Playing With Fire
The Controversial Career of
Hans J. Eysenck

Playing With Fire
The Controversial Career of Hans J. Eysenck

Roderick D. Buchanan
Honorary Fellow
School of Philosophy,
Anthropology and Social Inquiry
University of Melbourne
Australia

OXFORD
UNIVERSITY PRESS

Great Clarendon Street, Oxford OX2 6DP

Oxford University Press is a department of the University of Oxford.
It furthers the University's objective of excellence in research, scholarship,
and education by publishing worldwide in

Oxford New York

Auckland Cape Town Dar es Salaam Hong Kong Karachi
Kuala Lumpur Madrid Melbourne Mexico City Nairobi
New Delhi Shanghai Taipei Toronto

With offices in

Argentina Austria Brazil Chile Czech Republic France Greece
Guatemala Hungary Italy Japan Poland Portugal Singapore
South Korea Switzerland Thailand Turkey Ukraine Vietnam

Oxford is a registered trade mark of Oxford University Press
in the UK and in certain other countries

Published in the United States
by Oxford University Press Inc., New York

© Oxford University Press, 2010

The moral rights of the authors have been asserted
Database right Oxford University Press (maker)

First published 2010

All rights reserved. No part of this publication may be reproduced,
stored in a retrieval system, or transmitted, in any form or by any means,
without the prior permission in writing of Oxford University Press,
or as expressly permitted by law, or under terms agreed with the appropriate
reprographics rights organization. Enquiries concerning reproduction
outside the scope of the above should be sent to the Rights Department,
Oxford University Press, at the address above

You must not circulate this book in any other binding or cover
and you must impose this same condition on any acquirer

British Library Cataloguing in Publication Data
Data available

Library of Congress Cataloguing in Publication Data
Data available

Typeset by Glyph International Private Ltd., Bangalore, India
Printed in Great Britain
on acid-free paper by
CPI Litho

ISBN 978–0–19–856688–5

10 9 8 7 6 5 4 3 2 1

The views expressed in this book are not shared by me.

Sybil Eysenck, June 2008

Portrait of Hans Eysenck
Source: Centre for the History of Psychology, Staffordshire University.

Preface and acknowledgements

This book is not about whether Hans Eysenck was right or wrong. I was always more interested in what made Hans Eysenck such a compelling figure, in what made people pay attention to him in his pomp and still care about him after he was gone. Following the inclinations of my subject, this book is aimed at a broad readership. I have attempted to accommodate the different interests and expertise of the diverse groups Eysenck engaged. Nevertheless, readers may still need a modicum of tolerance: what may seem obvious and old hat to some may be revelatory and fresh to others. I have written with historians of science in mind, particularly scholars specializing in the history of the behavioral sciences. I have also sought to address the scientific community, especially those working in psychology—the budding students and the leading figures, Eysenck's progeny and former peers. I hope that lay readers will find this book rewarding too.

Since the late 1970s, biographical studies of scientists have been seen as a little passé. While biography remained a staple of the general public's non-fiction choices, it was repeatedly dismissed by serious historians as a hackneyed genre, full of cliché and error. The general view was that it was incompatible with the appropriate philosophical and sociological categories of analysis. This may have been true in practice, but not in principle. The indifferent quality of some work, all those hagiographies and hatchet jobs, need not have deterred us. Now it seems they no longer do, for science biography has made a comeback. Just as well, because modern-day science study—the history, philosophy, and sociology of science—has been built on case studies, on sophisticated analyses of episodes and innovations, themes, and events.

The idea of a biography of Hans Eysenck was conceived and organized by Maarten Derksen and Trudy Dehue. Without funding from the Wellcome Trust it would hardly have been possible, a generosity matched by the project's host, the Heymans Institute of the Psychology Department at the University of Groningen. The unstinting support and assistance of Trudy Dehue and Maarten Derksen was backed up by all the other members of the Groningen Theory and History group: Douwe Draaisma, Stefan Petri, Yvette Bartholomée, and Sarah de Rijcke. I look back on the time my family and I spent in the lowlands with great fondness. I never thought I would feel partial to the colour orange, never thought I would feel nostalgic when the weather turned cold and grey.

Writing a biography is an unavoidably personal affair. I respectfully acknowledge the courtesy and forbearance Sybil Eysenck and Darrin Evans extended toward me throughout the project. Many thanks must also go to Graham Richards, heading up the Centre for the History of Psychology, at Staffordshire University, Stoke-on-Trent. He and his staff went above and beyond the call of duty. The Hans Eysenck library, the newspaper cuttings, and the memorabilia at Staffordshire should prove a valuable resource for anyone interested in things Eysenckian. Many of the pictures appearing in this volume have been culled, with kind permission, from this source. Graham Richards also performed a similar role at the British Psychological Society's History of Psychology Centre in London, as well as commenting on draft sections of the manuscript.

Any historian is only as good as his or her sources. I came to rely on the knowledge and skill of many archivists and librarians in the course of the project, notably Martin Guha at the Institute of Psychiatry library, David Baker and his staff at the Archives of the History of American Psychology, Colin Gale at the Bethlem Royal Hospital Archives and Museum, Adrian Allan at the Special Collections and Archives at the University of Liverpool, and the staff at Special Collections and Western Manuscripts at Oxford University.

Eysenck's life and career left a vast network of friends, colleagues, and adversaries. Many took the time to be interviewed and/or provide additional information. Those kind enough to speak or write to me are listed are in the primary sources section at the end of this volume. A few went out of their way to be of assistance, particularly Arthur Jensen, Michael Eysenck, Phil Rushton, Frank Farley, Hermann Vetter, Michael Rutter, and Manfred Amelang. Sadly, several interviewees are no longer with us, including Jeffrey Gray, A.R. Jonckheere, and Victor Meyer. My fellow historians were generous enough to provide pointers and advice, among them Andrew Winston, Raymond Fancher, John Hall, Geoff Bunn, Sandy Lovie, Rhodri Hayward, Alexandra Rutherford, Wade Pickren, Bill King, Ron Borland, and the late Alison Turtle. A range of people also gave up their time and critical skills to appraise various draft chapters and thus deserve special mention. They included Eric Hobsbawm, John Burnham, Manfred Amelang, Patrick Rabbitt, Arthur Jensen, Paul Barrett, Michael Eysenck, and Michael Rutter. The Theory and History group at the University of Groningen did all this and more; *dank u wel* for your constructive criticism, suggestions, and encouragement. Thanks also to those in the History and Philosophy of Science Department (now part of the School of Philosophy, Anthropology, and Social Inquiry) at the University of Melbourne, especially Howard Sankey and Neil Thomason, who supported the project in its latter stages.

Family comes first, and in this case, last. To my daughter, Casey, thank you for the many encouraging post-it notes on my PC saying: 'finish the book'. To my son, Quin, thank you for setting a good example by writing countless books of your own while I worked on mine. To my partner, Sue, thank you for the patience and the trust. Little in this book will surprise her, since she read every draft. Val and Alan Finch also provided excellent proofing skills, for which I will be forever indebted. Finally, I would like to express my gratitude to my parents, Alan and Marilyn. Sadly, very sadly, both passed away during the final stages of this project. They were always fond of telling me that hard work has its own rewards. Here's to you then.

Rod Buchanan
September 2008

Contents

1 Introduction *1*
2 Presenting a German past *13*
3 The accidental psychologist *37*
4 Dimensions of personality *73*
5 The biology of personality *117*
6 Clinical partisan *181*
7 Mr. Controversial: the psychology of politics *241*
8 Mr. Controversial: race and IQ *271*
9 Smoking, cancer, and the final frontier *361*
10 Conclusions *409*

Primary sources *429*
Bibliography *433*
Index *469*

Chapter 1

Introduction

'RACISTS OUT! RACISTS OUT!' the ragtag activist group chanted in unison. They had taken up a position at the front of the dark, wood-panelled lecture theatre and they were not about to move. The grainy video footage shows Hans Eysenck being ushered in protectively, having been invited there by the Psychology Department of the University of Sydney. He is due to speak on personality topics rather than the explosive issue of race and IQ. Never mind, nothing could be heard above the din anyway.

These sorts of political protests were nothing new to Eysenck. During the early 1970s, this scene could have occurred at any number of campuses in England. But this is Australia, and the year is 1977. It was assumed the issue had simmered down. The demonstrators have caught university officials by surprise. They have made no contingency plans.

The chanting does not abate and there is a growing threat of violence in the air. Eysenck is suddenly nowhere to be seen. He has been secreted somewhere. Eysenck spends the next twenty minutes locked in a storeroom with Sydney academic Alison Turtle, waiting till the coast is clear. According to Turtle, he remains calm throughout, not the least bit worried or chastised.[1]

The race and IQ controversy defined Eysenck in the public imagination. It is usually the first thing anyone outside the scientific community can recall whenever his name is mentioned. Yet race and IQ was a side issue for Eysenck and the brouhaha he helped create a bit of a sideshow. It is not how Eysenck himself would like to be remembered; he tended to emphasize his more substantive contributions to his chosen field. Likewise, his peers had an appreciation of him that went beyond the headlines. However, as anyone interested in his legacy quickly discovers, the nature of this appreciation greatly depends on whom you ask.

---

[1] Video footage was taken for the purpose of recording Eysenck's talk. Now an interesting historical artefact, it was kindly shown to me, with narration, by the late Alison Turtle in December 2002.

Eysenck or Eysencks?

Hans Eysenck was clearly one of the great visionaries of 20th century psychology... He led the way in defining the structure of human personality... he was the British equivalent of B.F. Skinner in the extent of his influence...The American Psychological Society gave him its highest award... (Frank Farley, former student of Eysenck and ex-president of the American Psychological Association 1998)

There was never the slightest likelihood that he would be elected a Fellow of the Royal Society... he was viewed as having a very narrow approach to psychology and it would not be easy to point to any discoveries that were of great consequence... Of course, it is not that there isn't the odd mistake in the writings of even the best scientists... The charge in the case of Eysenck is that some were substantial and they certainly were not random... (Michael Rutter 2002)

Like Freud, Eysenck was a conquistador... (paraphrasing Tony Gibson 1981)

I worked with Piaget too and Eysenck could never be compared with him... (A.R. Jonckheere 2002)

He was constantly being misconstrued...many of his critics never read his work closely anyway... but he never held grudges... (Gisli Gudjonsson 2001)

People were afraid of him, afraid to take him on in public... (Ann Clarke 2002)

Scientific psychology in Britain really started with Hans. Before him there was nothing... (Sybil Eysenck, Eysenck's second wife and surviving widow 2001)

As his career progressed he seemed to develop a kind of 'God complex', that is, because he said it, it must be true... Michael Eysenck (Eysenck's eldest child with first wife Margaret Davies 2001)

I can't think of an unkind word to say about him... (Arthur Jensen 2002)

Down with Eysenck... Shaft Eysenck! (Paraphrasing Richard Christie and Milton Rokeach 1955)

Hans Eysenck? I love the guy... I owe him so much... I could talk about Eysenck all day... (Farley again 2002)

Talk about Eysenck? Only if I have to and only off the record... (Anon)[2]

This illustrates the spectrum of opinion about Eysenck, the most prominent British psychologist of the post-war era and easily the most paradoxical.

---

[2] The first quote from Frank Farley derives from a tribute Farley organized and moderated at the 1998 American Psychological Association Annual Convention, 14 August 1998. The session included Nicholas Cummings, Martin Seligman, and Charles Spielberger, and an edited version of it appears at http://freespace.virgin.net/darrin.evans/hjemf.htm. At this session, Farley reiterated material submitted for an obituary in the *New York Times*. See Honan, W.H. (1997), Hans J. Eysenck, 81, a heretic in the field of psychotherapy, *New York Times*, 10 September 1997. The second Farley quote comes from an interview

The biographical landscape

The basic details of the life and career of Hans Jürgen Eysenck are relatively easy to summarize. He was born in Berlin in 1916, during the Great War, the only child of German film and stage performers. As an academic psychologist in Britain, he became famous for the audacity of his theorizing, the expansive scope of his empirical research, and the forthright, often controversial views he expressed.

For young Hans it was never clear it would be thus. His parents had ambitions for him in the theatre, and he himself wanted to be a physicist. After graduating from secondary school in 1934, he fled Hitler's Germany and arrived in London the following year. He enrolled in psychology at University College, London, where he was taught by J.C. Flügel, S.J.F. Philpott, and Cyril Burt. He took a degree, and then rapidly completed a Ph.D. supervised by Burt focusing on the experimental analysis of aesthetic preferences.

The outbreak of war saw Eysenck declared an enemy alien and he found it difficult to get work. After a spell as a firewatcher, he landed a job at the Mill Hill Emergency Hospital in 1942. Headed by the imposing figure of psychiatrist Aubrey Lewis, Mill Hill functioned as the relocated Maudsley Hospital. Soon after the war, plans were drawn up for a new Institute of Psychiatry (IoP), a training and research facility to be affiliated with the Maudsley and Bethlem hospitals. Eysenck turned down offers at several other universities to head psychology there. He was given an unusual degree of bureaucratic freedom to organize research around his programme for a unified experimental psychology of individual differences. The IoP provided a stable institutional environment where he spent the rest of his career. By 1955, Eysenck had become a full professor in an independent psychology department. After he retired in 1983, his research activities and writing were undiminished, and his interests even more diverse. Eysenck developed a distinctive science of personality psychology that married descriptive statistics with physiological experimentation, and collapsed any firm distinction between pure and applied science. The heavily

---

with the author, 12 July 2002. The Tony Gibson paraphrasing comes from Gibson, H.B. (Tony) (1981), *Hans Eysenck: the man and his work*, Peter Owen, London, in which Gibson spent an entire chapter comparing Eysenck with Freud. The Richard Christie and Milton Rokeach paraphrasing comes from their April 1955 correspondence regarding their critique of Eysenck's political research. See Milton Rokeach Papers, Archives of the History of American Psychology. Provenance for the other quotes: Michael Rutter, personal communication, 9 May 2002; A.R. Jonckheere, interview, 22 April 2002; Gisli Gudjonsson, interview, 25 October 2001; Alan and Ann Clarke, interview, 25 April 2002; Sybil Eysenck, interview, 21 October 2001; Michael Eysenck, interview, 28 October 2001; and, Arthur Jensen, interview, 1 July 2002.

cited author of 85 books and over 1000 scientific papers, Eysenck was also renowned as a popularizer of psychological science.

There can be little argument about the importance of Eysenck as an historical figure. His name ranks alongside American contemporaries such as Gordon Allport, B.F. Skinner, and Raymond Cattell for recognition value, and he hardly had a rival on the UK scene during his lengthy heyday.[3] Perhaps only Donald Broadbent could be mentioned in the same breath when discussing his British peers. Venerated by a raft of students and influential peers, he was held in the highest esteem both personally and intellectually. But he was also reviled by just as many others as some kind of arrogant know-it-all, a 'ratbag' figure one simply loved to hate.[4] So what was it about Eysenck that made him so loved and yet so loathed?

Eysenck so divided opinions that a privileged, singular viewpoint seems impossible. Thus I invite the reader to hold in mind two markedly different visions of the man. Here was a good scientist who was incisive and fearless, well-intentioned and kind and, conversely, here was a bad scientist who was dishonest and destructive, dogmatic and vain. The moral tone in all this is unmistakable. Are we in for a sermon about science and its ethical responsibilities? How wrongheaded that would be and how tedious. Instead, I intend to characterize divergent perceptions of Eysenck, and identify what repelled or endeared him to people and why. This will surely take a whole book. Hopefully a good book.

Doing a new biography of Eysenck

Why do another biography of a man whose personal voyage has already been well-charted? Eysenck supplied versions of his life story for dictionaries of biography and short autobiographical collections, as well as putting out two editions of a full-length autobiography. Provocatively titled *Rebel with a cause*,

---

[3] See Haggbloom, S.J., *et al.* (2002), The 100 most eminent psychologists of the 20th century, *Review of General Psychology* **6**, 139–52. Using a method that was biased toward Americans, this otherwise exhaustive survey ranked modern era psychologists according to various indices of prominence and influence. Eysenck came in at number 13, behind Skinner, Bandura, Festinger, and Rogers *et al.* He was the highest British entry, and third-highest non-American behind Piaget and Freud. Eysenck was ranked even higher in terms of journal citations. Even so, Eysenck's eminence was still probably underrated. See Black, S.L. (2003), Cannonical [sic] confusions, an illusory allusion, and more: a critique of Haggbloom *et al.*'s List of eminent psychologists (2002), *Psychological Reports* **92**, 853–7.

[4] This phrase comes from media images of Eysenck. The *New Scientist* twice ran pieces on him with the title: 'The psychologist they all love to hate'.

they paint a picture of the fearless outsider, unintimidated by established orthodoxy. The personal details that colour memories of his childhood and early adulthood give way to lengthy accounts of the various 'battles' he engaged in. This format gave him the opportunity to take up the cudgels once again.[5] Moreover, the one other biographical account that was not written by Eysenck came from an acolyte of sorts, H.B. (Tony) Gibson. Eysenck cooperated fully with Gibson's retirement project, and approvingly dubbed the finished product 'a good read'. Written in 1981, *Hans Eysenck: the man and his work* was necessarily incomplete, since Eysenck was still a couple of years away from embarking on an extraordinarily energetic retirement.[6]

Eysenck's widow, Sybil Eysenck, has done her best to continue in this vein. She and youngest son Darrin monitor commentary about Eysenck and maintain a website in respectful homage.[7] Sybil still co-edits some of the journals he helped found. When interviewed, she reiterates the key aspects of the Hans Eysenck story in accordance with the public record, no more, no less. Perhaps unintentionally, she also placed an immense obstacle in the way of any new biography. She engaged the services of a professional disposal firm to destroy the extensive personal and professional records of her husband in the months after his death in September 1997. The more intimate details of his working life, his motivations, and the development of his thought will no doubt remain a mystery—a pity.

One cannot ignore the biographical materials gathered by others, for they can still serve as valuable secondary resources. These issues are especially acute with the existence of an autobiography. It would be equally mistaken to completely avoid or to slavishly follow my subject's public testimony. Instead, I think it appropriate to critically embrace it, to enter into an explicit dialogue with its more important aspects, as well as use it as a foil. For some periods of Eysenck's life, his memoirs are almost all that is available.

These historical concerns have an interesting flip-side, however, for there are other sources and other views on Eysenck's career. Suppressed by the volume and power of Eysenck's own version of the story, these unofficial fragments derive from the often rather snide anecdotes of some of his rivals, from *ad hominen* criticisms in scholarly commentaries, and from the differing recollections of contemporary figures. It is an underground narrative owing much

---

[5] See Eysenck, H.J. (1990*b*), *Rebel with a cause*, W.H. Allen, London, and Eysenck, H.J. (1997*b*), *Rebel with a cause* [revised and expanded], Transaction Press, New Brunswick, New Jersey.

[6] Gibson, H.B. (Tony) (1981), *Hans Eysenck: the man and his work*.

[7] See http://freespace.virgin.net/darrin.evans/hjemf.htm.

to the kind of incessant rumouring surrounding any famous figure in the twentieth century. However, the existence and persistence of this narrative, and the authority of some of its origins, makes it worth exploring. By addressing it explicitly, I hope to pre-empt any accusation that there was something about the man I was avoiding.

Special considerations

Whatever one's standpoint on the biographical genre it is clear that, as a subject, Eysenck presents some unique problems and opportunities. Perhaps the most immediate challenge would be to take the project out of the partisan realm so that it could not be dismissed as just another spade for the ideological trenches dug by supporters and opponents alike.

As well as being controversial, Eysenck was extraordinarily prolific. New publications by and about him continue to appear, a beyond-the-grave influence that would no doubt please him. No biography could possibly cover all that he waded into. What *is* covered necessarily becomes a matter of practicalities like time constraints and access to materials, as well as historical judgement.

Another key question revolves around how one treats the personality and private life of such a notorious figure. Should it be left out, treated as separate and immaterial to his work? In most respects the choice has already been made. A lack of primary source material means this will essentially be the story of a professional life as it played out in public. Most of the historical material for this project comes from published sources, supplemented by interviews with Eysenck's contemporaries and limited archival records.

And what of my personal views on Eysenck: do I sympathize with his social and intellectual viewpoint and does this matter? That old chestnut—that biographers either love or loathe their subject—has followed me since the project began. One view is that to understand the famous and write analytically about them, one has to hate them first. But, while fame may overwhelm scholarly detachment, what sense of balance is at risk when dealing with the famously loathed? Or, as Michael Eysenck asked me incredulously, how could *anyone* be neutral about his father?

However, it would be bumptious in the extreme for any dedicated historian to presume a greater capacity for judging the merits of a particular scientific viewpoint than the scientists themselves. After all, it is their business, their long and arduous training put to good use. Moreover, any kind of advocacy standpoint would make it impossible to fully understand what they were actually arguing about—if, in fact, a genuine old-fashioned slanging match is being had—just as any good analysis requires sympathetic symmetry. If advocacy is your bag, then the issue of why scientific disputes arise and persist would remain a conundrum. Moreover, victories and defeats could only be accounted for by a kind of retrospective sleight of hand: *x* got the prizes and *y* got the shaft because we *now* think that *x* was right and *y* wrong. If only things were so clear at the time. Far from being a risk to the project's balance, I see the very divergence of opinion about Eysenck's work as something utterly fascinating, something that should be, and can be, explained.

A related question concerns the author's place in the work. To cite two well-known examples, you can adopt what *Darwin* co-author James Moore called the *cine* approach and melt into the background.[8] At the other extreme, you can insert yourself up front not only as the storyteller but even into the action itself, as Edmund Morris once did in his biography of Ronald Reagan.[9] Faced with such choices, my response will be mixed and pragmatic. Signalled by the active voice, the author will make self-conscious appearances on occasions, but only when it serves the interests of the analysis. Conversely, it might be admirable to try to collapse what Thomas Söderqvist characterized as the ironic distance between subject and author that invites the reader to sit in judgement suffused with smug hindsight.[10] One can achieve this by spiriting all into the shoes of the subject, by carefully reconstructing the complex contingencies that operated then and there. However, a lack of sources again militates against this. Only in a few key episodes will this be attempted, most occurring in the recent past where memories, protagonists, and documentation are more freely available. Only then will the separation of life and work start to blur a little as more information about the way in which the personal shaped the professional can be conjured up. And perhaps only then will the constructively agnostic stance on scientific truth shift focus, since greater access to information has a way of leapfrogging hard-won professional wisdom.

Eysenck and the construction of objectivity

There is much to chew on in the Eysenck story. Detailing Eysenck's impact within academic psychology and the social conditions of twentieth century Britain would be a meal in itself. Then there were his attempts to marry the

---

[8] See Desmond, A. and Moore, J. (1991), *Darwin*, Michael Joseph, London, and Moore, J. (1996), Metabiographical reflections on Charles Darwin, in *Telling lives in science: essays on scientific biography* (ed. M. Shortland and R. Yeo), pp. 267–81, Cambridge University Press, Cambridge.

[9] See Morris, E. (1999), *Dutch: a memoir of Ronald Reagan*, Random House, New York.

[10] Söderqvist, T. (1996), Existential projects and existential choice in science: science biography as an edifying genre, in *Telling lives in science: essays on scientific biography* (ed. M. Shortland and R. Yeo), pp. 45–84, Cambridge University Press, Cambridge.

biological and the psychological in personality theory, his contribution to the nature versus nurture debate, his anti-psychoanalytic campaigning, his promotion of behaviour therapy, and on and on. Important lessons in the philosophy of science could be gleaned by following Eysenck through the many twists in his career as he formed and broke professional alliances, crossed disciplinary boundaries, and divided political and public interest groups. He was only following the facts, he said, only being objective when he disputed the efficacy of psychotherapy, questioned the link between smoking and cancer, or seriously considered the merits of astrology.

In recent years, science studies has sought to re-conceptualize objectivity as a *constructed* notion rather that some *a priori* ideal—potent, slippery, and historically situated at that. An energetic band of scholars has set about tracing the emergence of 'objectivity', its transformations and its functions, particularly its role in evincing trust from those who matter. These analyses have also had the additional benefit of often focusing on the social sciences. Perhaps as a consequence, the social sciences have been treated on their own terms rather than as poor cousins to the natural sciences.[11]

Hans Eysenck presents a very obvious example (perhaps misleadingly so) of a psychologist *qua* natural scientist. Eysenck's work can be used to explore the production of scientific objectivity in psychology. One may take a leaf out of Ted Porter's *Trust in numbers*, a study that elevated the tension between seasoned expert judgement and formalized empirical rules as a central historical theme. Eysenck's commitment to what he saw as rigorous science, to quantifiable concepts and measurement, would seemingly place him squarely on one side of this divide. One could ask why he adopted such a strategy and why it apparently paid off so handsomely, as it did for many others in modern psychology.

Exploring the production of objectivity in this sense doesn't dispel the contradictions surrounding him; rather, it heightens them in a potentially instructive manner. Eysenck constantly advocated a no-nonsense, empirically verifiable rigour. He asked us to trust the numbers not the scientist. Nevertheless, much of his output was extremely breezy, even cavalier. This was not just a

---

[11] See Solomon, J. (1998), *Objectivity in the making: Francis Bacon and the politics of inquiry*, Johns Hopkins University Press, Baltimore; Porter, T.M. (1995), *Trust in numbers: the pursuit of objectivity in science and public life*, Princeton University Press, Princeton; Dehue, T. (1995), *Changing the rules: psychology in the Netherlands, 1900–1985*, Cambridge University Press, Cambridge; Daston, L. (1992), Objectivity and the escape from perspective, *Social Studies of Science* **22**, 597–619; and, Levine, G. (2002), *Dying to know: epistemology and narrative in Victorian England*, University of Chicago Press, Chicago.

flow-on from the role he took on as a popularizer. Once Eysenck's institutional position had been consolidated by the mid-1950s and his intellectual framework put in place, a surprisingly high percentage of his publications could be classed as 'subscientific', works aimed at a broad audience.[12] For example, nearly half his books showcase accessible argumentation backed by a sprinkle of key data, facts, and references. Where is the eye-glazing detail, one wonders, that his official reputation promises? One can find some of that, especially in his journal articles, but one can find proportionally much more in the narrower, more pedestrian *oeuvre* of those he relies upon for support. Those who observed him at close hand describe him as more theoretician than experimentalist.[13] In fact, he spent much of the latter part of his career as a kind of intellectual tourist. He collected facts, information, and data from the programmes he helped set in train, from his avid consumption of obscure books and foreign-language journals. Much of this touring was used to pick the eyes out of a range of painstaking, dryly reported research carried out by scientific workers in highly specialized areas. In this capacity he was no more impersonal or rigorous than the people he appropriated or borrowed from—quite the reverse in some cases. However, Eysenck was no mere butterfly collector pinning items in static compilations. He made his booty work for him. These facts, information, and data were repeatedly used—cleverly, creatively, and at times quite slyly. They served to bolster, amplify, and extend an *existing*, ambitiously *broad* theoretical framework.

---

[12] Consider Eysenck's 80 or so books. The uncertainty depends on whether you count new editions, translations, test manuals, and so on. Over a third could be classed as popular or quasi-popular works, directed at an educated lay audience. Eysenck published proportionally more of this material in the middle and latter part of his career. Almost a third more of his books amount to edited volumes, collecting the journal papers and data of other researchers. Most of these edited works were put together in the middle part of his life. Then there are the numerous test manuals. Finally, there is a small set of books (which sometimes run several editions) that make up the core of his scientific output. However, to say that these core works are all that count and dismiss the rest as 'dumbed-down' pastiches is to miss the point and ignore what made Eysenck *Eysenck*. While his journal contributions contain proportionally less popular material, a sizeable chunk was made up of replies, rejoinders, and corrections. In addition, he still found time to write for middle-brow periodicals like *Encounter* and *The Listener*, as well as the likes of *Reader's Digest* and *Penthouse*. He also wrote many newspaper articles and letters to the editor—writings that seldom appear in any Eysenck bibliography. For a reasonably complete bibliography, as well as various other tribute pieces, see the special issue of *Personality and Individual Differences* (2001), Bibliography: Hans Eysenck, Ph.D., D.Sc., 1939–2000 [Sybil Eysenck *et al.*], *Personality and Individual Differences* **31**, 45–99.

[13] Alan and Ann Clarke, interview, 25 April 2002.

The message here is: don't let Eysenck's reputation and the rhetoric of rigour fool you. He was a special case of what he and the late Paul Meehl and the even later William James called tough-minded. He was a quantitative hard-nut, but that was hardly all. For one thing, the very power of his elite, celebrity status provided a strange counterpoint to his crusade for detached rigour. In very visibly promoting impersonal science, he was often attacked in a very personal manner. Eysenck's emphasis on objectivity was a red herring in at least one important respect. It just doesn't come to grips with the way people thought about him. Critics saw anything but the kind of virtues such a stance supposedly conferred. With his fondness for the interpretation of other people's results, Eysenck inevitably stood on some often quite distinguished toes. Many took issue with his presumption of a higher authority in his attempts to arbitrate scientific disputes. In their indignation, it was Eysenck's impartiality, his expert *judgement* that was being called into question. The notion of objectivity was both a means and an end for Eysenck; it was a resource underwriting his stated aims *and* the often quite partisan tactics he used to achieve them. However, he was hardly alone in this.

Playing the field

So what did make Eysenck special? Perhaps it was his enormous capacity to link diverse elements and ideas, including the concept of objectivity, and deploy them in an instrumental manner. Perhaps too this arose from the looping effects of *being Hans Eysenck* and all that that entailed. His reputation reinforced his unique approach to his work, and vice versa. This amounted to more than just being argumentative. While Eysenck occasionally acknowledged the strategic way in which he exploited his intellectual gifts to score points, he always ennobled his efforts under the banner of 'high science'. Some of his supporters liked to hive off his serious theoretical and empirical expositions from the 'propaganda'—the cleverly slanted critiques, the adversarial rhetoric that blew off opponents and changed the face of debates he intervened in. However, I doubt whether such a distinction can be sustained.

One might opine that Eysenck played his career like a game and that he was, in turn, caught up in this game of science like no other contemporary psychologist. His immediate goal—as in the games of chess, squash, or tennis that he loved—was to win. It encouraged and even necessitated a kind of competitive, advocacy style that threatened to become his consuming *raison d'être*. Combined with his bid for a broad audience and his penchant for hot topics, it made it hard for him to walk away. Whether he intended to or not, Eysenck ended up playing for keeps. He liked to win and, increasingly, he *had* to win—to maintain his scientific status, to save his moral and political reputation, to protect his celebrity brand name.

What do you get when a scientist operates in this manner? One wonders how an overriding need to win squares with old-fashioned notions of scientific progress and capital 'T' truth. I am not about to claim Hans Eysenck was simply self-interested. Yet his particular way of doing science was born of an ambitious individualism that made him appear less and less interested in collectively generated achievement and sharing credit. And, even though Eysenck was a master at subtly modifying his theories without conceding that he had done so, shoring up entrenched positions would inevitably became a priority. It made for the rejection of new areas and alternative approaches, with more and more of his attention directed toward the past rather than the future.

Eysenck as game-player is a rich enough metaphor to help explain some of the odd juxtapositions of his work. For a start, his seemingly compulsive output, with its high to lowbrow appeal, becomes much more intelligible. 'Mr. Psychology' has to have an across-the-board strategy. He was one in a growing tradition of the public intellectuals in Britain, like C.P. Snow, Julian Huxley, and J.B.S. Haldane. He was a star in the new media and the old—all of which tended to intensify the demands and effects of his celebrity. While the mass media may dazzle and dangle glittering prizes, scientists have generally affected a disdainful attitude toward those who yield to its temptations. Mixed with this snobbery is a kind of cautionary wisdom that suggests there is no surrender, no control and no closure in the court of public opinion. However, these negative characterizations of popular forums surely derive from invidious comparisons with 'proper' scientific channels of communication.

What is needed, then, is the kind of analysis that can get a purchase on scientists' interactions with the media. It should treat public debates on their own terms, rather than as an emotive and corrupt spill-over from 'internal' scientific debates. Popular forums have their own particular rules—different, perhaps, from those of conventional science, but rules just the same. Moreover, public debates do not take place independent of internal debates; the scientific community is certainly not unaffected by what goes on in popular media outlets. In any case, whether particular media outlets are considered inside or outside the scientific beltway partly depends on just how the scientists themselves draw the line. Clearly, Eysenck chose an expansive strategy that included a prominent public profile. Why he did so, and how it affected his work and reputation, thus become key questions.

However, if I had to single out one aspect about Eysenck that made him special it would be this: he could be very, very aggravating. Eysenck could make enemies without even trying, even though he was never personal and seldom overtly aggressive. It seemed he was only too pleased to get up certain people's noses, even though he denied this was ever deliberate. As his career progressed you get the unmistakable impression that he defined himself in terms of what he was against as much as what he was for. If he was not provoking or offending those he disagreed with—psychoanalysts and psychiatrists, left-leaning social theorists, medical epidemiologists, narrow experimentalists, and so on—he felt he was just not doing his job. For Eysenck, divide and rule *was* the rule.

While Eysenck seemed to possess a remote, cerebral outlook not uncommon amongst intellectuals, it came with an almost irresistible, impish urge to stir the pot. He never seemed to lose that sense of intellectual cheek, even as an elderly man. But Eysenck was more than just a naysayer, some kind of crank easily dismissed or laughed at. He was part of the academic psychological establishment (awkwardly at times), and he had real power and real influence. He was an empire-builder, intellectually and institutionally, vigorously policing boundaries he helped draw up. Aligning yourself with him could and did pay off. For those he nurtured, his gladiatorial reputation was a strange misnomer. For those he rubbed up against, it was the sum of him. Thus we see a fractured vista: Eysenck cutting a swathe through a crowd of hapless opponents, all the while cheered on by an enthusiastic band of loyal followers. Here was a man with a real talent for generating disagreement *and* consensus, distrust *and* good faith.

I hope to understand the professional man as well as account for the success he achieved in Britain for, in many respects, Hans Eysenck was a stranger in a strange land. The intellectual legacy he created was testimony to his remarkable self-belief and the stunning breadth of his intellectual gifts, as much as it was to the happenstance of opportunity and good fortune.

The key areas of Eysenck's work will be traced in a series of chronologically overlapping chapters. The narrative structure is deliberately episodic, following each thematic tread as Eysenck took on new challenges and a fresh set of adversaries. This is, above all, the story of a partisan leader, a cut-and-thrust debater, a man never far from the limelight. It is the biography of a *controversialist*.

Chapter 2

Presenting a German past

Documenting the early of life of Hans Eysenck is somewhat problematic. Few records from pre-war Germany exist, especially those pertaining to everyday civilian life such as schooling and work.[1] The passage of time has ensured that only the hardiest of souls who grew up with Eysenck in Berlin are still with us today. Very little additional information is available to supplement what has been written by Eysenck, given by him in interviews, and recounted by others with Chinese whispers-style variations.[2] However, Eysenck's published memoirs have the latter-day man written all over them; they are an active reading back for whatever purpose was at hand. His recollections are so imbued with a conscious sense of self-presentation that they would confound any attempt at psychobiography.[3] They could never be separated from what they were supposed to explain.

---

[1] Exceptions to this generalization are the military archives and the extensive records pertaining to National Socialist 'eugenic' programmes. With a ghastly kind of efficiency, the Nazis kept meticulous accounts of their operations and their victims.

[2] I am referring to Eysenck's autobiography, Eysenck, H.J. (1990b), *Rebel with a cause*, W.H. Allen, London and Eysenck, H.J. (1997b), *Rebel with a cause* [revised and expanded], Transaction Press, New Brunswick, New Jersey, and his other autobiographical pieces, Eysenck, H.J. (1980b), Hans Jürgen Eysenck, in *A history of psychology in autobiography*, vol. 7 (ed. G. Lindzey), pp. 153–87, W.H. Freeman, San Francisco, and Eysenck, H.J. (1991e), Maverick psychologist, in *The history of clinical psychology in autobiography*, Vol. 1 (ed. C. Eugene Walker), pp. 39–86, Brooks/Cole, Pacific Grove, California. Commentary about Eysenck's career includes: Gibson, H.B. (Tony) (1981), *Hans Eysenck: the man and his work*, Peter Owen, London; Cohen, D. (1977), Interview with Hans Eysenck, in *Psychologists on psychology* (ed. D. Cohen), pp. 101–25, Routledge and Kegan Paul, London; and, Richards, G. (2004), Eysenck, Hans Jürgen (1916–1997), psychologist, in *Oxford dictionary of national biography* (ed. C. Matthew and B. Harrison), Oxford University Press, Oxford. One of the first detailed versions of his career is contained in *Current biography* (1972), Biographical entry, in *Current biography* (ed. M.D. Candee and C. Moritz), pp. 18–21, H.W. Wilson, London. One can also find various biographical accounts, presumably culled from the sources above, on the internet. These are notable for the small variations in their detail.

[3] A taste of this was conveyed in the BBC's documentary on Eysenck as part of their *Heretics* series. The programme aired in mid-1994.

Nevertheless, this is exactly why these stories can be moved centre stage. While my overriding focus is on Eysenck's career as a combative adult scientist, Eysenck's distant memories are made more immediate, and more relevant, by Eysenck himself. He used his past as an argumentative resource to counter dire accusations and vouchsafe his credibility. His Berlin childhood, his reasons for leaving Germany, and his family background became especially important from 1971 onward, when he became embroiled in the race and IQ controversy. Critics made sure the personal became political, and Eysenck took the opportunity to return fire. The stories of his youth became key elements of the image-making process of this very public intellectual, allowing Eysenck to fashion himself as the outsider, the rebel truth-teller. They are also intriguing tales in their own right, for they touch on some of the most powerful moral tropes of the twentieth century.

Berlin, 4 March 1916

Hans Jürgen Eysenck was born during World War I in the German city of Berlin. His father, Eduard Eysenck, was born in 1889 (d. 1972) in Bergisch Gladbach in western Germany. Eduard Eysenck came from a Catholic family in the Rhineland, the youngest of three children. Most of the paternal side of Eysenck's family were solidly middle-class, with established professions like customs officer, restaurant owner, and various kinds of businesses.[4]

The maternal side of Eysenck's family was a little more intriguing. Eysenck's mother, Ruth Eysenck (*née* Werner), was born an only child circa 1892 (d. 1986) in Königshütte in Upper or Eastern Silesia, part of the Prussian empire of the time. Königshütte has since been renamed Chorzów and is now part of southern Poland. Silesia was largely Germanic in terms of language and customs, although it contained a sizeable number of Poles, especially in the east, as well as many who identified themselves as ethnically Austrian or Czech. In the nineteenth century, most Silesians were practising Roman Catholics, though Protestants tended to predominate in the more Germanic west. In addition, Silesia was home to a small number of Jews (less than 10%), torn between cultures and traditions.[5]

By Eysenck's account, Ruth Werner's mother came from 'a Jewish family in Silesia'.[6] This was an important detail, which Eysenck waited until near the end

---

[4] Eysenck (1990*b*), *Rebel with a cause*, p. 10.

[5] Vital, D. (1999), *A people apart: the Jews in Europe, 1789–1939*, Oxford University Press, Oxford.

[6] See Eysenck, H.J. (1990*b*), *Rebel with a cause*, p. 80.

of his life to reveal. Although Eysenck did not name her, it appears his maternal grandmother was Antonia Werner (*née* Sachs), born in 1863 in Hamburg.⁷ It is possible that Ruth Werner's father was also Jewish, though little is known of him. He was a doctor who died of tuberculosis in his thirties while Ruth was still a young girl. The Werner family had moved to the boom town of Berlin in the late 1890s, attracted by the professional opportunities it offered. Turn-of-the-century Berlin had a bustling sense of impermanence, as if distracted by the process of becoming something else.⁸ It had a restless, itinerant population that made it a welcoming destination for outsiders—especially Jews, intellectuals, and artists. Even so, the Werner family apparently made some attempt to assimilate to German culture once in Berlin, for Eysenck says that his mother was brought up as a Lutheran Protestant. Ruth Werner wanted to study law, but as a good-looking young woman she was encouraged to train in the theatrical arts. She became a successful stage actress and soon married the dashing actor–singer Eduard Eysenck around the beginning of World War I.

The Eysencks' marriage was not destined to last long, given that they were seldom together. Eduard Eysenck toured Germany and later took part in the war effort, leaving his new wife to look after young Hans. She continued to do stage engagements, leaving the boy in the care of his maternal grandmother. Adopting the stage name of Helga Molander at the end of the war, Ruth Werner became a star of the German silent screen era. She featured in over 30 films in

---

[7] While Eysenck did not name his maternal grandmother in the text of his autobiography, the index did cite her as 'A. Werner'. Since she died in a concentration camp, I have otherwise derived this information from records of the Holocaust, specifically Zentralinstitut für Wissenschaftliche Forschung, Freien Universität Berlin (1995), *Gedenkbuch Berlins der Jüdischen Opfer des Nationalsozialismus*, Berlin edition. Hentrich, Berlin. Assuming these lists are exhaustive, Antonia Werner (*née* Sachs) appears the most likely candidate based on the details Eysenck supplied in his autobiography. Nonetheless, I could not absolutely rule out another name—that of Anna Werner (*née* Lissner) of similar age (b. 1868). Anna Werner also lived in the Schöneberg district of Berlin and also died in a concentration camp—although this occurred, crucially, in the middle of the war. Anna Werner was born in Posen (or Poznań as it is now known in Polish), part of the eastern Prussian province of that name. How either Antonia or Anna Werner came to be in Königshütte remained unclear and would be especially intriguing if Antonia were in fact Eysenck's grandmother. While it would seem more likely that Eysenck's grandmother was born in nearby Posen, more direct evidence would be needed to tip the weight of evidence in favour of Anna Werner. An additional search of census records by staff of the Berlin Landesarchiv yielded very little additional information that would help resolve nagging doubts about her identity. Eysenck's surviving family could not shed any light on this matter either.

[8] See Ladd, B. (1997), *The ghosts of Berlin: confronting German history in the urban landscape*, University of Chicago Press, Chicago.

the late 1910s and 1920s.⁹ She also had a few speaking roles as the talkies era dawned in the early 1930s, although age and politics were to count against her. Hans's mother and father did unite briefly on screen, for example, in the 1919 film *Cagliostros Totenhand* (*Cagliostros Dead Hand*) during the dying embers of their relationship. According to various accounts supplied by Eysenck, his parents separated and divorced when he was two, or separated then and divorced another two years later.

Eysenck's father later remarried an actress called Tilly but they had no children. His mother also married again and likewise had no more offspring. Ruth Werner's second husband, Max Glass, was an entrepreneurial Austrian with a Ph.D. in philosophy from the University of Vienna. Glass had quickly worked his way up in the Berlin film industry, starting out as novelist and scriptwriter, before turning to production. By the mid-1920s, he headed Terra Film. In 1928 he founded his own production company, Max Glass Film Produktion GmbH, and another, Kristall Film, the following year. As Helga Molander, Ruth Werner starred in many films produced and directed by Max Glass, such as the 1922 *Der Mann mit der Eisernen Maske* (*The man with the iron mask*) and the 1923 *Bob und Mary* (*Bob and Mary*). Max Glass's first wife, Helene Münz, blocked his divorce application and Max and Ruth were obliged to keep their relationship under wraps. Even in the liberal climate of Berlin of the 1920s, they were still considered to be living in sin.

But there was danger ahead. In 1933, the Weimar dream died when Hitler became chancellor. Prominent figures in all fields suddenly felt pressured to conform to the National Socialist line. The Nazis had quickly grasped the power the new media had for persuading the masses. For those in the business of shaping popular opinion, the options were especially stark. Before Hitler took over, dissenting journalists were hounded and leant on; now they were taken away into 'protective custody'. From 1936, they also had to prove their Aryan ancestry dating back to 1800. Recent conversions from Judaism to

---

⁹ At the time the German film industry rivalled that of Hollywood and was technically superior, if anything. Helga Molander appeared in many silent films of this golden era, though not necessarily in starring roles. Probably the most notable film she appeared in was the 1919 *Anders als die Anderen* (*Different from the Others*), a pioneering work that attacked the criminalization of homosexuality. Made in conjunction with sex researcher Magnus Hirschfeld, it attracted large audiences and much controversy within German psychiatry and the wider public. See Mildenberger, F. (2007), Kraepelin and the 'urnings': male homosexuality in psychiatric discourse, *History of Psychiatry* **18**, 321–35. Eduard Eysenck also made at least one film that led to censorship from the state, the 1919 *Die Nakte: Ein Sozialpolitischer Film* (*The Naked Ones: A Sociopolitical Film*).

Christianity would have no bearing on racial designations formulated by the Nazis.

Those in the film industry likewise felt the heat. When unemployment levels had reached a crisis point in the last days of the Weimar Republic, foreign film companies were required to employ a minimum of 75% German nationals. The Nazis took these regulations several steps further, however, targeting the local industry as well. They quickly issued a decree in 1933 spelling out their preference for German films made by those with a 'German heritage'. As film historian Brigitte Berg noted, this spelt trouble: 'Not only was Max Glass a Jew (his conversion to Catholicism dating back to his youth in Vienna and his efforts at assimilation were to no avail), but several of the actors [who] appeared in his films were too, like... Szöke Szakall, Helga Molander and Grete Mosheim.'[10] Both of Max Glass's production companies were immediately shut down. Many of the films he had a hand in were eventually barred from commercial distribution as racial laws tightened. With no future in Germany, Max Glass went into exile in France. A similar cloud hung over Ruth Werner's career. Even though she was yet to become Max Glass's wife—as Eysenck implied she was—she had virtually no choice but to accompany him to Paris.

Max Glass recovered to set up production companies in the French capital with his sons from his first marriage. It was not a particularly successful professional transition for Ruth Werner, however. Her heavy German accent proved an insurmountable barrier to a film career in France. Even worse, Paris proved to be hardly far enough from the Nazis. When they overran Paris, Ruth Werner found herself in an internment camp. The next stop for German refugees handed over to the Nazis was usually the transit camp of Drancy, then on to Auschwitz. Max Glass had to spend a fortune in bribes to secure her release. But he was stripped of his French citizenship by the Vichy government in 1942 and his production companies were broken up. He and Ruth fled to South America. They spent the rest of the war in Brazil and the US, before returning to Paris afterwards. Max Glass and Ruth Werner would finally marry in 1957.

In contrast, Eduard Eysenck's career appeared to be barely inconvenienced by the new regime operating in Berlin. From ageing *Schauspieler* (actor), he

---

[10] Berg, B, Les indépendants du ler siècle—biographie de Max Glass, http://www.lips.org./Bio_GlassM_GB.asp. According to Berg, Ruth Werner's Jewish status was confirmed in a latter-day interview with one of Max Glass's cousins. What remains in doubt is just how widely known this was at the time. Brigitte Berg, personal communication, 3 December 2003.

became a *Conférencier* (cabaret host).[11] If anything, he enjoyed even more professional success under the National Socialists—partly because many rivals had been barred because they were Jewish. He was able to trace his family back through the centuries and was reassured by his impeccable Aryan pedigree. Even though he apparently hobnobbed with those in power, Eduard Eysenck did not immediately rush to join the National Socialists. He was not quite the enthusiastic Nazi Hans implied he was. Eysenck senior did not become a member of the Nazi Party until 1 May 1937, when it became virtually compulsory for those in his line of work.[12]

Growing up in Berlin

Hans Eysenck had an unusual childhood. Although he was never neglected, his parents had little to do with him. Eysenck remembered them as mean and stingy, and not particularly attentive. Of his mother he recalled:

> I saw very little of her, except occasionally on holidays, and she never managed to treat me as a child, or show much interest in what I was doing. Conversation with her was on strictly adult lines ... about the theatre, plays, literature and poetry, and cultural topics of that kind...[13]

He became genuinely interested in German literature as a result, he said, so 'perhaps her method of upbringing was not entirely mistaken'.[14]

Eysenck hardly spent any more time with his father. Instead he was palmed off to various members of his extended family. He spent time with his paternal grandmother in Rodenberg, near his father. However, for the best part of his boyhood he stayed with his maternal grandmother in Berlin. His tales of the city revolved around her flat on the Kaiserallee, its middle-class drabness home to some singular individuals in the making.[15] Weimar was a brief but extraordinary period in the arts and sciences in Germany, Berliners in particular. This was the city for cabaret and *The three-Penny opera*; it had universities

---

[11] According to his Actors Registration record, Eysenck senior did a long stint in the last years of the war at the Kabarett Wintergarten, the premier Berlin nightspot for varieté at the time. File held in Personenakten der Reichskulturkammer, Bundesarchiv, Berlin.

[12] This information was gleaned from Nazi party membership files. BA (ehem. BDC), NSDAP-Gaukartei, Bundesarchiv, Berlin.

[13] Eysenck (1990*b*), *Rebel with a cause*, p. 7.

[14] Ibid.

[15] Ibid., Chapter 1, and Gibson, H.B. (Tony) (1981), *Hans Eysenck: the man and his work*, Chapter 1.

boasting Albert Einstein, Max Planck, and, briefly, Max Born.[16] The streets of Weimar Berlin were captured in the work of artist George Grosz and writer Christopher Isherwood. The youthful Eysenck walked these streets, along with another future émigré, the historian Eric Hobsbawm.

Eysenck's father had wanted him to follow in the family footsteps into the theatrical arts. His mother was reportedly less keen on this idea. The world of facts and knowledge held more appeal to Hans himself, and he soon had his heart set on becoming a scientist. Eysenck later claimed that his parents, and his maternal grandmother, had little guiding influence on him—implying his adult character was very much the product of his constitution and his own actions. However, his parents' absence from his life may well have been telling. As Eysenck's first son Michael pointed out, not having one's parents around marks one out as different in a quite visible way.[17]

Eysenck entered Bismarck Gymnasium around 1925, moving to Prinz-Heinrichs-Gymnasium for his secondary education. Situated in Schöneberg, the Prinz-Heinrichs was a good school for some of the best and the brightest of the moneyed classes of Berlin. The school does not exist now, for it did not survive the war. Few details of Eysenck's education there can be found.[18] Nevertheless, Eysenck's account of his years at the Prinz-Heinrichs can be compared and contrasted with an account supplied by his fellow pupil Hobsbawm. Around 15 months Eysenck's junior but two grades behind, Hobsbawm was 'ein Engländer' as well as 'ein Jude'. Hobsbawm remembered the Prinz-Heinrichs as:

> A Prussian school with military connections … naturally Protestant in spirit, deeply patriotic and conservative. Those of us who did not fit this pattern—whether as Catholics, Jews, foreigners, pacifists or leftwingers, felt ourselves as a collective minority, even though in no measurable way an excluded minority.[19]

---

[16] For an anecdotal background on Weimar Berlin, see Friedrich, O. (1972), *Before the deluge: A portrait of Berlin in the 1920's*, Harper and Row, New York.

[17] Michael Eysenck, interview, 28 October 2001.

[18] As well as checking the Bundesarchiv, I have searched for Eysenck's school records in the Bibliothek für Bildungsgeschichtliche Forschung, Landesarchiv, Berlin, with the able assistance of German historian Stefan Petri. Unfortunately, this research yielded few new details of Eysenck's schooling. For an account of the history of the Prinz-Heinrichs written by old boys, see Stallman, H. (ed.) (1966), *Das Prinz-Heinrichs-Gymnasium zu Schöneberg, 1890–1945*, privately published, Berlin.

[19] Hobsbawm, E. (2002), *Interesting times: a twentieth century life*, Allen Lane, London, p. 52.

Yet Hobsbawm argued that the Prinz-Heinrichs was by no means a Nazi school. However, Hobsbawm left for England very soon after Hitler came to power.

Eysenck said most of the teachers at the school were right-wing nationalists, although few favoured the Nazis before the advent of Hitler. The majority of his classmates were gullible supporters of the new regime. In hindsight, he thought it a wonder that he was never beaten up, given that he was 'almost the only non-Jewish boy in an almost entirely Nazi school who strongly and vocally dissented from the majority'.[20] This dissenting standpoint inevitably brought him many enemies. In passages that echo the combative style of later life, Eysenck instructed the reader on the psychological tactics he used to face them down. Eysenck also noted the advantage he had in these school-yard battles: he was tall, athletic, and very good at sport. It provided him with a confidence and fighting capability that deterred the bullies.

According to Eysenck, he was not a terribly conscientious student. Bored by the syllabus, he was equally uninspired by his teachers, many of whom he characterized as stupidly pedantic. Eysenck found little excuse for their pedagogical limitations or nationalistic cupidity, and felt no reason to hide his superior intelligence and political acumen:

> I had one particularly cogent reason for not thinking very highly of my teachers, whom in retrospect I must say were excellent examples of their profession ... It was clear to me fairly early on that I was a good deal brighter than any of them ... that even in their specialities I probably *knew* more than most of them ... I simply couldn't get on with the emphasis on Prussian history ... and found the teaching singularly uninspired and ignorant of wider issues. [original italics][21]

Hobsbawm paints a slightly more sympathetic portrait of those 'misunderstood adults'.[22] Still, Eysenck's and Hobsbawm's accounts tally in many other respects and contain some striking parallels. These two misfits came of age in 'a world that was not expected to last'.[23] Both convey the sense of impending disaster as Weimar chaos segued into Nazi terror. Both dwell on the intense politicization of everyday life that was impossible to ignore. And both spell out how these social upheavals reached into their school: the replacement of the respected *Oberstudiendirektor* (Headmaster) Dr. Walter Schönbrunn with a

---

[20] Eysenck (1990b), *Rebel with a cause*, pp. 20 and 26.

[21] Ibid., p. 30.

[22] Hobsbawm (2002), *Interesting times: a twentieth century life*, Chapter 4.

[23] See Hobsbawm's memorable description of life in the last days of Weimar in ibid., Chapters 4 and 5. Quote comes from p. 47.

'bitterly resented', Nazi-appointed *Kommissarischer Leiter* (temporary external director), the dismissal of the popular Jewish maths teacher Professor 'Sally' Birnbaum, the fellow students who sported swastikas and mouthed racial propaganda ...

Eysenck's tales of his school days were framed in terms of the man he became, his youthful self-certainty bucking a tide of rigidity and ignorance. He recounted with relish the time he refused to sing on request and bit his teacher, the occasions he stood up to his classmates, the way he habitually defied authority. Here was the rebel, it seemed, already formed.

Eysenck's schooling was not restricted to Berlin, however. His teenage years were interspersed by several visits to the British Isles: for a holiday in Folkestone in 1929, for a term at a public school on the Isle of Wight in 1930, and for a term at Exeter College in 1932 where he studied English language, history, and literature.[24] This pattern of holidaying and education is left unexplained. Perhaps his mother, in particular, harboured the kind of Anglophilia common amongst European Jewish intellectuals. Eysenck liked to recount his youthful infatuation with English language literature and poetry, but emphasized that this was a self-led discovery. Virtually all Eysenck said about these trips was that they helped his English along nicely and smoothed the way for his subsequent immigration.

The records show Eysenck passed his *Abitur* (school leaving certificate) in Easter, 1934, around the time of his eighteenth birthday. Even though Eysenck says he was promoted from the third to second last year of school, he was not much younger than many of his fellow students. However, he was not to see much more of them.

Leaving the fatherland

By the time he left school, Eysenck said that he had already developed a reputation as something of a renegade. He had many friends in the German Communist Party, he recalled, 'some of them reasonably high up'.[25] He had many Jewish acquaintances and often publicly took issue with anti-Jewish propaganda and social measures. Eysenck also had several friends with National Socialist connections, though he seemed to want to excuse or play down these links. His first girlfriend, Ilsemarie, was a *Hitlermädchen*, and they remained

---

[24] There is some confusion about the dates of these trips, especially Eysenck's spell at Exeter College. Gibson says it was the following year, 1933. See Gibson (1981), *Hans Eysenck: the man and his work*, Chapter 1.

[25] Eysenck (1990*b*), *Rebel with a cause*, p. 32.

friendly even after the war. 'I argued with her frequently, pointing out what Hitler really stood for but she was so persuaded of his rhetoric and his promises ... that she simply could not see the dark side of his nature.'[26] His 'old friend' Egon Borgerhof was a member of the Strasser faction of the National Socialist movement. It was an association that almost saw both of them packed off to the concentration camps that were initially constructed to hold political dissidents. Eysenck says his politics were halfway between those of the Social Democrats and the Communists. Even so, he said he was not a member of any party.

Looking to further his education, Eysenck sought entry to Berlin University to study physics. According to Eysenck, university authorities made it clear that he had to join the *Schutzstaffel* (SS) to be accepted. He refused. Eysenck explained:

> This was not a general rule for all students, of course; I imagine the university authorities had heard of my political attitudes and behaviour, and wanted me to toe the line ... I didn't have to make any decision; I knew I could not live in that uniform, and with those people, and that emigration was the only possibility for me.[27]

On the face of it, the stipulation to join the SS seems unusual. The SS was Hitler's original security service—very prestigious and, reputedly, very selective. While it was growing rapidly in terms of its manpower and functions, the SS was not said to press-gang reluctant recruits. One wonders whether Eysenck actually meant the *Sturmabteilung* (SA), the brownshirted storm troopers. More broadly based, the SA was certainly less selective. Thus it might have easily accommodated someone like Eysenck. This suspicion is strengthened by the fact that Eysenck appeared to confuse the SA and SS on occasions, referring to both as the 'storm troopers'.[28] In earlier interviews he had been vaguer, claiming that his attendance at university was made conditional upon 'joining the Nazi Party or the storm troopers'.[29] However, even the SS was relatively lax about selection requirements in the early years of Hitler's chancellorship. Up until 1938, membership of the Nazi Party or one of its youth organizations was not a condition for joining the SS, although it was an advantage. One had to be under 30 years old, be reasonably tall, of good health, and 'Nordic appearance'.

---

[26] Ibid., p. 36.

[27] Ibid., p. 39.

[28] Ibid., pp. 22 and 39.

[29] See his notorious interview published in the first issue of the right-wing National Party's *Beacon*. *Beacon* (1977), Interview with Hans Eysenck, *Beacon*, February 1977, 8. For more on this see the discussion of the race and IQ controversy in Chapter 8.

One also had to pass a simple intelligence test. Sympathy with Nazi ideology was a criterion, but even this was handled rather flexibly at first. SS leaders took the view that any young man of good stock could be quickly re-educated if necessary. Apart from his somewhat left-leaning political inclinations, Eysenck may well have been regarded as an excellent SS recruit.[30]

Eysenck also claimed that when his father Eduard heard about all this he attempted to curry favour with Göring in order to get his son admitted to the SS at officer level. Whether Eduard Eysenck was in a position to obtain such a favour, and whether Göring was in a position to effect it, is difficult to say. It might also sound unlikely that university authorities would make such a stipulation, especially to study a course that was not yet officially race restricted. However, university entry was being tightened considerably at this time to combat what Nazi officials termed 'overcrowding'. Part of the National Socialist agenda was to purify the student body and bring it into line with Nazi ideology. Jews were being successively barred from leading courses like medicine, veterinary science, business studies, pharmacy, and the law, and had been purged from the academic faculty in most institutions—including Berlin University.

Student intakes dropped markedly in the mid-1930s as Nazi-style entry criteria and quotas were progressively imposed. The criteria laid down in February 1934 were of three sorts: intellectual; physical; and moral–political. A fourth racial criterion was added the following year. Each applicant's 'political reliability' had to be vouched for by his school's *Kommissarischer Leiter*, and approved by the provincial ministry's director of the division of higher education in consultation with local party leaders. These measures proved so successful in cutting intakes that quotas were subsequently dispensed with. By the time Eysenck finished school, previous membership of National Socialist organizations like the SS, SA, or Hitler Youth was a definite advantage, if not a must, when it came to getting into courses. Moreover, the entry process left ample room for interpretation, and ample opportunity for various officials to intervene in particular cases. There is some evidence that troublesome individuals *were* actually singled out for special treatment or barred from university entrance altogether.[31]

---

[30] Wegner, B. (1982), *Hitler's politische Soldaten: die Waffen SS, 1933–1945*, Schöningh, Paderborn, pp. 135–49. See also Gelwick, R.A. (1971), Personnel policies and procedures of the Waffen-SS, Ph.D. thesis, University of Nebraska at Lincoln.

[31] See Hartshorne, E. (1937), *The German universities and National Socialism*, George Allen and Unwin, London, Chapter 3.

Fig. 2.1 Eysenck's father (Anton) Eduard Eysenck.

Source: Centre for the History of Psychology, Staffordshire University.

One may never be able to ascertain the precise nature of the conditions set for Eysenck and from where they originated. However, the fact that conditions were set for him is not implausible. If he was half the rebel he said he was, then his troublesome 'political attitudes and behaviour' may well have attracted special conditions. Of course, his family's mixed ancestry may also have been a factor in all this, a factor Eysenck never acknowledged.

Given a terrible ultimatum by university officials in the summer of 1934, Eysenck turned émigré. His passage out of Germany was greatly facilitated by his mother and stepfather. Nevertheless, he did not immediately join them in Paris, electing to study in Dijon for a short time before moving on to England. Eysenck left behind his father, his stepmother, and his maternal grandmother who had cared for him. His father and new wife fared well. However, his maternal grandmother was deported and died in a concentration camp—not because she was old and crippled but almost certainly because she was regarded as a Jew.[32] She had sought sanctuary by converting to Catholicism

---

[32] Eysenck made this clear in his memoirs. See Eysenck (1990*b*), *Rebel with a cause*, p. 80.

Fig. 2.2 Eysenck's mother Ruth Werner a.k.a. Helga Molander.

late in her life. Despite being sheltered by two sisters of the faith, the Gestapo came for her.

Eysenck's political stand, his refusal to join the SS (or the SA or the Nazi Party), became a defining episode in his life. Leaving for a life of exile would cost him dear, he said, implying that it would have been easier to stay. But it gave Eysenck a valuable kind of leverage in later controversies. It enabled him to portray his decision to leave as a principled political act, a protest against fascism rather than just an understandable act of self-preservation.

The Jewish question

Being Jewish matters because people think it matters. Conflicting definitions of Jewishness have been woven into a long history of persecution. Claiming to be or not be Jewish can thus be a highly political act. However, in Hitler's

Germany, self-ascription was almost irrelevant. What counted were Nazi definitions. Eysenck's mixed ancestry mattered to the Nazis authorities, and this made it likely to matter in nearly all aspects of his life in Berlin: at home and at play, at school and at any university he might seek entry to. And like it or not, his ancestry still mattered in England, for it implicitly framed his identity as a German émigré.

Eysenck was always slightly evasive about the religious aspects of his background and several inconsistencies appear in the accounts he supplied over the years, adding to the uncertainty. For example, in his 1977 interview with David Cohen, Eysenck was quoted as saying his father was Lutheran and his mother Catholic.[33] Later he said this was a misunderstanding and it was the other way around, a version he subsequently stuck to.[34] When Eysenck first arrived in the UK it was widely assumed he was Jewish, an assumption noted by his biographer Gibson as well as his long-time clinical colleague, Bob Payne. 'Why else was he here?' the reasoning went.[35] Prior to the war, however, Eysenck specifically denied it. This may well have been the truth as he knew it at the time. He said he first learnt of his maternal grandmother's Jewish background when informed of her grisly fate at the end of the war. But it was not until 1990, when his full-length autobiographical account appeared, that he made these details public. As Payne noted, 'the only mystery is why he suppressed them for so long.'[36]

So Eysenck *was* part-Jewish, certainly by National Socialist definitions. He was at least one-quarter Jewish in terms of his religious lineage—a borderline case for special treatment by the Nazis. Was this also a factor in his decision to leave? Eysenck claimed he knew none of this at the time, implying it had nothing to do with his own thinking. Perhaps this was the case. Up to 1935, only some university courses required proof of Aryan descent. But these did not include the physics course Eysenck applied for. Some jobs did so as well—which was why his father was required to attest to his own German blood, dutifully tracing the Eysenck half of the family tree back several centuries.

Eysenck left Germany just before his part-Jewish status would have to have become obvious and official to everyone, including himself. Even so, the April decrees of 1933 had forced many people—two million civil service employees,

---

[33] Cohen, D. (1977), Interview with Hans Eysenck.

[34] See Gibson (1981), *Hans Eysenck: the man and his work*, p. 17.

[35] Ibid., p. 21, and Bob Payne to Alan R. Dabbs, October 1994. A copy of this private letter was kindly supplied to me by Alan Dabbs.

[36] Bob Payne to Alan R. Dabbs, October 1994.

doctors and lawyers, academics, students and their parents—to examine their ancestry. These laws sought to strategically divide Aryan from non-Aryan. The category non-Aryan included several 'racially alien' groups, including the Roma and Sinti people (a.k.a. gypsies), but it mostly covered Jews. A non-Aryan was defined as any person with at least one Jewish parent or grandparent, while a Jew was any person belonging to the Jewish religion. The manner in which this highly inclusive definition was interpreted depended on practical and political considerations. For instance, all those with any Jewish ancestry whatsoever were removed from the civil service, since there were relatively few such cases left by the time the Nazis took over. Almost all those classed as non-Aryans were likewise dismissed from public schools and university faculties.

Legislation drawn up in April 1933 to prevent the 'overcrowding' of German schools and universities stipulated maximum percentages of non-Aryan students, but made exceptions for children of mixed marriages, war veterans, and so on. This meant that schools were obliged to tally up students' backgrounds. Many school children found it a shock to be classed as non-Aryan status, since their parents had led them to believe otherwise. They found that being officially branded a Jew set them apart. In the classroom and the playground, Jewish students were singled-out, marginalized, often traumatized.[37] These effects were deliberate. Early racial laws had an 'educational' intent; they were designed to foster public racial consciousness in tandem with anti-Semitic propaganda.[38]

The fact that Eysenck's grandmother came from a Jewish family meant that Eysenck was non-Aryan according to the 1933 decrees. It is quite possible, likely even, that Eysenck had to make a declaration on his ancestry either in his last year at school or when presenting for university in the middle of 1934. If not, he may have been able to avoid being publicly labelled non-Aryan. Eysenck's maternal grandmother had few ties with the Jewish community, making her status less than obvious. His mother and stepfather were a different story, however, even though they were no longer around.

---

[37] For an affecting account of the impact these new laws had in German schools see Kaplan, M.A. (1998), *Between dignity and despair: Jewish life in Nazi Germany*, Oxford University Press, New York, Chapter 4, and Abrahams-Sprod, M.E. (2007), Life under siege: the Jews of Magdeburg under Nazi rule, Ph.D. thesis, University of Sydney, Chapter 5, at http://hdl.handle.net/2123/1627.

[38] Nazi propaganda depicted Jews as an amalgam of evil, claiming that the German people were locked into life and death struggle with these parasitic but powerful foreign conspirators. Jews, especially Jewish men, were caricatured as hideously deformed and subhuman on posters and in pro-Nazi publications such as Julius Streicher's rabid *Der Stürmer*.

After Eysenck left for France, the Nürnberg laws of September 1935 sought to clarify ambiguous cases and bring definitions of Jewishness in line with Nazi eugenic objectives. They divided non-Aryans Jews into full-blooded Jews (those with at least three Jewish grandparents) and part Jews, termed *Mischlinge*—literally 'mixed ones' or 'mixed breed', but with the derogatory connotations of 'mongrel'. According to these 1935 edicts, Eysenck was *Mischlinge* of the first, second, or even third degree, depending on whether his grandmother was regarded as fully Jewish and on whether her late husband was regarded as Jewish as well.

In his 1990 autobiography, Eysenck admitted that he knew his 'life in Germany would encounter grave problems'.[39] But he still preferred to emphasize the sacrifice he made. Yet, had he stayed, things would have been grim. It would have partly depended on how he was classified. For example, the 1935 Nürnberg laws set out different marriage restrictions for first and second degree *Mischlinge* with the aim of converting half-Jews to full-Jews and assimilating quarter Jews as Aryans.[40] But as the Nazi era rolled on, racial measures were made progressively more draconian and murderous, and Judaic links became more difficult to conceal. These measures came to affect *Mischlinge* in unpredictable ways, since they were applied in an inconsistent, discretionary fashion—often by Hitler himself. While many *Mischlinge* were persecuted and deported, others were allowed to continue their work or serve in the armed forces, and several attained very high rank. Even with the advent of the Final Solution midway through the war, the Nazis were still unsure about what to do with the various degrees of *Mischlinge*, proposing but not necessarily enacting policies of sterilization and extermination. However, *Mischlinge* who looked or acted Jewish were always at risk of deportation and death.

Even if Eysenck had been willing to toe the political line, a university education might have been difficult. Up to 1940, *Mischlinge* were still allowed to study most subjects, except in key areas like dentistry and law. However, Eysenck would have faced a hostile environment that saw most Jewish students

---

[39] Eysenck (1990*b*), *Rebel with a cause*, p. 39.

[40] The Nürnberg laws stripped Jews of their citizenship, but not those of *Mischlinge*. *Mischlinge* might or might not have been discriminated against under the various measures that were introduced, and this might depend on whether they were first or second degree cases. For instance, those with only one Jewish grandparent (second degree *Mischlinge*) were still able to sit university exams in the otherwise restricted university courses up to the time Eysenck departed. There is a huge literature on this sad period of German history See, for example, Hilberg, R. (1985), *The destruction of the European Jews*, Holmes and Meier, New York, and Friedländer, S. (1997), *Nazi Germany and the Jews*, Harper Collins, New York.

disappear from the secondary school system and universities by the late 1930s. Obtaining rewarding and responsible employment after university would have been tough. With the German military effort escalating, the armed forces would have been his most likely destiny, and any refusal to join would not be looked upon kindly. And even if he did not end up in the army, his life would have remained in danger. Late in the war the Nazis would often round up *Mischlinge* to make up the numbers on transports or to get rid of 'troublesome' individuals. With his non-Jewish appearance and a slice of luck, Eysenck might have got through the war. But given the compromises he would have had to make, the aftermath of allied victory might not have been good to him.

Such speculation merely serves to make a point. We can allow Eysenck his mantra of a principled political stand. There is no reason to doubt his objections to the politicization of his education, no reason to think that his abhorrence of Nazism was anything but sincere. Yet in the full knowledge of hindsight there are other ways to read his actions. Political principles become something of a luxury; for Hans Eysenck, *Mischlinge*, leaving the Fatherland was a matter of personal survival.

There is little direct evidence to suggest Eysenck knew he had to leave because he was partly Jewish. He did mention in his autobiography that he was termed a 'white Jew' by his fellow students because, he said, he was friendly with and overtly sympathetic to those of that faith.[41] Bob Payne suggested that he must have known and that such a label was probably based on an 'awareness of his mixed ancestry'.[42] The racial consciousness generated by the new education laws of 1933, and the fact that his famous film star mother and her high-profile industry partner had to flee the country because they were both widely regarded as Jewish, make this conclusion hard to resist. It is quite possible that the conditions set for his attendance at Berlin University originated for this very reason. His 'radical' politics might have been the least of it. Eysenck might have been asked to join the SS (or SA or the Party) to absolve his partial Jewishness. His racial impurity could then be overlooked in favour of his obvious intelligence and tall, *blond* good looks.

When Eysenck enrolled in London University in October 1935, he put down Eduard Eysenck at an address in Berlin as his responsible parent. Perhaps it was his father's side of the family that Eysenck preferred to emphasize at the time, for it would sidestep the issue of why his mother had fled to France. But, then again, if Eysenck did know about his part-Jewish ancestry before the war,

---

[41] Eysenck (1990*b*), *Rebel with a cause*, p. 23.

[42] Bob Payne to Alan R. Dabbs, October 1994.

he never confided this to his first wife, Margaret Davies. Eysenck made the last of his annual visits to Germany to see his father and grandmother in 1937. He and his mother had already managed to smuggle their family fortune out of the country; he was fortunate to make it out on this occasion given that he was a male of military age without the necessary special travel permit. Margaret accompanied him on at least one of these trips to Germany, probably this last one. According to Michael Eysenck, his father gave no indication to Margaret of any problems his grandmother might face in the future, nor the dangers he faced in visiting Germany. It seems possible that Hans might have been extremely cavalier about his own safety, full of youthful bravado. But perhaps this reflected a genuine ignorance of his ancestry and the risks it conferred.

Nevertheless, Eysenck conceded that both his mother *and* father must have known. When he first learnt of his grandmother's death, it suddenly dawned on him why Eduard had tried to get him into the SS at officer level: 'to protect me from persecution'.[43] In principle, Jews and part Jews were barred from obtaining officer ranks in the German armed forces, especially in the SS. Anyone serving in such a position would be assumed to be beyond reproach. It might not have been possible, however. As Payne noted, Eysenck would have been 'wise not to rely on Göring to suppress the fact that he was half-Jewish'.[44]

Irrespective of whether Eysenck was fully informed at the time, he had several good reasons for denying he was Jewish both before and after immigrating. For a start, he probably genuinely felt non-Jewish. From his accounts, religion had played little role in his upbringing—even though he was made acutely aware of the racial designations of his peers and extended family. Many of those living in Germany at the time had their sense of ethnic and religious identity challenged by Nazi edicts, and by the assumptions made when they emigrated. As Eric Hobsbawm put it:

> Several of my student contemporaries in Cambridge, Oxford and the LSE were Germans. One assumed that they were all Jews, though hardly any came from religious families and some from converted ones—but even this we neither knew nor cared about. Very few at the time took a specific interest in their own Jewishness, and I cannot remember discussing Jewish identity with them, though we were, presumably, all aware that we were Jews by Nazi standards.[45]

Second, Eduard Eysenck may have suffered had it become known his émigré son was considered a Jew. Eysenck may also have known that the safety of his

---

[43] Eysenck (1990*b*), *Rebel with a cause*, p. 80.

[44] Bob Payne to Alan R. Dabbs, October 1994.

[45] Eric Hobsbawm, personal communication, 21 September 2004.

grandmother depended on his non-Jewish status. In his memoirs Eysenck claimed that, at the time, he had a clear vision of the horrific future ahead for 'Jews, gypsies, and all sorts of cripples …' although he did not directly connect these thoughts to his grandmother's fate.[46]

Third, a non-Jewish identity was probably the simplest, most expedient one for Eysenck to take on once he arrived in England—even when he did become aware of his mixed ancestry after the war. It probably helped smooth his passage through the education system. While many successful psychologists and psychiatrists in England were Jewish, they still encountered subtle forms of prejudice. Eysenck's future mentor Cyril Burt embodied the casual anti-Semitism of some English dons, being in the habit of calling the young Eysenck a 'boastful … German Jew' behind his back.[47] When informed of this in 1979 by Ann Clarke, Eysenck's response was worth recording:

> Burt knew perfectly well of course that I wasn't Jewish … it suggests that his hold on reality was much weaker than I had given him credit for … the passage certainly establishes his insane racism, which I had always doubted, but which apparently he did show. There was the same flaw in Galton of course.[48]

One wonders whether Burt knew something others didn't or was simply making prejudicial assumptions. Conversely, Eysenck's denial can probably be attributed to a sincere sense of disconnection to Judaism. It was probably just easier to not have to explain. Over the decades he continued to deny his partial Jewishness, and then came to equivocate about the category itself.[49]

In the mid-1970s, when he became embroiled in the race and IQ controversy and questions about his own background intensified, he took up Arthur Koestler's bizarre and inflammatory line of argument.[50] Eysenck would tell anyone who asked that the term Jewish was meaningless. In the first place, not

---

[46] Eysenck (1990*b*), *Rebel with a cause*, p. 33.

[47] The racial consciousness of Burt was noted by others—including his biographer Leslie Hearnshaw, and Tony Gibson and Leon Kamin. See the Burt correspondence files in the Liverpool University archives, especially Cyril Burt to Marion Burt, 4 February 1949.

[48] Hans J. Eysenck to Ann Clarke, 19 November 1979, p. 1. A copy of this letter was kindly supplied by Ann and Alan Clarke.

[49] For another example of his flat denial of his partial Jewishness, see *Beacon* (1977), Interview with Hans Eysenck.

[50] Koestler, A. (1976), *The thirteenth tribe: the Khazar empire and its heritage*, Hutchinson, London. Koestler's odd book caused disbelief and anger amongst his fellow Jews, and became a standard resource of the anti-Semitic right. Eysenck wrote to John Baker—author of the book *Race* that Eysenck had reviewed—to discuss Koestler's viewpoint. Despite his conservative inclinations, Baker could not agree with Koestler's conclusions. See John R. Baker to Hans J. Eysenck, 9 July 1977, Papers and Correspondence of John

all those descended from Jews followed the Jewish religion, he claimed; many were Christians or atheists. Moreover, the term did not apply to any particular race as he saw it, given most European Jews were not semitic. Following Koestler, Eysenck suggested the Jews of Europe were not descendants of those originating in the region of modern-day Israel. Eysenck argued that a tribe of Aryan-speaking Caucasians, the Khazars, had converted to Judaism while living on the north-east side of the Mediterranean after the birth of Christ, and later migrated to Russia and eastern Prussia.

It was a line Eysenck repeated in his 1990 autobiography, when Eysenck was ready to reveal more details of his heritage. This information came with some strong caveats about Jews as a race, especially European Jews. Eysenck argued that 'Jewishness is a religious faith, and now maybe a national entity, but not a race.'[51] He rejected any genetic implications this concept carried, thus making his genealogy irrelevant to his current status. And, since he was not a follower of the Jewish faith and not a citizen of any Judaic nation, he was not a Jew. By this reasoning, 40 years of flat denials could be made consistent with the realities of his grandmother's death.

In trying to make sense of Eysenck's autobiography, Bob Payne maintained that he had 'never known Hans to say something that was not true'.[52] For the most part, Eysenck wanted to retain the right to define his own status—it was all about the religion of the individual concerned, not about race, not even about religious ancestry. He had, he felt, no religious inclinations or baggage. While his mother was nominally a Protestant and his father a lapsed Catholic, Eysenck said little about any religion his grandmother may have practised during his childhood. Yet Eysenck freely ascribed the identity of 'Jew' to Max Glass, despite his conversion to Catholicism.

Despite rejecting a Jewish identity for himself, Eysenck still wanted to emphasize an affinity with Judaic values and virtues. He noted that many of his friends and work colleagues were Jewish, as was his second wife. Thus 'it is obvious that Jews have an attractive quality for me.' Pointing to laudable Jewish achievements in the arts and sciences he added, 'perhaps a little of that creative genius had touched my life!'[53]

Randal Baker, CSAC 114.5.86/E.99, Special Collections and Western Manuscripts, Bodleian Library, Oxford University.

[51] Eysenck (1990b), *Rebel with a cause*, p. 34.

[52] Bob Payne to Alan R. Dabbs, October 1994.

[53] Eysenck (1990b), *Rebel with a cause*, p. 81

Taking the personal from the past

The past is not a fiction, but any reconstruction of it can't be separated from the act of storytelling. As an unwanted only child who never knew a stable, two-parent home, Eysenck comes across as precociously clear-eyed. Here was a young man who grew up in a fabled city during a turbulent period of history. He saw the rise of an evil dictatorship disguised as populist salvation. If only his fellow Germans had seen through this masquerade and managed to avoid the horror ahead. Consider, for example, the remarkably prescient vision he said he had at a Hitler rally in 1933:

> To see and hear him speak ... made an indelible impression on me—an impression of naked evil, of original sin, if you like, of unimaginable viciousness and cruelty ... I knew at that moment that war was inevitable ... I knew exactly how Cassandra must have felt—able to foresee the future, unable to change it.[54]

Eysenck's tales of his early life illustrated his remarkable capacity to draw a line between the personal and the political. A recurring theme of these tales centres on rejection: his rejection of Hitler, of Nazism, and of much of his German upbringing. But did he reject his homeland, or did it reject him? Even if Eysenck did know of his mixed ancestry, he had to eventually concede that those close to him certainly did.

It is impossible to say how Eysenck really felt about his fated trek. Perhaps a clue can be gleaned from a comparison with the reactions and recollections of his counterpart from the Prinz-Heinrichs, Eric Hobsbawm. While Hobsbawm was nearly two years Eysenck's junior, they shared some of those 'interesting times' in Germany. Yet each claimed to take something quite different from these experiences. Hobsbawm's family guardians moved him to Britain in 1933 for a variety of reasons; even though they were Jewish, fear of persecution was not the only consideration. However, Hobsbawm recalled that his relatively brief education at the Prinz-Heinrichs and on the streets of Berlin was the making of him as an activist. The intensity of this experience translated into direct political action, to protest marches and leafleting for the Communist Party. It was a first-flush experience of youth that imprinted him for life, he said. Hobsbawm wanted to hold on to that feeling, to the heady atmosphere of 1933 Berlin, for as long as possible. He found England a tame, sleepy backwater in comparison and wrote his private diaries in German for years to come.

For Eysenck, it was quite the opposite. He found the politicization of everyday life in Germany repellent and oppressive. You sense him turning up his nose as he tries to negotiate a way through it then, and 50 years later.

---

[54] Ibid., p. 33.

To Eysenck, pre-war Berlin was weird and dangerous. He used terms like 'Kafkaesque', 'insane', and 'surreal' to describe an ever-worsening situation. Eysenck's tales repeatedly focus on the debasing force of politics, the way partisan appeals could overwhelm rationality and dupe the innocent. For example, when arguing with girlfriend Ilsemarie or SS man Mathias Steinberg about Hitler's real intentions, he learnt of the 'utter impotence of reason when faced with rhetorical and emotional hogwash'.[55] Far from wanting to hold on to the memory of Berlin, Eysenck exorcized much of it and reputedly refused to speak German for years.[56] He took to English life with enthusiasm. English literature, politics, and sport became great passions, even if experienced from the vicarious perspective of a foreigner. Like his mother, Eysenck did not rush to visit his homeland after the war, despite the fact that his father and relatives still lived there.[57]

Caught between a rock and a hard place in Berlin, Eysenck opted to disengage. He remained defiantly out of step as a German refugee in England, still just as detached. Nonetheless, the way the Nazis magnified and manipulated popular racial consciousness left indelible marks—literally and figuratively—on those they persecuted. It made it hard to escape from this kind of thinking even as one tried to oppose it. Eysenck did not, he said, bridle at Nazi anti-Semitic propaganda because it hurt him or his family directly. This kind of racial typing was wrong because it was factually wrong—irrespective of whether it was offensive to those it targeted. Then and now, he put the accent on evidence that showed Jews were actually brave, intelligent, and not 'Jewish looking'. Moreover, such propaganda was anathema to Eysenck because it represented irrational prejudice promulgated by undeserving authority figures and repeated by unthinking, ignorant plebes. And Eysenck repeatedly linked the debasing power of political commitment with the rise of history's biggest villain.

Eysenck could not quite claim the moral authority bestowed upon those of the wartime resistance and the survivors of the camps. He came close on occasions, talking of taking on 'Nazi bullies' at school. But he didn't face Nazism

---

[55] Ibid., p. 36.

[56] The articles Eysenck published in German journals were generally translated by others. A rapprochement with things German occurred late in Eysenck's life when he began collaborating with Ronald Grossarth-Maticek.

[57] According to his second wife, Sybil, Eysenck's final break with his father came in the early 1950s, when he announced plans to re-marry. Being the daughter of well-known Jewish musicians Max Rostal and Sela Trau, Sybil was not warmly received because Eysenck's father Eduard was quite anti-Semitic. She recalled that he refused to accept her as part of the family and cut off contact. Sybil Eysenck, interview, 13 September 2004.

firsthand for very long; instead he went though the Blitz with the rest of London, which was terrifying enough. The few dissident intellectuals who did stay were dealt with severely by the Nazis, and were eyed with suspicion by the victorious allies after the war. Even though he characterized himself as a left-wing radical in his youth, it did not amount to any deep or permanent commitment to any particular set of political values. His allegiances were ultimately insular, tied to a trust in his own rational judgement that often put him at odds with the received wisdom of a sheepish majority. This was what he was always 'rebelling' against, he implied. Political engagement was not for him in Berlin, and was not for him ever since. Perhaps it can help resolve one of the biggest paradoxes of Eysenck's career, and explain why his politics were so hard to read later in his career. Why didn't Eysenck seem to care that neo-Nazis, white supremacists, and anti-Semites took succour from his views on race and IQ? Why didn't he denounce them as loudly as many liberal social scientists thought he should? Selective disengagement was the key, and it started at an early age.

Unfinished business

Leaving Germany behind, the young Hans Eysenck had to tread a fine line between conflicting loyalties and personal expedience. Mostly painfully, he had to forsake his elderly grandmother. Antonia Werner was deported to Theresienstadt concentration camp on the 19th of April 1943 and died there some time in March 1944.[58] No exact date of death was recorded. Since Theresienstadt was not a death camp, she probably died as result of old age and the horrific living conditions.[59]

This unfinished business haunted Eysenck for the rest of his life. With a self-conscious effort of repression, he would not discuss her nor would he countenance any kind of historical examination of the Holocaust. Eysenck's second wife, Sybil, shares his reluctance to look back. While Jewish herself, she insisted

---

[58] The other plausible candidate, Anna Werner, was also deported to Theresienstadt on 1 September 1942, and died there just over a month later, 11 October 1942. The key detail given by Eysenck that made Antonia Werner the more likely candidate was that Eysenck said his grandmother was deported and died in a camp 'about a year before the end of the war...' Eysenck (1990*b*), *Rebel with a cause*, p. 80.

[59] Theresienstadt was the usual destination for *Mischlinge* brought up as Jews, for Jews in mixed marriages who remained in the Jewish community, and for Jews in mixed marriages that had been dissolved by death or divorce. Kaplan (1998), *Between dignity and despair: Jewish life in Nazi Germany*, p. 190.

Hans was not.[60] However, his first wife, Margaret, felt bitterly misled. According to Michael Eysenck, she was convinced Hans concealed his Jewish heritage from her before the war. And she thought that if he had made this known at the time, more might have been done to get Frau Werner out.[61]

Even if one assumes Eysenck was fully informed from an early age, one must be wary of reading post-Holocaust certainties back too far. One should not fault him for not anticipating what only became absolutely clear after 1938 with the annexation of Austria and Kristallnacht.[62] Moreover, there is no evidence to suggest that anyone else in the Eysenck/Glass family tried to get her out, even though they must have been aware of her dangerous racial status. As a frail old woman, it would have been a daunting trek.

Many other émigrés also left unfinished business. Many of those they left behind suffered a similar fate to that of Eysenck's grandmother. For example, four of Sigmund Freud's sisters died in the camps too. Despite the Freud family's connections, they were unable to secure visas.[63] Few countries readily accepted elderly Jews and few in Germany foresaw the full extent of the atrocities to come.

---

[60] Sybil Eysenck, interview, 13 September 2004.

[61] Commenting on this delicate family issue Michael Eysenck wrote: 'my mother was liberal-minded, so it was entirely the fact of him concealing something important that angered her, and not at all the fact that he was partly Jewish.' Michael Eysenck, personal communication, 3 December 2003.

[62] It is an easy historical trap to fall into, as Eric Hobsbawm himself pointed out when commenting on a draft of this chapter. Eric Hobsbawm, personal communication, 21 September 2004.

[63] The parallels with the fate of Eysenck's grandmother are striking. While Freud had managed to get out, his four aged sisters remained behind in Vienna. Despite the family's valiant efforts to evacuate them to the south of France, they were deported to the concentration camps. Dolfi was taken to Theresienstadt on 29 June 1942, where she perished on 29 September 1942. Paula and Mitzi were also taken to Theresienstadt on 29 June 1942 and then on to Treblinka on 23 September 1942. Paula and Mitzi were probably murdered in Treblinka some time late in 1942. Finally, Rosa was taken to Theresienstadt on 28 August 1942 and was probably exterminated some time late in 1942 in Treblinka, rather than in 1944 in Auschwitz as some sources suggest. This more recently discovered information came from the (now defunct) website http:/www.javari.com/FreudPsa/prod04.htm (accessed 23 October 2002. See also Gilbert, M. (1987), *The Holocaust: the Jewish tragedy*, Fontana, London, p. 476 and www.holocaustresearchproject.org/ghettos/freud.html.

Chapter 3
The accidental psychologist

Hans Eysenck arrived at University College in London in 1935, having just turned 19 years of age. He was a confident but unformed *savant*. While reasonably well-versed in the classics and physical sciences, Eysenck had learnt little psychology before he came to England. He barely knew the discipline existed. Eysenck was free to absorb the intellectual traditions of University College, London, and did so quickly and deeply. German, Russian, and American researchers might have influenced him, but only after he had been thoroughly inculcated into the 'London school' of individual differences psychology.

Eysenck's decision to enrol at University College dashed a boyhood dream of a career in physics. Instead it meant he would train as a psychologist, and train as a certain *sort* of psychologist. He entered an academic world riven with understated but profound divisions. By the time Eysenck got there, the kind of psychology practised at University College contrasted sharply with that practised elsewhere in Britain. Eysenck became a protégé of departmental leader Cyril Burt. He took on Burt's methodological outlook and most of his philosophical assumptions and, through Burt, those of Spearman and Galton as well. These key figures shaped Eysenck's thought: what to look at, how to look at it, and how to use it. It seemed that much of the mature scientist was formed in these early days. But what also comes through is the young Eysenck's sheer hubris, his unshakeable self-belief in the face of wartime hardship and uncertainty.

Enrolling as a psychologist

Eysenck's summer sojourn at Dijon University in 1934 was a short one. Better acquainted with English culture and literature than its French counterparts, he moved on to London in August that year. Eysenck's German school credentials were not recognized in England, so he was obliged to study for and sit the matriculation examination. Through the winter of 1934–35 he read mathematics, English, and several other arts subjects at Pitman's College, London as preparation. Sitting his exams at the University of London's buildings in South Kensington, Eysenck easily qualified for admission by coming 'near the top on all the subjects [he] took'.[1] However, he made an unfortunate and naive assumption. In the German university system Eysenck was familiar with, students were relatively free to choose subjects and courses once they had been accepted. This was not the case in Britain, and turned out to be a momentous misunderstanding. Eysenck had taken only one science subject in his matriculation, not the minimum two required to study physics (or any other science subject). Eysenck was furious at the time, and laid the blame at the university's doorstep, 'one of the worst places in the world as far as bureaucracy and red tape are concerned'.[2] His only option was psychology, then in the Arts faculty. It was only thing 'on the science side' he could take, the nearest thing to physics.[3] 'Try it, you'll like it', the university admission officers assured him. Eysenck could have re-sat his matriculation exam with the required subjects. However, he could not afford to wait another year. He was short of money and in a hurry.

Much has been made of this twist of fate by commentators, and by Eysenck himself. His first biographer, Tony Gibson, interpreted Eysenck's interest in all things rigorously scientific as a rebellion against his parents' aspirations for him in the arts. Gibson, amongst others, suggested that Eysenck became a psychologist styled as a frustrated or *de facto* physicist.[4] Yet this was surely based on a secondhand appreciation of what a science like physics was, for he never actually studied it at a formal, tertiary level. Eysenck's idealization of the natural sciences was based on his reading of the philosophy of science, which he immersed himself in from early on. His was a 'philosophy of physics envy' rather than 'physics envy' *per se*—but it still reflected a widely held mid-twentieth century view of the pecking order of scientific disciplines.

The natural sciences, especially physics, had achieved a prestige and influence that put them at the apex of human intellectual achievement. They were the best, the most authoritative form of knowledge. A roll-call of scholarly writers—Mach, Carnap, Popper, Merton, even Kuhn and Lakatos—all took this 'obvious' truth as a starting point. Criteria derived from natural sciences were habitually extended to evaluate other disciplines, the social

---

[1] Eysenck, H.J. (1990*a*), A sanguine veteran of psychological warfare, *Sunday Telegraph*, 11 March 1990.

[2] Eysenck, H.J. (1980*b*), Hans Jürgen Eysenck, in *A history of psychology in autobiography*, Vol. 7 (ed. G. Lindzey), pp. 153–87, W.H. Freeman, San Francisco, p. 156.

[3] Eysenck, H.J. (1997*b*), *Rebel with a cause* [revised and expanded], Transaction Press, New Brunswick, New Jersey, p. 47.

[4] Gibson, H.B. (Tony) (1981), *Hans Eysenck: the man and his work*, Peter Owen, London, Chapter 1.

sciences included. Psychology, not surprisingly, got short shrift. For the logical positivists, psychological theories usually lacked coherence, logical organization, and explanatory power. For Popper, much of psychology was unfalsifiable and therefore not scientific at all; for Lakatos, it was non-progressive; and for Kuhn, it was an immature protoscience. Rarely addressed was the kind of epistemic privilege granted to the natural sciences that, combined with a form of intellectual imperialism, made the backwardness of psychology seem like a no-brainer.[5]

For Eysenck, such put-downs were never cause for dismay; rather, they were a source of empowerment. For a start, they helped generate plenty of room for new approaches. Throughout his career he found the philosophy of science had many other uses as well. For example, it provided him with methodological blueprints and a critical tool-bag. Eysenck's early work took its prescriptive cues from English analytical philosophers like Susan Stebbing.[6] His anti-Freudian attacks drew upon the criticisms of Karl Popper and Frank Cioffi.[7] His latter-day theoretical integrations utilized aspects of Kuhn's influential formulations.[8] And he repeatedly mined the history of science to bear out the narrow hypothesis-testing approach of his research and the essential incompleteness of evidence.[9] Above all, Eysenck tried to emulate what these scholars said science was or should be, and he had plenty of precedents. Many of the first modern psychologists evoked the ideal of the natural sciences for promotional purposes, for claiming psychology as a separate speciality that was also

---

[5] Such put-downs tend to be more qualified and less imperialistic these days. See, for example, Dehue, T. (1997), Managing distrust, Review of *Trust in numbers: the pursuit of objectivity in science and public life* by Theodore Porter, *Theory and Psychology* **7**, 417–20. Despite Dehue's criticism that Porter invoked a Kuhnian yardstick when taking a dim view of psychology's methodological obsessions, his work remains a key example of the social sciences treated as authentically powerful; see Porter, T.M. (1995), *Trust in numbers: the pursuit of objectivity in science and public life*, Princeton University Press, Princeton.

[6] See Eysenck's appropriation of Stebbing, S. (1930), *A modern introduction to logic*. Methuen, London, in Eysenck, H.J. (1952a), *The scientific study of personality*, Routledge and Kegan Paul, London, pp. 9–12.

[7] See Eysenck, H.J. and Wilson, G.D. (eds.) (1973), *The experimental study of Freudian theories*, Methuen, London, especially their jointly written introduction.

[8] See Eysenck's attempt to justify the unification of Cronbach's two schools of correlational and experimental psychology in Eysenck, H.J. (1997a), Personality and experimental psychology: the unification of psychology and the possibility of a paradigm, *Journal of Personality and Social Psychology* **73**, 1224–37.

[9] See Eysenck, H.J. (1959a), Scientific methodology and *The dynamics of anxiety and hysteria*, *British Journal of Medicine and Psychology* **32**, 56–63.

clearly distinguishable from the amateur fringes. At the same time, such promises usually carried the implicit idea of (temporary) inferiority, for they inevitably drew upon non-native criteria for what a proper science should be.[10]

Eysenck made it abundantly clear that he was unimpressed by the field he was forced into.[11] Yet even though he dismissed the psychology of time as 'the hope of science', he had no regrets:

> Competition in the hard sciences is much fiercer than in psychology, and the really successful practitioners are quite something. I have known many of the leading psychologists … but none have impressed me half as much as the leading physicists and astronomers. It turned out to be quite easy to be a big fish in a small pond …[12]

Eysenck's dismissive attitude toward his fellow psychologists could thus be understood as clearing a space for his personal authority, rather than as just arrogance. Encouraging a perception that psychologists were collectively thick-headed not only implied comparatively sharper analytical skills, it also hinted at an independent, de-mystifying perspective. Here was an expert insider prepared to present his science to the public in manner that stripped it of its self-protecting, elite pretensions. Even amongst psychologists, those offended or bemused by Eysenck's stance were counterbalanced by allies galvanized by the promise that he, and they, were the scientific future. Eysenck's philosophy of physics envy largely played out as a potent leverage tool. It gave him discursive traction when numbers alone might not do. It also lent a romantic

---

[10] I would argue that the historical effects of this kind of appropriation are worth studying, regardless of one's stance on whether psychologists *should* listen to philosophers and historians. For example, the crisis talk it encouraged may look like weakness, but it also hides a functional strength. Psychologists arguing for new theories or for the uptake of alternative applied techniques could almost always lambaste the present state of affairs, and draw upon aspects of the philosophy of science to do it. One could trace a line from John Watson and William McDougall, through Eysenck and Skinner, to the contemporary tales of woe of Cosmides and Tooby. One result has been to produce a more open and dynamic discipline, even if it cost psychologists a reassuringly coherent self-image. The oft-bemoaned absence of broad disciplinary consensus or overarching paradigms can thus be seen as an effective form of pluralism. Modern psychology has become hugely successful in the West because it appealed to and infiltrated key institutions and bureaucracies, the government and the military, the media and 'everyday life'. To do so, it had to be responsive; it had to be many things to many people. The trade-off, of course, was the high maintenance costs of insecure, permeable borders. As the discipline was constantly remade, expertise had to be re-defined and consumer trust re-negotiated. Public controversy and 'pop' psychology are the symptoms of this trade-off; they are hardly the *unambiguous* threat or sad reflection that critics of the discipline have imagined.

[11] See Eysenck (1997b), *Rebel with a cause*, pp. 52–3 and p. 66.

[12] Ibid., p. 47.

tinge to the way he characterized his early career that was not without its contradictions—contradictions I will return to later.

London calling: individual differences and the factors of the mind

British psychology had been slow to develop as an academic speciality and just as slow to professionalize. Up to World War II, there were only a handful of psychology departments in England, one of the most important being that at University College, London (UCL). University College was the third oldest higher education institution in England, the first of a series of such institutions making up the administrative federation known as 'London University'.[13] Oxford and Cambridge had virtually monopolized higher education for centuries, propped up by the twin pillars of the clergy and land-owning gentry. UCL was the secular counterpoint to Anglican Oxbridge. Established in 1826, it was to be no mere finishing school for the refinement of aristocratic young minds. This 'godless college of Gower Street' was far more inclusive, catering to the free-thinkers and the new merchant–professional classes of Britain's industrial age. Only the more traditional King's College rivalled it amongst the growing number of colleges of London University of the late nineteenth century. UCL was founded on utilitarian principles. As such, it came to be associated with a kind of technocratic liberalism that reflected the changing role of science and society. The practice of science had ceased to be the exclusive province of a self-selected group of gentleman amateurs. German universities had already demonstrated that scientific research could be harnessed to social and economic goals. UCL's early history as primarily a teaching college ensured that many of its faculty members were sympathetic to the emerging ideal of practical, professional (not to mention secular) science.[14] Not coincidentally, UCL had been a key institutional launch pad for evolutionary thought. It was home to the biometric school of statistician Karl Pearson and biologist Walter Weldon, both heavily indebted to the teachings and material legacy of Francis Galton.

---

[13] There is some debate about this claim, however, since UCL did not receive legal charter until 1836, post-dating the founding of Durham University in 1832 and the charter of King's College in 1829.

[14] Many of those who benefited still had some misgivings about the professionalization of science. For example, even Karl Pearson feared that professional careerism could distort disinterested inquiry. See Porter, T.M. (2004), *Karl Pearson: the scientific life in a statistical age*, Princeton University Press, Princeton, Chapter 9.

Galton set the agenda for psychology at UCL. Galton took variation in form and function as equally, if not more, important than its intrinsic, compositional aspects. His cousin Charles Darwin had argued that deviations from the typical or the average were hardly Nature's mistakes. Variation across the natural world was ubiquitous; it was a pre-condition for evolution by natural selection. Galton and a growing number of contemporaries were determined to follow the implications for human history, to naturalize human nature. Individual differences in almost any quality carried an intriguing set of questions. For example, how were these qualities distributed, and how could they be measured? Moreover, how were they related from person to person, and from generation to generation? Hereditary mechanisms had become perhaps *the* pressing question for the new evolutionary biology of the late nineteenth century. Galton took the radical step of supposing that it was not just physical features or biological properties that were passed on. Galton suggested that attributes of mind might be as well, including an individual's capacity for higher thought.

The attributes of mind that most preoccupied Galton were *talent* and *character*.[15] Galton argued that eminence in Victorian public life was dictated more by pedigree than by upbringing and education. In the terms he coined, it was *nature*—rather than *nurture*—that apparently determined achievement. Bolstering these claims with an enthusiastic programme of anthropometric testing, Galton concluded that intellectual capacities varied enormously. Even Darwin was pleasantly surprised by this revelation.[16] Most significantly, Galton guessed that the distribution of these unequal gifts could be modelled mathematically. He fitted the distribution of talent to a bell-shaped curve that reproduced itself across successive generations according to statistical laws of inheritance.

Subsequent developments made Galton's work look prophetic. Galton's intellectual heirs at UCL—especially Burt—took up where he left off. With Galton's help, James Sully had equipped UCL with a psychological laboratory

---

[15] Gillham, N.W. (2001), *A life of Sir Francis Galton: from African exploration to the birth of eugenics*, Oxford University Press, Oxford. See also Hearnshaw, L. (1979), *Cyril Burt: psychologist*, Hodder and Stoughton, London.

[16] On reading Galton's *Hereditary genius*, Darwin remarked: 'You have made a convert out of an opponent in one sense, for I have always maintained that, excepting fools, men did not differ much in intellect, only in zeal and hard work...' Quoted in Gillham (2001), *A life of Sir Francis Galton: from African exploration to the birth of eugenics*, p. 169.

in 1897.[17] After Sully departed, the psychology programme had been taken over by Carveth Read, with W.H.R. (William Halse Rivers) Rivers and William McDougall taking care of experimental laboratory work for a time around the turn of the century. The department was later built up by redoubtable ex-army man, Charles Spearman. Spearman was made Grote Professor of Mind in 1911 and finally became Professor of Psychology in 1928. Burt took over this chair in 1932.

Spearman and Burt both favoured and became leading experts in a kind of correlational psychology that was both quasi-naturalistic and firmly quantitative. Much of their work involved extracting psychological meaning out of daunting arrays of numbers, deriving latent constructs from more directly assessed, manifest measures. These numbers were typically gathered from samples of important life tasks (e.g. school results, professional achievements) or, better still, from new standardized measures that *resembled* and/or *dissected* such tasks (e.g. anthropometric tests of memory and reaction time, psychometric intelligence tests). The methodological bond between Spearman and Burt did, nevertheless, hide some deeper level philosophical differences.

Spearman had been educated for a spell in Leipzig and hoped to take Wundtian experimental psychology into the world outside the laboratory.[18] He wanted to build a systematic psychology using correlational methods to 'search for laws and uniformities' rather than describe individual differences *per se*.[19] From the outset, Spearman focused on intelligence, particularly the question of whether there was a 'single subjective activity' as opposed to 'no discernible structure'. He noted how measures of intellectual function tended to vary together; moreover, matrix tables of the correlations between tests could be arranged as a hierarchy, that is, to decrease *smoothly* along the rows and down the columns. For Spearman, this suggested there was essentially one faculty of intelligence that each test was imperfectly assessing. Spearman developed various statistical means for adducing what he hoped would be the necessary proof. He devised a complicated statistical procedure for reducing the

---

[17] Gurjeva, L.G. (2001), James Sully and scientific psychology, in *Psychology in Britain: historical essays and personal reflections* (ed. G.C. Bunn, A.D. Lovie, and G.D. Richards), pp. 72–94, BPS Books, Leicester. Sully is remembered as an important founding figure in the child-study movement of the latter part of the nineteenth century, a practical interest prompted in part by his insecure place in academia.

[18] Burt, C. and Myers, C.S. (1946), Charles Edward Spearman, *Psychological Review* **53**, 67–71.

[19] See p. 207 of Spearman, C.E. (1904), 'General intelligence' objectively determined and measured, *American Journal of Psychology* **15**, 201–93.

number of variables necessary to account for this correlation matrix. It was a prototype version of a set of techniques that came to be known as factor analysis.[20] These techniques would become a primary methodological tool for Hans Eysenck.

In general, factor analysis summarizes the co-variation between a set of variables by representing this co-variation as a smaller set of hypothetical equations called 'factors'.[21] Each factor can be expressed, in part, as a weighted sum of these original variables. These weights reflect the contribution (i.e. the 'saturation' or, in more modern terms, the 'loading') each of the original variables makes to this new factor. Spearman's initial version of factor analysis strongly presupposed that a more succinct description of the data was possible, in line with his starting hypothesis about the general nature of intelligence. In effect, Spearman asked whether one general factor common to all his intelligence tests, and a second (necessarily smaller) factor specific to each test, could account for the observed matrix of correlations. Since it only allowed for the extraction of two factors, it was about as succinct as one could get—beyond suggesting that all psychological phenomena were indivisibly universal. Armed with results showing that such a re-description was indeed possible, Spearman proposed a two-factor theory of intelligence. The first factor was interpreted as general intelligence, or *g*, with the second factor, *s*, seen as an index of the specific contribution of each test. For Spearman, the *g* concept was a decisive construct that he hoped would dispose of the debate about distinct intellectual faculties and components. Not everyone was convinced. Critics at home—William Brown at King's College, Godfrey Thomson at Edinburgh, and Karl Pearson—questioned Spearman's assumptions. Thomson in particular wondered whether Spearman's one general factor actually demonstrated the existence of a general cognitive structure or was merely a function of the statistical procedure itself. Spearman's arch-enemy Karl Pearson also took exception, slamming Spearman's grand 1927 opus, *The abilities of man*, in an

---

[20] Spearman instructed Burt in the application of this new procedure as he developed it, which allowed Burt to later mischievously claim priority as its originator. See Lovie, A.D. and Lovie, P. (1993), Charles Spearman, Cyril Burt, and the origins of factor analysis, *Journal of the History of the Behavioral Sciences* **29**, 308–21. The Lovies' research strongly confirmed Hearnshaw's account of Burt's duplicity in this instance.

[21] Factor analysis should not be confused with the related but philosophically distinct technique of principal component analysis developed by Pearson and Hotelling. Principal component analysis seeks to exhaustively re-describe rather than summarize the co-variation between a set of variables by producing an orthogonal set of components equal in number to the variables originally considered.

anonymous review. Yet these were quarrels within the one broad methodological church, however venomous the tone.

Although Burt was Oxford educated, he was more practically inclined than Spearman. Burt had worked outside academia for a lengthy period before taking up the reins at UCL. His outlook was explicitly Darwinian. Burt went as far as to claim that his entire academic career was an effort to preserve (even rescue) the tradition founded and developed there by Galton of 'individual' or 'differential' psychology. Burt was entering some choppy waters, however. The London school had been both bolstered and challenged by developments overseas, especially in the US. Part of this trans-Atlantic confrontation reflected the changing use of factor analysis in the mid-1930s. The technique had come to be seen as generally applicable to any multivariate problem, for it had been more than just tinkered with by a range of mathematical and statistical thinkers. Now it was possible to overcome the limitations of Spearman's original procedure to extract multiple factors, rather than just two. Factors could also be rotated in geometric space, exchanging the amount of common variance a factor may summarize for saturations that were more in line with theoretical suppositions. For example, any set of intelligence tests would be expected to generally correlate positively. Negative saturations on a factor made it less intelligible as a summary construct, though some zero saturations might be expected if the tests were thought to measure relatively distinct abilities. Factor rotation could make the necessary adjustments in the hope of enhancing the substantive interpretation of results. Factor analysis had played a deductive, confirmatory role in Spearman's theorizing. However, more sophisticated versions came to be used as inductive, exploratory heuristics—ones that could be used to construct measures of almost any ability or attribute.[22] Thus the g concept came to be vigorously attacked by several notable American psychological researchers whose instrumental usage of factor analysis was less encumbered with *a priori* theory. Both Truman Kelley and L.L. (Louis Leon) Thurstone put up alternative schemes of multiple abilities (e.g. verbal, spatial, and numerical). Even William Stephenson in Spearman's own department suggested that substantive factors other than g could be ascertained in correlation matrices of tests.[23]

---

[22] Lovie, A.D. (1983), Images of man in early factor analysis—psychological and philosophical aspects, in *Studies in the history of psychology and the social sciences* (ed. S. Bem, H. van Rappard, and W. van Hoorn), pp. 235–47, Leiden University, Leiden.

[23] Hearnshaw (1979), *Cyril Burt: psychologist*, Chapter 9. See also Lovie, P. and Lovie, A.D. (1995), The cold equations: Spearman and Wilson on factor indeterminacy, *British Journal of Mathematical and Statistical Psychology* **48**, 237–53.

Burt took up Spearman's *g* concept but steered a middle course between the warring factions. It was a characteristically artful compromise directly expressed in the statistical techniques he refined. Burt proposed his own brand of group factor analysis. Instead of rotating factors to remove negative factor saturations and maximize the number of zero saturations (as the pragmatic Thurstone did), Burt partialled out the effect of the first big factor and then factored residual intercorrelations to obtain secondary 'group' factors. Thus Burt held on to *g*, but went on to emphasize its secondary components (e.g. arithmetic, language, memory, and composition) more than Spearman ever did.

Burt also sought evidence for his long-standing conviction that general cognitive ability was both innate and highly inherited. Utilizing the twin study method that Galton pioneered, Burt had apparently identified and kept tabs on a large set of identical and non-identical twins, although he had not yet fully analysed the data this collection would provide.[24] Burt's compromise position in the debate over *g* and the structural features of his version of factor analysis reflected his basic assumptions about how the mind worked. Fond of elegant theoretical orderings, Burt followed Sherrington and McDougall in proposing a series of levels between the psychological and the biological. Unlike Spearman, Burt did not subscribe to realist interpretation of the factors he obtained. They were nominal, descriptive constructs that merely confirmed that there was a dispositional structure to the mind without entirely prejudging what that structure was.

London and Oxbridge: practical and political differences

There was probably only one other psychology department in England, besides UCL, that really mattered prior to World War II. It was at Cambridge. UCL and Cambridge were rivals in terms of research and student output, but their ideas about what psychology was and how it should be conducted were diametrically opposed. Cambridge psychology was characterized by a relatively narrow experimentalism—even by the standards of the day—a purist English version of the Fechner–Wundt tradition. Differences between and even within individuals were no more than an annoyance. Rather than being a sign of progress, of evolution in action, such variability was seen as error surrounding

---

[24] The long-running debate over Burt's alleged fraudulence need not concern us here. It is discussed in more detail in Chapter 8. For those unacquainted with this vast argumentative literature, a good place to start would be Mackintosh, N.J. (ed.) (1995), *Burt: fraud or framed?* Oxford University Press, Oxford or, most recently, Tucker, W.H. (2007), Burt's separated twins: the larger picture, *Journal of the History of the Behavioral Sciences* **43**, 81–6.

a true value or process. Part of Cambridge's narrowness stemmed from an entrenched, theologically based opposition to the study of mind or soul. Oxbridge intellectuals had seen the emerging discipline of psychology of the late nineteenth century as more a threat than an opportunity.[25] Even so, Cambridge did agree to tolerate this 'upstart subject', although only if its distasteful empiricism was restricted to the relatively narrow study of the senses and perception. Thus James Ward and W.H.R. Rivers managed to achieve limited facilities at Cambridge by the early years of the twentieth century. A more elaborate laboratory of experimental psychology was founded by Charles S. Myers in 1913. Only after Myers proposed, designed, and put up the money himself, did the university grudgingly agree to it.[26] The readership Myers was promised was still restricted by the university to experimental psychology, ruling out more socially oriented, applied psychology. Convinced of psychology's wider scope, Myers soon resigned to take up industrial work. He handed the readership over to his assistant Frederic Bartlett, who maintained the requisite narrow focus.

By the mid-1930s, Cambridge had become a leading site for research on visual and auditory perception, and memory. Bartlett argued that the experimental approach of Cambridge was entirely unpretentious. There were, he said, no theoretical preconceptions or prescribed set of methodological rules. The measurements taken involved no 'contrived' scaling procedures, and analyses contained few or no statistics. Prior to World War II, inferential statistics—particularly the work of Ronald Fisher—had yet to make a big impact on the design and analysis of psychological experiments anywhere, and certainly not at Cambridge. Results were often reported as simple comparisons of percentage differences or in the form of graphical displays. The Cambridge approach was geared to studying fundamental or common mechanisms, deducible from controlled adjustment of factors thought relevant and the exclusion of factors thought extraneous. Representational schemes were constructed to account for the limits, the strengths, and weakness of perceptual and cognitive processes. Objectivity was underwritten by deductive logic,

---

[25] Richards, G. (2001), Edward Cox, the Psychological Society of Great Britain (1875–1879) and the meaning of an institutional failure, in *Psychology in Britain: historical essays and personal reflections* (ed. G.C. Bunn, A.D. Lovie, and G.D. Richards), pp. 33–53, BPS Books, Leicester.

[26] See Costall, A. (1992), Why British psychology is not social: Frederic Bartlett's promotion of the new academic discipline, *Canadian Psychology* 33, 633–93, and Costall, A. (2001), Pear and his peers, in *Psychology in Britain: historical essays and personal reflections* (ed. G.C. Bunn, A.D. Lovie, and G.D. Richards), pp. 188–204, BPS Books, Leicester.

by the elimination of alternative explanations. Wider applicability was ensured by the stepwise deconstruction of psychological processes that enabled their reconstruction in other contexts.

In Bartlett's world, the laboratory was a place of unprejudiced, practical enquiry. Bartlett's successor at the applied psychology unit, Kenneth Craik, saw the laboratory as a place of 'infinite possibilities and far horizons', a place where phenomena can be isolated and manipulated, and effects created and studied.[27] Collectively, Cambridge psychologists projected an image of reflective enquiry—exemplified in part by Bartlett's rich experimental work on memory between the wars. However, much of the work Cambridge psychologists undertook had specific human engineering applications in industry and the military—particularly after Bartlett set up the applied psychology unit at Cambridge during World War II.[28] It was a tradition geared to the improvement of human performance, for solving specific problems, for tailoring bodies to machines. It made great use of the contrivances like perceptual illusions, paradoxes, and conflicts, as well as gadgets that mechanically modelled or improved on aspects of mind.[29] Bartlett's 'no theory, no method, no statistics' dictum may have amounted to a self-serving caricature that downplayed the assumptions of experimentalism and hid the work necessary to translate this work into non-laboratory applications. However, it was quite in tune with the not entirely baseless Oxbridge conceit of effortless superiority. And it did serve to make a contrast with the approach taken down the road in London.[30]

Just as it did in terms of pure fundamentals, the London school took its applied cues from Galton. Much of Galton's later career was motivated by an interest in eugenics. This amounted to assisted or artificial selection and was prompted by *fin de siècle* fears of degeneration. Galton had set up a eugenics laboratory at UCL and founded the *Eugenic Review*, designed to promote good breeding patterns. Galton opened the imagination to the possibility of

---

[27] See Bunn, G.C. (2001), Introduction, in *Psychology in Britain: historical essays and personal reflections* (ed. G.C. Bunn, A.D. Lovie, and G.D. Richards), pp. 1–29, BPS Books, Leicester.

[28] Collins, A. (2001), The psychology of memory, in *Psychology in Britain: historical essays and personal reflections* (ed. G.C. Bunn, A.D. Lovie, and G.D. Richards), pp. 150–68, BPS Books, Leicester.

[29] See Hayward, R. (2001), 'Our friends electric': mechanical models of mind in postwar Britain, in *Psychology in Britain: historical essays and personal reflections* (ed. G.C. Bunn, A.D. Lovie, and G.D. Richards), pp. 290–308, BPS Books, Leicester.

[30] This is an unreconstructed version of the Cambridge approach that Leslie Hearnshaw gave in Hearnshaw, L. (1964), *Short history of British psychology, 1840–1940*, Methuen, London, Chapter 11, still the last major attempt to give British psychology a grand narrative.

managing human resources for improving society by improving human stocks. Moreover, the exploration of the hereditary basis for psychological qualities paved the way for decidedly contemporary versions of meritocracy. Talent, irrespective of social station, should be nurtured and rewarded. In keeping with the socialist leanings of his protégé Pearson, Galton was hardly enamoured with the ruling class aristocracy. Both advocated a meritocratic course guided by the probabilistic nature of inheritance that sought to avoid the opposing perils of blind egalitarian democracy and elite hereditary peerage.[31]

The intellectual tradition Galton founded would be less about adjusting the individual and coping with new technology, more about coping with each other and our inherent differences. It produced knowledge that lent itself to applications at an administrative and social policy level—in schools and universities, prisons, clinics, and mental hospitals, and in bureaucracies of one form or another. It was the science of governmentality best exemplified by that most practical of psychological tools, the standardized test. Psychologists had begun devising various tests of intellectual functioning in the early part of the twentieth century. Such tests effectively repackaged the old school exam as a precision instrument. Clear and uniform content made for much broader applicability, and quantitatively standardized scaling promised a universal basis for comparison. Extending on this model, psychologists developed similar tests of non-cognitive attributes and attitudes. Standardized testing gave these attributes and attitudes a newly refined scientific status, even though their meaning still partly derived from popular usage.

The application of psychological testing had the potential to cut across existing orderings and procedures. It provided for novel social interventions that, *at the time*, tended to have a liberal, even radical, edge—not something that could be said about the perceptual modelling and 'fitting work' at Cambridge. Even the cyber technologies that Cambridge experimentalists later helped to develop were more socially remote and ambiguous, politically harder to read than the direct policy implications of the London tradition.

It is precisely on the issue of tests that the greatest contrast between London and Cambridge can be drawn. The London school made great use of tests, while those at Cambridge made practically none, and actually led attacks on testing in several instances.[32] Those in London tended to be more inclusive.

---

[31] See Porter (2004), *Karl Pearson: the scientific life in a statistical age* and Gillham (2001), *A life of Sir Francis Galton: from African exploration to the birth of eugenics*.

[32] See, for example, the piece by Cambridge psychologist Chambers, E.G. (1932), Statistical psychology and the limitations of the test method, *British Journal of Psychology* **33**, 189–99.

While they saw the Cambridge approach as unduly restrictive, they were usually willing to entertain the methods and results of experimentation—facilities and training permitting. Eysenck himself came to embody this inclusive attitude, regarding experimental research as an important extension to his correlational methods. However, the reverse was much less true, with the blue-bloods to the north openly dismissive of the London school's output.[33] For example, Bartlett once said he would never let a statistician loose in his laboratory, while his successor Oliver Zangwill slammed the factor analytic and testing tradition as a thoroughly misleading waste of time and energy.[34]

It was not a trifling point, however, given that testing was perhaps *the* central means for connecting psychology with life. It must be remembered that, in almost all instances (e.g. in France, the US, the UK, and in Germany to some extent), the introduction of standardized testing in the first decades of the twentieth century represented a progressive move—a bid to adapt, mould, or even overturn the status quo. Contrasting with latter-day critiques, testing was largely welcomed as a way of making opportunity more meritocratic.[35] It was just the sort of new technology that modernizing governments and bureaucracies liked to rely on. It was empirical and thus scientific, impersonal and thus fair. Many of psychological testing's most enthusiastic proponents overseas—Binet, Stern, Goddard, Terman, Yerkes—became the applied face of psychology. Indeed, Spearman's earliest work had focused on fundamental statistics issues

---

[33] The arguable point, however, is the extent to which this asymmetry of attitudes exhibited by London and Oxbridge psychologists can be attributed to home-grown politics rather than the historical rift between the 'two schools' of psychology that cut across the discipline as a whole.

[34] See Zangwill, O. (1951), *Introduction to psychology*, Methuen, London. See also Bartlett's politely critical remarks concerning Spearman's penchant for grand generalizations in Bartlett, F.C. (1930), Experimental method in psychology, *Journal of General Psychology* **30**, 49–66.

[35] The passage of time changes everything, however. Through mid-century, applied psychologists achieved a level of institutional integration that made them victims of their own success. For example, once the practice of psychological testing became embedded as a yardstick of educational achievement and the basis for selective streaming, it took on a very different political hue. By the 1960s, applied psychologists were vehemently criticized as the 'servants of power', and tests attacked as rigidly narrow measures of 'class-conditioned good manners'. See Buchanan, R. (2002), On *not* 'giving psychology away': the MMPI and public controversy over testing in the 1960s, *History of Psychology* **5**, 284–309. This re-figuring was directly reflected in the career of Hans Eysenck (and Burt too), with the kind of psychology he represented increasingly seen as an expression of conservative interests, out of step with the radical politics of the 1960s and 1970s. Again see Chapter 8 for more on these points.

that laid the basis for modern psychometrics. Hearnshaw dubbed Cyril Burt the 'first professional psychologist' in Britain, a title especially relevant here given that Burt was an important pioneer of intelligence testing in his own right. His educational testing work at the London County Council in the 1913–32 period combined clinical assessment of school children and the screening of the subnormal with test research and development. This and Burt's later work with the progressive English Board of Education provided a key example of the impact psychology could have for child guidance, and for educational reform based on selective streaming according to largely innate general ability and specific aptitudes.

In this between-the-wars English context Burt was, if anything, politically to the left. He was never particularly friendly to the British 'establishment', as Hearnshaw put it, and he felt he owed his achievements and position to his own wits rather than fortuitous social circumstances. Burt was actually knighted by a Labour government.[36] He was, and remained, a detached scholar who favoured the progressive welfare programmes and the managed knowledge economy of tripartite schooling he helped install. Eysenck's education at UCL must thus be read in these historical terms. The intelligent and ambitious young German eagerly took to this quantitative science of innate hereditary individual differences that was as consciously practical as it was liberal.

Eysenck's undergraduate years

Eysenck started out at UCL in the autumn of 1935. At the time, UCL psychology was still a relatively small department, headed by Burt. Eysenck took liberal arts subjects in his first year of undergraduate study—Latin, English, German, economics, and ethics—before starting to specialize in his major of psychology in years two and three. Eysenck did suffer from an understandable deficiency in English at first, evidenced by the fact that he got only a third in this subject in his special intermediate BA exam in 1936. However, he was quickly able to correct this to become an all-round first rank student.[37] He was taught ancient and modern philosophy. He also had time to take in lectures on statistics, from Karl Pearson's son Egon, and genetics, from J.B.S. Haldane. Experimental psychology and psychophysics were covered by Stanley J.F. Philpott. Social psychology was taught by the only other member of staff, J.C. Flügel. Former UCL junior staff member William Stephenson gave instruction in psychometrics while seconded at the time from Oxford. Eysenck took a minor in sociology,

---

[36] Hearnshaw (1979), *Cyril Burt: psychologist*, p. 127.

[37] Gibson (1981), *Hans Eysenck: the man and his work*, p. 32.

a field he developed scant respect for. Like Burt, Eysenck remained only dimly appreciative of much of the continental tradition in sociology and anthropology.[38]

Despite the quantitative emphasis at UCL, not all that went on there could be squarely located within the London tradition. As Eysenck liked to point out, many in the department were conversant and sympathetic with psychoanalytic ideas. Eysenck saw it as an uncritical engagement with what he already regarded as a pseudo-science. Flügel had authored several influential works, notably psychosexual accounts of masculine dress codes and family life, as well as being an analyst of sorts. Flügel had also authored a history of psychology and was fluent in German. He and Ernest Jones had founded the British Psycho-Analytical Society in 1919, an exclusively Freudian group. Philpott was interested in the notion of the collective unconscious put forward by Jung. Even Burt was attracted to Freud, and he too was a founding member of the British Psycho-Analytical Society.[39]

Eysenck obtained a first-class BA degree in July 1938, and immediately enrolled in a Ph.D. By this time, he had also met and married Margaret Davies. They met as students at UCL. Margaret Davies was Canadian, six years older than Eysenck, with a first degree in mathematics. She must have impressed the 21-year-old Eysenck, for they were quickly married in 1938 just as he got his BA. Their only child, Michael William, was born on 8 February 1944.

Davies had embarked on a master's degree in psychology, and was also supervised by Burt. Being very mathematically literate, she was in a position to assist Eysenck as he cut his teeth on factor analytic techniques and his early publications refer to her work repeatedly. In fact, they must have worked very closely together. Completed in 1938, her MA could be considered a dry run for his Ph.D. Eysenck focused on the same kind of problem, employed the same factorial methodology, and probably used some of the same subjects. No doubt they shared the onerous task of hand-cranking the complex calculations necessary in even the most rudimentary factor analyses.

Davies's master's research looked at the perception of smell, factor-analysing to derive a general factor of olfactory preference.[40] She also weighed into a contemporary controversy over factor analysis, namely, the merits of factoring across persons rather than tests. Burt was a pioneer of this kind of factorial

---

[38] See Hearnshaw (1979), *Cyril Burt: psychologist*, Chapter 2.

[39] Richards (2001), Edward Cox, the Psychological Society of Great Britain (1875–1879) and the meaning of an institutional failure.

[40] Davies, M. (1938), A statistical study of individual preferences for olfactory stimulus, MA thesis, University of London.

technique, apparently being the first to factor analyse across persons when analysing children's subject preferences at school, in aesthetics, and, later, examiners' marks.[41] It was, Burt said, a technique most appropriate 'when we possess no external or impersonal standard', no right answer or objective means of ascertaining a true order.[42] Burt also suggested a reciprocity principle applied, that the conclusions reached when correlating persons or tests would be essentially identical. His colleagues could not agree. Godfrey Thomson doubted the benefits of factoring across persons and attacked Burt's reciprocity principle as naïve.[43] Conversely, William Stephenson argued that it represented a different, intrinsically valuable approach and one-upped Burt with a 1936 paper that applied the technique to the analysis of personality.

Margaret Davies's review of a number of previous studies suggested that factoring across persons usually resulted in one major factor from correlation tables that tended to display hierarchical arrangement—as when factoring across tests.[44] In her view, these were complementary rather than opposing techniques. The one big factor produced by an across-persons analysis appeared to limit its applicability to personality. However, the technique was quite appropriate, Davies felt, when looking for evidence of reliability across persons (e.g. in academic grading) or an 'objective' aspect to individual preferences (e.g. in aesthetic, sensory, or ethical judgements).[45]

Burt the mentor

As a student, Eysenck was quite the intellectual sponge, soaking up the ideas of the environment he found himself in. However, Cyril Burt was Eysenck's chief intellectual influence and professional model, especially during Eysenck's doctoral work. Eysenck also apparently assisted with some of Burt's BBC media

---

[41] Hearnshaw (1979), *Cyril Burt: psychologist*, p. 164.

[42] See pp. 62–3 of Burt, C. (1937), Correlations between persons, *British Journal of Psychology* **28**, 59–96.

[43] Thomson argued that the results of factoring across persons and tests would only be equivalent if the tests were of average difficulty and the people sampled were of average ability.

[44] Davies, M. (1939), The general factor in correlations between persons, *British Journal of Psychology* **29**, 404–21.

[45] Davies's MA was originally to be supervised by Stephenson, but he left for Oxford. For whatever reason, her MA thesis results took some to time to see the light of day, appearing after Eysenck's doctoral articles. See Davies, M. (1944), An experimental and statistical study of olfactory preferences, *Journal of Experimental Psychology* **39**, 246–52. Margaret Davies later went into the psychology of ageing.

presentations in this period.[46] Burt thought highly of the young German who was able to master statistical and psychometric techniques so quickly. According to Eysenck, Burt told him he had produced one of the most outstanding performances in his BA examination he had seen. Nevertheless, Burt later claimed Eysenck's handwriting was so bad he had not been able to read anything Eysenck had written and only gave him a first because he knew the quality of Eysenck's work.[47] Burt specialized in this kind of backhanded insult, although I can attest that Eysenck's handwriting could be far from legible.

Burt took on the supervision of Eysenck's Ph.D. Burt had instigated a collaborative relationship with Eysenck, apparently while Eysenck was in his second year of undergraduate study.[48] Burt asked Eysenck to factor analyse the data Thurstone had presented in 1938 supporting the notion of multiple intellectual abilities.[49] Burt wrote up the results supporting his compromise position, but the paper was published in the *British Journal of Educational Psychology* in 1939 under Eysenck's name only—presumably an attempt on Burt's part to create the impression of widespread opposition to Thurstone's views.[50] This was Eysenck's first published article. It was a not altogether happy experience, however, sullied by Burt's habitual tinkering with the proof copies of manuscripts. According to Eysenck, the final copy was far more critical of Thurstone's position than the draft Eysenck was shown.[51] Eysenck let it go—not wanting to confront Burt perhaps—giving the impression the views expressed in this paper were entirely his own in publications that appeared soon after.[52] The episode also helped Eysenck get to know 'Leo' Thurstone, as well as establish his credentials in this heavily quantitative area.

---

[46] According to Gibson, Eysenck's first TV appearance came when he assisted Burt in the BBC programme 'Experiments in science', broadcast 4th November 1937. See Gibson (1981), *Hans Eysenck: the man and his work*, p. 46 (note). If this were indeed the case, it would have been in the very earliest days of TV. Burt also presented numerous BBC radio broadcasts throughout his career.

[47] See Eysenck (1997b), *Rebel with a cause*, p. 57.

[48] Eysenck remembered this collaboration as occurring while he was still an undergraduate, although the dates of Thurstone's monograph and Eysenck's response suggest this work was done a little later. See Eysenck (1997b), *Rebel with a cause*, p. 55.

[49] Thurstone, L.L. (1938), *Primary mental abilities*, Psychometric monographs, no. 1, University of Chicago Press, Chicago.

[50] Eysenck, H.J. (1939b), Critical notice of 'Primary mental abilities' by L.L. Thurstone, *British Journal of Educational Psychology* **9**, 270–5.

[51] Eysenck (1997b), *Rebel with a cause*, p. 55.

[52] See p. 91 of Eysenck, H.J. (1941c), The empirical determination of an aesthetic formula, *Psychological Review* **48**, 83–92.

Burt's original thesis proposal was for Eysenck to re-standardize the Binet intelligence scale for British subjects, but the project did not excite Eysenck. Instead they agreed that Eysenck should extend on Burt's work on aesthetic preferences. It was a topic that did not require much equipment or extensive logistical support—an important consideration even before war made elaborate laboratory research impossible. One could simply use postcards of art and cut-out figures for subjects to evaluate and rate.

Eysenck's interest in aesthetics had been engaged by Bedford College philosopher Susan Stebbing's lectures suggesting certain theoretical laws could be applied to aesthetic composition. His published papers also pointed to older philosophical traditions proclaiming the aesthetic principle of 'unity in variety', the modern significance of which Eysenck attributed to Fechner.[53] But his decision to pursue this line of research, he later said, also grew from a conviction that there must be a value hierarchy in aesthetics that crossed social and cultural divides—that a Rembrandt must be better than a child's scribble or, for that matter, 'African or Bushman or Red Indian art'.[54] Wife Margaret had already conducted a number of related studies. Moreover, his former flatmate John Butler Parry joined him in similar research, Eysenck claiming to have brought Parry over to psychology from the 'wasteland' of philosophy. Both were able to draw and extend a great deal of work in aesthetics that Burt had done in the 1930s, much of which Burt had not taken the time to publish.[55]

Experimental aesthetics

At Burt's suggestion, Eysenck sought to adjust American mathematician G.D. (George David) Birkhoff's formulation that aesthetic pleasure depended on conceptual order divided by complexity of elements. Margaret Davies had already found minimal support for this contention. In Burt's formulation (and Eysenck's reworking of it), aesthetic pleasure was better considered a product of these two terms. Eysenck used subjects' rankings of the aesthetic value of various polygon figures and compared the pleasure rating given by Birkhoff's formula and a regression product equation based on component features of the polygons. Lo and behold, Eysenck found a much better fit with his

---

[53] See Eysenck, H.J. (1942a), The experimental study of the 'good gestalt'—a new approach, *Psychological Review* **49**, 344–64. As mentioned earlier, Eysenck cited Stebbing as a formative influence on what he considered to be the proper scientific approach. See Stebbing, S. (1937), *Philosophy and the physicists*, Dover Publications, New York.

[54] Eysenck (1997b), *Rebel with a cause*, p. 68.

[55] See Hearnshaw (1979), *Cyril Burt: psychologist*, pp. 215–21.

product formulation. However, Eysenck went much further, exploiting as many possibilities as Burt's factorial techniques allowed with an interlocking series of studies.

Eysenck set about isolating aesthetic judgement as a psychological quality free from culturally conditioned rules. Factor analysing across persons for the polygon rankings had yielded one big factor, and a smaller but significant second factor, suggesting that there were general and particular aspects at work in an individual's judgements. In an effort to describe these aspects more generally, Eysenck got his subjects to rank many other kinds of art. The examples of art he presented were unfamiliar, which ensured no learnt cultural cues would intrude. He argued: 'No tradition or teaching should point to one of them as superior ... they should be roughly of the same degree of excellence ... they should be equally unknown to the subjects ... Within reason, the pictures selected for the present research comply with these three conditions.'[56]

Initially factoring across persons, Eysenck found one large general factor for each set of art work. This justified the simpler procedure of calculating a subject's agreement with the average rank order for each set of art. An average correlation of $r = 0.34$ (adjusted for the contribution each subject makes to the average order) was taken to be indicative of a common basis of aesthetic judgement, also providing a comparative value to each subject's aesthetic capacity. Taking each subject's level of agreement with the average order for each art set as akin to a test score, Eysenck factor analysed them in the manner Burt recommended. Eysenck extracted what has been described as a relatively modest first factor, accounting for around 20% of the variance that loaded on most sets of art work.[57]

Eysenck had statistically summarized the consistency in his subjects' capacity to rank art work. The question was: how did one interpret this consistency? In re-describing his data in this manner, Eysenck wanted to demonstrate that this consistency was not just a function of a particular data collection; the consistency across each set of art work was taken to imply a common quality in subjects' preferences. Burt had shied away from ontological realism for most of his career, preferring to interpret the factors he derived as nominal constructs. As Burt cautioned in 1937, 'factors themselves, of course, need have no

---

[56] See p. 96 of Eysenck, H.J. (1940*b*), The general factor in aesthetic judgements, *British Journal of Psychology* **31**, 94–102.

[57] See David Nias's description of this factor as 'rather small' and his discussion of the substantive import of this work in the context of 'humour and personality', in Nias, D. (1981), Humour and personality, in *Dimensions of personality: papers in honour of H.J. Eysenck* (ed. R. Lynn), pp. 287–313, Pergamon, Oxford.

more real existence that the lines of latitude and longitude on a map.'[58] However, his bold German student came close to such a position. Eysenck interpreted his first factor as a general tendency for people to agree on what constituted good art. He claimed he had quantified the 'core of reality behind what is generally called "good taste"'[59] Eysenck dubbed it the T factor, and made an explicit analogy to Spearman's g. Nevertheless, he was disappointed to find that his general aesthetic factor T did not correlate very highly with g.

Most art materials yielded only one primary factor. Only portraits and statues produced evidence suggesting a secondary factor played a significant role. Using Burt's group factor technique, Eysenck interpreted it as a bipolar dimension, with positive and negative loadings on some forms of art and not others. Eysenck compared this second factor with measures of personality (the Heidbreder test of introversion), political attitudes (Vetter's radicalism–conservatism test), and across gender and age groups. This secondary K factor, provisionally characterized as 'brightness', seemed to divide the population in their preference for modern art, correlating with extraversion, radicalism, and age.[60]

Not all of Eysenck's doctoral work could be seen to be restricted entirely within the London tradition, hinting at a willingness to combine this approach with experimental methods. The most ambitious aspect of Eysenck's thesis project sought to explain the 'good gestalt'—the features of a stimulus underlying the pleasing and sudden perception of wholeness. Eysenck's aim was to clarify the perceptual factors underlying this process. He had the foresight to do follow-up experiments with a tachistoscope, examining the features of subjects' reproductions of rapidly presented figures. Eysenck's results reworked Birkhoff's formula: aesthetic pleasure was a *product* of order and complexity. Complexity seemed to enhance rather than cancel out pleasure, vindicating Kurt Koffka's dictum that 'perception is artistic'.[61] Developed from the last chapter of his doctorate, the theoretical sweep of this paper was striking.

---

[58] Burt (1937), Correlations between persons, p. 62.

[59] Eysenck (1940*b*), The general factor in aesthetic judgements, p. 100.

[60] Ibid.

[61] Eysenck linked his findings to the concept of neural energy and the second law of thermodynamics to derive a general law of aesthetics and a number of testable corollaries. According to Eysenck's law, perceptual pleasure was 'directly proportional to the decrease in energy capable of doing work in the total nervous system, as compared with the original state of the system.' Eysenck (1942*a*), The experimental study of the 'good gestalt'—a new approach, p. 358.

It illustrated Eysenck's talent for systematizing, as well as a precocious self-assurance. He was weighing into an already well-developed field, dominated by German research.[62]

Eysenck's doctoral work demonstrated a masterly command of the literature. He was especially adept in pulling diverse elements together and integrating results published in several different languages. The research materials he used were the stuff of everyday life, his subjects distinguished only by their willingness to make structured, comparative judgements. It was the statistical techniques that were the heavy artillery. Eysenck used factor analysis to extract commonalities from his subjects' responses that could then be represented as fundamental (and novel) psychological constructs. Factor analysis enabled Eysenck to get a grip on an ephemeral, culturally loaded property like aesthetic taste—defining what it was by describing how it varied. This oblique form of explanation could be filled in and extended by correlating factorial constructs with other measures, and by treating them as variables in new experimental designs. It demonstrated the kind of interlocking correlational and experimental methodology that would become synonymous with the Eysenck name, a synthetic expansion of the individual differences approach that was almost entirely in evidence while Eysenck was still a student. And he achieved this rapid intellectual maturation amidst a backdrop that was hardly serene.

Wartime trials and uncertainty

When Britain declared war in September 1939, University College was quickly evacuated to University College of Aberystwyth in Wales. Burt moved with it. Eysenck stayed in London, however, living in a flat with Margaret on Howitt Road in Hampstead. Margaret had a job she could ill afford to leave and Eysenck apparently worked at home but depended on resources at the library of the British Museum, amongst others.

Once war was declared, Eysenck's life became even more complicated. He applied for duty in the Royal Air Force, and other branches of the armed forces, but was knocked back because he was still a German national. Thus he had no choice but to return to his Ph.D. studies.[63] However, the threat of

---

[62] Eysenck's acknowledgements thanked Koffka, Burt, and Spearman, adding 'that it must not be assumed, however, that views here advanced would necessarily be endorsed by them.' Eysenck (1942*a*), The experimental study of the 'good gestalt'—a new approach, p. 344.

[63] Eysenck (1997*b*), *Rebel with a cause*, p. 78.

internment of 'enemy aliens' would soon have Eysenck 'in a panic'. He narrowly managed to avoid being taken away on several occasions.[64] In these uncertain times, Eysenck worked single-mindedly, even feverishly. Obtaining a Ph.D. was not just professional qualification, it could also help avoid being locked up.

Burt continued to supervise Eysenck from a distance, giving great attention and feedback in the form of detailed letters commenting on thesis drafts. As Burt admitted, the situation was far from ideal. Because of this, Burt gave Eysenck (and Parry) more attention than they otherwise might have received. The intelligence and promise both seemed to possess would no doubt have encouraged Burt's mentoring efforts as well. Burt provided painstaking advice about exposition and style, and on how to frame arguments and moderate tone in order to make a position readily defensible. On more than one occasion Burt gently rebuked Eysenck for appearing overconfident and intellectually bumptious. 'I myself would prefer a tentative understatement rather than an emphatic overstatement', Burt cautioned, 'particularly when one is criticising rather eminent authorities'.[65] He instructed Eysenck in how to insulate his work from criticism by anticipating points of attack and removing opportunities for counterarguments. For example, Burt warned that examiners 'often attack … not so much on the candidate's analysis of his own experimental data, but upon his knowledge of the literature. A candidate therefore should be careful not to leave the examiner a chance for tripping him up on these matters'.[66] More substantive content was dealt with in detail elsewhere, polishing the total package. Here was the master mentor at work, listened to closely by an attentive and respectful student.

In the process of factoring across persons, Eysenck had developed a version of the Spearman–Brown prophecy formula for persons instead of items. It enabled estimation of the subject sample size necessary to achieve a certain degree of fidelity with a hypothetical 'true' population order. It led to Eysenck's first authentic publication.[67] However, it drew a sharp response from Bernard Babington Smith, then at St Andrews and later at Oxford, who suggested Eysenck's use of an approximate formula was misleading. Smith went on to

---

[64] See, for example, Hans J. Eysenck to Cyril Burt, 14 July 1940, Cyril Burt Papers, D.191/14/2, Special Collections and Archives, University of Liverpool.

[65] Cyril Burt to Hans J. Eysenck, 21 May 1940, Cyril Burt Papers, D.191/14/2, Special Collections and Archives, University of Liverpool, p. 2.

[66] Ibid.

[67] Eysenck, H.J. (1939a), The validity of judgements as a function of the number of judges, *Journal of Experimental Psychology* **25**, 650–4.

argue that the central claim of Eysenck paper 'had overreached itself' by conflating the concepts of reliability and validity.[68] Eysenck typed a draft response, convinced that Babington Smith was attacking a 'man of straw' with 'particularly little resemblance to author!'[69] Burt was not so sure. Burt's response was a long and delicately worded treatise, full of great erudition and mathematical proofs. It was virtually a primer in argumentative technique.[70] Burt was at pains to point Eysenck toward a defensible position that blunted the potentially valid points of Babington Smith's attack. Burt commented: 'your proper line of reply is to admit that you were using "validity" in a rather broad sense and to agree that … "it is unsafe to discuss validity in the absence of a known criterion".'[71] Since Eysenck had originally stated that he was dealing with a problem (i.e. aesthetic judgement) where no such criterion existed, Burt urged him to emphasize this as the key justification. In such a case, the validity of small group judgements could only be compared with the benchmark judgements of much larger groups. Burt also helped Eysenck nuance other points; for example, he recommended that Eysenck make a clear distinction between chance and systematic error (implicit in Eysenck's original article) to avoid another of Babington Smith's criticisms. Burt added: 'Of course, your claims are stated with youthful enthusiasm … I sometimes fancy that your effort to be definite and clear makes other readers think you are claiming more than you mean to claim.'[72] Eysenck adapted almost all Burt's suggestions in his published reply. Careful to cite his supervisor's research generously, he used the pronoun 'we' to suggest a more general UCL viewpoint. Armed with Burt's

---

[68] Babington Smith, B. (1941), Discussion: the validity and reliability of group judgements, *Journal of Experimental Psychology* 29, 420–6. The distinction between reliability and validity was more contestable at this time than some readers might imagine. The clear separation of these two concepts that contemporary psychologists would be familiar with owes much to the more formalized treatments of psychometric issues in the 1950s and 1960s. See in particular, Lord, F.M., and Novick, M.R. (1968), *Statistical theories of mental test scores*, Addison-Wesley, London.

[69] Hans J. Eysenck to Cyril Burt, 10 September 1940, Cyril Burt Papers, D.191/14/2, Special Collections and Archives, University of Liverpool, p.1.

[70] Eysenck made an edited copy of this letter for Burt. Even so, this typescript ran to 11 pages. See 'Copy of Letter, 12 September 1940, concerning Mr. Babington-Smith's Criticism of H.J. Eysenck's Article on Validity etc.' Cyril Burt Papers, D.191/14/2, Special Collections and Archives, University of Liverpool.

[71] 'Copy of Letter, 12 September 1940, concerning Mr. Babington-Smith's Criticism of H.J. Eysenck's Article on Validity etc.' Cyril Burt Papers, D.191/14/2, Special Collections and Archives, University of Liverpool, p. 6.

[72] Ibid., pp. 8–9.

tips, Eysenck made strategic concessions look like steadfast defence, casting his opponent in the role of the irrelevant nit-picker with little understanding of the psychological issues involved. Burt's guidance provided a lesson in rhetoric that Eysenck undoubtedly took to heart, for this was Eysenck's first public sparring match. It was a challenging, perhaps even thrilling experience that he would repeat throughout his career.[73]

Burt was more than impressed by his protégé's output. Burt wrote to Eysenck's prospective examiner, Spearman's ex-student Wynn Jones:

> Eysenck you may remember as a sharp young German who obtained a First last year. The chief problem that I suggested to him was to compare Birkhoff's aesthetic measure with a rough formula of my own ... Eysenck, however, is an original youth and has attempted something much more ambitious and suggestive.[74]

Characteristically though, Burt still could not resist comparing Eysenck unfavourably to another of his Ph.D. students at the time, Eysenck's friend John Parry. Burt added: 'Parry is the stabler and I think the abler of the two.'[75] It was an intriguing judgement. Perhaps Burt underestimated the difficulty of being a German émigré at that time and how this might affect even the most eventempered. Perhaps he was already beginning to perceive Eysenck as a threat, as Eysenck said Burt came to do when he was later trying to explain Burt's spoiling efforts. Parry remained an important contact for Eysenck, even though their lives were to take quite separate paths.

In his letters to Burt, Eysenck showed himself to be extremely studious, as well as politely deferential to his professor's authority and expertise. The correspondence between them also conveyed some of the problems Eysenck faced, as the Battle of Britain started in July 1940. He certainly depended on Burt for more than just intellectual guidance. At the beginning of July 1940, a new order to intern all German- and Austrian-born men had come into effect. Only those who could show reason why they should not be interned would be granted exemptions. Eysenck mentioned to Burt that he was writing to the Home Office to this effect. His unselfconscious tone is touching:

> I should be most grateful if you would be good enough to write them too... You might say that my research has given some useful results, and if allowed my freedom I should be capable of rendering good service to the country, given the opportunity. My name is on the Central Register of Aliens, and they have advised me that only those with

---

[73] See Eysenck, H.J. (1941a), Reply: the validity and reliability of group judgements, *Journal of Experimental Psychology* **29**, 427–34.

[74] Cyril Burt to Wynn Jones, 11 June 1940, Cyril Burt Papers, D.191/14/3, Special Collections and Archives, University of Liverpool, p. 1.

[75] Ibid., p. 2.

special qualifications were selected for the list. I have always been anxious to help in any way I could...[76]

This was a very stressful time. Eysenck had applied for naturalization in August 1939, as soon as the required five years of residence was completed. However, that seemed a distant prospect now. His wife Margaret had her British nationality taken back at first and then restored, but as he wrote Burt: 'she is deeply distressed over the imminent possibility of me being taken off for internment, without warning and for no one knows how long.'[77]

Eysenck submitted his thesis, which ran to 420 pages, on the 2nd of June 1940. He included four published papers as appendices, a feat not many would have been able to boast of at the time.[78] Eysenck faced and passed his oral the following month. In the lead up to the completion of his dissertation, he put in for several scholarships—including the John Stuart Mill Scholarship, and the Perrault Scholarship for Psychical Research—so that he might continue to do research on humour and hypnosis that he and Burt had mapped out. He did receive a Mill scholarship, and this supported work on humour and other research. However, Eysenck also wrote several unsuccessful job applications leaving him to wonder whether his nationality had counted against him yet again. Having heard that University College would be re-established in London, Eysenck even asked Burt to help secure him a position of any kind there. However, the College was to remain in Wales until the autumn of 1944, so nothing came of this.

Eysenck's biographer Gibson suggested that Burt was actively unhelpful to Eysenck's efforts to get a job in this period. However, I could find no evidence for this. Times were tough and opportunities in London scarce. By the same token, Cambridge was out of the question. As Eysenck himself admitted, not having a Cambridge degree was a fatal obstacle to obtaining a job there. It seems Burt did all he could. At the end of the war Burt was able offer Eysenck part-time teaching at UCL and strongly supported Eysenck's application for the position of senior psychologist at the Maudsley Hospital in 1946.[79]

---

[76] Hans J. Eysenck to Cyril Burt, 4 July 1940, Cyril Burt Papers, D.191/14/2, Special Collections and Archives, University of Liverpool, p. 1.

[77] Ibid.

[78] Eysenck, H.J. (1940c), Processes of perception and aesthetic appreciation, Ph.D. thesis, University of London. The papers included as appendices were on Thurstone's primary abilities, the validity of group judgements, the general factor in aesthetic judgements, and one on hypnosis.

[79] See Gibson (1981), *Hans Eysenck: the man and his work*, p. 36; Eysenck (1997b), *Rebel with a cause*, p. 53, and Minutes of the Medical Advisory Committee, Maudsley Hospital, 3 October 1946, Box 12, Aubrey Lewis Papers, Bethlem Hospital Archives.

As with his doctoral studies, finding work was not just about career advancement. As Eysenck noted at the time, 'new regulations exempt refugees of scientific and academic standing when important work is available to them. Lecturing would be regarded as such.'[80] Eysenck had been receiving financial support from his mother and stepfather. When the Nazis overran Paris in June 1940, however, this source of funds was cut off. He had no job in prospect, severely restricted by the kind of work he could take on as an enemy alien. Even though he was strapped for cash, Eysenck continued to work on psychological research through the latter half of 1940 and early 1941, during the worst days of the German bombing. He was able to gain a meagre income vetting propaganda broadcasts to Germany, but the couple's main source of income in this period came from Margaret's office job. On September 1940 he wrote to Burt:

> As you say, we have had some excitement—about five bombs within a few hundred yards! Your house is still standing, but there was a bomb quite near in Adelaide Rd, and another in Primrose Hill Rd. Ever [sic] now and then you can hear a time bomb going off somewhere around. Two of my best and most loyal subjects have been bombed out of their home twice! This sort of thing makes it rather difficult to conduct experiments.[81]

Eysenck added that his friend Parry was now an air raid warden and 'spends most of his days extinguishing fire-bombs and guarding bomb-craters. He is grateful that the Blitz did not start till he finished his thesis.'[82] With some restraint Eysenck complained that 'there are some difficulties in the way of writing anything—the British Museum for instance seems to be full of time bombs, and they don't let anybody in to consult journals etc.'[83]

Finding his voice

There was a growing sense of confidence in Eysenck—not to mention a remarkable degree of composure in the face of adversity—in the period immediately following his Ph.D. examination. With a well-regarded doctorate in the

---

[80] Hans J. Eysenck to Cyril Burt, 7 September 1940, Cyril Burt Papers, D.191/14/2, Special Collections and Archives, University of Liverpool, p. 4.

[81] Hans J. Eysenck to Cyril Burt, 29 September 1940, Cyril Burt Papers, D.191/14/2, Special Collections and Archives, University of Liverpool, p. 1.

[82] Ibid.

[83] Ibid.

bag, Eysenck began to strike out on a more independent intellectual path.[84] He now had many published papers already in print, on the way, or planned. Eysenck began to openly disagree with Burt on the implications of some of his findings. For example, Eysenck defended his draft article on the general aesthetic factor by asserting:

> You say elsewhere that an average correlation of 0.49 between six judges ranking 16 objects is hardly significant, while I regard it as very significant ... As a single correlation, the chances of it being obtained by chance would be about one in twenty; but as all fifteen correlations of which this is the average would by chance occur once in twenty, the chances of such an average occurring would be 15 times 20 ... That surely is very significant?[85]

Spin-off work from his Ph.D. yielded several articles on humour and appreciation of poetry.[86] More important were articles on colour measurement that overlapped with Parry's research. Eysenck suggested that the same explanation of a general 'good taste' factor could likewise be applied to colours.[87] Eysenck took issue with the research literature suggesting that colour preferences were too varied and chaotic to be 'native'. Re-analysing a long line of results from German, English, and American studies, he derived a consistent preference order for the colours they all employed. Eysenck obviously felt pleased with this corrective; in an aside to Burt he suggested this paper could serve as a useful 'illustration of how wrong even the most careful workers, such as Allesch, can be if they don't use the appropriate statistical methods.'[88] It was also the

---

[84] Eysenck's thesis examiner, Wynn Jones, had only minor quibbles. Jones questioned the extent to which the results indicated Eysenck's general aesthetic factor was unrelated to intelligence or *g*, whether age effects were accounted for, and whether the personality and political preference tests used were good enough to warrant the conclusions reached. Hans J. Eysenck, 'Notes in reply to examiner', c. September 1940, Cyril Burt Papers, D.191/14/2, Special Collections and Archives, University of Liverpool.

[85] Hans J. Eysenck to Cyril Burt, 7 September 1940, Cyril Burt Papers, D.191/14/2, Special Collections and Archives, University of Liverpool, p. 2.

[86] Eysenck, H.J. (1940*a*), Some factors in the appreciation of poetry, and their relation to temperamental qualities, *Characteristics of Personality* **9**, 160–7; Eysenck, H.J. (1942*b*), The appreciation of humour: an experimental and theoretical study, *British Journal of Psychology* **32**, 295–309.

[87] Factor analysis of colour preferences yielded a first factor as large that for intelligence tests, and it correlated highly (at 0.53) with the T factor he found for art work. Eysenck, H.J. (1941*e*), A critical and experimental study of colour preferences, *American Journal of Psychology* **54**, 385–94.

[88] Hans J. Eysenck to Cyril Burt, 14 July 1940, Cyril Burt Papers, D.191/14/2, Special Collections and Archives, University of Liverpool, p. 1.

first (and almost only) time Eysenck utilized the category of race as a variable in his research.[89]

Research on colour preferences led to Eysenck's first publication in the prestigious journal *Nature*—a notable achievement for such a young researcher, especially in psychology. Sidestepping the units of measurement problem that had bugged psychophysicists for years, Eysenck was content to show that Ostwald's colour charts were not psychologically equidistant. Burt was not entirely impressed with Eysenck's treatment of the measurement issue, however, and Eysenck would avoid such fundamentals for much of his subsequent career.

Burt continued to offer advice and support in Eysenck's post-Ph.D. period, even though his formal responsibilities as a supervisor had been discharged. More subtly, Burt remained a moderating influence on his eager charge, tempering Eysenck's tendency to overstate his conclusions. For his part, Eysenck was sure he had something important to say, but was unsure how to say it. It seemed the young researcher was having trouble finding his voice. In his article on the aesthetic appreciation formula, Eysenck omitted criticism of noted Harvard psychologist J.G. (John Gilbert) Beebe-Center, and Birkhoff himself, apparently in deference to Burt's warnings. Noting that criticism needs 'a particularly fine touch to be effective and acceptable', Eysenck admitted that he had been 'putting his feet down very heavily' at the moment. The point was, though, that he was beginning to engage with many noted figures in the field. When arguing a case by citing his previous research Eysenck admitted he found it 'difficult to refer to myself without appearing either intolerably conceited or painfully shy'.[90]

Eysenck's early articles look very assured, in spite and perhaps because their methodological details are hard to follow. For example, in his key article on the general factor in aesthetic judgements, it is not clear at which point Eysenck is factoring across persons and what saturations (weighted or/and unweighted) are being used when factoring across sets of art work.[91] Eysenck later came to criticize Burt for his habit of overreferencing his work. If anything, Eysenck appeared to edge toward the other extreme, even in these early days. Critics would later complain of a lack of careful description and documentation that

---

[89] Eysenck found no difference between 'white' and 'coloured' subjects' colour preferences. See Eysenck (1941*e*), A critical and experimental study of colour preferences.

[90] Hans J. Eysenck to Cyril Burt, 7 September 1940, Cyril Burt Papers, D.191/14/2, Special Collections and Archives, University of Liverpool, pp. 1 and 3.

[91] Eysenck (1940*b*), The general factor in aesthetic judgements.

conveyed a take-it-from-me style more intent on persuasion than on open reporting for the purposes of evaluation and replication.

Few contemporary graduate students hit the ground so full of running. By the close of 1941, just over a year after his Ph.D., Eysenck had 11 research articles published in some of the most prestigious psychological and scientific journals in the world. The combined correlational–experimental aspects of Eysenck's approach spun out maximum output and offered a template for things to come. He began his habit of publishing almost compulsively and effortlessly. There seemed little agonizing prevarication or self-censorship. He appeared to consult quickly and decisively with his key advisors and collaborators before putting his work out in the public domain.

Eysenck promptly left the cul-de-sac area of experimental aesthetics to focus on more meaty questions. He did not return to this topic until the mid-1960s.[92] It remained one of his many sideline interests, the subject of a 'mere' twenty or so later research and review articles. Still, even work in such a relatively esoteric area had obvious practical applications. For example, his doctoral research could provide a psychological guide to why some people prefer certain types of paintings. More practically, it could lead to the construction of a measure of aesthetic sensitivity or aptitude, useful perhaps in a vocational guidance context. This is exactly what Eysenck later did, following the work of other researchers who had constructed similar such tests for various purposes.[93] Somewhat disingenuously, Eysenck said he was always surprised about the hostility he encountered to this work from those in the art world.

Burt also steered Eysenck on to the study of personality, or 'temperament', as it still tended to be termed at the time. He had engaged Eysenck partly through his own work on mental factors and the dispute he was having with various rivals. Burt had just published his momentous work *The factors of the mind* that summarized most of his factorial work to date. In it, Burt reviewed various types of analyses, including factoring across persons and tests, as well as applications to temperamental types. Through Burt, Eysenck took issue with American Joy P. Guildford's surprising failure to affirm introversion–extraversion as an important temperamental type. Again using Burt's group factor technique, Eysenck's re-analysis suggested there was such a factor. In the

---

[92] Eysenck, H.J. (1968a), An experimental study of aesthetic preference for polygonal figures, *Journal of General Psychology* **79**, 3–17.

[93] Eysenck, H.J., *et al.* (1979), A new visual aesthetic sensitivity test: (VAST. 1. Construction and psychometric properties, *Perceptual and Motor Skills* **49**, 795–802.

process, he verified his secondary K factor that separated aesthetic preferences according to personality type using this more modern questionnaire.[94]

In addition, Burt had wanted Eysenck to take Stephenson to task over the value of factoring across persons when analysing personality. Pressured in the last days of his Ph.D., Eysenck had to ruefully decline. Eysenck told Burt that his wife Margaret had made notes on one of Stephenson's articles with the view to criticizing it, but

> we feel that really it is rather a delicate task, and that neither of us is at the moment well enough fitted to undertake it ... We discussed the possibility of repeating St.'s experiment, but have come to the conclusion that not even between us could we hope to find 46 people we know sufficiently well to rate them for 176 traits with any degree of accuracy. Preconceived notions of introversion and extraversion of the individuals, or liking them etc, would tend to vitiate the ratings.[95]

Teasingly, Eysenck added:

> For some time I have been thinking of a plan to get ratings done so that the problem of definite temperamental types among more or less normal people could be tackled. My plan was to enlist the assistance of several psychoanalysts who would be able to rate their patients for a number of traits ... After the thesis is submitted, I hope I can do something along this line, as the subject interests me very much.[96]

But reality had a way of intruding on his intellectual aspirations. Out of work and with no professional options in sight, Eysenck did a short stint in the Air Raid Precaution Service. His account of this episode was hardly flattering to his fellow plane-spotters, whom he depicted as the misfits and miscreants of London town.

Burt in hindsight

The period ending with Eysenck's induction into the Air Raid Precaution Service represented the end of his apprenticeship with Burt. Eysenck recalled that he was determined to do more than just follow in Burt's footsteps, for he saw many weaknesses in his mentor's work. Nevertheless, Burt's general approach was 'invulnerable', Eysenck said. 'That I took over. Not because I was

---

[94] Eysenck, H.J. (1941*b*), Personality factors and preference judgements, *Nature* **148**, 346.

[95] Hans J. Eysenck to Cyril Burt, 16 April 1940, Cyril Burt Papers, D.191/14/2, Special Collections and Archives, University of Liverpool, pp. 1 and 2. It was not clear which of Stephenson's articles Eysenck was referring to, but most likely it was Stephenson, W. (1939), Methodological considerations of Jung's typology, *Journal of Mental Science* **85**, 185–205.

[96] Hans J. Eysenck to Cyril Burt, 16 April 1940, Cyril Burt Papers, D.191/14/2, Special Collections and Archives, University of Liverpool, p. 2.

his student, but in spite of being his student.'[97] While he also nominated Spearman and Galton as major influences, Eysenck remembered Burt as 'probably the most intelligent person I ever met'.[98]

Eysenck remained in contact with Burt over the next decades and still held his work in great esteem, but Eysenck's relationship to his intellectual father figure grew more distant and complex. The posthumous scandals surrounding Burt saw Eysenck leap to his mentor's defence, then rapidly distance himself when Burt's guilt seemed to be well-established. Burt's biographer, Leslie Hearnshaw, characterized his subject as the practised impostor, and Eysenck followed suit by harping on Burt's 'psychopathic' tendencies.[99] Eysenck also remarked on Burt's disarming but duplistic style, how he would show

> his aggressiveness and his dominance behind the backs of the people involved, never to their faces. He never said an unkind word to me in person, but his letters and comments to others were dripping with poison, as Hearnshaw makes clear. In public debates he had a dangerous technique of praising the contribution of his opponent, only to insert the stiletto at the last minute.[100]

Still, Eysenck paid Burt the ultimate compliment. Burt was, he said, the only fellow psychologist he ever feared, the only one he would think twice about taking on in a public argument. Moreover, the thrust of the correspondence in this mentoring period throws a slightly different light on Burt's subsequent (real and imagined) efforts to thwart Eysenck's career.

As well as being more helpful to Eysenck's early career than others have assumed, Burt may not have been as critical as was supposed. For example, a trenchant review of Eysenck's second book *The scientific study of personality* was widely assumed at the time to be Burt writing under the alias 'WLG'. This was not quite the case.[101] Eysenck later recalled how he caught up with WLG, a well-known statistician, who told Eysenck that Burt had rewritten his review to be more critical of Eysenck.[102] Prompted by Hearnshaw's negative verdict

---

[97] Cohen, D. (1977), Interview with Hans Eysenck, in *Psychologists on psychology* (ed. D. Cohen), pp. 101–25, Routledge and Kegan Paul, London, p. 109.

[98] Ibid.

[99] Eysenck's latter-day attitude to Burt was probably best summed up by the title of his article in Eysenck, H.J. (1983*h*), Sir Cyril Burt: polymath and psychopath, *Journal of the Association of Educational Psychology, Centenary Issue* **6**, 57–63.

[100] Eysenck (1997*b*), *Rebel with a cause*, p. 116.

[101] See WLG (1953), Review of *The scientific study of personality* by Hans Eysenck, *British Journal of Psychology (Statistical Section)* **5**, 208–12.

[102] Eysenck did not identify this prominent statistician, however. When asked for the name by Ronald Fletcher some years later Eysenck said he could not remember—which seems

on Burt and Ann and Alan Clarke's antipathy to their former supervisor, the latter-day Eysenck went out of his way to highlight the more aberrant aspects of Burt's behaviour, lumping many incidents together as devious back-stabbing.[103] Eysenck remembered Burt warning him off publishing in certain areas because Burt claimed he was supervising students with allegedly similar projects. While Eysenck saw this as another example of Burt's propensity for sinister interference, these were not totally groundless concerns nor were they obviously self-interested on Burt's part. Moreover, when Burt retired from UCL in 1950, he tried to interest Eysenck in taking over the chair. Eysenck dismissed these efforts as grudging at best. Both Burt and Aubrey Lewis failed to support Eysenck's subsequent proposal that he remain head of psychology at the Institute of Psychiatry after taking up the reins at UCL—an opposition Eysenck put down to unreasonable institutional pride on both their parts.

Burt's put-down of Eysenck as a 'boastful Jew', his veiled criticism of Eysenck's work, and his attempts to stall plans for a clinical psychology programme at the Maudsley may be seen as the calculated behaviour of an ageing don anxious to protect his intellectual legacy. Nonetheless, most accounts of Burt's malfeasance reflected the interested judgements of those he had made enemies of. It was hardly likely that Burt was a schemer amongst innocents. Burt's apparent anti-Eysenck spoiling may have also stemmed from a perverse sense of responsibility as a disciplinary standard bearer. Not only did he see Eysenck as a threat, he saw Eysenck's style as overly ambitious, potentially disastrous for the kind of differential psychology Burt championed.

While Eysenck defended much of the substantive content of Burt's research, he still had reservations about Burt's interpretative habits. Despite his sophisticated statistical skills, Eysenck recalled, Burt was surprisingly indifferent to the manner and circumstances in which his data was collected.[104] He was too 'concerned with theories, and the mathematical and statistical elaboration and

---

odd for a man with such good recall. See Fletcher, R. (1991), *Science, ideology and the media: the Cyril Burt scandal*, Transaction, New Brunswick, New Jersey, p. 186 (footnote 8).

[103] See Eysenck, H.J. (1995a), Burt and hero and anti-hero: a Greek tragedy, in *Burt: fraud or framed?* (ed. Nicholas J. Mackintosh), pp. 111–29, Oxford University Press, Oxford and Eysenck (1997b), *Rebel with a cause*, pp. 231–2.

[104] Eysenck opined: 'To analyse in detail the results of tests administered by largely untrained teachers, to unwilling pupils, in any old school that would allow the test to be done did not, in my view, provide results that could be relied upon, however inspired the statistical treatment.' Eysenck (1997b), *Rebel with a cause*, p. 55.

testing of models', Eysenck explained.[105] Yet several people who worked with Eysenck implied much the same about Eysenck himself.[106]

The big break: Aubrey Lewis and Mill Hill

Just as Eysenck despaired over how his work with the Air Raid Precaution Service was distracting him from his new-found mission in life, things took a turn for the better. In the middle of 1942, the treatment of enemy aliens relaxed. No longer threatened with the lock-up, work opportunities opened up. During the war, Eysenck's friend John Parry worked in the psychology department of the Air Force, collaborating with Philip Vernon who had a similar role in the Army. It was Vernon's recommendation to influential and imposing psychiatrist Aubrey Lewis that helped secure Eysenck's first job. Burt's support no doubt helped. In June 1942, Eysenck was hired by Lewis as senior research officer at the makeshift Mill Hill Emergency Hospital in northern London, replacing the more psychoanalytically oriented Eric Trist who had taken up a military selection post. Mill Hill was now functioning as the relocated Maudsley Hospital. Eysenck spent the remainder of the war there. After the war, most of the Mill Hill staff returned to the Maudsley. If one considers Mill Hill to be the wartime equivalent of the Maudsley, it was Eysenck's first and only job.

With research support from the Rockefeller Foundation, Eysenck's position at Mill Hill gave him the freedom to define his own role, that of research scientist working in a clinical setting. He continued to publish at an astonishingly proficient rate. In his early research there Eysenck studied hypnosis and suggestibility, continuing a sideline interest from his University College days.[107] He also made his first forays into personality assessment and research, a new field he would soon stake out as his own.

Rose-coloured memories

The war years were hard for almost everyone in Britain. One could argue that in many respects Eysenck was lucky. The risks of military service, the difficult adjustment afterwards, and the lengthy career interruption were not things he had to worry about. Even so, the immediate post-Ph.D. period was one of the

---

[105] Ibid., p. 116.

[106] Sidney Crown, interview, 22 April 2002; Alan and Ann Clarke, interview, 25 April 2002.

[107] Eysenck said this research had occurred when he was still an undergraduate, though the article he refers to was not published until 1941, three years after his BA. See Eysenck, H.J. (1941*d*), An experimental study of the improvement of mental and physical functions in the hypnotic state, *British Journal of Medical Psychology* **18**, 304–16.

trickiest and most uncertain in Eysenck's life and he got through it with barely a hint of self-doubt. The unselfconscious testimony that is available from this period clearly illustrated his ability to cope with whatever was thrown at him. Equipment shortages, lack of money, and wartime constraints made for a rather British kind of improvisation—making the best of the people, the resources, and the found objects at one's disposal. Even with no immediate job prospects, Eysenck continued to do research. Even though the bombs were going off around him, Eysenck clearly believed he had something more important to deal with. Even before he graduated, he was already a player in the international scene. He clearly identified his interests with those of the discipline itself. Psychology had a future, in part owing to the contributions he had already made and those he imagined he could make in the future.

This introduces a fascinating contrast with Eysenck's latter-day accounts of his student years. In hindsight, Eysenck made it clear he thought psychology was in a parlous state when he first entered it, a discipline badly in need of new ideas and rigorous thinking. Yet the young Eysenck appeared to give it more respect, taking to the psychology he was presented with like a duck to water. The London school gave him an almost complete scientific world view that encompassed the type of questions asked, the methodology favoured, the knowledge produced, and the social interventions made or envisaged. Beyond an impatient ambition, there was little hint of the romantically rebellious standpoint he later adopted. As a fresh-faced graduate, it was business as usual, as he sought to adjust, extend, or correct the work of his more senior contemporaries. Eysenck's philosophy of physics envy may well have functioned as a multipurpose leverage tool in his subsequent career, but it obscured the obvious continuity he maintained with his intellectual heroes. Despite Eysenck's claims of a brave new scientific approach, one does not have to look far for his initial inspirations. Burt, Spearman, Galton *et al.*—please stand up.

Chapter 4

Dimensions of personality

Hans Eysenck's personality research can be thought of as the core of his career, beginning with the research that led to his first book, *Dimensions of personality*.[1] Published in 1947, *Dimensions of personality* outlined two main personality factors of neuroticism and introversion–extraversion. It represented a brash act of coordination and integration that took its cues from American and continental work but was said to be unprecedented in Eysenck's adopted homeland. In 1952, *The scientific study of personality* introduced a third dimension, that of psychoticism.[2] Eysenck spent the remainder of his career defending these three independent dimensions as the essential underlying structure of personality.

Eysenck felt that these two books, particularly *Dimensions of personality*, were his most original and important. It was a judgement even his harshest critics came to agree with. But there are still some intriguing contradictions surrounding this work and the foundational reputation it acquired. From a present-day perspective, *Dimensions of personality* comes across as somewhat piecemeal empirical work. It reads like a psychologist just beginning to grapple with the problems of the area; it is full of half-digested data and empirical asides. Some of its organizing ideas also look quaintly old-fashioned, having been long since discarded. Moreover, the book appeared to deliberately shy away from any pretence to novelty.

Given these impressions, it is worth reconstructing the context for Eysenck's *Dimensions* to gain a historically sensitive appreciation of its significance. The way Eysenck took to mid-century personality psychology, the assumptions he made, and the insights he gained were as much a function of circumstances as intellectual clarity. One might sum this up as: 'opportunity meets ambition'— two things that defined Eysenck's career.

---

[1] Eysenck, H.J. (1947a), *Dimensions of personality*, Routledge and Kegan Paul, London.

[2] Eysenck, H.J. (1952a), *The scientific study of personality*, Routledge and Kegan Paul, London.

Making the Maudsley

The Maudsley Hospital took its name from Henry Maudsley, perhaps the most prominent British 'alienist physician' (doctors working in psychiatric institutions) of the nineteenth century. The origins of the Maudsley Hospital lay in the professional cul-de-sac of the county asylums movement in Britain.[3] Stigmatized by their association with the insane and isolated from the medical mainstream, Britain's 'mad doctors' were poorly trained and lacked effective professional representation.[4] Maudsley and his allies saw a glaring need for an institution situated within the community it drew patients from. Early intervention was desirable; research and teaching capabilities a must. Encouraged by London County Council asylums pathologist Frederick Mott, Maudsley donated 30, 000 pounds to the Council for a new psychiatric facility. The Council acquired a site on Denmark Hill in south London opposite King's College hospital and started building.[5]

From the very outset, the hospital was envisaged as a university hospital. Maudsley was particularly anxious that the old designation 'asylum'—with its negative custodial connotations—should be avoided. Other English county asylums were not re-badged as 'hospitals' until 1930, just as 'mental illness' was officially re-labelled as such. With inpatient and outpatient services, the Maudsley was closely modelled on Emil Kraepelin's clinic, which Mott had specifically visited in Munich. German psychiatry had enjoyed a productive alliance with neurology; to British eyes it looked like the future.

---

[3] As a more appropriate alternative to the prisons and workhouses, the asylums often represented a step up in living standards for those they housed. Although the asylums were relatively closed worlds, they were also isolated from the disease-ridden hovels of urban life, from typhus, cholera, and rising crime. They were headed by a typically imperious medical superintendent and staffed by live-in and consultant doctors and support staff. As Kathleen Jones concluded, these institutions were quite insular worlds, with their own economy, regulations, and social mores. However, she concluded, the asylum served its times. 'We should neither deny its usefulness in its own day nor regret its passing.' See p. 27 in Jones, K. (1991), The culture of the mental hospital, in *150 years of British psychiatry, 1841–1991*, Vol. 1 (ed. G.E. Berrios and H. Freeman), pp. 17–28, Royal College of Psychiatrists, London.

[4] Turner, T. (1991), 'Not worth powder and shot:' the public profile of the Medico-Psychological Association, c .1851–1914, in *150 years of British Psychiatry, 1841–1991*, Vol. 1 (ed. G.E. Berrios and H. Freeman), pp. 3–16, Royal College of Psychiatrists, London.

[5] Allderidge, P. (1991), The foundation of the Maudsley Hospital, in *150 years of British psychiatry, 1841–1991*, Vol. 1 (ed. G.E. Berrios and H. Freeman), pp. 79–88, Royal College of Psychiatrists, London.

The Maudsley Hospital was completed in 1915, but the demands of the war delayed its full opening until 1923. The following year the hospital and central pathological laboratory became the Maudsley Hospital Medical School, a recognized part of London University. Maudsley's vision was taken over by the first superintendent of the hospital, Edward Mapother, who linked up with US philanthropic bodies like the Commonwealth Fund and the Rockefeller Foundation for research support.[6] Nevertheless, up to World War II the Maudsley's inpatient capacity remained small and teaching capabilities relatively limited.

Aubrey Lewis became clinical director in 1936 and would eventually succeed Mapother as professor of psychiatry after the war. Lewis was an unlikely mandarin figure. He had been born to a relatively humble Jewish family in Australia and educated at the University of Adelaide. Only in the special circumstances of war, Lewis confided to a colleague, could an outsider like himself come to head such a prestigious institution.[7] After graduating from medicine in 1923, Lewis won a Rockefeller Fellowship and joined the Maudsley staff in 1929.

In August 1939, the Ministry of Health evacuated the Maudsley and reallocated staff to deal with the mental health casualties expected from impending air warfare. Two large emergency services hospitals were set up on the northern and southern outskirts of London—one at Mill Hill, with Lewis as clinical director, and one at Belmont in Sutton, headed by Eliot Slater. The Maudsley was gutted of equipment and closed. Both hospitals played major roles in the inpatient treatment of neuroses, serving the army and Greater London civilian populations. Both hospitals continued the Maudsley's teaching responsibilities and both continued to support research.

Eysenck was hired in 1942 to continue the Maudsley group's research brief and to create a small subdepartment in psychology. Part of this brief was to help diagnose and tabulate the expected increase in neuroses due to the war. Eysenck had come to an institution where those at the top made a point of employing German-Jewish intellectuals looking for a new professional home. Mapother had prevailed upon the Rockefeller Foundation to provide fellowships for

---

[6] The means to fully realize Maudsley's original vision would not become available until after the war, when the Maudsley amalgamated with the venerable and rich Bethlem Hospital and the Institute of Psychiatry was created. See Waddington, K. (1998), Enemies within: post-war Bethlem and the Maudsley Hospital, in *Cultures of psychiatry and mental health care in postwar Britain and the Netherlands* (ed. M. Gijswijt-Hofstra and R. Porter), pp. 185–202, Rodopi, Amsterdam.

[7] Gibson, H.B. (Tony) (1981), *Hans Eysenck: the man and his work*, Peter Owen, London, pp. 64–5.

refugee psychiatrists in the mid-1930s and a number of eminent German refugees had joined the Maudsley before Eysenck arrived. This included psychiatrists Wilhelm Mayer-Gross and Eric Guttman and neuropathologist Alfred Meyer. In 1949, Austrian psychiatrist Erwin Stengel joined the staff, a noted advocate of psychodynamic psychiatry.

Both Mapother and Lewis had a deep sympathy with the plight of war émigrés. For Lewis, especially, it was a matter of scientific, political, and personal principles. *The Lancet* of 5 August 1933 had run a strong, unsigned leader condemning the 'extremist' racial laws enacted by the Nazis as a 'gross overstatement of our present knowledge of hereditary' that exhibited 'a disregard for the individual human being, and a willingness to act upon racial prejudice'.[8] The author of this piece, it later emerged, was Aubrey Lewis.[9] In the mid-1930s, Lewis had also sent Eliot Slater to Europe to study the genetic aspects of psychiatry, where he came face to face with Nazi eugenics. While Slater became a lifelong advocate of genetic research, he too was at pains to distance himself from such perverse 'misuse'.[10]

Eysenck recalled that Lewis asked him few pertinent academic questions in what for him was his first and only job interview. Eysenck assumed Lewis simply took Philip Vernon's word that he was a good psychologist, and his fine academic pedigree spoke for itself. Unspoken assumptions relating to Eysenck's status as a German émigré may have also counted in his favour. However, it would soon become clear that Eysenck had little in common with most of his former countrymen at the Maudsley.

Like Mott and Mapother, Lewis was an admirer of German psychiatry. Lewis had been educated in a liberal arts tradition; he was multilingual and had a particular interest in anthropology and genetics. Lewis was also quite taken with the dynamic, community-based psychiatry of Adolf Meyer after visiting him at Johns Hopkins in Baltimore.[11] Meyer's eclectic approach focused on developmental and environmental factors in the aetiology of mental distress, and encouraged input from paramedical groups.

---

[8] See p. 298 in *Lancet* (1933), Eugenics in Germany [unsigned editorial; later found to be by Aubrey Lewis], *The Lancet* **ii** (August 1933), 297–8.

[9] See Gottesman, I.I. and McGuffin, P. (1996), Eliot Slater and the birth of psychiatric genetics in Great Britain, in *150 years of British psychiatry, 1841–1991*, Vol. 2 (ed. H. Freeman and G.E. Berrios), pp. 537–48, Athlone Press, London.

[10] See Slater, E. (1971), Autobiographical sketch, in *Man, mind and heredity* (ed. J. Shields and I. Gottesman), pp. 1–23, Johns Hopkins Press, Baltimore.

[11] Shepherd, M. (1977), A representative psychiatrist: the career and contributions of Sir Aubrey Lewis. *American Journal of Psychiatry* **134**, 7–13.

More than anyone else in Britain, Lewis put in place his version of Meyerian psychiatry. He insisted on careful and systematic case notes, conducted admission and discharge case conferences attended by the professional team, and founded multiple research departments that included the non-traditional (in Britain) areas of physiology and psychology. According to Eliot Slater, Lewis's greatest contribution was 'teaching us to define our operational terms, to define problems so that they would be subject to empirical attack ... and to proceed in an orderly and purposeful way'.[12] Lewis also shared some of his American hero's shortcomings, however. He expanded the hospital's services and research facilities enormously in the post-war years. But for those not sharing his Meyerian optimism, it looked rather indecisive and haphazard, especially in hindsight.[13] Moreover, Lewis was not a particularly gifted theorizer. And, while his political inclinations were liberal, his managerial style was resolutely authoritarian.

An academic psychologist in a wartime hospital

Eysenck was an ambitious, quietly driven young man. Sensing this, Lewis hoped Eysenck could position psychology as a basic science for psychiatry, as basic as physiology was to disease medicine. Lewis gave him the security of a permanent job, and the freedom, to make his mark. Temperament was one topic Eysenck already had earmarked for attention well before he ever sighted a psychiatric ward—but he needed expert ratings and measures, research assistants, and lots of subjects. Mill Hill provided him with just that.

Eysenck's new position was unusual in several respects. It was still relatively uncommon for a psychologist to be working in a psychiatric context in Britain in 1942. Up to then, the academic corps of British psychology was exceedingly small. Psychologists had had to fight hard for a foothold in England, not quite so hard in Scotland. Only six chairs had been created. Three were in London—at King's College, University College (UCL), and Bedford College. The total number of psychologists in academia numbered around 30.[14] For the most

---

[12] Slater (1971), Autobiographical sketch, p. 15.

[13] As psychiatrist–historian John Cramer commented, Lewis added research department to research department in the hope that, left to their own devices, they would achieve some sort of breakthrough. See Cramer, J.L. (1996), Training and education in British psychiatry, 1770–1970, in *150 years of British psychiatry, 1841–1991*, Vol. 2 (ed. H. Freeman and G.E. Berrios), pp. 209–36, Athlone Press, London.

[14] Outside the capital, the only other long-standing academic sites for psychological research in England were in Cambridge, Manchester, and, at various times, Reading, Bristol, and Liverpool. Only a subset of these could be described as fully fledged teaching

part, these university-based scholars were ambivalent about connecting their work with the world outside. Applied psychological work was mainly undertaken in some kind of educational context or as part of child guidance services. Only a few worked in psychiatric institutions, and when they did so it was typically on a part-time basis and often with children.[15] While some psychological researchers used psychiatric patients as subjects, their primary affiliation was usually academic. For example, around the beginning of World War I Charles Spearman and Bernard Hart had begun testing the effects of dementia, and whether it affected the g or s components of intelligence. After the war, William Stephenson led a research group on a general investigation of Spearman's factors in psychiatry.[16] Working at the Bethlem, Horton, and Maudsley hospitals, the group included Murdo MacKenzie, Constance A. Simmins, and Grace L. Studman. Wynn Jones performed similar research, and Burt's work with the London County Council could also be included under the same heading. The remaining psychologists who worked in psychiatric hospitals and outpatient departments were a heterogeneous group, less visible than those mentioned above. Most were professionally unaffiliated and relatively poorly trained. Many were women, whose low-status, 'helping' orientation made their contribution easily overlooked. During the 1930s, the Maudsley employed several psychologists on a part-time basis. These included Grace Studman, J.W. Pinard, Nancy Samuel, and Gertrude Keir, either working in the children's department or doing research in the adult sections of the hospital.

Eysenck was one of the first highly trained psychologists in Britain let loose in an adult psychiatric institution—along with Philip Vernon, J.M. Blackburn, Eric Trist, and John Raven.[17] In addition, Eysenck was given a priceless gift: he

---

departments. Up north, the Scots had been more receptive to psychology, with reasonably well-equipped and staffed departments in Aberdeen, Glasgow, and Edinburgh, and, more latterly, St Andrews. See Hearnshaw, L. (1964), *Short history of British psychology, 1840–1940*, Methuen, London, Chapter 14, and Kenna, J.C. (1966), Some aspects of the development of psychology departments in British universities, unpublished manuscript, December 1966.

[15] Many of the first generation of British psychologists engaged in a little private therapeutic practice. However, most still had at least one foot firmly planted in academia. C.S. Myers, W.H.R. Rivers, R.J. Bartlett (not to be confused with Frederic), and William Brown could all be cited as prominent figures who adopted this kind of dual role. See Hearnshaw (1964), *Short history of British psychology, 1840–1940*, Chapter 15.

[16] Stephenson, W. *et al.* (1934), Spearman factors and psychiatry, *British Journal of Medical Psychology* **14**, 101–35.

[17] Both Philip Vernon and J.M. Blackburn worked at the Maudsley prior to the war but moved on before Eysenck arrived. John Raven's psychiatric experience began in 1934

was able to do research *full-time* without any obligation to meet any service needs. Here was an individual differences psychologist coming face to face with the intractable problems of clinical psychiatry, opportunity meeting ambition.

Eysenck began with little in the way of research funds, equipment, or laboratory facilities at Mill Hill. Few tests were available either. His reading material also appears to have been remarkably limited during his time. Moreover, he was almost totally ignorant about psychiatry. Unimpressed and unintimidated by what he found, he decided to:

> act on the wise words of Lord Rutherford ... 'we have no money so we will have to think!' I decided that to begin with I would take a standard textbook psychiatric statement, and try to test it empirically.[18]

Eysenck's de-bunking impulses would make him many enemies amongst Maudsley psychiatrists.[19] Nevertheless, there were many aspects of clinical psychiatry he trusted and used. However, this input did not come directly from his medically trained co-workers; it came from Eysenck's appraisal of the psychiatric literature. Henderson and Gillespie's 1943 textbook provided one important source, as did psychodynamic psychiatrist John Bowlby's *Personality and mental illness*.[20] Eysenck also looked to the psychological literature for inspiration. He drew from Ross Stagner's 1937 text on personality, and discussed material mentioned in it at length in subsequent publications. He may well have had Gordon Allport's influential compendium *Personality: a psychological interpretation* on hand as well, though this is less clear.[21] His friend Philip Vernon's research was a key reference point, as was J.P. Guilford's. From the continent, Jung, Ernst Kretschmer, and Ivan Pavlov were also on his

---

when he was hired as psychological assistant to Lionel S. Penrose at the Royal Eastern Counties Institution in Colchester—developing his famous progressive matrices test there. During the war, Raven worked with Eysenck at Mill Hill and then moved on to the Crichton Royal in Dumfries.

[18] Eysenck, H.J. (1997*b*), *Rebel with a cause* [revised and expanded], Transaction Press, New Brunswick, New Jersey, p. 92.

[19] A.R. Jonckheere suggested that this was partly because Eysenck questioned their authority *per se* and partly because they thought he had no familiarity with the problems of clinical work. A.R. Jonckheere, interview, 22 April 2002. Numerous other interviewees attested to the frosty relationship that began to develop between Eysenck and his fellow psychiatrists at the Maudsley, a sentiment proudly repeated by Eysenck himself.

[20] Henderson, D. and Gillespie, R.D. (1943), *Textbook of psychiatry*, Oxford University Press, Oxford; Bowlby, J. (1940), *Personality and mental illness*, Kegan Paul, London.

[21] Stagner, R. (1937), *Psychology of personality*, McGraw-Hill, New York; Allport, G. (1937), *Personality: a psychological interpretation*, Holt, New York.

reading list. However, his most important port of call was the Joseph McVicker Hunt edited volumes on *Personality and behavior disorders*—particularly the first chapter, 'The structure of personality', written by Bryn Mawr psychologist Donald MacKinnon.[22] These volumes read like a template for Eysenck's future research programme, with MacKinnon's chapter in particular an eerie rehearsal of the central ideas of his first book.

Into the breach

> One might imagine that English psychologists would be especially well fitted for investigations of personality which would combine the advantages of, or avoid the disadvantages of both the American and German approaches ... But apart from applied psychology our contributions to the experimental study of temperament and personality are meagre. (Vernon, P.E. (1933), The American v. the German methods of approach to the study of temperament and personality, p. 171)

Psychologists customarily date the birth of personality psychology as 1937, pointing to the work of Gordon Allport, Henry Murray, and Ross Stagner.[23] Even if the efforts of these scholars represented some kind of watershed, the concept of personality had a much longer history. Similar concepts like 'character', 'temperament', and 'the self' had been around for even longer, products of markedly different intellectual and social circumstances.[24] In Depression-era America, static surface trait models vied with more dynamic depth conceptions; additive component schemes vied with those stressing that

---

[22] See Hunt, J.McV. (ed.) (1944), *Personality and the behavior disorders*, Vols. 1 and 2, Ronald Press, New York and, in particular, MacKinnon, D.W. (1944), The structure of personality, in *Personality and the behavior disorders*, Vol. 1 (ed. J.McV. Hunt), pp. 3–48, Ronald Press, New York.

[23] While this may only amount to a historical convention, it still conveys the unfortunate suggestion that a whole subfield emerged immaculate and new in the hands of a select group of American researchers. See, for example, Monte, C.F. (1991), *Beneath the mask: an introduction to theories of personality*, Holt, Rinehart, and Winston, New York, p. 637, and various chapters in Craik, K.H. et al. (eds.) (1993), *Fifty years of personality psychology*, Plenum Press, New York.

[24] Danziger, for example, located the roots of personality as a naturalized entity in nineteenth century French psychiatry, a medicalization of selfhood that was picked up by William James at the turn of the twentieth century. Other writers have alluded to a variety of antecedents in France, Germany, and Russia, making for a confusing and contradictory genealogy. See Danziger, K. (1997), *Naming the mind: how psychology found its language*, Sage, London, Chapter 7. Lombardo and Foschi extend on Danziger's historical view by examining the contradictions that emerged as different traditions intersected. See Lombardo, G. and Foschi, R. (2003), The concept of personality in 19th century French and 20th century American psychology, *History of Psychology* **6**, 123–42.

selfhood was more than the sum of its parts. Meanwhile, standardized testing had demonstrated how psychologists could get a grip on human diversity, a practical feat quite in tune with a home-grown positivism of observable behaviour and operational definitions. Allport's work was particularly crucial for stressing the integrative potential of the personality concept. Allport managed to structure a rag bag field of atomistic, component traits by incorporating German wholism. In the process, he banished the elitist moral tone of 'character' and ignored the strongly innate, biological reductionism of 'temperament'.[25]

American psychologists still exhibited a pronounced conceptual and methodological pluralism up to World War II. But they converged on the category of personality itself, and they did so just as they began to assume a dominant international position. The war consolidated this new disciplinary order, with the US established as the centre of the discipline across the English-speaking world and much of Europe. After the war, America became the chief exporter of psychological personnel, ideas, and techniques.[26]

So when Eysenck arrived at Mill Hill in 1942, most of the personality research that mattered was either American or German.[27] Local efforts were, in comparison, very sparse. Academic departments in Britain tended to be smaller and more specialized than those abroad, with UCL and Cambridge occupying two well-developed positions in an incomplete spectrum.[28] In his 1933 review

---

[25] Nicholson, I. (2003), *Inventing personality: Gordon Allport and the science of selfhood*, APA Press, Washington, DC.

[26] See Danziger, K. (2006), Universalism and indigenization in the history of modern psychology, in *Internationalizing the history of psychology* (ed. Adrian C. Brock), pp. 208–25, New York University Press, New York. This meant that formerly distinct European traditions would be progressively assimilated, marginalized, even erased. For example, prior to the war, German work emphasized wholistic conceptions of character and much of this work had a constitutional, biological slant. But it was a different story after the war, as German psychology progressively yielded to New World priorities. Conversely, Russian research retained much of its independence courtesy of the Cold War, but would later be selectively mined by Western researchers.

[27] For a survey of pre-war German work on personality, see Maller, J.B. (1933), Studies in character and personality in German psychological literature, *Psychological Bulletin* **30**, 209–32.

[28] As Hearnshaw rightly noted, British higher education was exceedingly conservative. Faced with the reactionary and cautious attitudes of many university leaders, especially amongst the dreaming spires of Oxford, psychologists had to justify their inclusion in rigidly traditional university curricula and stake a claim for resources more established sciences already had dibs on. See Hearnshaw, L. (1979), *Cyril Burt: psychologist*, Hodder and Stoughton, London, p. 17.

of American and German approaches to temperament and personality, Philip Vernon ruefully remarked that 'important English work in the field is conspicuous by its absence'.[29] Only the London school had made much of an effort. One of Spearman's students, E. Webb, had performed perhaps the first factorial study of temperament in 1915. Webb isolated the factor 'will'—defined as 'deliberate volition'—from subject ratings.[30] Grace Studman had come up with a similar factor, along with a factor labelled mental fluency.[31] There was a smattering of similar other such work on motor performance and 'perseveration' ('will' re-labelled).[32]

In 1915, Cyril Burt had presented a talk to the British Association for the Advancement of Science on 'The general and specific factors underlying the primary emotions'.[33] He reported a general factor of 'emotionality', plus two specific bipolar factors, derived from interview and performance test estimation data. Burt returned to the topic of emotions in the mid-1930s at UCL. In a 1938 paper on the 'Analysis of temperament', Burt elaborated on temperament factors. Besides that of general emotionality, he proposed three further factors. Of these three, 'by far the most conspicuous was a bipolar factor making for aggressive or extraverted behaviour when positive and for inhibited or introverted behaviour when negative'.[34] While Burt would later overstate the coherence and rigour of this early work, it was still a significant British precedent for the mapping of personality through factorial means.

Eysenck obviously saw an opening, especially since his appreciation of the research literature was greatly aided by his capacity to read German, English, and even French. The fact that personality was emerging as a superordinate

---

[29] From p. 171 of Vernon, P.E. (1933), The American v. the German methods of approach to the study of temperament and personality, *British Journal of Psychology* **24**, 156–75.

[30] Webb, E. (1915), Character and intelligence, *British Journal of Psychology, Monograph Supplement* **1**, no. 3.

[31] Studman, G.L. (1935), The factor theory in the field of personality, *Character and Personality* **4**, 34–43.

[32] See, for example, Pinard, J.W. (1932), Tests of perseveration, 1. Their relation to character, *British Journal of Psychology* **23**, 5–19. See also K.H. Rogers' article bemoaning the lack of attention given to Spearman's approach to the subject: Rogers, K.H. (1935), The study of personality, *Journal of Abnormal and Social Psychology* **29**, 357–66.

[33] Burt, C. (1915), The general and specific factors underlying the primary emotions, *Report to the British Association for the Advancement of Science* **69**, 45.

[34] Burt, C. (1940), *Factors of the mind: an introduction to factor-analysis in psychology*, University of London Press, London, p. 374. Criticisms from Stephenson in particular put Burt on the defensive in this methodological stock-take.

conceptual category no doubt attracted him as well, and Burt had already oriented him to the field.

Eysenck took on Burt's factorial methodology and aspects of his hierarchical structure of mind, a debt he openly acknowledged. However, Eysenck would later single out Webb as the founding factorialist of temperament.[35] Eysenck would likewise credit other factorialists (though not Stephenson) as often as he cited Burt. In contrast, Burt suggested that Eysenck took up his work lock, stock, and barrel, merely renaming his 'general emotionality' factor as neuroticism and appropriating his bipolar introversion–extraversion factor.[36] These were both very much self-interested accounts. While the similarities between Burt and Eysenck were obvious enough, there were also some important differences. Eysenck wanted to improve upon the 'subjectivity' of Burt's ratings and interviews, and he wanted to anchor his factors with non-psychometric data. Above all, Eysenck wanted a more significant status for his factors than the more cautious Burt ever contemplated.

Philip Vernon was just as important as Burt as a role model. Although Vernon had a Ph.D. from Cambridge, he had become one of the most authoritative English psychometricans of the period. Vernon had worked at the Maudsley before joining the army selection programme during the war, and had long been interested in German personality research. Moreover, he collaborated with Gordon Allport in the early 1930s while on a Rockefeller Fellowship, which led to their *Study of values: a scale for measuring the dominant interests in personality*.[37] Like Allport, he refused to come down decisively on either side of qualitative versus quantitative approaches to personality.[38]

---

[35] Eysenck always tended to emphasize the priority of Webb's factorial research (which he cited as published in 1914), adding that Burt's 1915 work was dogged by problems of interpretation. See Eysenck (1947a), *Dimensions of personality*, pp. 40 and 55 (footnote). See also Eysenck, H.J. (1992b), A hundred years of personality research, from Heymans to modern times, Undelivered lecture, University of Amsterdam, 12 February 1992.

[36] See the claims and counterclaims of a review assumed to be at least partly written by Burt, and Eysenck's response. See WLG (1953), Review of *The scientific study of personality* by Hans Eysenck, *British Journal of Psychology (Statistical Section)* **5**, 208–12; Eysenck, H.J. (1954b), A note on the review, *British Journal of Psychology (Statistical Section)* **6**, 44–6; and, WLG (1954), A reply, *British Journal of Psychology (Statistical Section)* **6**, 46–52. See also Hearnshaw (1979), *Cyril Burt: psychologist*, Chapter 9.

[37] Allport, G. and Vernon, P.E. (1931), *Study of values: a scale for measuring the dominant interests in personality*, Houghton Mifflin, Boston; Nicholson (2003), *Inventing personality: Gordon Allport and the science of selfhood*, p. 177.

[38] Hearnshaw (1964), *Short history of British psychology, 1840–1940*, pp. 252–3.

Vernon provided an empirical bridge between German and American work and Eysenck was to cite him approvingly and often.

Apart from singling out the contributions of Webb and Burt, Vernon's 1933 review suggested that 'controlled empirical techniques might be developed which would aim at validating manifestations of personality as concrete *Gestalten*, rather than as abstracted statistical ciphers'.[39] Vernon suggested connecting test results and psychometrically derived personality factors with other measures—different levels of expression, including anatomical types, endocrinal and chemical changes, motor performance and reaction time, handwriting and gestures, artistic expression, and aesthetic preferences. He also canvassed the possibility of enlisting the help of psychiatrists; the different measures of expression could be linked with hereditary and environmental factors of normal and abnormal subjects. As we shall see, Eysenck took Vernon's suggestions very seriously. Vernon also remained an important source of advice and was kind enough to critically examine a draft copy of the *Dimensions of personality*.

The programme begins

Eysenck's initial research work at Mill Hill was supported mainly by grants from the Rockefeller Foundation. As he built up his programme of projects, he was able to expand this grant backing to fund the positions of a number of co-workers, among them Hilde Himmelweit and one of Burt's ex-students Asenath Petrie. Lewis also encouraged Eysenck to work with psychiatrists, in particular Linford Rees.

Eysenck's first project focused on suggestibility and neurosis, putting a more psychiatric angle to work as he had already done at UCL with hypnosis. He set about testing the idea that hysterics were highly suggestible, an accepted tenet of psychiatric textbooks and one prominently mentioned in MacKinnon's chapter.[40] These studies led to a useful distinction between primary and secondary suggestibility based on a factor analysis of test results. Expressive movement tests—such as the body sway test—best tapped primary suggestibility, while secondary suggestibility was more effectively measured by Binet's progressive lines and weights tests. Importantly, these performance tests were direct and 'objective'; according to Eysenck they avoided the unreliable 'subjectivity' of observer ratings and the wilful misrepresentations of self-report

---

[39] Vernon (1933), The American v. the German methods of approach to the study of temperament and personality, p. 171.

[40] MacKinnon (1944), The structure of personality, p. 17.

questionnaires.[41] Surprisingly, hysterics were found to be no more suggestible than non-hysterics, and even men were shown to be more suggestible than women.[42] It was a finding Lewis valued highly; Eysenck's boss asked him to check his calculations carefully and then encouraged him to publish. It illustrated an important component of Eysenck's approach—the harvesting of accepted wisdom with a view to adjusting or undermining it.

Beyond defining himself in these negative terms, however, Eysenck embarked on a far more ambitious track. Lewis organized a standard form of patient accounting, recording a wealth of information on the personal history and circumstances, signs and symptoms of new patients. These patient case notes also contained the diagnoses given by the psychiatrist the patient had been assessed by, often on more than one occasion by different psychiatrists. During the war they were modified to be recorded on a single sheet of stiff cardboard, with holes punched down one side to indicate the presence or absence of a particular factor. There were about 160 such holes and they made for easy data extraction. Lewis' data sheets coded the Meyerian approach in a shorthand tabular fashion (Fig. 4.1). They were just what Eysenck had been waiting for.

At Sutton hospital, Eliot Slater and his statistician brother Patrick had already begun to analyse similar standardized case study material in terms of diagnostic variables, particularly neurosis. Eysenck's first thought was to calculate the reliability of psychiatric diagnoses, for the sheets often gave more than one diagnosis. He found a surprisingly low level of agreement but was, he said, blocked from publishing this result by the administrative head at Mill Hill, Walter MacLay.

---

[41] Eysenck placed great store on such performance tests at this time for this very reason, even though most were blindly empirical. There were often no *a priori* reasons why tasks like discriminating weights or sizes should show differences between particular criterion and control groups (e.g. neurotics vs. normals), hardly adding to the theoretical exposition of the trait or type in question. Moreover, the apparent unreliability of such tests eventually took a toll. Even Eysenck's flagship test, his adaptation of Hull's body sway test, was apparently greatly influenced by extraneous variables like the style of instruction and testing circumstances. Alan and Ann Clarke, interview, 25 April 2002; A.R. Jonckheere, interview, 22 April 2001. See also Alan Clarke's dissertation done under Eysenck's supervision, Clarke, A.D.B. (1950), The measurement of emotional stability by means of objective tests: an experimental inquiry, Ph.D. thesis, University of London.

[42] For a historical analysis of gendered assumptions in British psychiatry, see Showalter, E. (1987), *The female malady: women, madness and female culture*, Virago Press, London.

The most difficult data to record in this scheme are the patient's symptoms and signs. The following is the proposed list:-

Anxiety	3 grades.
Depression	3 "
Irritability	2 "
Elation.	
Paranoid	2 "
Apathy, stupor, Depersonalisation.	2
Perplexity.	
Tremor	3 "
Sweating, flushing, ejaculatio praecox.	
Palpitation.	
Dyspnoea.	
Headache.	
Dizziness.	
Vomiting.	
Pain: Precordial, other. Enuresis and other, Dysuria.	
Diarrhoea	2 "
Anxiety dreams, nightmares.	
Paresis and paralysis:	before this illness, only during this illness.
Stammer:	before, only during illness
Spasm:	" " " "
Blindness:	" " " "
Deafness:	" " " "
Dyspepsia:	" " " "
Dysmnesia: organic, psychogenic.	
Fugue.	
Pseudo-Dementia.	
Sleep walking.	
Sexual: impotence, homosexuality, other perversion, masturbation worries. Willing or unwilling to be in the Army.	
Hypochondriacal attitude:	3 grades.
Worries, Preoccupation: With war, with domestic, financial etc.	
Delusion, Hallucination.	
Obsession: impulse to action, ideas, rumination and folie de doute.	
Phobias.	
Sucidal:	thoughts, attempts.
Schizophrenic features.	
Loss of weight.	
Poor appetite.	
Lassitude, fatigue.	
Definite signs of organic nervous disorder.	
Signs of organic disease other than of nervous system.	

There are many other categories to be included, they are given on the attached sheet.

Fig. 4.1 Draft version of a key portion of Aubrey Lewis's patient data sheet.
Source: Bethlem Royal Hospital Archives and Museum.

Research reports from the period hint at a bigger project emerging from diverse and piecemeal research. In 1943, Eysenck and his team were joined by a number of others working on projects concerned with the further study of:

> suggestibility and hypnotisability in relation to personality traits and clinical syndromes. The personality traits have been investigated by means of questionnaires of various types, biographical data, clinical interview, and studies of expressive movements and of projection tests.[43]

Olga Marum was analysing expressive movements in handwriting, P.M. Yap focusing on conditioned salivary output, Petrie and Himmelweit examining test performance, and Desmond Furneaux was looking at hypnosis. That year, Eysenck also embarked on a 'statistical analysis of clinical data... the study of the general and specific factors which appear in the syndromes for which the patients are admitted; this ... has not hitherto been used for the neurosis. Factorial analysis was employed on similar clinical material by Slater and Slater at Sutton.'[44]

New wine in old bottles: vintage 1944 personality factors

In 1944, Eysenck's landmark factorial study of personality appeared. Eysenck took a subset of variables contained on Lewis's data sheet and factor analysed them for 700 soldiers classed as neurotic. Eysenck had planned to use 1000 subjects but to simplify the factors he excluded cases complicated by epilepsy, head injury, organic central nervous system disease, and other physical diseases and injury. While diagnoses were:

> highly complex cognitive processes, based on a variety of information, some of which can interpreted on a subjective basis. The symptoms themselves are fairly objective ... I decided to look at the correlations between the symptoms and try out a factor analysis to see whether they might not give rise to superordinate concepts derived directly from the facts.[45]

By 'objective' Eysenck meant reliable, not subject to personal interpretative variation. He selected 39 variables, whittled down from the over 70 on the original sheet. In order to make them comparatively equivalent, these 39 variables

---

[43] AJL/JMT, 'Report: research work, under the Rockefeller Grant, at Mill Hill Emergency Hospital', 21 February 1944, Aubrey Lewis Papers, Box 9, Bethlem Royal Hospital Archives and Museum, p. 1.

[44] Ibid., p. 2.

[45] Eysenck (1997b), *Rebel with a cause*, p. 96.

were transformed as dichotomous variables even though some had multiple response options (e.g. age and intelligence).

Eysenck produced a 39 by 39 matrix of 741 intercorrelations. Consistent with the structural assumptions of Burt's general factor summation method that he used, Eysenck derived a positive general factor and a number of smaller bipolar group factors. The first general factor accounted for 14% of the variance, while the next three bipolar factors accounted for 12%, 8%, and 6% of the variance, respectively. Eysenck was thus able to statistically summarize a significant proportion (i.e. 40%) of the information on the data sheet with just a few factors.[46] Nevertheless, he wanted to do much more than provide a practical, descriptive shorthand.

Surveying the variables loading highly on his first factor (e.g. 'badly organized personality', 'dependent', 'abnormal before illness') Eysenck concluded that 'clearly, the factor is one of "neuroticism" or "lack of personality integration".'[47] In this first article he settled on the label 'integration', but would return to the more common appellation 'neuroticism' in later work. He dubbed his first general factor the obverse of Webb's 'will' factor and identical with the 'neuroticism' factor that emerges from so many American questionnaire studies, 'often falsely labelled "introversion"'.[48]

The second bipolar factor was more complex. Its loadings contrasted 'anxiety', 'depression', 'obsessiveness', and 'apathy' on the one hand, with 'hysterical conversion symptoms', 'hysterical attitude', 'narrow interests', and 'sex abnormalities' on the other. In a significant interpretative leap, Eysenck argued that this dichotomy bore out C.G. Jung's 'well-known statement' that there are two large groups of functional nervous disorders—one embracing forms designated hysteria, the other those the French school designated psychasthenia. In Eysenck's reading of Jung, the hysteric belonged to the extraverted type, the psychasthenic to the introverted type.[49] Similarly, Eysenck drew on McDougall's 'two great categories of disorder'—the hysteric and the neurasthenic. According to Eysenck, McDougall used the term 'neurasthenic' in the same way as French writers like Janet used the term psychasthenic,

---

[46] Mind you, taken together, these four factors still left 60% of the variance unaccounted for. Eysenck attributed this low community to the fact that some variables (e.g. 'age', 'exposure to enemy attack', 'alcohol intake') showed little or no correlation with these four factors. Eysenck, H.J. (1944a), Types of personality: a factorial study of seven hundred neurotics, *Journal of Mental Science* **90**, 851–61.

[47] Ibid., p. 854.

[48] Ibid. Here Eysenck was taking aim at J.P. Guilford especially.

[49] Ibid., p. 855. Eysenck gave the quote in English but cited the German language edition.

thereby reinforcing his interpretation. Nevertheless, he regarded both terms as obsolescent and instead suggested the more modern appellation 'dysthymia' for the anxiety–depressive–obsessive group. This second bipolar factor, Eysenck concluded, corresponded with the introvert–extravert distinction that so many scholars had pointed to.

Eysenck's factorial interpretation depended on a fairly sweeping survey of some old debates over psychiatric diagnoses. The broad nineteenth century category of neurasthenia—a general malaise or nervous fatigue of possible organic origin—had never been totally accepted in Britain. Alternative explanations for some of its symptomatic features saw this category broken up in the early years of the twentieth century. Some of these features were included within the bridging concept of psychasthenia, which for a time included the anxiety-related phobic and obsessional neuroses. Hysteria, conversely, was exemplified by organically fictitious 'conversion' symptoms. It was the classic disease of blameworthy women. Hysteria was disappearing as a diagnostic category, however, until it was temporarily revived during World War I. Freudian psychiatrists, in particular, took to classifying trench warfare pathology as either phobic neurosis or hysteria. Such a distinction still had a particular resonance across the channel during the next war.[50]

Eysenck's third bipolar factor was characterized by 'hypochrondriasis', 'dyspepsia', 'hypochrondriachical personality', 'fainting', and 'pain', as opposed to 'sex abnormalities', 'wartime separation', and 'unsatisfactory home'. Eysenck deemed this factor more psychological than the first two, and labelled it 'hypochrondriasis'. He later came to dismiss this third factor as relatively unimportant, while he discarded his fourth factor on the spot. Eysenck wanted to concentrate attention on his first two factors. They had summarized 26% of the variance of data sheet intercorrelations. They were, he hoped, 'fundamental vectors in the field of personality and temperament'—especially so if they can be shown to emerge in other studies of young and old, normal and abnormal, human and animal.[51]

Sourcing the big picture

> MacKinnon (1944) ... presents a Table in many ways similar to our own ... (Eysenck, H.J. (1947a), *Dimensions of personality*, p. 12)

---

[50] Berrios, G.E. and Porter, R. (eds.) (1995), *A history of clinical psychiatry: the origin and history of psychiatric disorders*, Athlone, London.

[51] Eysenck (1944a), Types of personality: a factorial study of seven hundred neurotics, p. 857.

Eysenck's autobiography gave the impression that he factored the data sheets before he read the MacKinnon chapter in the 1944 McVicker Hunt volumes.[52] This was probably true, for he must have launched this study just before he set eyes on these books. However, he certainly read these volumes before publishing his 1944 paper—for he cited them—and did so just as he was trying to make sense of what these results meant. Particular noteworthy is the 'Table of two-fold typologies' that appears on page 18 of MacKinnon's chapter (Table 4.1).[53] Eysenck reproduced it in full in *Dimensions of personality* with a few additional elements, notably the ideas of McDougall (see Table 4.2). MacKinnon had drawn up this table from the literature in psychiatry, and from abnormal and personality psychology. It sets out a range of contrasting, opposite types that had been suggested in relation to the psychoses, the neuroses, and normal personality. It was a table that no doubt fired Eysenck's imagination, as did MacKinnon's presentation of the seven personality factors that had been reported in a least three factorial studies of personality to date. MacKinnon was himself summarizing a 1942 review paper by another American psychologist, Dael Wolfle.[54]

Eysenck clearly saw ample opportunity for integration and simplification in the summary of factors and in the table. MacKinnon's review suggested many researchers had

> seen somewhat similar constellations to the one I had stumbled upon, but had never quite put the pieces together ... Much of the work that I have done within the ensuing forty years has been devoted to the pursuit of this apparition.[55]

While Eysenck was able to compare these oft-reported factors with his own, MacKinnon's table provided the more significant resource. *Dimensions of personality* could be read as a straightforward attempt to empirically connect the dichotomous traits and types of this table. Eysenck was clearly taken by the possibility of constructing a dimensional link between opposing terms. Two notable factor-analytic findings made several terms stand out. Introversion–extraversion had already become a well-known factor of temperament, although its psychometric unity was still contested. While this factor had been adduced amongst normal populations, it had yet to be extensively studied within psychiatric populations. Conversely, the Slater brothers had just

---

[52] Eysenck (1997b), *Rebel with a cause*, pp. 96–7.

[53] MacKinnon (1944), The structure of personality, p. 18.

[54] Wolfle, D. (1942), Factor analysis in the study of personality, *Journal of Abnormal and Social Psychology* **37**, 393–7.

[55] Eysenck (1997b), *Rebel with a cause*, p. 97.

Table 4.1 MacKinnon's table of two-fold typologies

Psychotic Types	
Manic-depression	Dementia Praecox (Kraepelin)
Manic-depression	Schizophrenia (Bleuler)
Psychoneurotic Types	
Hysteria	Psychasthenia (Janet)
Psychotic Personality Types	
Cycloid	Schizoid (Kretschmer)
Psychoneurotic Personality Types	
Hysteroid	Obsessoid (Janet)
Normal Personality Types	
Shallow-Broad	Deep-Narrow (Gross)
Extraverted	Introverted (Jung)
Nonperseverative	Perseverative (Spearman)
Objective	Subjective (Stern)
Cyclothymic	Schizothymic (Kretschmer)
Syntropic	Idiotropic (Wertheimer & Hesketh)
Color-Type	Form-type (Scholl)
Extratensive	Introversive (Rorschach)
B-type	T-type (Jaensch)
Integrate	Disintegrate (Jaensch)
Personality Traits	
Suggestibility	Nonsuggestibility (Janet)
Hypnotizability	Nonhypnotizability (Janet)
Short secondary function	Long secondary function (Gross)
Extraversion	Introversion (Jung)
Nonperseveration	Perseveration (Spearman)
Color-abstraction	Form-abstraction (Külpe)
B-type eidetic imagery	T-type eidetic imagery (Jaensch)
Integration of psychic processes	Disintegration of psychic processes (Jaensch)
Morphological Types	
Pyknic	Leptosomic (Kretschmer)
$\frac{\text{Height}}{\text{Chest volume}}$ low	$\frac{\text{Height}}{\text{Chest volume}}$ higt (Wertheimer & Hesketh)

Source: MacKinnon, D.W. (1944), The structure of personality, in *Personality and the behavior disorders*, Vol. 1 (ed. J.McV. Hunt), pp. 3–48, Ronald Press, New York.

Table 4.2 Eysenck's adaptation of MacKinnon's table

Psychotic Types		Author
Manic-depressive	vs. Dementia Praecox	Kraepelin (1899)
Syntonic	vs. Schizophrenic	Bleuler (1924)
Neurotic Types		
Hysteric	vs. Psychasthenic	Janet (1894)
Hysteric	vs. Neurasthenic	McDougall (1926)
Personality Types		
Extraverted	vs. Introverted	Jung (1923)
Objective	vs. Subjective	Binet (1900)
Sthenic	vs. Asthenic	Burt (1937)
Cyclothymic	vs. Schizothymic	Kretschmer (1926)
Extratensive	vs. Introvertive	Rorschach (1942)
Surgent	vs. Desurgent	Cattell (1933)
Inhibitory	vs. Excitatory	Pavlov (1941)
Explosive	vs. Obstructive	James (1890)
Shallow-broad	vs. Deep-narrow	Gross (1902)
Syntropic	vs. Idiotropic	Wertheimer et al. (1926)
B-Type	vs. T-Type	Jaensch (1926)
Adient	vs. Avoidant	Holt (1931)
Viscerotonic	vs. Cerebrotonic	Sheldon (1942)
Manic	vs. Melancholic	Heymans et al. (1908)
Personality Traits		
Suggestibility	vs. Non-suggestibility	Babinski (1918)
Short secondary function	vs. Long secondary function	Gross (1902)
Fluency	vs. Lack of fluency	Cattell (1933)
Dissociation	vs. Anxiety	McDougall (1926)
Plastic eidetic imagery	vs. Rigid eidetic imagery	Jaensch (1926)
Colour-attitude	vs. Form-attitude	Scholl (1927)
Non-perseveration	vs. Perseveration	Spearman (1927)
Slow oscillation	vs. Quick oscillation	McDougall (1926)
Synthetic ability	vs. Abstractive ability	Kretschmer (1926)
Careless	vs. Careful	Downey (1923)
Slow personal tempo	vs. Quick personal tempo	Kretschmer (1926)
Lacking in persistence	vs. Persistent	Downey (1923)
Ascendent	vs. Submissive	Allport (1928)
Sociable	vs. Unsociable	Guilford (1936)
Emotionally demonstrative	vs. Non-demonstrative	Guilford (1936)
Constitutional Types		
Digestive	vs. Respiratory-cerebral	Rostan (1828)
Sympatheticotonic	vs. Vagotonic	Eppinger (1917)
Megalosplanchnic	vs. Microsplanchnic	Viola (1933)
Pykinc	vs. Leptosomatic	Kretschmer (1926)
Endomorph	vs. Ectomorph	Scheldon (1940)

Synoptic table showing sample of current dichotomous typologies, arranged according to psychotic type, neurotic type, personality type, personality trait, and constitutional type. A more detailed table of constitutional types is given later in the book.

Source: Eysenck, H.J. (1947a), *Dimensions of personality*, Routledge and Kegan Paul, London.

demonstrated that a general factor of neuroticism could be extracted from such hospital samples. The dots were there to be joined. Might the two main types of neurosis be separated according to an underlying temperamental dimension; might introversion characterize the psychasthenic type, and extraversion characterize the hysteric type? How might these neurotic types be related to other personality traits and tests, and to morphology? This was the central organizing idea of *Dimensions of personality* and the starting point for decades of work. When Linford Rees claimed that Eysenck based his entire career on Lewis's data sheets he was only half right.[56] With the addition of this table, Eysenck had the programmatic mother lode, a recipe to make qualitative spirit quantitative flesh.

MacKinnon's chapter also served as model for the literature review for *Dimensions of personality*, with extensive quotation and paraphrasing from it in Eysenck's 'Methods and definitions' chapter. While the empirical data in *Dimensions of personality* might have been new, the central concepts were almost completely borrowed. Perhaps in light of this, Eysenck's introduction to *Dimensions of personality* is a study in politic humility:

> Little novelty is claimed for most of the experimental procedures adopted, or the theories advanced … No claim is made that we have been able to do more than advance a very small distance toward the goal which we set ourselves.[57]

Pinpointing the inspiration for *Dimensions of personality* highlights the kind of strategy Eysenck had hit upon. He did not and could not pursue all the primary literature, and thus he did not risk getting caught up in minute subspecialisms. One wonders how many of the secondary sources Eysenck cited he was actually able to get hold of, given the dearth of up-to-date reading material during the war. So, perhaps understandably, the 'Concepts and definitions' chapter in *Dimensions of personality* (pp. 22–5) paraphrases various chunks of MacKinnon (pp. 4–10) and included most of the same references.[58] Instead, Eysenck appeared to assess the field via general reviews, and read around the secondary literature. His aim was to cut through the confusion by drawing together various strands and attempting to bear them out empirically. To do so, he had to simplify and integrate previous results. He read to confirm, to support and extend these connections, and adjust them when necessary.

---

[56] Linford Rees, interviewed by H.B. Gibson, 28 February 1979.

[57] Eysenck (1947a), *Dimensions of personality*, pp. ix–x.

[58] One also wonders how many of the French writers Eysenck discussed—like Janet—he actually was able to read at the time, given his limited resources.

Already he had his eye on the Big Picture, and would continue to do so for the rest of his career

Dimensions of personality was a call to arms, an argument for a higher-order perspective and for programmatic research. 'The time has come', Eysenck wrote, 'when preliminary surveys of isolated traits, and the exploratory study of small groups must give way to work planned on an altogether larger scale.' Moreover, the disparate theories that had emerged from small scale research have been 'so divorced in the main from operational definition and experimental control that ruthless discarding appeared more necessary than an attempt to add to the confusion.'[59] This first book set out a framework for systematic research that required just a few sweeping assumptions to render it a basis for the exploration of universal human nature. Surprisingly too, the book aimed at a wide audience, targeting psychologists of all stripes and then some. Mundane empirical detail had been removed and readers were referred to journal articles for the full tables of intercorrelations and the like.

Demonstrating his facility for grand theorizing, Eysenck combined and integrated trait and type conceptions within a hierarchical model. With a behaviourist–operational turn on the ideas of Gardner Murphy and Friedrich Jensen, Eysenck defined his personality framework from the bottom up.[60] Observable acts or specific responses (S.R.s) could be grouped as a set of habitual response tendencies (H.R.s). These tendencies could then be grouped as traits (e.g. 'persistence' and 'rigidity'), and then as types. The notion of type was made equivalent to highest 'factor' level, of 'introversion', for example. It was a structuring move that got around the already vexed question of the general versus situational specificity of traits. In Eysenck's scheme, traits were a bit of both, an observed constellation of behavioural tendencies. His dimensional types were the most general personality variable, an observed constellation of traits. Thus he was able to claim that he was defining his higher order concepts in strictly operational, observable form; it also enabled him to upwardly integrate the meaning of a range of experimental results, and to downwardly generate many testable hypotheses. It was an all-encompassing framework, flexible yet precise.

With the theoretical elements parcelled up, the rest was easy. After a snappy summary of his original factorial study, Eysenck supplemented it with some illuminating discussions of theories of neuroticism and introversion–extraversion. The remainder of *Dimensions of personality* recounts experimental results of a

---

[59] Eysenck (1947a), *Dimensions of personality*, p. ix.
[60] Murphy, G. and Jensen, F. (1932), *Approaches to personality*, Coward-McCamus, New York.

number of researchers and students working under Eysenck and how they might fit into this kind of framework. Eysenck's gift was to give this improvisation the look of forward-planned research. The research coverage was deliberately 'shallow–broad'.[61] No apology was made for the uneven attention given to various traits; this was an inevitable function of 'external circumstances', as well as how adequately particular traits had been investigated in the past. If these factors are 'really *personality* factors', Eysenck argued, 'they should be expected to cover all the diverse features' of an individual's behaviour. [original italics][62]

Everyday madmen: continuity between the normal and the abnormal

Perhaps the most audacious move Eysenck made was to draw a line from the abnormal to the normal. American personality psychologists had made their object of study the ordinary person. Their conclusions were derived from, and were applied to, just such subject samples. Not surprisingly, they had evolved a habit of explaining the 'abnormal' in terms of deviance or difference from the 'normal', rather than the other way around. For Eysenck, though, circumstances dictated otherwise. While he was drawn to the psychometric description of everyday psychological functioning, Eysenck's work at Mill Hill placed him in the context of the odd and aberrant. He had only limited access to non-psychiatric samples, although this would change after the war.

The subjects that Eysenck had at his disposal were the psychiatric casualities of war, defined by institutional fiat as abnormal.[63] They were a captive subject pool, amenable to study in ways that others were not. For example, one could check the effect of various drugs and utilize experimental procedures that might have been more problematic with non-institutional participants. Most were classed as neurotic in some respect. There were also many con-men, misfits, and malingerers; many patients 'acted out' for reasons well understood at the time.[64] The psychiatric taxonomy employed at the Maudsley aimed at

---

[61] Eysenck (1947a), *Dimensions of personality*, p. 20.

[62] Ibid.

[63] As the Blitz ended but hostilities wore on, military needs tended to overwhelm civilian demands. War casualties came to predominate at Mill Hill. For example, in 1945, staff at Mill Hill treated 427 civilian cases and 1655 military cases; nearly half were women. 'Mill Hill Emergency Hospital, January 1st 1945 – September 3rd 1945', *c.* late 1945, Aubrey Lewis Papers, Box 11, Bethlem Royal Hospital Archives and Museum.

[64] Alan and Ann Clarke, interview, 25 April 2002. See also Kendrick, D.C. (1981), Neuroticism and extraversion as explanatory concepts in clinical psychology, in

capturing and describing the particulars of each 'case'. Severity of disturbance was handled in relatively crude categorical terms, and there was no concept of continuous measurement, only vague hints of more or less disturbance.

Eysenck had to make the best of his lot; he had to construct some sort of bridge to the everyday to make his work broadly relevant. Assuming the abnormal subjects he had at hand were an extreme or less muted version of normal subjects would do the trick. And he had some notable precedents. For example, Emil Kraepelin and Alfred Binet had made attempts to reconstruct the workings of the normal mind from experiences in psychiatry and mental deficiency.[65] A handful of psychiatrists Eysenck was familiar with—including German Ernst Kretschmer, Austrian Otto Gross, and Frenchman Pierre Janet—had also tried to connect psychiatric categories with normal personality functioning. Even Freud and his fellow analysts had started with the odd and unusual.

Eysenck was still up against it at Mill Hill. His medically trained colleagues were not going to be much help. Psychiatric practice focused on those regarded as candidates for special treatment. Intellectual understandings reflected and justified the social separation these individuals experienced. Psychiatry's task was to document manifest varieties and suggest possible causes and treatments. As Danziger pointed out, their natural object of investigation was not abstracted qualities of the person within the group, but the clinical observation of the individual case.[66] But for the London-trained Eysenck, patient variety *per se* was almost useless. It was more a matter of how, and by how much, these patients differed from the normal. The same attributes or variables that could be used to describe mental patients could be used to describe the person in the street. The difference would be merely one of degree, rather than kind.

Eysenck's 1944 factor analysis presumed that dimensionality could be found amongst the Mill Hill checklists of clinical signs and tests if—and this is a big 'if'—they were treated as indicative of continuous variables. The factorial dimensions he derived from his abnormal sample suggested that similar descriptive dimensions might be applied to normal populations. Moreover, the dimensions of neuroticism and introversion–extraversion might be

---

Dimensions of personality: papers in honour of H.J. Eysenck (ed. R. Lynn), pp. 253–62, Pergamon Press, Oxford.

[65] In 1894, Kraepelin founded a journal devoted to this kind of 'psychological work', *Psychologische Arbeiten*, but it was discontinued after only nine volumes soon after his death. See also Binet, A. and Henri, V. (1895), La psychologie individuelle, *L'Psychologique* 2, 411–65.

[66] Danziger (1997), *Naming the mind: how psychology found its language*, p. 125.

coextensive from normal to abnormal samples, though this was still not clear. In his 1944 article, he wrote of introverted and extraverted 'types' but quickly added that he did not wish to imply a bimodal distribution consistent with a qualitatively distinct dichotomy—suggesting he thought (or hoped) otherwise. Distributional characteristics remained to be determined by further investigation. His only other comment at this stage was that his results were not in conflict with the Slater brothers' view that neuroticism was distributed in a normal Gaussian manner.

In *Dimensions of personality*, both the neuroticism and introversion–extraversion factors showed a distinctly normal distribution of scores, a particularly surprising result for neuroticism. If such a factor displayed a normal distribution across the whole population, Eysenck's sample of hospitalized neurotics might be expected to represent the tail of the high end of this bell curve. Eysenck explained this finding by arguing that his sample probably did represent a good cross-section of the population. Even some of those low in neuroticism had wound up in psychiatric care due to the extreme stresses of war. Eysenck otherwise pleaded agnostic on the continuity issue. His results were compatible with the notion of a continuous distribution for these factors, but they did not prove it, he said. Nor did he rule out that at the extremes there might exist a qualitatively different pathological type that overlay an otherwise continuous bell-shaped distribution.[67]

Eysenck's *Dimensions* pulled together a great variety of original and second-hand results to characterize the neurotic soldier, a somewhat sicker version of all of us. Such a man was defective in mind and body, low in intelligence, persistence, and emotional control, and he was suggestible, unsociable, and denied unpleasant facts. Neurotic introverts were anxious and depressed, obsessive, retiring, self-conscious day-dreamers. They tended to be slight of stature, highly intelligent, and very persistent. In contrast, neurotic extraverts tended to be hysterical, lazy, unreliable, and accident-prone, and frequently off work through illness. They were heavy set, low in intelligence, and easily bored. Eysenck thus concluded that: 'If "g" or intelligence is a general factor in the cognitive sphere, so "neuroticism" is a general factor in the conative (integrative) sphere, while "introversion" is a general factor in the affective sphere.'[68]

Graphically, Eysenck's initial two-dimensional scheme was an inverted-T rather than a cross. Only one factor had a strictly pathological anchor-point.

---

[67] Eysenck (1947a), *Dimensions of personality*, Chapter 2.
[68] Ibid., pp. 261–2.

The bottom plane of this inverted-T was represented by the introversion–extraversion dimension, all normal, all good. Neither extreme could be considered a sign of pathology. Stretching upward into the darker reaches of aberration was the neuroticism dimension. Excessive levels of neuroticism were a bad thing, suggesting a propensity to breakdown. In fact, no amount of neuroticism was good. At this early stage, only the neuroticism dimension provided a link between the normal and the abnormal. However, Eysenck would soon develop a psychoticism dimension that made the same link. The assumption of continuity was the untested leap of faith underpinning all Eysenck's early work, one that he knew would be questioned. If it held, if he was right, each of us could be placed along these dimensions, each of us had something fundamentally in common with the humanity-in-a-box Eysenck first encountered at Mill Hill.

Fig. 4.2 Eysenck giving a presentation early in his career.
Source: Centre for the History of Psychology, Staffordshire.

Fig. 4.3 Eysenck listening to a presentation early in his career.
Source: Centre for the History of Psychology, Staffordshire.

Fig. 4.4 The young researcher.
Source: Centre for the History of Psychology, Staffordshire.

Psychiatry, psychology, and factor analysis

> Eysenck and his collaborators have been most searching in checking their results ... Their mass of evidence is formidable but welcome: it brings into psychiatry the method of mathematics and gives hope that we can be more accurate in our assessments, more certain in our pronouncements. (*Lancet* (1947), Review of *Dimensions of personality* by Hans Eysenck, *The Lancet* **249**, 713)

Dimensions of personality was not pitched as an *overt* challenge to psychiatry. Mindful of his circumstances, Eysenck touted his dimensional scheme as a way to improve diagnostic practices. The idea was to get away from the 'subjectivity' involved in psychiatric judgements by abstracting the main dimensions that underlay these ratings. By doing this, it would then be possible to develop tests to reliably characterize and measure these underlying dimensions, to re-present clinical observation in an empirically distilled, standardized form. Eysenck's 1944 article and research reports emphasize these practical aims. Even though Eysenck said Aubrey Lewis barely made any comments on his work, Lewis's foreword to *Dimensions of Personality* was quietly effusive, displaying a great deal of sympathy for Eysenck's ambitions to analyse:

> by reliable statistical techniques, of experimental and clinical data, so that measurement may be possible and a sight obtained of the promised land where mental organisation will be as well understood as the physical organisation of human beings...[69]

For psychiatrists, Eysenck offered a kind of partial vindication of their hard-won wisdom. His early work bore out as much as challenged psychiatrists' 'unreliable' diagnoses. For example, the variable 'psychiatric diagnosis' had the highest loading (at 0.71) on the first 'neuroticism' factor in Eysenck's factor analysis of a battery of tests characterizing his newly minted dimensions. Eysenck also looked back on psychiatric lore to interpret his dimensions, especially when trying to resolve problematic contradictions. For instance, much American questionnaire research had tended to equate neuroticism with introversion, itself a potentially interesting reflection of American cultural values. How then could these be two independent dimensions? Eysenck resolved this paradox by arguing that US psychologists had paid too much attention to Freud; instead one should look to Jung who clearly distinguished between these two concepts. Eysenck drew on the clinical portrait of the (presumably extraverted) hysteric. According to Henderson and Gillespie, the hysteric was 'reserved and peculiar', thereby reinforcing Eysenck's surprising conclusion that extraversion had little to do with sociability. By extension, 'absence of

---

[69] Lewis, A. (1947), Foreword to *Dimensions of personality*, by Hans Eysenck, p. vii, Routledge and Kegan Paul, London.

sociability' was more a feature of neuroticism than introversion. It was a partition at odds with most definitions of these concepts before and since, and Eysenck was later to modify this view substantially.[70]

Eysenck's reluctance to explicitly challenge psychiatric authority at this stage might have lain in part on the epistemic position he had put himself in. His work was, in the first instance, dependent on psychiatrists' clinical observation. It effectively halted this process of first-hand data collection to re-describe it in factorial terms. Any claim to superiority would turn on whether these factors could be corroborated and extended by other means. Thus *Dimensions of personality* represented only a veiled future threat; more prominent was the promise of efficiency, of a renewed confidence and trust in diagnostic categories potentially revised in light of this kind of re-description.[71]

For psychologists, it was a somewhat different story. 'The whole field of personality and temperament study is in a state of acute conflict and dissociation.'[72] Cleaning up this mess was paramount. Eysenck took care to correct, adjust, and re-direct personality psychologists' attention. He wanted to extend the power of factor analysis as the pre-eminent heuristic technique for the study of the emotions and temperament. Much of this was aimed at the Americans. Eysenck had come up with a powerful simplification that contrasted sharply with the speculative trait lists and checklist inventories of personality psychologists in the US. Eysenck's theoretical project went beyond the practical, social management goals embodied by the American approach. Personality psychology was to be more than just the measurement of non-cognitive attributes, it should aim to describe the basic structure or set of such attributes accounting for most of the variation in human personality. Eysenck's model was hardly superficial or atomistic; it was a tightly structured, inclusive hierarchy. At the top of this structure were his two new—or at least empirically enhanced—personality factors. However, Eysenck still faced a big question: what ontological status did his factors have?

Far from being naïvely assumed to be a faithful representation of psychological nature, the status of factor analysis was hotly debated from the technique's inception—more hotly, if anything, than it is today. Despite the great

---

[70] For more on this modification, see Eysenck, H.J. and Eysenck, S. (1969), *Personality structure and measurement*, Routledge and Kegan Paul, London.

[71] For an indication of the cautiously positive medical response to Eysenck's early work, see *Lancet* (1947), Review of *Dimensions of personality* by Hans Eysenck, *The Lancet* **249**, 713.

[72] Eysenck (1944a), Types of personality: a factorial study of seven hundred neurotics, p. 851.

leap forward in factor analytic sophistication between the wars, those championing the technique found themselves on the cusp of some serious philosophical dilemmas. Factor analysis could propose underlying hypothetical structure but it could not prove it. It could detect communality but not what caused it. It could provide multiple mathematical solutions but no definitive way of choosing between them. Factors were indeterminate constructs, their generality beyond particular data sets uncertain. When Eysenck first trained as a psychologist in the late 1930s, the use of the technique was at a crisis point, assailed from within and without. In Britain, the categorical opposition of Cambridge psychologists was entrenched. However, even those with considerable expertise like Godfrey Thomson doubted the reality of factors and preferred more direct regression calculations for predicting particular behaviours. In the US, prominent specialists like Allport and Anastasi also viewed factors as essentially statistical artefacts.

Did a strongly naturalistic interpretation of factor analysis have a chance? As the leading factorialist in the UK, Cyril Burt hoped his moderate views would hold sway. Up to a point they did. MacKinnon and Wolfle, for example, quoted Burt chapter and verse. For Burt, factors were principles of classification. Further advances in parallel fields, in physiology and neurology, for example, might reveal them to be more than this, but for this we would have to wait. Nevertheless, as long as factor analysis only tells the truth, Burt argued, 'we need not condemn it for failing to telling the whole truth.'[73]

Eysenck was explicitly aware of the interpretative pitfalls he faced, a tension lurking in the background throughout his career.[74] While factor analysis was undoubtedly a sophisticated tool, the problems of indeterminacy and data-set dependence were especially apparent when it came to the basic question of factor interpretation. Eysenck toyed with several different names for his factors and the implications they conveyed. While he made comparisons with the factor analytic work of others, these comparisons were rule-of-thumb. There was no agreed standardized yardstick available for such a task. Conversely, other researchers could inspect his loadings and the relevant literature and make up their own minds. Or they could simply take his word for it. Factor naming was

---

[73] Burt (1940), *Factors of the mind: an introduction to factor-analysis in psychology*, p. 138.

[74] For example, Eysenck characterized internal means of choosing factors, like the 'principle of simple structure' and 'proportional profiles', as a 'gigantic game of tautological hunt-the-slipper, in which artificial statistical rules applied to a matrix ... are supposed to give reliable and valid information about real psychological influences'. See pp. 41–2 of Eysenck, H.J. (1950b), Criterion analysis: an application of the hypothetico-deductive method to factor analysis, *Psychological Review* **57**, 38–53.

tied up with preferences for particular factorial solutions, and it inevitably invoked deeper-level philosophical commitments and a great deal of judgement.

Eysenck's discussion of the contrast his results made with those of American J.P. Guilford's was a case in point. Using Thurstone's methods, Guilford had not extracted a general factor akin to Eysenck's neuroticism and had seemingly broken up introversion–extraversion into a number of rotated components. Eysenck's re-analysis using Burt's iterative summation technique demonstrated that Guilford's results could be equated with his.[75] In other words, one set of results could be seen to be a mathematical transformation of the other, a position that Burt had often expressed.

Eysenck was committed to a London world view but needed a way to overcome its limitations. Factor analysis was the way forward; the task was to show it was.[76] Eysenck wanted to overcome his mentor's caveats; he wanted to regard his factors as *real*. Factor analysis might not be sufficient to prove 'unitary' or 'primary' abilities or dimensions, Eysenck allowed, but it was more than just a summarizing device. For example, if it showed the existence of a factor of introversion, and if

> later on introversion could be shown to be due to demonstrable Mendelian factors ... then our factor would surely deserve a higher status scientifically than a mere principle of classification; it could rightly be regarded as a fundamental dimension of mind.[77]

---

[75] Eysenck (1947a), *Dimensions of personality*, Appendix A.

[76] In 1944, Eysenck had even tried to encompass philosophy in a trait dimensional space. He took up William James's observation that philosophy was largely the clash of human temperaments and ran it through the factor-analytic mill. Philosophical questions were correlated with personality attributes resulting in ... two factors. The first (taking up 51% of the variance) divided idealists from materialists, the second (taking up 13% of the variance) divided monists from dualists. Thus the highly nuanced arguments of Western philosophy, all hard thought and fought, could be re-described in terms of personality tendencies like introversion, social shyness, nervousness, and desire for success. What was this other than an attempt to demonstrate the superordinance of the factor analytic method as a tool for providing powerfully reductive explanations of human complexity? It put solipsistic philosophy in its place and underlined the pre-eminence of psychology over its 'ugly sister'. See Eysenck, H.J. and Gilmour, J.S.L. (1944), The psychology of philosophers: a factorial study, *Characteristics of Personality* **12**, 290–6. James's dictum was itself the first line of MacKinnon (1944), The structure of personality. Eysenck may have also taken a lead from MacKinnon in other more trivial ways. Note the snappy alliterative title of Donald MacKinnon's article, MacKinnon, D. (1953), Fact and fancy in personality research, *American Psychologist* **8**, 138–46, reminiscent of Eysenck's popular Penguin paperbacks, e.g. Eysenck, H.J. (1965c), *Fact and fiction in psychology*, Penguin Books, London.

[77] Eysenck (1947a), *Dimensions of personality*, p. 17.

This was where Eysenck took his leave from Burt, a move characterized by his more realist philosophical inclinations and greater ambition. Eysenck reasoned that his factors could be seen to be *the* solution amongst many possible solutions if they could be anchored down, rather than rotated away or broken down into components. It also signalled Eysenck's departure from a whole gamut of American personality researchers.

The Americans had, by and large, worked hard to render 'personality' as a wholly *psychological* concept. In doing so, they tacitly assumed that the way we talk about ourselves reflected important underlying personality structure. Analysis at this level would therefore tell us something about this structure, even though 'personality' was otherwise assumed to be a natural entity situated outside descriptions of it.[78] It tended to lock US researchers down in a statistically defined semantic space. Eysenck was willing to entertain this kind of reflection hypothesis. But he wanted to do more, to move beyond the restrictions of ordinary language and the assumptions of self- report assessments. *Dimensions of personality* makes some long and painful points about the difference between the scientific and everyday use of a term like 'intelligence'.[79] Mixing Vienna Circle positivism with the operationalism of Percy Bridgman, Eysenck declared that scientific usage of such terms was 'connotative' rather than 'denotative'. The significance and meaning of these terms was dependent on their theoretically embedded function, upon operational definitions and testable implications. It was a position exemplified by Eysenck's search for collateral evidence—via experimental tests of suggestibility, salivary output, and night vision, and from the measurement of physique and exercise response. Eysenck was not (and could not afford to be) all that fussy about the nature of this evidence. Any and all corroborating data would do. Eysenck did not want to depend on conservative assumptions about everyday language. He wanted to go beyond discursive description and he was more than willing to reconnect with the biological discourses his New World contemporaries had roped off.[80] While Eysenck was hardly alone when it came to factor analytic explorations of

---

[78] For a discussion of these points in relation to the history of personality testing in the US see Buchanan, R. (1997), Ink blots or profile plots: the Rorschach versus the MMPI as the right tool for a science-based profession, *Science, Technology and Human Values* 21, 168–206.

[79] See for example, Eysenck, H.J. (1947*a*), *Dimensions of personality*, Routledge and Kegan Paul, London, pp. 17–18.

[80] Thus Eysenck felt at liberty to take issue with Eliot Slater—now together with Eysenck at the Maudsley after the war—on the hereditary basis for neuroticism. See Prins, A. (1998), Ageing and expertise: Alzheimer's disease and the medical professions, 1930–1990, Ph.D. thesis, University of Amsterdam, pp. 93–111.

personality—J.P. Guilford, Raymond Cattell, and Truman Kelley were already prominent names in the US—none were so eager to venture into the biological realm. Cattell, for example, restricted his avenues for factor verification to ratings, tests, and clinical description. Eysenck's 1944 paper and *Dimensions of personality* were notable for their inclusion of Pavlovian ideas.[81] This made him, according to some of his critics, just an old-fashioned theorist of temperament. I will discuss these points at greater length in the following chapter.

The scientific study of personality

Much of Eysenck's output in the years immediately following *Dimensions of personality* could be seen as an attempt to transform his more tentative early proposals into a formalized system. For example, he produced several articles outlining the basic role of factor analysis in generating hypotheses and theory building. Eysenck moved further away from Burt's 'principles of classification' viewpoint to argue that factors represented potential causal properties. Gone too were his reservations about rotation. He came to wholeheartedly agree with Thurstone on this score; only rotation could supply the unique and invariant factors that went beyond mere description. Only rotation provided the basis for factor purification that might enable the isolation of underlying hypothetical causes for dimensional traits or types.[82]

In 1950, Eysenck developed a new statistical technique that he hoped would answer potential criticisms. It was designed to serve the dual purpose of directly assessing the question of continuity, as well as combating the problem of rotational indeterminacy by adding substantive backbone to factor interpretation. Eysenck called this technique 'criterion analysis'.[83] It involved comparing a factor like neuroticism with a column of various tests' (biserial or tetrachoric) correlations with the criterion of interest (e.g. 'normals' versus hospitalized neurotics). When the correlation between a factor and the criterion column was at a maximum, the factor was correctly positioned. Eysenck was tweaking the more usual practice of correlating factor measures and criterion tests. Instead he proposed correlating the pattern of test loadings on the factor with

---

[81] Eysenck (1947*a*), *Dimensions of personality*, p. 37.

[82] Eysenck, H.J. (1953*c*), The logical basis of factor analysis, *American Psychologist* **8**, 105–14. See also Eysenck, H.J. (1952*f*), Uses and abuses of factor analysis, *Applied Statistics* **20**, 345–84.

[83] Eysenck (1950*b*), Criterion analysis: an application of the hypothetico-deductive method to factor analysis.

measures of discrimination of the criterion of interest. The idea was that tests (like his new Maudsley medical inventory and performance measures such as the body sway test) would show varying capacities to discriminate the criterion. A correctly rotated factor should exhibit high loadings for highly discriminating tests and low loadings for poorly discriminating tests. Moreover, Eysenck proposed that criterion analysis should be carried out separately for normal and abnormal samples. This offered the chance to check for continuity between them. If the correctly rotated factors in each sample lined up, it suggested they were co-extensive dimensions. It was a complicated procedure that tried to do two things at once, straining Eysenck's (and everyone else's) understanding of the conceptual limits of factorial techniques. Still, Eysenck recalled that it was 'probably the most original idea I ever had as far as statistics [was] concerned' and it paved the way for his second book.[84]

The scientific study of personality picked up where *Dimensions of personality* left off. Published in 1952, it employed the same factorial methodology and corroborative experimentation as his first book. As Eysenck said, this book 'finalized my system of personality description'.[85] However, it was a far more self-conscious work, much more preoccupied with defending a position already taken. As if emboldened by the positive reception to his first book, *The scientific study of personality* was more confident in tone, grander in its claims. *The scientific study of personality* was a bigger, flashier public relations effort. It contained lots of photos of the Maudsley, of experimental equipment, procedures, and technicians, quite unusual and costly in a book of this type at the time. These pictorial elements appear to be aimed at non-psychologists; apart from reinforcing the neo-objectivity of Eysenck's psychological science, their instructive purpose was unclear. The book also contained more data than Eysenck's first book, more numbers, more tables.

The scientific study of personality drew heavily on the history and philosophy of science to justify his methodological standpoint, particularly the work of Susan Stebbing. Eysenck contrasted common sense and properly scientific thinking, leaving the reader in little doubt as to whose side he was on. A scientific approach to personality should be hypothetico-deductive, a logical arrangement of higher-order theory and lower-order observable, testable consequents. He rejected the notion of the unique personality championed by the likes of Allport and Windelband as useless. Everybody is unique, Eysenck said,

---

[84] Eysenck (1997*b*), *Rebel with a cause*, p. 104.

[85] Ibid.

'so is my old shoe.'[86] Personality research should be rigorously and unapologetically nomothetic.[87] Eysenck's second effort also challenged psychiatric practices much more overtly. He dismissed categorical taxonomies (as opposed to dimensional continua) as characteristic of more primitive, pre-scientific thinking. The final chapter of *The scientific study of personality* spelt out the uses of his dimensional framework for evaluating psychiatric treatments like leucotomies, as an aid to making administrative decisions on mentally defective patients, and for assessing unskilled worker adjustment and so on.

Most importantly, *The scientific study of personality* introduced a third dimension, that of psychoticism. In the final chapter of *Dimensions of personality*, Eysenck had speculated whether the same kind of analysis he had applied to the neuroses could be applied to the psychoses. At the time, he had difficulty answering such a question due to the lack of appropriate samples. However, towards the end of the war, the dilapidated Maudsley buildings were reopened and soon after the professional staff regrouped. Late in 1947, a merger was set up with the Bethlem Royal Hospital, Britain oldest psychiatric hospital, which had moved to its present location at Monks Orchard. The Bethlem had the space, the buildings, and the inpatient capacity that the Maudsley lacked. It was a marriage of convenience: one was very old and very rich, the other very young and very poor.[88] While the Bethlem took more of the chronic and difficult patients, Lewis saw to it that early and recoverable cases were directed to his domain on Denmark Hill. Even so, Eysenck now had access to a greater range of patients than he did during the war.

Eysenck drew heavily on the work of Ernst Kretschmer, who had proposed continuity from the normal to the psychotic. Could psychoticism be isolated as a factorial dimension, and might the two major types of schizoid and cycloid (manifested as schizophrenia and manic-depression) also relate to introversion–extraversion? Eysenck reviewed the contradictory evidence linking introversion and extraversion with either schizophrenia or manic-depression, before dismissing this idea. Instead, Eysenck choose to focus on psychoticism as an entirely separate dimension, and investigate the possibility of a further

---

[86] Eysenck (1952a), *The scientific study of personality*, p. 18.

[87] See also Eysenck's reply to Samuel Beck, Eysenck, H.J. (1954d), The science of personality: nomothetic! *Psychological Review* **61**, 339–42.

[88] This observation came from Lewis's successor, Denis Hill. The Bethlem had been stripped of its teaching status, unable to attract psychiatric trainees. Those at the Bethlem feared it would lose its autonomy and be absorbed within the new National Health Service as just another struggling public hospital. See Waddington (1998), Enemies within: post-war Bethlem and the Maudsley Hospital.

dimension marked out by schizothymia and cyclothymia at each pole. Based on evidence from factor analyses, and criterion and discriminant analyses, Eysenck argued that there was, in fact, a distinct dimension of psychoticism. However, he claimed to find no evidence for a bipolar schizothymia–cyclothymia dimension. Like neuroticism, Eysenck's psychoticism dimension had one pole anchored in the normal and the other anchored in the aberrant.

Eysenck's proposition of a universal factor of psychoticism had far fewer precedents in the psychological literature than neuroticism and introversion–extraversion. Few psychologists had attempted to provide psychometric descriptions of the severely disturbed, much less suggest that the characteristics reflected extreme versions of normal personality. Such patients were always difficult to deal with. The inappropriate test performance of psychotics and the common complications of organic factors tended to restrict the study of such patients. It made Eysenck's work on psychoticism groundbreaking, much braver than his earlier work. It also made it more problematic. Lacking interpretative signposts, psychoticism was constantly revised and refined over the years. It proved difficult to independently accommodate with his first two dimensions, whose substantive reality was always assumed.[89]

What the periodicals said

> Eysenck's main achievement has been the valuable but limited one of refining existing useful classifications. It has, however, few of the far-reaching implications for psychology that he would believe it to have. (Albino, R.C. (1953), Some criticisms of the application of factor analysis to the study of personality, p. 168)

Aubrey Lewis's foreword to *The scientific study of personality* gently rebuked Eysenck for taking too negative a view of psychiatric diagnostics. 'Not as good as the first book' was his private opinion, Eysenck recalled. In addition, the authority of *The scientific study of personality* was diminished by uncertainty over the value of criterion analysis. It was very technical and difficult to implement. One needed to find a 'large and varied' set of criterion-discriminating tests. Just how many tests were needed and what level of discrimination was

---

[89] This topic was as much the project of second wife Sybil as it was Eysenck's, for they collaborated closely on it. See Eysenck, H.J. and Eysenck, S.B.G. (1971), The orthogonality of psychoticism and neuroticism: a factorial study, *Perceptual and Motor Skills* **33**, 461–2; Eysenck, H.J. and Eysenck, S.B.G. (1976), *Psychoticism as a dimension of personality*, Hodder and Stoughton, London; Eysenck, H.J., Eysenck, S.B.G., and Barrett, P. (1985), A revised version of the psychoticism scale, *Personality and Individual Differences* **1**, 21–30; and, Eysenck, H.J. (1992c), The definition and measurement of psychoticism, *Personality and Individual Differences* **11**, 757–85.

acceptable, was left up in the air.[90] These difficulties alienated psychologists and psychiatrists. Worse, the statisticians hated it. Ardie Lubin argued that criterion analysis did not provide the kind of independent verification Eysenck claimed it did. Given that Lubin worked in Eysenck's own statistical section at the Institute of Psychiatry (IoP), he might have felt a need to pull his punches. In a *very* carefully worded, abstract critique, Lubin pointed out difficulties in directly calculating the correlations Eysenck's technique demanded. Moreover, at the limit, criterion analysis amounted to a factorial re-description of test-battery correlations with the criterion. According to Lubin, it would be completely successful when there were no unique factors, when 'all of the criterion variance that can be predicted from the tests is a function of the common factor scores of the test battery'.[91] The implication was that the technique simply shifted the burden of statistical proof: the validity of a factor thus positioned would depend on the validity of the criterion—usually based on observer ratings and judgements—that could be just as difficult to establish and improve.

Eysenck must have found it difficult to answer Lubin's criticisms, for he did not attempt to. Others pointed to logical problems with the technique, and these were broadsides Eysenck *did* respond to.[92] However in the years to follow, he tended to de-emphasize the technique's value for the positioning of factors. Even more tellingly, he turned to other techniques in an attempt to provide independent evidence for a continuous dimensional framework. In his 1955 article that re-framed the discrete categories of psychiatric diagnosis as a 'psychological and statistical problem' he took up Lubin's recommended

---

[90] When pressed by later criticism, Eysenck suggested that in 'due course it will prove possible to give a more operational definition of "varied"' and that the number of discriminating tests 'should not be below 20, with the standard of significance to be taken at the $p = 0.01$ level ...', p. 430 of Eysenck, H.J. (1958*a*), The continuity of abnormal and normal behaviour, *Psychological Bulletin* **55**, 429–32. Obviously such a procedure was not going to be an easy one to implement.

[91] See p. 54 of Lubin, A. (1950), A note on 'criterion analysis', *Psychological Review* **57**, 54–7. It was not the only occasion on which those within Eysenck's statistical section were publicly critical of the way he interpreted his data. According to Lubin's statistical section colleague, A.R. Jonckheere, Eysenck would often ask them to do things that were mathematically difficult or impossible. In frustration, they would suggest that Eysenck try doing such calculations himself. A.R. Jonckheere, interview, 22 April 2001.

[92] See Beezhold, F.W. (1953), On criterion analysis, *Journal of the National Institute for Personnel Research* **5**, 176–82, and Eysenck, H.J. (1953*d*), On criterion analysis: a reply to F.W. Beezhold, *Journal of the National Institute for Personnel Research* **5**, 183–7

(canonical variate) discriminant analysis instead.[93] The results separated the normal, the neurotic, and the psychotic on various tests. According to Eysenck, it provided indisputable evidence for distinct neurotic and the psychotic dimensions. Arguably Eysenck was attacking a straw man; Eysenck crudely interpreted Freud's notion of pathogenic regression as implying continuity between neurosis and psychosis. Such an interpretation was at odds with contemporary analytic and non-analytic wisdom.[94]

Eysenck's work began to attract more attention, indicative of the impact he had made as a new kid on the block. The incorporation of the experimental and the biological marked out his approach as unusual. Moreover, his commitment to precise measurement, quantifiable entities, and rigorous testability made him the über-positivist of British psychology. Factorialists of personality—Burt, Cattell, Guilford—found themselves challenged by an even more ambitious proponent of the technique. Conversely, those already sceptical about such an approach—especially the blue-bloods to the north—found a new reason to dismiss the factorial approach as little more than a statistical contrivance. Eysenck did not often tangle directly with the Oxbridge set, but when he did the sparks flew. For example, at the second meeting of the Experimental Psychology Group (later Society) in 1947, Eysenck read a paper on 'The measurement of personality'. Oliver Zangwill's minutes record that 'his paper was followed by a lively discussion, and there was considerable difference of opinion as the interpretation of the "neuroticism" factor revealed by this work.'[95] Eysenck liked to characterize Zangwill and his Oxford colleagues as remarkably ignorant of basic statistics, and remembered Zangwill's criticisms at this meeting as 'childish'.[96] However in his autobiography, Eysenck was a little more respectful. According to Eysenck, Bartlett's position on factor analysis was that 'you only get out what you put in', while Zangwill would always argue

---

[93] Eysenck, H.J. (1955c), Psychiatric diagnosis as a psychological and statistical problem, *Psychological Reports* 1, 3–17. See also Loevinger, J. (1955), Diagnosis and measurement: a reply to Eysenck, *Psychological Reports* 1, 277–8, and Eysenck's response, Eysenck, H.J. (1956b), Diagnosis and measurement: a reply to Loevinger, *Psychological Reports* 2, 117–18.

[94] It was a point of Freudian interpretation that Loevinger, amongst others, challenged Eysenck on. See Loevinger (1955), Diagnosis and measurement: a reply to Eysenck.

[95] One must read this quote with an awareness of the distinctly English propensity for understatement. Minutes of EPG Meeting, 4 January 1947, in Mollon, J.D. History of the EPS: Meetings. http://www.eps.ac.uk/society/meetings.html.

[96] Hans J. Eysenck, interviewed by H.B. Gibson, 1 March 1979.

that such 'statistics could never show the structure of psychological reality'.[97] Here was a clash of outlooks that made for an implacable but increasingly disengaged hostility.

Things were hardly more convivial amongst the psychiatrists at the Maudsley. Many had begun to dislike Eysenck, even if they did not take him on in print. He was very much the opposite of many of these psychiatrists—Erwin Stengel, for one. Yet they were ill-equipped or not interested in taking on Eysenck on his terms. For example, if Stengel was presented with a chi-square that contradicted his ideas, it just illustrated to him the problem with that kind of quantitative approach.[98] Eliot Slater occupied the middle ground, pushing for a multifaceted, qualitative and quantitative approach.

Public confrontation was for psychologists. Eysenck's approach was criticized by Neil O'Connor, then at the IoP's Occupational Psychiatry Research Unit, for being too distant from the object of its investigation, for worshipping the 'false god' of the statistical table above the 'human element'.[99] Eysenck was attacked by the University of Natal's R.C. Albino for not acknowledging that clinical experience provided the primary guide for what to factor analyse, and for not being convincing on the continuity issue.[100] Most notorious was the review he received from 'WLG' in the *British Journal of Psychology*, a review already discussed in the previous chapter. Assumed at the time to be ghost written by Burt, this review took Eysenck to task on just about every major assumption and claim Eysenck had to offer. The review doubted the priority of Eysenck's factors, citing their comparability to Burt's work in particular; it questioned Eysenck's heavy reliance on detached quantitative methods as opposed to clinical observation; and, it faulted Eysenck's appropriation of the philosophy of science as backing up his model of psychological science. Most particularly, WLG took Eysenck to task on the notion of continuity.

---

[97] Eysenck (1997b), *Rebel with a cause*, p. 124.

[98] Jenner, F.A. (1991), Erwin Stengel: a personal memoir, in *150 years of British psychiatry, 1841–1991*, Vol. 1 (ed. G.E. Berrios and H. Freeman), pp. 436–44, Royal College of Psychiatrists, London.

[99] O'Connor, N. (1952), Review of *The scientific study of personality* by Hans Eysenck, *Bulletin of the British Psychological Society* **3**, 115.

[100] See Albino's critique and Eysenck's reply: Albino, R.C. (1953), Some criticisms of the application of factor analysis to the study of personality, *British Journal of Psychology* **44**, 164–8, and Eysenck, H.J. (1953a), The application of factor analysis to the study of personality: a reply, *British Journal of Psychology* **44**, 169–72. Albino argued that quantitative evidence of a continuous distribution of a syndrome (e.g. neuroticism) was not conclusive evidence of continuity of cause. Eysenck countered that he had never made such an argument, and otherwise called for more data to resolve the issue.

Rather than adjudicating even-handedly on this issue, Eysenck appeared to regard it as a 'foregone conclusion'. By:

> applying tests that yield measurements on a continuous scale, and subjecting the measurements to a factor analysis, he reaches results that are in harmony with a hypothesis of continuity, and therefore claims to have 'shown' such continua exists...[101]

Eysenck's empirical evidence was all necessary but not sufficient. WLG faulted Eysenck for not canvassing the opposite possibility of *discontinuity*, and then checking for confirming and disconfirming evidence for that.

Eysenck called it the worst review he had ever received, a 'complete hatchet job'.[102] It was 'so biased and professionally damaging' that he felt obliged to reply.[103] His riposte cited a 'baker's dozen' list of distortions, misrepresentations, and lack of understanding on the part of the critic.

He implied this was just a sample of the review's problems that he could have tackled, with space not permitting a complete and exhaustive correction.[104] Questions of priority were likewise a scientific red-herring, Eysenck maintained. If his neuroticism factor had anything in common with Burt's notion of 'emotionality', for example, it was because the work of both could be situated in a venerable tradition dating back to Heymans and Wiersma at the turn of the century.

Which parts of this review (and follow-up reply) were written by Burt is still a matter of conjecture. At the time, though, Eysenck clearly *believed* he was facing Burt. It was thrilling spectacle: the up-and-coming former student facing his mentor, the ageing master of rhetorical combat. Eysenck emerged bruised but not beaten, more wary of Burt than ever. Eysenck also became highly sensitized to the accusation that he was assuming rather than demonstrating continuity. Several other critics would also latch on to Eysenck's 'unjustified' desire to bridge the abnormal–normal divide,[105] an impression he

---

[101] WLG (1953), Review of *The scientific study of personality* by Hans Eysenck, p. 210.

[102] Eysenck, H.J. (1995*a*), Burt and hero and anti-hero: a Greek tragedy, in *Burt: fraud or framed?* (ed. Nicholas J. Mackintosh), pp. 111–29, Oxford University Press, Oxford, p. 116.

[103] Eysenck gave the impression he seldom made such direct responses to criticism. This seems a little at odds with his publication record, however. By my rough count, Eysenck published *at least* 40 pieces reacting to poor reviews and negative commentary. In this sense, he was nothing if not combative, never one to take criticism lying down.

[104] Eysenck (1954*b*), A note on the review and WLG (1954), A reply.

[105] See Pearson, J.S. and Kley, I.B. (1957), On the application of genetic expectancies as age-specific base rates in the study of human behaviour disorders, *Psychological Bulletin* **54**,

helped sustain by sometimes conflating the ideas of dimensionality and continuity.[106] In an effort to claim the intellectual high ground on the issue, Eysenck would point to the results he had achieved with criterion analysis. But he was forced to concede that the technique had its problems. He never came up with a better one. Years later, Eysenck sadly remembered criterion analysis as 'misunderstood by the critics' and 'widely disregarded'.[107] This was as close as Eysenck ever came to an admission of failure—at any time, on any issue. Eysenck would return to the continuity issue sporadically over the years, attacking, for example, the categorical framework of later DSM revisions.[108]

Foundational if not original

Eysenck cited *Dimensions of personality* as his most original and worthy contribution, and his intellectual heirs have tended to follow suit.[109] However, Eysenck's early work carries the halo of immaculate conception because the context in which it emerged has been wiped away. As contemporary trait theorist Ian Deary remarked, going back to *Dimensions of personality* makes one realize how much personality research there already was at the time.[110] It is a revelation quickly confirmed by a perusal of the McVicker Hunt volumes.

406–20, and the subsequent exchange Eysenck (1958*a*), The continuity of abnormal and normal behaviour, and Pearson, J.S. and Kley, I.B. (1958), Discontinuity and correlation: a reply to Eysenck. *Psychological Bulletin* **55**, 432–5.

[106] See, for example, Eysenck, H.J. Wakefield, J.A., and Friedman, A.F. (1983), Diagnosis and clinical assessment: the DSM-III, *Annual Review of Psychology* **34**, 167–93. For a latter-day critique of Eysenck's position, see Wakefield, J.C. (1997), Diagnosing DSM-IV—part II: Eysenck (1986) and the essentialist fallacy, *Behaviour Research and Therapy* **35**, 651–65.

[107] Eysenck (1997*b*), *Rebel with a cause*, p. 104.

[108] Eysenck, H.J. (1986*b*), A critique of contemporary classification and diagnosis, in *Contemporary directions in psychology: toward the DSM-IV* (ed. T. Millon and G.L. Klerman), pp. 73–98, Guilford Press, New York. For a latter-day critique of Eysenck's position on diagnosis, see Wakefield, J.C. (1997), Diagnosing DSM-IV—part II: Eysenck (1986) and the essentialist fallacy, *Behaviour Research and Therapy* **35**, 651–65.

[109] See Eysenck (1997*b*), *Rebel with a cause*, pp. 97–9, and celebratory volumes such as Lynn, R. (ed.) (1981), *Dimensions of personality: papers in honour of H.J. Eysenck*, Pergamon Press, Oxford, especially Lynn's preface, and Nyborg, H. (ed.) (1997), *The scientific study of human nature: tribute to Hans J. Eysenck at eighty*, Pergamon, Oxford. Numerous obituaries also expressed similar sentiments, for example, Gray, J. (1997*b*), Obituary: Hans Jürgen Eysenck (1916–97), *Nature* **389**, 794.

[110] Deary, I. (1999), The origins of Eysenck's dimensions, *Contemporary Psychology* **44**, 318–19.

Most of Eysenck's empirical research had methodological precedents in the contemporary literature, even if some of the performance tests and experimental set-ups were novel.

Many of Eysenck's inspirations have been subjected to a form of enforced forgetting by a self-consciously progressive discipline. The past is read in terms of current truths and the steps necessary to get there, with the archaic and embarrassing all but erased. As Deary also noted, *Dimensions of personality* was replete with Victorian ideas of the separation of the physical expression of emotional distress into neurasthenia, hypochrondriasis, and hysteria.[111] These terms and many of the implications they conveyed have no place in contemporary psychiatric nomenclature. The correlation of physical morphology and personality—the suggestion that dysthymics, for example, tend to be tall and skinny 'leptomorphs'—looks just as old fashioned.

Dimensions of personality proclaimed the 'discovery of two main factors, both of which bore a close relation to similar factors previously discovered in normal subjects by numerous investigators'.[112] There was very little shock of the new. Much of what Eysenck did was utterly mainstream within a particular tradition, distinguished only by a bower-bird style inclusiveness and a willingness to carry out psychophysiological experimentation. *Dimensions of personality* was diverse, but the literature it addressed was even more so. There were many scattered studies, an overabundance of traits, attributes, and factors, and a notable absence of theoretical convergence. Allport and Vernon had identified at least five different *classes* of personality definitions alone.[113] Into the breach rode Eysenck, combining much of this material in a promisingly programmatic way and finishing the job with a theoretical spit and polish. With few psychologists in Britain so much as dabbling in the area, there was never going to be a rash of competing titles. Even so, the impact *Dimensions of personality* made illustrated Eysenck's amazing capacity to get noticed, to make a splash. Eysenck managed to distinguish himself from esoteric and scholarly contemporaries like Burt by pitching this first book to a broad audience.

---

[111] Ibid., p.319.

[112] Eysenck (1947a), *Dimensions of personality*, p. 244.

[113] Allport, G. and Vernon, P.E. (1930), The field of personality, *Psychological Bulletin* **27**, 677–730. If any further support for this point is needed, consider this. In 1927, A.A. Roback put together his compendium *Psychology of character*. Since the text was over 600 pages long, Roback decided to publish the bibliography separately. This bibliography ran to 340 pages! See Roback, A.A. (1927), *A bibliography of character and personality*, Sci-art, Cambridge.

Here was an easy-to-digest synthesis on an emerging topic, its weighty, unifying title carrying an implicit 'the' at the front.

Reconstructing the context for Eysenck's early work illustrates just how dependent it was—ontologically and methodologically—on the work of others. Eysenck would no doubt have argued that this was exactly the point. The personality dimensions he came up with were not novel, only the basis for believing them. Eysenck wanted to upgrade factorial descriptions of personality to the level of causal laws. To do so, he had to make several leaps of faith—from statistical factors to personality structure, from discrete classes to continuous measurement, from the abnormal to the normal. These leaps of faith were dictated by his inclinations, his training, and his institutional circumstances. Having made them, he would spend the rest of his career demonstrating their utility while assuming their validity.

Dimensions of personality acquired its foundational status because the tireless Eysenck made it so. He and a host of followers used it as a basis for an extensive research programme that, in turn, dovetailed with other research developments in personality psychology.[114] In the late 1980s and 1990s, a trend toward consolidation—at a semantic, psychometric level at least—led to a surprising consensus in factorial descriptions of personality. The first two factors of the now dominant 'big five' (i.e. extraversion, emotional stability/neuroticism, agreeableness, conscientiousness, and openness/culture) are regarded as equivalent to Eysenck's.[115] Latter-day proponents of Eysenck's 'even bigger three' have also argued that two of the remaining three factors of the big five (i.e. agreeableness, conscientiousness), could be subsumed under Eysenck's psychoticism dimension, and have suggested that openness is more a cognitive construct.[116] It seems that the young Eysenck anticipated much of what was to come in his chosen field, and he did so using methods and conceptual suggestions that already had a relatively long history.

---

[114] See Nathan Brody's discussion of these developments in Brody, N. (1987), Controversy and truth: can scientists' contributions be evaluated by studying the controversies they have generated? Review of *Hans Eysenck: consensus and controversy*, ed. Sohan Modgil and Celia Modgil, *Personality and Individual Differences* **8**, 983–4.

[115] See, for example, Costa, P.T. and McCrae, R.R. (1992), The five factor model of personality and its relevance to personality disorders, *Journal of Personality Disorders* **6**, 343–59, and Matthews, G. and Deary, I. (1998), *Personality traits*, Cambridge University Press, Cambridge, Chapter 1.

[116] Revelle, W. (1995), Personality processes, *Annual Review of Psychology* **46**, 295–328.

Chapter 5
The biology of personality

A descriptive taxonomy of personality was never going to be enough for Hans Eysenck. Inspired by the eugenic concerns of Galton and German physical morphology, Eysenck set about laying down what he thought of as the 'causal' basis for his dimensions. This meant anchoring introversion–extraversion, neuroticism, and psychoticism in corporeal biology. These enhanced biodimensions could then be correlated with characteristically different patterns of conditioning, performance, and problem solving. Eysenck's learning paradigm was classically behaviourist and never strayed far from its physiological underpinnings. While Hull was taken on board and then discarded, Skinner was barely considered at all. Latter-day cognitive approaches were also largely ignored. But Eysenck's affinity with Russian work remained, especially with that of a Nobel laureate who considered himself primarily a physiologist. Eysenck made the misunderstood Ivan Pavlov his muse—a distant but compelling guide linking material body with immaterial mind.

Cued by Pavlov, Eysenck took the radical step of situating behavioural differences within the functional economy of the nervous system. Personality differences were directly related to differences in the brain, differences held to be largely determined by heredity. Utilizing small-scale animal breeding and behaviour research and the limited techniques of functional neurology of the day, Eysenck was able to produce the landmark 1957 book *The dynamics of anxiety and hysteria*. His famous 'typological postulate' mapped introversion–extraversion on to a simplified version of excitation and inhibition, while neuroticism was linked to an anxiety-related notion of drive. Prompted by criticism and further research, these ideas were revised a decade later in *The biological basis of personality*. Eysenck's dimensional grid both reflected and predicted differential learning patterns that, in turn, were the result of differential hereditary propensities. Biologizing personality also enabled Eysenck to claim that his dimensions were culturally universal, with a continuum from humans to animals.[1]

---

[1] See Eysenck, H.J. (1997b), *Rebel with a cause* [revised and expanded], Transaction Press, New Brunswick, New Jersey, pp. 199–203.

Eysenck achieved a remarkable feat by becoming a 'biological behaviourist' and 'personality learning theorist'. His breadth and vision set him apart from a wash of American trait psychologists. No one else—not even his closest counterpart Raymond Cattell—had the gumption to connect psychometric descriptions with the control structures of the brain. Eysenck was now a disciplinary boundary-crosser. It was a high risk, high yield strategy. Each of these subfields had their native experts, none of whom had a neutral attitude to encroachment. Yet Eysenck treated traditional divisions and intellectual niceties as an opportunity for transgression, the better to make bolder, more novel claims.

While many researchers were intrigued, few could contemplate such a programme. It was not just an intellectual bridge too far; it ran up against daunting financial and institutional constraints. So how did Eysenck do it? How did he generate and manage the kind of resources, the equipment, and the personnel required? As we will see, he had a knack for making a little go a long way. Those attempting to follow in his footsteps also claimed Eysenck's approach could unite the discipline as never before by reconciling Cronbach's 'two disciplines' of experimental and correlational psychology. Eysenck's final biological model laid out a blueprint for research in the neurophysiology of temperament differences in the decades to follow. However, Eysenck's integrative efforts encountered resistance from those he tried to unite and left traditional boundaries still in place. In part this could be represented as entrenched resistance to what was seen as an attempted takeover. But it was not just that.

Eysenck's ambitious efforts provoked bad critical notices from some of his former colleagues at the Institute of Psychiatry. Beautiful numbers also failed to convince many farther afield, even those willing to give his conjectures a sympathetic hearing. Eysenck's style was all detachment, all overt formalism, all public performance; it sacrificed consensus-building for partisan didacticism. Eysenck did not, and perhaps could not, do the behind-the-scenes work to allay the doubts he evoked. Out of this emerged a conspicuous dissonance. Eysenck's escalating disciplinary profile was accompanied by increasingly toxic cynicism amongst his peers—highlighting the importance of what is left out of public scientific discourses for making knowledge claims and breaking reputations.

Creating an Eysenckian department

The multisection research department Eysenck organized at the Institute of Psychiatry (IoP) was a crucial ingredient to the breadth of his research. While Eysenck was lucky to be in the right place at the right time, he cleverly adapted the circumstances to fit his broad intellectual vision. It was set up to both research and serve Eysenck's interests, embodying a set of choices not without trade-offs. The young researchers and students who came to the department joined an expansive, multipronged programme, but found themselves working for Eysenck as much as with him.

At the end of World War II, plans were put in place for a new system of postgraduate medical teaching centres or institutes geared to training in eight clinical specialties. The Maudsley Hospital was always the leading contender as the base for psychiatry.[2] As well as the power and prestige value it would confer, a new institute at the cash-strapped Maudsley made a lot of financial sense. With it, the hospital could tap government funding set aside for medical education. Likewise the merger with the Bethlem Royal Hospital made sure the Maudsley would have access to 'mystical Bedlam's' annual endowment of over 36, 000 pounds—a huge amount in those days. The Bethlem would also take the strain of psychiatric service demands, leaving the Maudsley comparatively free to pursue teaching and research.

These promises came just in the nick of time for Eysenck, since support from the Rockefeller Foundation had begun to dry up immediately after the war.[3] This was Eysenck's main source of funds. Eysenck said he continued to earn a 'meagre' salary from his lectures to medical students at the Maudsley and his guest lectures at his alma mater University College (UCL), but debts began to mount.[4] Worse, it seemed his research programme might have to be cut back. Aubrey Lewis pleaded with the Rockefeller Foundation for bridging finance to keep Eysenck and his assistants afloat until the association with British Postgraduate Medical School had been effected.[5] In October 1946, Eysenck was appointed senior psychologist at the Maudsley with special responsibility for teaching and research, ahead of other applicants F.S. Grimwood and K.J. Saalfeld. His annual salary was 700–800 pounds.

---

[2] See Hall, J. (2007b), The emergence of clinical psychology in Britain from 1943 to 1958, Part II: Practice and research traditions, *History and Philosophy of Psychology* **9**(2), 1–33.

[3] The Rockefeller Foundation had been giving research grants to the Maudsley since 1935. However, the Foundation gave notice that grants would probably not be extended past 1945. Eysenck recalled that the termination of Rockefeller funding occurred a year later. See Eysenck (1997b), *Rebel with a cause*, p. 106.

[4] In February 1946 Burt had offered Eysenck a part-time position at University College teaching personality psychology to budding undergraduates.

[5] AJL/JMT, Untitled draft of funding proposal, 6 November 1945, Aubrey Lewis Papers, Box 9, Bethlem Royal Hospital Archives and Museum.

The following year consolidated the quick reversal in Eysenck's professional fortunes. He was granted British citizenship and his first book, *Dimensions of personality*, made quite an international impact. Suddenly in demand, he had to weigh up two job offers from Duke University in the US and from Edinburgh University with Godfrey Thomson. Faced with losing one of his prized protégés, Lewis assured Eysenck of a significant role in the new institute. In October 1947, Eysenck was appointed reader in the Maudsley's Medical School. The following year the school was renamed the Institute of Psychiatry (IoP) and became one of the constituent institutes of the British Postgraduate Medical Federation, with the University of London assuming administrative control. Psychologists at the Institute were paid by the Department of Education, through the university. Hospital wards, bathrooms, and kitchens were converted to laboratories, testing rooms, and offices. Later, purpose-built facilities were provided when a new IoP building was constructed in the 1960s.[6] With the Maudsley and Bethlem hospitals officially merged in 1948, research funds once again became available. Bethlem coffers provided upwards of £15, 000 per year for this purpose. The IoP quickly became the centre of the English psychiatric universe, a research powerhouse and the preferred posting for psychiatric trainees.[7]

When Eysenck took up his readership in 1947, psychology was officially listed as a subdepartment within the department of psychiatry. By 1955, it had become an independent department and Eysenck was given a full professorship.[8] The way it was organized reflected Eysenck's keen eye for matters of power and autonomy. While the clinical course had special service requirements Eysenck could delegate to others, the rest of the department was up to him. As a result, it had a 'God-professor' structure reminiscent of university departments in Eysenck's native land. Everyone within the department was, in principle, answerable to Eysenck. They were not formally accountable to the medical hierarchy of the hospital, avoiding a lot of potential problems given that there were 'more psychiatrists per square inch at the IoP than anywhere

---

[6] Martin, I. (2001), Hans Eysenck at the Maudsley—the early years, *Personality and Individual Differences* 31, 7–9.

[7] Waddington, K. (1998), Enemies within: post-war Bethlem and the Maudsley Hospital, in *Cultures of psychiatry and mental health care in postwar Britain and the Netherlands* (ed. M. Gijswijt-Hofstra and R. Porter), pp. 185–202, Rodopi, Amsterdam. The centre of the universe phrase comes from p. 229 of Scull, A. (2004), The insanity of place, *History of Psychiatry* 15, 417–36.

[8] As well as psychiatry and psychology, the IoP came to house separate departments of biochemistry, clinical neurophysiology, neuroendocrinology, and neuropathology.

in Britain'.[9] If anyone at the hospital had an issue with someone from psychology they could take it up with Eysenck, a man whose daunting reputation only grew throughout his tenure at the IoP.

Programmatic research was as much Lewis's idea as Eysenck's.[10] With Lewis's blessing, Eysenck was able to put in place a broad but relatively autonomous research agenda.[11] A small statistical section was formed to assist with design and analysis.[12] A technical workshop was set up to make the experimental apparatus, test equipment, and so on. In 1949, visiting American Fulbright scholar Roger W. Russell also established an animal psychology laboratory at Bethlem.[13] In the early days of the IoP, most research was otherwise carried out on site—even if this necessitated the use of some fairly makeshift facilities. This made it easier for Eysenck to keep research focused, and to direct his team according to his needs. Until the mid-1950s, funding sources were fairly limited. The only research support received was from joint hospital coffers—although Desmond Furneaux's research on hypnosis, cognition, and intelligence in the psychiatry department was supported by the Nuffield Foundation. From 1956 onward, more diverse sources of funds were procured, including the Medical Research Council, the Ford Foundation, and more.

---

[9] William Yule, interview, 22 July 2003.

[10] Eysenck liked to portray his output as the theoretical integration of a vast and diverse programme of research. For example, Eysenck's 1948 public relations piece in *Acta Psychologica* extolled the virtues of the programmatic research model advocated by American psychologist Donald Marquis and his success in fulfilling just such a model at the IoP. Eysenck, H.J. (1951*b*), Psychology Department, Institute of Psychiatry, Maudsley Hospital, University of London, *Acta Psychologica* **8**, 63–8.

[11] In practice, however, the various research arms of the IoP worked independently—although some collaborative research was undertaken at an individual level. Eysenck was later to lament this situation, claiming that he tried to convince Lewis of the value of the multidisciplinary approach to specific psychiatric problems. Yet the social isolation that the psychology department in particular was to experience was in part a direct result of Eysenck's personal style. Eysenck, H.J. (1983*g*), Forty years at the Maudsley: changes and otherwise, *Bethlem and Maudsley Gazette* **31**, 4–5.

[12] Ardie Lubin and A.R. Jonckheere first served here, as did A.E. Maxwell and Patrick Slater later on.

[13] Russell quickly departed, however. Russell replaced the retiring Burt as chair at UCL, but only after Eysenck's proposal for a joint chair at UCL and the IoP was rejected by both Lewis and Burt. See Chapter 3.

Bringing back the biological

Eysenck's earliest work on personality had been notable insofar as it included ideas from outside the psychological discourses of his day, including decidedly unfashionable figures like Pavlov, Kraepelin, Gross, and McDougall. *Dimensions of personality* came with a chapter on the relation of body-build to personality. His second book, *The scientific study of personality*, included a study on the hereditary basis of his personality dimensions. It made him seem both innovative and nostalgic, looking forward to a new hybrid science of personality as well as looking backward to a time and place wherein the divisions of mind and body were more permeable. It made Eysenck's work stand out from the kind of psychological research typical in Britain at the time.

By the end of the nineteenth century, secular science's confrontation with organized religion was all but over in Britain. A conservative antipathy to mechanistic materialism still lingered, however, especially amongst Oxbridge-educated souls. The first generation of modern British psychologists had to engage in an awkward process of separation and specialization.[14] While they took on the experimental methods of continental psychology, they still incorporated aspects of 'mental philosophy' to buttress their choice of subject matter (i.e. ideas, feelings, and the conscious self) and put some distance between themselves and adjoining biological disciplines. Mental phenomena were not to be treated as merely the 'special complications of processes that were not mental'.[15] The Darwinian impetus—which had driven the comparative research of George Romanes, Lloyd Morgan, and Thomas Hobhouse—had dissipated by the turn of the century. The British lead in comparative psychology was squandered, and animal research almost disappeared. Before he left for Harvard after World War I, McDougall was one of the few notable advocates of a biologically based psychology in Britain. However, the neo-Lamarckian animism he came to espouse was a decidedly atypical marriage of mind and body, and he quickly went from respected leader to ignored outsider.

Very little psychological research in Britain in the first decades of the twentieth century made reference to physiological research. Across the Atlantic, first generation behaviourists like John Watson, Edward Thorndike, and E.C. Tolman had quickly discarded the evolutionary framework in favour of

---

[14] See various contributions in Bunn, G.C. *et al.* (eds.) (2001), *Psychology in Britain: historical essays and personal reflections*, BPS Books, Leicester.

[15] G.F. Stout, quoted by Hearnshaw, L. (1964), *Short history of British psychology, 1840–1940*, Methuen, London, p. 230.

short-term, experimental study of animal learning linked to practical social engineering. However, American behaviourism made very little impact in Britain prior to World War II. Moreover, there was an almost complete rejection of Russian work, of 'laboratory mind' and that 'simpleton Pavlov' in particular.[16] Even though Pavlov had made a lecture tour to London in 1906, his work on conditioned reflexes was not translated for British audiences until 1927.

Eysenck, for one, was to change all that. As with the factorial description of personality, he saw a yawning gap in home-grown research with its exclusion of the biological level of explanation. It was, he remembered, always clear to him that human nature was 'biosocial'. Any notion of a dualism of mind and body should be rejected in favour of a continuum. This was, Eysenck recalled, 'too obvious to require supporting argument' and he seldom gave one.[17]

One must remember that Eysenck was mindful of the scepticism surrounding the methods and constructs bequeathed to him by the London school. He had to convince everyone—including himself—that what he was measuring was some sort of natural quality. Factorial constructs could be made more *real*—that is, more stable and robust, mobile and universal—if they could be correlated with measurements whose status was less questionable. More than operational definitions were required, Eysenck argued. The 'interpretation of a factor ... should try and relate the factor causally to relevant parts of general and experimental psychology.'[18] Eysenck had no qualms about crossing the ontological divide separating psychological and biological discourses. Indeed, he quickly sharpened his focus. Not content with relating his dimensions to various physiological indices like physical endurance and body build, he began to search for the executive structures that dictated these personality differences. This meant going directly after the brain.

---

[16] These phrases come from George Bernard Shaw, quoted on p. 244 of Wilson, D. (2001), A 'precipitous *dégringolade*'? The uncertain progress of British comparative psychology in the twentieth century, in *Psychology in Britain: historical essays and personal reflections* (ed. G.C. Bunn, A.D. Lovie, and G.D. Richards), pp. 243–66, BPS Books, Leicester. See also Smith, R. (2001), Physiology and psychology, or brain and mind, in the age of C.S. Sherrington, in the same volume, pp. 223–42.

[17] See Eysenck (1997b), *Rebel with a cause*, p. 64 See also Eysenck, H.J. (1980a), The biosocial nature of man, *Journal of Social Biology Structure* 3, 125–34. Eysenck suggested latter-day sociobiologist E.O. Wilson overemphasized biological determinants but criticized American behaviourists like Watson and Skinner for ignoring biochemical and physiological factors.

[18] Eysenck, H.J. (1970a), *The structure of human personality*, 3rd edn, Methuen, London, p. 49.

Eysenck's first move to ground his personality factors in biology was via genetics, a move he foreshadowed at the beginning of *Dimensions of personality*. His first research proposal upon returning to the Maudsley from Mill Hill had emphasized this as a priority, as part of a full-scale, Galtonian-style project directed at the developing personality. The proposal stated that 'work at Mill Hill ... resulted in the isolation of two dimensions of personality.'[19] Three main channels of future research were envisaged: the isolation of a possible schizoid–cycloid dimension; the discovery of the influence of heritable factors on the genesis of the personality dimensions by studying twins; and the discovery of the influence of environmental factors on the personality dimensions by a longitudinal study of children. While the longitudinal study was never to be fully realized, Eysenck soon came up with a prototype psychoticism dimension.[20] In addition, he rapidly put in place a collaborative programme of research that would address the question of the heritability of his personality dimensions.[21] But it wasn't until 1953 that IoP reports start to mention any attempt to link his personality dimensions with neurophysiological functioning.

Pavlov re-conditioned

Without diaries, notebooks, and letters it is difficult to get inside Eysenck's head and trace the development of this thinking. However, his publications suggest that Pavlov was the clearest pointer to where he would go. By the early 1950s, Eysenck had read up on Russian research and had begun to seriously consider the implications for individual differences.

Eysenck started work on the experimental analysis of learning and behaviour, a field he regarded as the most scientifically developed in psychology. The behavioural perspective allowed psychologists to study learning in quantifiable terms without having to explicitly address the biological processes involved. But Eysenck wanted to make the link between the psychological and the neurological. Excitation and inhibition would provide the key. These two concepts—inhibition especially—had attained a prominence in several related

---

[19] Hans J. Eysenck, 'Programme of Research, Psychological Department', c. late1945, Aubrey Lewis Papers, Bethlem Royal Hospital Archives and Museum, p. 1.

[20] Ibid. Eysenck claimed the longitudinal study would take at least five full-time man years to complete, double that of the other proposals. A scaled-down form of this proposal was taken up by the IoP's child development research unit in the early 1950s. The unit was disbanded in 1954, having mainly developed tests for measuring children's emotional stability.

[21] Eysenck, H.J. and Prell, D.B. (1951), The inheritance of neuroticism: an experimental study, *Journal of Mental Science* **97**, 441–65.

scientific discourses in the early years of the twentieth century.[22] 'Over the whole family of uses', wrote Sheffield psychologists Pilkington and McKellar in 1960, 'loomed the dark, bearded figure of Ivan P. Pavlov.'[23] By mid-century, however, Pavlov's reputation and influence in the West was ambiguous to say the least. While his conditioning work had been central to the development behaviouristic psychology, neuroscience specialists had come to think of him as a crude physiologist whose work was both inaccessible and difficult to reconcile with the structural concepts and neuronal processes they worked with.[24]

In Pavlov's scheme, excitation and inhibition were two opposing forces of the central nervous system that direct the strength, equilibrium, and mobility of an animal's responses to stimuli. Pavlov clearly viewed excitation and inhibition as cortical processes, likening them to the functional increase and decrease of neural conductivity. The structural neurological basis for excitation and inhibition remained uncertain, however, as Eysenck was to later acknowledge.

Together, excitation and inhibition channelled learning, including the kind of associative conditioning that had made Pavlov famous. In the classical Pavlovian paradigm, learning occurred when an unconditioned stimulus like food was repeatedly presented with a conditioned stimulus like a bell. The unconditioned biological response became the conditioned or learnt response when elicited solely by the conditioned stimulus, for example, when salivation was triggered by only the bell. The concept of excitation allowed for the acquisition of conditioned responses; inhibition provided a check to their escalation and proliferation. Various forms of inhibition focused, displaced, or suppressed responses. For example, external inhibition explained why distracting stimuli might lower performance. Internal inhibition explained why a conditioned response might begin to disappear in spite of repeated reinforced performance. In such cases, the conditioned response had been met by the rise of an internally generated inertia. The active cortical cells had become exhausted (or 'satiated' in later terminology).

---

[22] Turn-of-the-century neurophysiologists like Charles Sherrington and Hughlings Jackson had employed the concept of inhibition to account for reflexes or coordinated behaviour that demanded the suppression of incompatible or antagonistic responses. Inhibition was also used in relation to memory research, and featured prominently in Freud's writings on selective repression. Smith, R. (1992), *Inhibition: history and meaning in the sciences of mind and brain*, Free Association, London.

[23] See p. 194 of Pilkington, G.W. and McKellar, P. (1960), Inhibition as a concept in psychology, *British Journal of Psychology* 51, 194–201.

[24] See Smith, R. (1997), *The Fontana history of the human sciences*, Fontana, London, Chapter 17, and Smith (1992), *Inhibition: history and meaning in the sciences of mind and brain*.

Pavlov had speculated about which psychiatric conditions might be compared with the different 'neurotic' behaviour patterns he and his co-workers had induced in dogs. 'Experimental neurosis' could be achieved by presenting the dogs with discrimination tasks that were too difficult for them—for example, by conditioning them to salivate in response to a circle but not an ellipse and then presenting intermediate shapes. Pavlov interpreted the odd behaviour this produced as a conflict between excitation and inhibition.

Eysenck reviewed Pavlov's writings but was hamstrung—and he knew it—by the 'execrable translations' the Nobel laureate had received.[25] Multilingual as he was, Eysenck did not read Russian. He only had the 1927 English edition of Pavlov's *Conditioned reflexes* and a 1955 translation of Pavlov's final summary on psychology-related topics, *Selected works*.[26] As Jeffrey Gray pointed out a decade later in his important re-interpretation of Russian research, one of the most alarming problems in early English-language versions of Pavlov was the inherent ambiguity in Russian terminology for 'strength of the excitatory process'. This meant that in Pavlov's writings, descriptions of strength could be used in a typological sense when applied to the nervous system as a whole (taking in both excitation and inhibition), as well in a neuronal sense when applied to the intensity of a specific reactive response. Eysenck quoted from a translation of Pavlov that appeared to founder on this very point: 'In dogs with the more resistant nervous system [canine experimental neurosis] leads to a predominance of excitation, in dogs with the less resistant nervous system, to a predominance of inhibition.'[27] Gray's improved translation would later untangle this generalization, effectively swapping the 'weak' and 'strong' labels. In most situations (i.e. low to moderate stimulation), a 'weak nervous system' responded with strong excitation, while a 'strong nervous system' responded with weak excitation.[28] Only at high stimulus intensity, when protective inhibition cut in to suppress the weak nervous system's excitation level, was the

---

[25] Eysenck, H.J. (1957*b*), *The dynamics of anxiety and hysteria*, Routledge and Kegan Paul, London, p. 109.

[26] Pavlov, I.P. (1927), *Conditioned reflexes: an investigation of the physiological activity of the cerebral cortex* [translator G.V. Anrep], Oxford University Press, Oxford, and Pavlov, I.P. (1955), *Selected works* [translator S. Belsky], Foreign Languages Publishing House, Moscow.

[27] See p. 34 of Eysenck, H.J. (1955*a*), A dynamic theory of anxiety and hysteria, *Journal of Mental Science* **101**, 28-51.

[28] See Gray, J. (ed. and translator) (1964), *Pavlov's typology: recent theoretical and experimental developments from the laboratory of B.M. Teplov*. Pergamon Press, London, pp. 160–1 (including footnote) and p. 275 (footnote).

weak system–weak response, strong system–strong response formulation an accurate representation of Pavlov's research. Below this stimulus threshold, Gray wryly observed, the ambiguity between the two senses of 'strength of the excitatory process' would prove 'particularly embarrassing'.[29]

Poor translation went a long way in explaining why Pavlov appeared to say one thing on one page, the opposite the next. For example, Eysenck noted that Pavlov linked the 'sanguine' type of dog with a predominance of excitation, and the 'melancholic' type of dog with a predominance of inhibition. In the context of Pavlov's other observations, Eysenck complained, it would make much more sense the other way round.[30]

Eysenck struggled to find a way through the fog, trying to 'reconcile the contradictions in Pavlov's theory'.[31] In the end, he was forced to take the great man's pronouncements as a series of hints. With his own preliminary data to guide him, Eysenck took those aspects of Pavlov that suited him and disregarded the rest. And it was the comparisons Pavlov made between canine 'neuroses' and human psychiatric conditions that Eysenck found the most useful. Pavlov had linked high excitability and low inhibition with neurasthenia (which Eysenck took to be roughly equivalent to the introverted condition of dysthymia) and low excitability and high inhibition with hysteria (which Eysenck took to be an extraverted condition). This suggestive nugget of an idea provided Eysenck with almost all he needed to connect personality with functional neurology.

The postulate Eysenck eventually arrived at might have oversimplified Pavlov. However, one could argue that Eysenck's 'variations' on the master's work were testament to his intellectual independence, to the creative, feel-your-way process he engaged in. Despite the impression Eysenck gave of an orderly progression from theory to prediction and then to empirical testing, his postulate was incubated in an exciting back-and-forth exchange of experimentation, data analysis, and explanation. It was a process that sociologists of science have more recently deemed to be characteristic of much scientific work—*good* scientific work—when deductive logic meets the mangle of practice.[32]

Eysenck's other key influence in the 1950s was Hull. Like Eysenck, Hull used 'objective' performance tests and had a penchant for expressing his hypotheses

---

[29] Ibid., p. 275.

[30] Eysenck (1957b), *The dynamics of anxiety and hysteria*, p. 113.

[31] Ibid.

[32] The phrase comes from Pickering, A. (1995), *The mangle of practice: time, agency and science*, University of Chicago Press, Chicago.

in terms of symbols and equations. This kind of formalist precision no doubt appealed to Eysenck, who also copied Hull's habit of italicizing postulates and deductions.[33] Eysenck equated Hull's notion of reactive inhibition with Pavlov's idea of internal inhibition, but rejected the peripheral 'tissue injury or fatigue' basis Hull gave it. Eysenck's version of reactive inhibition was a central nervous system process. As such, it could be used to account for a variety of learning phenomena. For example, the build-up of reactive inhibition during a repetitive motor task explained the spontaneous improvement in the performance of that task after a rest—a phenomenon that became known as 'reminiscence'.

Eysenck's thinking mirrored that of his first book, *Dimensions of personality*. The introvert–extravert distinction again provided the fundamental contrast. Eysenck was already aware of research literature equating inhibition with various measures of cortical satiation.[34] If perceptual after-effects were assumed to reflect cortical satiation, then their extent and duration should vary according to the different inhibition levels characteristic of introverts and extraverts. Eysenck looked at kinaesthetic after-effects, wherein subjects were asked to estimate the width of a wooden rail whilst rubbing another, broader rail. The more the subject rubs the comparison 'stimulus' rail, the smaller the 'test' rail feels. The after-effect was the extent and duration of this perceived shrinkage. Eysenck assembled subject groups from either end of his introversion–extraversion dimension, in this case hysterics and dysthymics. The hysteric group showed longer and stronger after-effects, as predicted. The conditioned eye-blink paradigm offered an even more telling demonstration of personality differences in conditioning, including response acquisition effects. Subjects acquired a conditioned 'blink' response to an auditory tone when it was presented in conjunction with a puff of air. Using the same contrasted group method, Eysenck presented data showing introverted dysthymics gave more conditioned responses than normals in acquisition trials, and even more so than extraverted hysterics, and their conditioned responses trailed off at a higher level across extinction trials.

---

[33] See Hull, C.L. (1952), *A behavior system*, Yale University Press, New Haven, Connecticut. For Hull, behaviour was a product of conditioned habit strength (i.e. acquired links between stimulus receptors and organ/motor effectors) and drive strength (i.e. physical and psychological motivations), minus reactive and conditioned inhibition. This was expressed as: $_sE_r = {_sH_r} \cdot D$.

[34] See, for example, Kimble, G.A. (1949), An experimental test of a two factor theory of inhibition, *Journal of Experimental Psychology* **39**, 15–23.

Although Eysenck discussed these results in terms of excitation and inhibition, his initial 1955 formulation incorporated the concept of inhibition only.[35] Eysenck first assumed that introverts and extraverts differed according to the speed and strength of their development and dissipation of inhibition. Following Pavlov, Eysenck then linked inhibition with general conditionability—'learning' in a behaviourist sense. Extraverts developed inhibition rapidly and strongly, and dissipated it slowly—making it harder for them to learn and easier to 'unlearn'. Conversely, introverts developed inhibition slowly and weakly, and dissipated it quickly—making it easier for them to learn and harder to 'unlearn'.[36] But Eysenck must have realized his initial postulate was only half the story; it did not explicitly theorize differential response *acquisition*, it did not explain what *helped* as well as hindered learning. The elaboration this would entail became Eysenck's first biological model.

The biology of personality mark I: *The dynamics of anxiety and hysteria*

Eysenck's revised version of his typological postulate was the centre-piece of his 1957 book *The dynamics of anxiety and hysteria*. Dedicated to Hull, it was his most densely written, technical work to date. It was an attempt to move from the descriptive to the biologically causal. This required experimentation rather than mere observation, much of it done between 1953 and 1957.[37] Backed by his data, Eysenck's postulate expanded to include differences in *excitation* as well as inhibition. Introverts were characterized by quick and strong excitation, along with weak inhibition, and were prone to dysthymic disorders when breakdowns occurred. They formed conditioned responses quickly and strongly, and preserved these responses longer. Extraverts were characterized by slow and weak excitation and strong inhibition, and tended to exhibit hysterical–psychopathic disorders when breakdowns occurred. They formed conditioned responses slowly and weakly, and preserved these

---

[35] Eysenck (1955a), A dynamic theory of anxiety and hysteria, and Eysenck, H.J. (1955b), Cortical inhibition, figural after effect, and a theory of personality, *Journal of Abnormal and Social Psychology* 51, 94–106.

[36] Note that this was a formulation opposite to that of everyday usage that suggests introverts are socially 'inhibited', extraverts 'uninhibited'.

[37] An IoP annual report of the period noted: 'Much new apparatus has been built, and a number of new phenomena investigated with the aim of clarifying ... Professor Eysenck's "personality postulate" linking inhibition with hysteria and psychopathy, and excitation with anxiety and reactive depression.' Institute of Psychiatry (1956–57), Annual Report, 1956–1957, Maudsley Hospital, p. 43.

responses for shorter periods. Eysenck's postulate proposed a balance between excitation and inhibition, but in an either/or fashion. Less of one meant more of the other, depending on where the individual was located on the introversion–extraversion dimension. Input stimuli differences were barely considered at all, implying consistency across situations.

While Eysenck's first biological model appeared relatively simple, it combined a variety of theoretically loaded terms that criss-crossed psychological and neurophysiological discourses. Their measurement was functional and indirect; adducing evidence for their law-like combination was a distinctly synthetic affair. Eysenck welded together masses of experimental, behavioural, observational, and anecdotal material. Out went the 'almost infinite variety' of atheoretical performance tasks—such as the body sway and leg persistence tests—and in came the more focused measures of conditioning, learning, and physiological processes.[38]

The central concepts of excitation and inhibition were certainly richly inclusive. Any physiological variable said to affect either cortical excitation or inhibition, or their balance, could be related to introversion–extraversion. For example, with their higher inhibition, extraverts tended to show larger reminiscence effects and more inertness in repetitive tasks.[39] Since inhibition correlated with cortical satiation, introverts and extraverts should also systematically differ on dark adaptation, perceptual rotation tendencies, and duration of after-affects. Pharmacology was an even more fertile area. Stimulant drugs like coffee that were thought to boost excitation and lower inhibition (e.g. benzedrine) and depressant drugs like alcohol thought to lower excitation (e.g. sodium amytal) were trialled for their effects on motor learning tests with generally expected results.

Since it covered a wide range of topics, *The dynamics of anxiety and hysteria* skimmed over some complicated experimental procedures and gave greatly condensed analytical summaries. Much of the data was presented as graphical displays, backed up by curve fitting and statistical significance testing of differences. While easy to digest, these graphs carried a heavy burden of persuasion. The eye-blink conditioning data was the most crucial. Much of this work came from Cyril Franks's doctoral research, which he completed in 1954.[40] Franks then went on to demonstrate characteristically different

---

[38] The quote comes from Martin (2001), Hans Eysenck at the Maudsley—the early years, p. 7.

[39] See, for example, Eysenck, H.J. (1956e), Reminiscence, drive and personality theory, *Journal of Abnormal and Social Psychology* **53**, 328–33.

[40] Franks, C. (1954), An experimental study of conditioning as related to mental abnormality, Ph.D. thesis, University of London.

patterns of conditioning for contrasted normal groups as well as for hospitalized neurotics—a finding that could be accounted for in terms of introvert versus extravert conditionability.[41] To make this explanation convincing, the various groups had to be seen to be appropriately matched (i.e. with respect to tone sensitivity, spontaneous blink rates, and a common initial baseline). Prior to conditioning, they must differ only in terms of a valid measure of introversion–extraversion.

In the first chapter of *The dynamics of anxiety and hysteria*, Eysenck briefly restated his conviction that introversion–extraversion was the key to distinguishing neurotic disorders. Although Eysenck cited his original *Dimensions of personality* factorial study and the evidence of other researchers, he also prominently featured as 'typical' a study done in his own department. Peter Hildebrand had completed a Ph.D. at the IoP in 1953, testing some of the basics of the dimensional framework with normal and neurotic groups. Most of his results were in line with Eysenck's theory, with three dysthymic groups (i.e. anxiety states, reactive depressives, and obsessives) testing as introverted, while psychopaths tested as extraverted. There was one curious anomaly, however. Hysterics, while more extraverted than dysthymics, were found to lie on the *introverted* side of Hildebrand's normal group. This was a problematic result, for it cast doubt on perhaps *the* most basic differentiation underlying Eysenck's dimensional framework. For Eysenck, hysteria had defined the extraverted pole of neurosis, just as dysthymia had defined introverted pole. Eysenck was no doubt perturbed by this finding and argued Hildebrand's normal sample was 'usually extraverted'. To prove the point, Eysenck went out and gathered his own less extraverted normal sample. Armed with this data, he was able to display a diagram of Hildebrand's results in *The dynamics of anxiety and hysteria* with hysterics in their 'proper', somewhat extraverted place on the dimensional grid.[42]

Eysenck also saw an opportunity to include a discussion of some of the first studies coming from the IoP animal laboratory done by Peter Broadhurst. Broadhurst had tried and failed to induce 'experimental neuroses' in rats, following the classical conditioning paradigm Pavlov pioneered with his dogs. Stalemated, Eysenck suggested he emulate American researcher C.S. Hall who had developed 'emotional' and 'non-emotional' rat strains by selective breeding. The resulting studies partly supporting the Yerkes–Dodson law were

---

[41] Franks, C. (1956), Conditioning and personality: a study of normal and neurotic subjects, *Journal of Abnormal and Social Psychology* **52**, 143–50.

[42] See Eysenck's discussion of these points in Eysenck (1957b), *The dynamics of anxiety and hysteria*, Chapter 1.

included because they potentially shed light on the influence of drives (i.e. basic biopsychological needs) on learning. The more emotional strains learnt better at low and medium levels of motivation and task difficulty, but not when intensely motivated to do difficult tasks. Eysenck speculated that anxiety had a kind of multiplicative influence that could be causally related to his neuroticism dimension, because it loaded highly on this dimension in factor analysis. While Eysenck linked anxiety with the autonomic nervous system, he did not get very far with this idea. Thus neuroticism was not given a detailed biological grounding of introversion–extraversion.

Despite its broad and complex subject matter, *The dynamics of anxiety and hysteria* contained some curious padding. Eysenck often broke off to consider topics with little direct bearing on his postulate, such as reaction-time studies of schizophrenics. He (again) slew a psychiatric dragon: neurosis and psychoses were continuous dimensions rather than discrete categories, though psychosis was not an extension of neurosis, as he (again) claimed psychoanalysis implied. To back this up, Eysenck presented D.S. Trouton and A.E. Maxwell's factor analysis of psychiatric admission data at the IoP that suggested two distinct neurotic and psychotic dimensions, and he repeated a summary of his 1955 study that statistically discriminated neurotics from psychotics. Eysenck's conclusions included a lengthy discussion of the potential applications of learning principles to psychotherapeutic practice, seemingly out of kilter with the laboratory focus of the previous chapters. Apart from demonstrating a tendency to tilt at windmills, the effect was to strengthen the impression of wide-ranging, programmatic research. The book contained 590 references, including over 100 to studies by Eysenck and his co-workers. He credited more than a dozen research staff and former students by name, including those in the new animal laboratory and other special units at the IoP, and collaborating psychiatrists. Eysenck also gratefully acknowledged the 60 additional student researchers, those 'mute and inglorious Miltons' whose data helped firm up the manuscript.[43] They would not all remain mute for long.

Student power

Eysenck attracted a lot of research students, particularly in the early years of the IoP. They were the engine room of his programme. All were postgraduates; there was no undergraduate instruction of any kind at the IoP. Other universities performed that particular time- and labour-consuming service. The IoP's students were certainly a diverse bunch. They came from all over the world,

---

[43] Ibid., p. 5.

though mostly from British Commonwealth countries (Australia, Canada, New Zealand, South Africa, and India), a few from the US. The IoP proudly advertised the international background of its trainees across all the fields it served. More importantly, overseas students paid full fees and were thus an important source of income.

Many of these first and second generation students would go on to illustrious careers, testimony to Eysenck's capacity to nurture talent. Although not all psychology students worked on Eysenck-related topics, most did. The clinical section had several senior appointees to teach the clinical course, along with a number of visiting psychiatrists. However, research degrees in psychology were essentially made up of thesis-based research. Outside the clinical area, Eysenck dominated as the most senior staff member. So unless a student chose a topic more appropriately supervised by one of the clinical course staff—Monte Shapiro in the first instance and later other clinical teachers like Gwynne Jones, Reg Beech, and Bob Payne—Eysenck was likely to be their supervisor.

Eysenck did not require his students to work on any particular aspect of his programme but it made practical sense to do so. Most students applied to IoP because they were already interested in personality topics and wanted to do that kind of work. Many also worked as research assistants during and after their candidature, mainly supported by hospital research coffers. A few graduated up the IoP ranks to more senior teaching and research positions. For example, Cyril Franks became an IoP lecturer for short period after completing his doctorate, while Irene Martin and Sybil Eysenck spent their entire careers there, the latter in a research-only position.

A few other research staffers came in at a more senior level, usually bringing in funding from outside sources. Roger Russell was one of the first of these, beginning a short stint in 1948 with support from the Fulbright scheme. As the department expanded through the 1950s, some of the more independent research staff Eysenck had appointed—fellow German J.C. 'Hans' Brengelmann working on perception and personality and Peter Broadhurst working in the animal lab—took on their own students. But even then Eysenck was a strong secondary influence over these students' research—a *de facto* co-supervisor.

At first the University of London would not allow Eysenck to officially supervise graduate students; they had to be registered through University College. However, this changed with the founding of the Institute so that by 1950 there were around 20 higher degree research students in Eysenck's subdepartment. Most had embarked on Ph.D.s, a handful on masters degrees. Two years later, this figure swelled to over 40 and would hold steady at around this level for nearly two decades. From mid-1951 to mid-1952, 34 students enrolled for higher research degrees—a big injection of personnel that would help ensure

Eysenck's research programme got into full swing. While intakes for subsequent years were not nearly as large, the clamour to get to Denmark Hill was testament to the prestige of the IoP brand and Eysenck's pulling power as the hot new thing in British psychology.

The 1950s and 1960s were the halcyon years of Eysenck's attempt to map the neurobiological terrain of personality. At any one time he had at least 25 Ph.D. students on the books. Around a half dozen, on average, completed their doctorate each year. Far from being a burden, these students were the key to Eysenck's productivity. Eysenck had perfected a relatively non-demanding (for him) and non-intrusive (for them) style of supervision. He would guide the students in their choice of topic, letting them dictate the level of consultation. He otherwise remained in the background, save for timely prodding to push them along. It all worked on implicit, barely articulated expectations generated by Eysenck's growing research profile and 'out-there' public reputation. Most students came knowing what they would encounter and were trusted to be clear-eyed about their research. The few students working on non-programmatic topics tended to get less attention and resources.[44] Thus Eysenck's student 'load' represented an enormous self-motivated resource; he had the luxury of up to 40 minds and pairs of hands at his disposal during these productive years.[45]

While it was not clear what dictated such high research student numbers—be it budget, intellectual policy, or IoP politics—it would seem that Eysenck's graduate student allotment far exceeded that of most other university psychologists in Britain. A more typical workload would entail the supervision of a handful of Ph.D. and masters students, combined with a share of undergraduate and postgraduate teaching and administrative duties. Eysenck's teaching responsibilities were not huge. In any case, he wore these responsibilities lightly and tended to publish much of his instructional material. For example, lectures to psychiatry trainees were turned into popular paperback books in the 1950s. Overall, this kind of student-led research programme was comparatively economical with resources and apparatus. The structure of the

---

[44] Arthur Jensen, interview, 1 July 2002.

[45] As the 1952–53 IoP Annual Report made clear, 'the work of research students is integrated with that of members of the department …and the general theoretical framework with which most work is being pursued is elaborated by Dr. H.J. Eysenck in his latest book, *The structure of human personality*' (p. 35 of Institute of Psychiatry, Annual Report, 1952–1953, Maudsley Hospital). *The structure of human personality* was first published in 1953 and went through many editions. Eysenck updated it several times, re-stating his dimensional framework in light of accumulating research.

department largely accounted for the way Eysenck's research programme worked and the enormous number of publications he was able to churn out.

Honey from the beehive

The IoP psychology department was very bottom-heavy in terms of staffing, the research section in particular. Eysenck sat alone atop a comparatively broad-based pyramid, first as reader then as professor. Next in line was the head of the clinical programme Monte Shapiro at senior lecturer level, followed by a handful of lecturers and researchers, and then a host of students and research assistants. Although the department grew through the 1950s and 1960s, this structure remained essentially unchanged in this period.[46] It was not until clinician Jack Rachman was given a professorship in the 1970s—ahead of Shapiro—that there was someone at Eysenck's level. By then, though, the clinical and research sections had gone their separate ways.

From the testimony of those at the IoP at the time, there was a tremendous sense of group solidarity in the department, especially amongst members of the research section. Here was an army of young scientists on a mission. Only in the clinical section, with its higher numbers of senior teaching appointments, was there a pronounced pluralism of aims and methods. This divergence had a personal as well as intellectual edge, the most notable example being Eysenck's chronically cool relationship with Shapiro. Despite this, Eysenck still managed to convince many clinical staffers to focus on an agenda he had mapped out, aided by the fact that, in the early days at least, this mixed band of students, researchers, and lecturers walked, talked, and ate in close proximity. It was all made possible by the unique aspects of the IoP as a postgraduate research and clinical teaching department. There was no obligation to cover the full purview of psychology, but no brake on pursuing a particular line of research beyond funding and logistical constraints. No wonder he turned down offers from other universities.

---

[46] In 1949–50 there was one reader (Eysenck), one senior lecturer (Shapiro), five lecturers, four assistant lecturers, three research or statistical assistants, and about 20 students. In 1955–56, there was one professor, one senior lecturer, 12 lecturers or assistant lecturers, 13 research assistants, and at least twenty more students. By the mid-1960s, Shapiro had been promoted to reader, and there were now two senior lecturers (G.W. Granger and A.E. Maxwell), a dozen lecturers or assistant lecturers, around 20 listed as research workers and guest workers, plus a similar number of additional students. This data has been put together from IoP annual reports. I also owe some of the insights contained in this section to interviews with ex-IoP staffers, especially Chris Frith, interview, 26 October 2001.

Nevertheless, this kind of research operation also had its disadvantages. The 'shallow-broad' research coverage Eysenck alluded to in *Dimensions of personality* went hand-in-hand with his huge student allotment. Eysenck had quickly realized that the opposite tack would not work. Deep-narrow coverage would require a swag of specialist researchers with long-term appointments. Certainly, students were a cheap and abundant form of intellectual labour. However, a deep-narrow approach would mean they would have to be recruited and trained to work in particular areas and encouraged to stay in order to maintain acquired knowledge and skills. It would be far easier to take them as they came, allowing them adapt their interests and experience to Eysenckian topics. A few stayed on for relatively lengthy periods, and their work tended to have a more focused quality. Peter Broadhurst could be put in this category, as could Irene Martin with her physiological and conditioning research. But much depended on this 'here today, gone tomorrow' cadre of graduate students and junior assistants, especially in the early years.

Psychological research at the IoP was thus quite diverse. Projects ranged over a broad set of topics—from rat running experimentation, to indices of brain damage, to learning amongst the mentally handicapped. Even so, student theses all tended to be very measurement-oriented, all highly empirical. Objective personality tests featured prominently, including Guilford's scales, forerunners to Eysenck's famous Maudsley personality inventory (MPI) that he developed in 1959. However, a surprising number of junior researchers and students utilized projective tests in the early 1950s, or did analytically inspired empirical investigations.[47] This diversity was a function of student diversity, heightened by Eysenck's own topic-hoping proclivities. He spread himself thinly for the most part because that was the most advantageous way of doing things, the way he could make the biggest impact. But it tended to produce a brief flurry of work in a particular area that was then left to go cold. It made for research that was more piecemeal than Eysenck implied it was, it made for a programme that was as much a function of happenstance as planning.

With Eysenck unrivalled at the top of the tree in the department's research section, intellectual exchange between equals was not a typical part of his working life. While he collaborated with several eminent scientists at other institutions, Eysenck lacked the positive, day-to-day critical spark this might have given him. Eysenck did not take criticism as constructive or worthy unless it came with the detailed knowledge and acceptance of the broad assumptions

---

[47] For example, A.W. Meadows, M. Malloy, Peter Keehn, Maryse Israel, and Joseph Sandler all did studies using the Rorschach test, and Peter Hildebrand's thesis explicitly tested Jungian ideas that Eysenck had used.

of his work. This was no better illustrated than during the informal discussions he held at his house on a Tuesday or Thursday night—Eysenck's version of Pavlov's 'Wednesdays'. Former colleagues such as Irene Martin spoke with reverence about the 'at-home' sessions.[48] Far from being free-for-alls, they had a relatively set format. A dozen or more from the Institute would gather after dinner at Eysenck's house in Herne Hill. Some would come with interesting data or provocative ideas to discuss. Typically only four or five would do much talking, with Eysenck presiding at the end of the room. These interactions were always directed to Eysenck, never across him. Eysenck enjoyed the discussion, but clearly didn't expect to have his mind changed. He would assess the figures and the arguments put to him, either dismissing strange results and criticism as irrelevant or showing how they could be incorporated within his thinking. These sessions had an important disciplinary function: they demonstrated the power and reach of the Eysenck programme, the benefits of staying within it, and the dangers of deviating from it. They also illustrated Eysenck's dominant status. Not only was this shyly charismatic figure the administrative leader, he was *the* singular intellectual influence, commanding enormous respect and deference. It tended to stifle new approaches and alternative thinking, a deficiency in Eysenck's work that would become more apparent as time went on. But, then again, having no one to really challenge him only gave him more time to do the things he really wanted to do.

A recent biography of Pavlov compared the Russian's institutional set-up to that of a factory.[49] There were elements of the organization of IoP psychology that were classically industrial. Certainly, Eysenck was the boss, suggesting projects and overseeing their implementation. However, he also put more than just a finishing touch on the final product. It made him more than a 'foreman' or 'owner'. So much of what came out of the IoP psychology department had Eysenck's unmistakable impression—his authorship.

Eysenck was a quietly compulsive writing machine. He once compared the writing process to having a baby—as though there was something inside that had to come out.[50] But his stunning profligacy outstripped the most enthusiastic human breeding; it put him, metaphorically speaking, somewhat lower in the great chain of being. Eysenck's productivity was akin to that of a *queen bee*. The IoP psychology research section resembled a *beehive* more than a factory, with a coordinated set of skilled drones (i.e. students and junior researchers)

---

[48] Irene Martin, interview, 27 October, 2001.

[49] See Todes, D.P. (2002), *Pavlov's physiology factory: experiment, interpretation, laboratory enterprise*, Johns Hopkins, Baltimore.

[50] Sybil Eysenck, interview, 25 July 2003.

serving their own interests and that of the collective's by providing for a single, central reproductive figure. Fed a royal jelly diet of empirical data, Eysenck was able to churn out ideas and theories, conference presentations and publications. He produced up to fifty journal articles and book chapters a year during his mid-career glory years of the 1950s and 1960s, as well as at least one book. He was also the sole author on a large percentage of his publications. More often than not he was the principal writer—usually indicated as the first named author—on joint publications. Only around a third of his 80 books and 1100 articles were jointly authored or edited, and most of these publications came in the latter half of his career. Moreover, wife Sybil was his most frequent co-author and collaborator, keeping it within the family.

Given he was head of a large research-oriented department, Eysenck's work practices were unusual. Executive scientists generally tend to avoid the mundane hack work of writing. Being short of time and having the power to delegate, they often restrict their authorial input to comments on drafts. In contrast, Eysenck took it all on: drafting, revising, and submitting in one quick bite. By the early 1970s, it appeared Eysenck had become the most published psychologist ever.[51] He relished writing, took pride in it, and made it a priority. In contrast, Eysenck delegated administrative duties wherever possible and avoided time-consuming committee work. He would habitually collect the minutes later, or send others in his place. His secretary's filing cabinets were reportedly filled with file-copy letters detailing why he could not come to this or that meeting. Only the prospect of a key decision affecting the things he cared about—his research programme, his students, and his department's functional independence—tended to attract Eysenck's full attention.

Stories of Eysenck's effortless efficiency abound. He could formulate articles, even whole books, in his head. This enabled him to dictate very effectively

---

[51] While Eysenck's publication rate may seem a little less remarkable in this collaborative, computer-driven era, it was exceptional given the constraints of the day. Comparisons with Eysenck's peers are thus illuminating. Of his British contemporaries, no one else came close to matching his output. Eysenck's bibliography exceeded that of Oxbridge experimentalists like Bartlett, Broadbent, and Zangwill by as much as a factor of ten, and he outstripped his mentor Burt quite early in his career. Only a handful of Americans remotely rivalled his publication rate. One of the most notable was Raymond Cattell, but even then Eysenck easily surpassed his output, as well as trebling that of another very prolific researcher, E.L. Thorndike. Eysenck regarded this as a good measure of his achievement and eminence. In his autobiography, he included graphs of his own publication and citation rates, quoting impact indices for his journals and displaying tables demonstrating the research output of the IoP. For more on these points see Rushton, J.P. (2001), A scientometic appreciation of H.J. Eysenck's contribution to psychology, *Personality and Individual Differences* 3, 17–39.

either to a secretary or on to a tape, although he also typed some scripts himself. He made very few mistakes and could apparently recite references from memory, minimizing back and forth consultation and additions. Eysenck could polish off a paper in a morning before heading out for an energetic game of tennis or squash at lunchtime, leaving the afternoons free for consultation with colleagues, advising students, reading, and other matters. He almost never worked late. And according to wife Sybil, Eysenck rarely brought work home with him, even when caught up in the most exciting of work phases or the hottest of controversies.

Dictating also allowed Eysenck to be more flexible with his time; he could be at work whenever and wherever he had his dictaphone handy. For instance, he composed his 1964 book *Crime and personality* on a short vacation while walking along the beach; he would often dictate papers, letters, and reviews in the garden outside his office when the weather was fine. He relied heavily on his formidable memory, keeping reference cards for books and articles but taking few notes. According to those who saw him up close, Eysenck also had the uncanny ability to produce lengthy quotes in a made-to-measure fashion. For example, Arthur Jensen recalled an incident that occurred while in Australia in the 1970s. Eysenck was approached by a journalist for a short feature; he was able to dictate it on the spot, with no backtracking, within a few words of the required length. Eysenck was practical too. He would generate a new paper simply by flipping the 'on' switch on the tape recorder at one of his lectures; reprise and update old manuscripts by literally cutting and pasting well before word processors made this more common.[52]

In routinely churning out all these books and papers, Eysenck did not get too fussed about the mistakes and inconsistencies that got through—a fault critics often seized on. His attitude was always to get it out, and then move on. Eysenck built his professional image around quantitative measures of scientific achievement; it both encouraged and justified his prolific output. He did not have the time or the inclination to look back. Contemplation and revision, polishing and checking, simply got in the way of writing the next piece. Any serious rethinks were included in future publications. It was a strategy based on speed and volume, of keeping one step ahead of critics while staying on message over and over and over again.

It made practical sense for Eysenck to write up the results of collaborative projects, and it also carried the advantage of ultimate control. Student theses

---

[52] Gisli Gudjonsson, interview, 25 October 2001; Jensen, A.R. (1997*a*), Eysenck as teacher and mentor, in *The scientific study of human nature: tribute to Hans J. Eysenck at eighty* (ed. H. Nyborg), pp. 543–59, Elsevier, Oxford; Arthur Jensen, interview, 1 July 2002.

and special projects were all grist for the mill, artfully incorporated within the grander theorizing of his books. Moreover, such in-house research was hard for outsiders to argue with. For example, Eysenck explicitly drew on at least 17 IoP unpublished doctoral dissertations in *The dynamics of anxiety and hysteria*. Nothing was left to waste. If a student was having trouble writing up their material or not interested in doing so, Eysenck would do it for him or her. It represented a marshalling of resources that was distinctly Eysenck-centric and would therefore require careful management.

While student research at the IoP was relatively diverse, Eysenck pulled it all together. He, almost alone, wove the relevance and meaning of these diverse projects into the larger whole of his first biological model of personality. Perhaps not unexpectedly, this opportune usage began to cause friction.[53] But when proprietorial disputes with his IoP minions did emerge, it had more to do with Eysenck's take on their results—misquoting those 'inglorious Miltons' if you like—rather than his decision to publish or speak for them *per se*. This is no better illustrated than by the notorious set of broadsides aimed at *The dynamics of anxiety and hysteria*.

Critical hit-and-run

Eysenck had initiated a new hybrid field integrating correlational statistics and experimental biology. As with most critical attention, it is those with a stake in the ideas and the necessary expertise that make a public stand. Early critical engagement with *The dynamics of anxiety and hysteria*, including some very sharp attacks, came from a very select group very close to home. A series of critiques from Vernon Hamilton, Lowell Storms, John Sigal, Kolman Star, and (surprisingly) Cyril Franks were arguably the most important of Eysenck's career for several reasons.

By the late 1950s, Eysenck had already been engaged in several high-profile skirmishes, most recently the heated dispute he had with American researchers over the computation and interpretation of data in *The psychology of politics*. However, the leading critics of *The dynamics of anxiety and hysteria* were as knowledgeable as they were negative. They were *insiders*; all were IoP researchers or students in the 1950s. Sigal and Storms were North Americans who had come to the IoP as graduate researchers—Sigal on a Canadian Research Council grant and Storms on a Fulbright—and had apparently become disenchanted

---

[53] Interviews with former colleagues suggested Eysenck was a surprisingly shy man who did not indulge in small talk or gossip. Even if he was a little distant—occasionally upsetting his secretaries—he was still more courteous and considerate than many holding such powerful positions. See Jensen (1997*a*), Eysenck as teacher and mentor.

with what they saw. Conversely, Franks, Hamilton, and Star had come through the system at the IoP as Ph.D. students, graduating in 1954, 1956, and 1957, respectively. All five had moved on to other jobs by the time their critiques were published.

While these renegades each had different axes to grind, it was possible to read a degree of convergence in their various assessments, and this affected the lessons onlookers subsequently drew. The most lengthy and detailed critique, and perhaps the most important, was jointly authored by Storms and Sigal.[54] They argued that Eysenck's presentation of the evidence was biased and his interpretations misleading. Both had been working in the bio-personality programme—and thus appeared to be in a good position to say just how selective Eysenck might have been. They presented a long list of Eysenck's sins, including the omission of unfavourable results, the non-reporting of statistically non-significant tests, and the misleading nature of some diagrams. For example, in his representation of Trouton and Maxwell factorial study, Eysenck left out variables loading on both the neuroticism and psychoticism factors, but retained items 'neatly and snugly' clustering around these axes. The result was a factor diagram more supportive of the notion of two independent factors, whereas Storms and Sigal suggested that 'oblique factors appear as reasonable as orthogonal ones'.[55] Storms and Sigal also disputed Eysenck's interpretation of the experimental evidence, as did Hamilton in a separate article, suggesting it was not as straightforwardly supportive as Eysenck seemed to think. Were the groups truly matched in the eye-blink conditioning experiments they wondered? And what of the predicted differing *rates* of conditioning and extinction, since the curves appeared to be parallel across groups? They even pointed out that a reproduction of one of Franks' eye-blink graphs contained an error in the theoretically expected direction. Where Eysenck saw clarity, Storms and Sigal saw ambiguity and doubt—an array of disparate results and findings that could be fitted to a number of different models of personality differences and brain function. It was all geared to the accusation that Eysenck did not use a balanced set of scales when weighing up the evidence.

A key point of issue was Eysenck's appropriation of other people's data especially, it seemed, that of IoP researchers like Franks, Peter Hildebrand, Peter Venables, and D.S. Trouton. Hildebrand's work was a particular bone of contention, partly because it drew attention to problems in Eysenck's

---

[54] Storms, L.H. and Sigal, J.J. (1958), Eysenck's personality theory with special reference to *The dynamics of anxiety and hysteria*, British Journal of Medical Psychology **31**, 228–46.

[55] Ibid., p. 231.

introversion–extraversion distinction of neuroses. Cyril Franks surprisingly joined with Sigal and Kolman Star to question whether dysthymics and hysterics could be used as criterion groups for introversion and extraversion.[56] They concluded that these groups could not be used to test Eysenck's typological postulate and/or that the introversion–extraversion scale of the new MPI used to select these groups was seriously flawed.

Eysenck's counter to this volley of well-informed analyses hardly missed a beat, though they must have tested his considerable skills. While dismissive of Hamilton, he took the critiques of the others more seriously.[57] In his lengthy riposte to Storms and Sigal, Eysenck drew on his reading of the history of science.[58] Eysenck suggested his critics were unreasonable, even foolhardy in implying that definitive data could ever be obtained for any particular set of conjectures. The facts were never in, he said, and that made partiality inevitable. Psychology was a young science following the divergent but important traditions initiated by Galton and Pavlov. The immature field of personality research had to be treated leniently—a caveat he continued to emphasize in later work. Eysenck argued that one must look at available options and follow the most promising avenues that the evidence suggested or no progress would be made. It was a form of special pleading with a sting in the tail. Even if one accepted the destructively naïve attitude of his critics, Eysenck said, specific criticisms could be countered in each case. He did not deal with all them, but created the impression he *could* have.[59] Likewise, he argued that the evidence in *The dynamics of anxiety and hysteria* might have been selective—for a book can only be so long, he said—but only in a neutral fashion. For instance, he claimed his reproduction of the Franks' eye-blink diagram had small 'artist's

---

[56] Sigal, J.J., Star, K.H., and Franks, C.M. (1958a), Hysterics and dysthymics as criterion groups in the study of introversion–extraversion, *Journal of Abnormal and Social Psychology* **57**, 143–8.

[57] See Hamilton, V. (1959a), Eysenck's theories of anxiety and hysteria—a methodological critique, *British Journal of Psychology* **50**, 48–63; Eysenck, H.J. (1959b), Anxiety and hysteria—a reply to Vernon Hamilton, *British Journal of Psychology* **50**, 64–9; and, Hamilton, V. (1959b), Theories of anxiety and hysteria—a rejoinder to Hans Eysenck, *British Journal of Psychology* **50**, 276–9.

[58] Eysenck, H.J. (1959a), Scientific methodology and *The dynamics of anxiety and hysteria*, *British Journal of Medicine and Psychology* **32**, 56–63. See also Storms and Sigal's final word, Storms, L.H. and Sigal, J.J. (1959), Misconceptions in 'Scientific methodology and *The Dynamics of anxiety and hysteria*', *British Journal of Medical Psychology* **32**, 64–7.

[59] This point was echoed in an interesting discourse analysis of Eysenck's argumentative tactics, along with those of several other psychologists, by John Soyland. See Soyland, A.J. (1994), *Psychology as metaphor*, Sage, London.

errors' both ways. (The error *unfavourable* to Eysenck's typological postulate was hard to spot, though.) He invited interested readers to compare his reproduction with Franks's original published graph, as he did with his version of the Trouton and Maxwell factorial diagram. If anything, it was the critics who were biased. They deliberately highlighted the most negative omissions. Thus Eysenck made his critics look doubly in error: misguided in approach and wrong in detail. While Storms and Sigal remained unbowed in their reply, Eysenck had conceded little ground.

The problems raised by the Sigal, Star, and Franks study of the MPI were a little harder to wave away. They had zeroed in on the statistical uncertainty surrounding Eysenck's personality dimensions. Like any such psychometric summary, these dimensions were a movable feast, an indeterminate, data-dependent solution. In lieu of the biological basis Eysenck was pursuing, these dimensions could only be tethered to normative markers. An individual was extraverted or introverted only in comparison to the distribution of scores of a relevant group. Eysenck's dimensions might not readily transfer across social contexts and investigators. Moreover, Eysenck's dimensions still drew a degree of conceptual coherence from their relation to psychiatric taxonomies they were originally derived from. Yet these psychiatric taxonomies (and judgements underlying them) were themselves inherently unstable. Dysthymia had never totally caught on as a diagnostic category and its symptomatic domain had been divided up into a variety of anxiety-related disorders. Conversely, hysteria, especially its conversion-symptom manifestations, was now beginning to disappear. Perhaps this was the result of changing diagnostic schemes and patterns of care. Or perhaps it was because patients had learnt to present differently—part of the 'looping effect of social kinds', to coin Ian Hacking's phrase.[60] But as long as medical yardsticks held some sway, changes in psychiatric diagnosis threatened the substantive touchstones of Eysenck's dimensions. Eysenck's solution was to look back—rubbishing contemporary psychiatric thinking as pre-scientific while quoting select historic figures for support.

Sigal, Star, and Franks reiterated the question Hildebrand had posed about the dysthymic–hysteric distinction. Like Hildebrand, their results with Eysenck's brand new MPI suggested that hysteria could not be characterized as neurotic extraversion. Hysterics, it seemed, were much nearer the midpoint in

---

[60] See Hacking, I. (1995), *Rewriting the soul: multiple personality and the sciences of memory*, Princeton University Press, Princeton. A good place to start for the disappearance of hysteria would be Micale, M.S. (1995), *Approaching hysteria: disease and its interpretations*, Princeton University Press, Princeton.

terms of introversion–extraversion. If anything, this pole was better exemplified by diagnosed psychopaths. Taken at face value, these findings implied that the interpretation of Eysenck's dimensions had to be changed, especially in the extraverted neurosis quadrant, or the MPI needed to be revised. While introversion–extraversion appeared to be a much less useful way of distinguishing neurotic disorders, the experimental data that contrasted dysthymic and hysteric groups in *The dynamics of anxiety and hysteria* was also rendered problematic.

Sigal, Star, and Franks had touched a sensitive spot. Why else would Eysenck repeatedly distance himself from the dysthymic–hysteric distinction underlying the derivation of his first two dimensions?[61] It was all Jung's idea, not his. While Eysenck claimed that psychopathy and hysteria were traditionally thought of as the same 'psychiatric complex', he made another debatable move. He argued that the dysthymic–hysteric distinction related to introversion–extraversion differences in psychiatric samples only, but had no implications for normals. Neuroticism and introversion–extraversion might also be correlated amongst psychiatric patients, rather than independent. The continuity assumption underlying his original work was put aside for the time being and Eysenck was forced to admit that 'Sigal et al. are correct in stating that *hysterics are not significantly more extraverted than normals ...*' [original italics][62] Sigal, Star, and Franks's final rejoinder suggested that Eysenck was misrepresenting Jung and his own work.[63] Both Jung and Eysenck portrayed hysterics as extraverted *per se*—which, they said, meant extraverted in comparison to normals. Hysterics were not merely 'less introverted introverts'. Sigal *et al.* argued that the full range of the continuum must be covered to provide complete measurement and valid explanations at the extraverted end of the spectrum. Eysenck had taken the 'least parsimonious' option, they noted sadly, by accepting the validity of his MPI introversion scale.[64] Obviously a rethink would be needed in order to save the idea of orthogonal dimensions, and he would soon revise the MPI.

In 1959, a symposium on inhibition was organized at the British Psychological Society's Cambridge conference, attesting to the interest generated by

---

[61] Eysenck, H.J. (1958c), Hysterics and dysthymics as criterion groups in the study of introversion–extraversion: a reply, *Journal of Abnormal and Social Psychology* **57**, 250–2.

[62] Ibid., p. 251.

[63] Sigal, J.J., Star, K.H., and Franks, C.M. (1958b), Hysterics and dysthymics as criterion groups in the study of introversion–extraversion: a rejoinder to Eysenck's reply, *Journal of Abnormal and Social Psychology* **57** 381–2.

[64] Ibid., p. 382

Eysenck's work. But others could not help but be highly critical of the theoretical underpinning of Eysenck's work and data he used to support it. Aberdeen's R.L. Reid conceded that the balancing concepts of excitation and inhibition had potential.[65] While doffing his cap to Eysenck's integrative ambition, Reid nonetheless suggested that a 'whole new theory' was called for. Eysenck's model was not sensitive enough to explain 'everyday life'. It implied introverts would learn faster and perform better at just about everything. Not only would they excel academically, they would talk earlier, play the violin better, achieve higher levels of socialization, and so on—hardly in accord with most people's experience, Reid opined.[66] Young IoP researcher Donald Kendrick presented new data adjusting the problematic relationship between reactive and conditioned inhibition that Reid had pointed to, while clinician Gwynne Jones summarized a range of the latest conditioning results achieved at Denmark Hill. But even Jones conceded that 'Eysenck's theory ... gains little support from recent work on conditioning. Relationships with neuroticism are observed at least as frequently as those with extraversion and the latter are seldom strong.'[67]

Rounding off the mixed reception to *The dynamics of anxiety and hysteria* was David Lykken from the University of Minnesota. Lykken had likewise spent time at the IoP as a visiting scholar; Eysenck had even utilized some of his research for support in his book. While admiring Eysenck's capacity to 'winnow kernels of supporting evidence from the chaff of complex experimental findings', Lykken ended with a devastating conclusion:

> Not only has Eysenck failed to make a reasonable case for his hypotheses but, hidden under a veneer of scientism and confident assertion, he has woven in this book a tangled skein of sophistry and tendentious scholarship calculated to mislead the gullible and to denigrate the whole factorial-experimental approach...[68]

Lykken's one reference was to the Storms and Sigal salvo. It seemed Eysenck's in-house critics had already set the tone, and would for years to come.

---

[65] See p. 230 of Reid, R.L. (1960), Inhibition—Pavlov, Hull, Eysenck, *British Journal of Psychology* **51**, 226–32.

[66] Ibid., p. 231.

[67] See p. 223 of Jones, G. (1960), Individual differences in inhibitory potential, *British Journal of Psychology* **51**, 220–5. See also Kendrick, D.C. (1960), Effects of drive and effort on inhibition with reinforcement, *British Journal of Psychology* **51**, 211–19.

[68] See p. 379 of Lykken, D.T. (1959), Turbulent complication, Review of *The dynamics of anxiety and hysteria* by Hans Eysenck, *Contemporary Psychology* **4**, 377–9.

The suggestion that Eysenck misrepresented his own and others' data would follow him for the rest of his career, a scientific sin that implied his conclusions should be taken with a bucketful of salt. Few, save Michael Rutter in recent times, have gone on to say so publicly.[69] However, only if one takes a very literal position of the sanctity of data do Eysenck's errors or fudges really matter, especially when viewed in isolation. It was never suggested that Eysenck's inaccurate reproduction of the Franks eye-blink graph, for example, significantly altered the interpretation of what were already published data. The importance of these incidents—and why some critics liked to bring them up when talking about Eysenck—lay elsewhere. They used them as evidential pointers in their unfavourable *moral* readings of Eysenck's character. These moral readings are a routine part of the informal, 'unofficial' yardsticks scientists use when they discuss each other's work.[70] Moreover, these readings tend to be *cumulative*. Thus the contentious interventions and dubious associations of Eysenck's subsequent career would intensify doubts about some of his earlier research. Yet no one clearly articulated what separated Eysenck from his supposedly more virtuous peers, no one could decisively arbitrate on whether Eysenck's 'pathological partisanship' was a matter of degree or kind. And no one could or would substantiate the wilder forms of these accusations when pressed by myself or by Eysenck's supporters.[71]

For those knowing little of Eysenck's research, these detailed critiques of *The dynamics of anxiety and hysteria* must have seemed authoritative. They had come from those in the know, from those who accepted the central assumptions of the programme. For a gathering crowd of Eysenck antagonists, it just fed the fire. For example, the Oxbridge set had their worst suspicions confirmed: not only was Eysenck an overly ambitious factorialist, he was a dodgy experimentalist undone by in-house insurrection. As Cambridge experimentalist John Mollon attested: 'My own information is third-hand, but I was

---

[69] Rutter recalled that Eysenck fired his imagination but was also responsible for his 'distrust of academic evangelists and my loathing of those who distort the evidence to support their own particular viewpoint'. P. 423 of Rutter, M. (2001), The emergence of developmental psychopathology, in *Psychology in Britain: historical essays and personal reflections* (ed. G.C. Bunn, A.D. Lovie, and G.D. Richards), pp. 422–32, BPS Books, Leicester; Michael Rutter, interview, 26 April 2002.

[70] For more on this point see the concluding chapter.

[71] This was a point that frustrated Jeffrey Gray, for example, when he heard his peers mutter about Eysenck's supposed lack of scientific honesty. But Gray said these critics could never back these accusations up, even when pressed to do so. Jeffrey Gray, interview, 25 July 2003.

brought up in a world where Eysenck was regarded with some scientific suspicion.'[72]

Conversely, Cyril Franks did not think of these critiques as a rebellion, just a close critique of 'sloppy, selective and not well-supported work'.[73] Franks maintained that Eysenck did not take this critique to heart. He remained on good terms with Hans and Sybil, and went on to do notable clinical work in the US. Star went to Newcastle University in Australia, and later returned to research work at the IoP. For the visitors Storms and Sigal and for Vernon Hamilton, the episode had a more moral edge. Sigal and Hamilton went to work in the British psychiatric system, and Storms went to UCLA. And despite shrugging it off publicly, some things obviously continued to rankle with Eysenck too. For one thing, the statistical foundations of his dimensions had been badly shaken. Over the next two decades, Eysenck faced the difficult task of re-jigging the content components of his dimensions—especially in relation to the psychoticism scale—and re-birthing the MPI.

The man whose results were the original *cause célèbre*, Peter Hildebrand, joined the 'other side' at the Tavistock Clinic soon after completing his 1953 doctorate. Hildebrand became a respected figure in psychoanalytic circles.[74] He was not the only one to follow such a career path. Joseph Sandler also crossed the divide and became a prominent leader in British psychoanalysis, and one of Eysenck's sworn enemies. Of these renegades, Eysenck said they could only be characterized as an 'extraverted' bunch—not exactly a compliment in Eysenckian terms, for it implied they were less morally developed. He otherwise claimed that allowing such students to study psychoanalytic theory and the Rorschach test demonstrated that he was not a dogmatic, doctrinal figure. Such diverse research was less evident in later years, however. By the late 1950s, Eysenck had begun insisting that projective tests had little empirical merit, and they ceased to be used at the IoP. He wrote a trenchant review of the Rorschach in the 1959 edition of Buros's *Yearbook*.[75]

---

[72] John Mollon, personal communication, 24 June 2003.

[73] Cyril Franks, telephone interview, 13 July 2002.

[74] Peter Hildebrand had taken a degree in psychology at the Sorbonne in 1951 before moving on to the IoP. He worked at the Tavistock Clinic from 1953 to 1993 and was for many years chief psychologist. He had become a psychoanalyst in 1958 and a training analyst for the British Psychoanalytic Society since 1967. Hildebrand died before he could be interviewed for this project.

[75] Eysenck, H.J. (1959c), The Rorschach test, in *The fifth mental measurement yearbook* (ed. O. Buros), pp. 276–8, Gryphon Press, Highland Park, New Jersey. It was a time when the weight of academic opinion and the priorities of clinical practice had begun to turn

Eysenck was, Franks said, still flattered by the critical attention given to *The dynamics of anxiety and hysteria*. According to Franks, he would have been disappointed if this major step forward did not attract substantial consideration. In the spirit of Oscar Wilde—the only thing worse would be to have been ignored. Eysenck even saw it as somewhat of a compliment to be attacked. Franks recalled Eysenck saying he would have been proud if an 'anti-Eysenck' journal was founded, and he increasingly styled himself as the *enfant terrible* of British psychology.[76] Yet how much of this was bravado, one wonders, a reaction to the tight spots he put himself in?

From the ashes: the biology of personality mark II

Eysenck's initial 1957 biological model was hardly a resounding success. It had been put together with some rather limited and dated intellectual resources, and it showed. Research on brain functionality offered much more than Eysenck had been able to incorporate in the marriage he had arranged with an individual differences perspective. Moreover, the classical learning perspective had all but run out of steam in the US. Worse, solid replications and empirical extensions of Eysenck's first model were not readily forthcoming from other labs.[77] At Cambridge, for example, Donald Broadbent had taken up the idea of personality differences in memory and attention in the late 1950s, particularly along the introversion–extraversion dimension. However, he and the young group of researchers under his wing at the applied psychology research unit found the implications of Eysenck's first biological model frustratingly difficult to work with. The impressive results Eysenck had reported proved exceedingly difficult to duplicate in their own work. As a result, some members of Broadbent's group began to explore other ways of conceptualizing the neurophysiological basis of individual differences in performance.

---

against projective tests in general. See Buchanan, R. (1997), Ink blots or profile plots: the Rorschach versus the MMPI as the right tool for a science-based profession, *Science, Technology and Human Values* **21**, 168–206.

[76] Cyril Franks, telephone interview, 13 July 2002.

[77] For example, researchers at Northwestern University could not duplicate Eysenck's results. See Rechtschaffen, A. (1958), Neural satiation, reactive inhibition, and introversion–extraversion, *Journal of Abnormal and Social Psychology* **57**, 283–91. Likewise, at the University of Michigan, Angus Campbell came to see Eysenck's work as difficult to reproduce, as too smoothly obedient to theoretical expectations.

Fig. 5.1 The Maudsley Hospital, *c.* late 1940s.
Source: Eysenck, H.J. (1952a), *The scientific study of personality*, Routledge and Kegan Paul, London.

Fig. 5.2 Dynamometer persistence test—a young Sybil Eysenck looks on.
Source: Eysenck, H.J. (1952a), *The scientific study of personality*, Routledge and Kegan Paul, London.

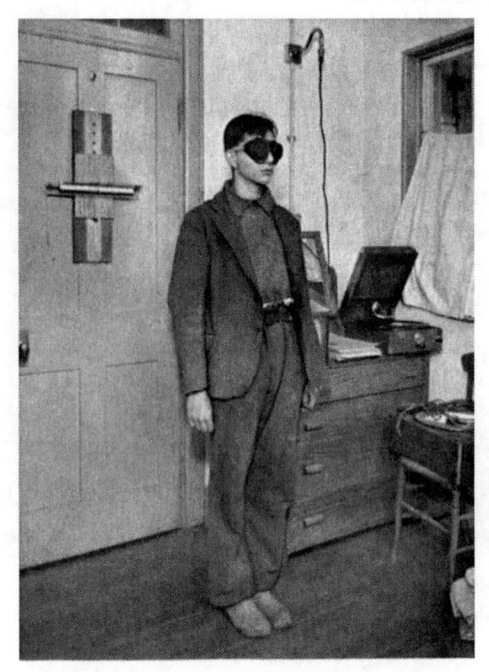

Fig. 5.3 Body sway test of suggestibility with portable recording apparatus.
Source: Eysenck, H.J. (1952a), *The scientific study of personality*, Routledge and Kegan Paul, London.

Fig. 5.4 Speed of decision test—Desmond Furneaux and Hilde Himmelweit demonstrate.
Source: Eysenck, H.J. (1952a), *The scientific study of personality*, Routledge and Kegan Paul, London.

Fig. 5.5 The statistical section of the IoP department.
Source: Eysenck, H.J. (1952a), *The scientific study of personality*, Routledge and Kegan Paul, London.

Things had got a lot more complicated than Eysenck had allowed for. Inhibition theory did not have the flexibility to explain the diverse performance differences of various personality groups. Conditioning seemed to be greatly affected by the parameters of the experimental trials. As Eysenck later acknowledged, shorter conditioned–unconditioned stimulus intervals, a stronger unconditioned stimulus, and higher proportions of reinforced responses tended to favor extraverts rather than introverts. In the early 1960s, Eysenck began to move beyond a classical learning model as he and his colleagues began to dabble in more complex verbal learning, memory, and intelligence tests.

While Eysenck's typological postulate was brash and brave, it became clear it could only be regarded as a prototype. 'Not entirely successful' was how Eysenck put it.[78] Gordon Claridge summarized the key problems: while the introversion–extraversion and neuroticism dimensions were said to be (mostly) independent, at a causal biological level they appeared quite interactive.[79]

---

[78] Eysenck (1997b), *Rebel with a cause*, p. 204.

[79] Claridge, G. (1986), Eysenck's contribution to the psychology of personality, in *Hans Eysenck: consensus and controversy* (ed. S. Modgil and C. Modgil), pp. 73–85, Falmer Press, Philadelphia.

The differences exhibited by different highly neurotic groups could not be explained in terms of introversion–extraversion mechanisms alone. The Pavlovian concepts of excitation and inhibition were a cul-de-sac in neurophysiological terms, defying elaboration. Furthermore, the causal biological basis of neuroticism still remained underdeveloped, psychoticism not at all.

Eysenck would revise the biological basis of his theory in the mid-1960s, casually remarking that this meant going the 'whole reductionist hog'.[80] A young Jeffrey Gray was indispensable in making this shift. Gray's input helped adjust Eysenck's 'cavalier' treatment of Pavlov and deepen understandings of other Russian researchers.[81] Gray had learnt to speak and read Russian at a very high level during his stint in the army. An Oxford graduate, Gray first came to the IoP in 1959 as a clinical trainee. He recalled that soon after arriving Eysenck handed him a two-volume set on Russian experimental work, in Russian. It had been sent to Eysenck by Pavlov's intellectual heir, B.M. Teplov, whom Eysenck had met earlier at a conference. Fatefully, Eysenck asked Gray: 'Do you think this is important?' Gray dutifully translated and annotated this material, for it indeed seemed important to him. Eysenck then suggested that Gray write a book about it and sent him to a maverick publisher with a track record in bringing Russian science to the West. This was Robert Maxwell—the colorful publishing magnate, nicknamed 'Captain Bob' by the English tabloids, then head of Pergamon Press. Maxwell was in the process of helping Eysenck found the key journal *Behaviour Research and Therapy*. Gray was able to secure a generous advance of £500 and the book finally appeared in 1964 as *Pavlov's typology*. In the meantime, Gray completed a Ph.D. on the Pavlovian notion of strength and in 1963 returned to Oxford. He later accompanied Eysenck to Russia as an interpreter for the 1966 International Congress of Psychology in Moscow.[82]

Gray suggested that Russian work could be made consistent with English-speaking research on cortical arousal, and had the potential to play a key role explaining personality differences. In two painstaking chapters, Gray reinterpreted Teplov's extensions of Pavlov in light of Western research on functional neurology. He straightened out misunderstandings of Pavlov's strong and

---

[80] Eysenck (1997*b*), *Rebel with a cause*, p. 204.

[81] The 'cavalier' label comes from p. 369 of Claridge, G. (1997), Eysenck's contribution to understanding psychopathology, in *The scientific study of human nature: tribute to Hans J. Eysenck at eighty* (ed. H. Nyborg), pp. 364–88, Pergamon, Oxford.

[82] See Eysenck, H.J. (1966*a*), Conditioning, introversion–extraversion and the strength of the nervous system, in *Proceedings of the 18th International Congress for Experimental Psychology*, pp. 33–44, Moscow.

weak nervous systems, he pointed to the importance of stimulus intensity and protective 'transmarginal' inhibition, and he linked nervous system types and arousal to the brainstem's reticular activation system. He introduced something far more interactive than Eysenck's first model: performance was a function of external environment input and internal individual settings. Gray kept to Pavlov's original weak/strong labels, but finished with a clear statement that significantly adjusted Eysenck's typological postulate. The weaker the nervous system, Gray concluded, the 'more pronounced' the effects of arousal and inhibition. Inhibition was identified as part of the same feedback loop as arousal. While arousal ascended from the brainstem, inhibition descended from the cortex to the brainstem to dampen performance in conditions of intense, protracted, and/or repeated stimulation.

Eysenck took up these ideas to completely overhaul his biological theory. First, he reapplied the explicit analogy between Pavlov's strong and weak nervous systems and introversion–extraversion. This time he got the Russian right. The weak nervous system was that of the introvert—more sensitive, more excitable, and less stable than the strong system of the extravert. Eysenck went on: 'A more fashionable way of making the same distinction would be to say that the weak nervous system is more easily and more highly "aroused"...'[83] Any past misapprehensions were left unmentioned. Following Gray's reinterpretation, the reticular activation system was said to regulate the stimulation input to the cortex like a homeostat. Eysenck incorporated the idea of optimal levels of stimulation matching up with optimal levels of arousal, and suggested that these bandwidths were set differently for introverts and extraverts. In a more independent move, Eysenck related neuroticism to limbic system activation. The limbic system, or Papez circuit, had been associated with emotional responsiveness since the 1930s. Various researchers had outlined the role of limbic system components, especially the hypothalamus, for initiating autonomic and hormonal changes associated with feelings.

The biology of personality mark II was more than just a change of terminology—a face-saving view Eysenck sometimes encouraged and others lazily followed.[84] As Jeffrey Gray observed, Eysenck was a master of absorbing new ideas and shifting tack as if no change had taken place. Eysenck's new theory did in fact represent a substantial overhaul, Gray was quick to point out,

---

[83] Eysenck, H.J. (1967a), *The biological basis of personality*, C.C. Thomas, Springfield, Illinois, pp. 241–2.

[84] For example, Tony Gibson's biography, Gibson, H.B. (Tony) (1981), *Hans Eysenck: the man and his work*, Peter Owen, London, tended to follow this line.

not least because 'it made different predictions'.[85] In this new theory, a given stimulus provoked higher levels of arousal, but only up to a certain threshold when transmarginal inhibition kicked in. The relationship would be interactive and curvilinear. Arousal should exhibit an inverted-U relationship with various learning and performance measures, and the parameters of this curve should be different for introverts and extraverts. For example, at high stimulation intensities, transmarginal inhibition ensured that it was the strong nervous system of the extravert rather than the weak system of the introvert that would be expected to exhibit higher arousal (or 'excitation' in Eysenck's original terminology)—opposite to that previously predicted. Introverts did not simply learn faster and better than extraverts, as Eysenck's first model implied. Differential learning was now much more a matter of horses for courses. With their higher levels of internal arousal, introverts should perform better in less stimulating environments, but would tend to be overwhelmed when things got very stimulating. Thus they tended to shy away from such situations. Conversely, extraverts would seek such high-input situations out, for that is where they performed at their best.

On top of a far greater awareness of differential stimulus input effects, other variables now came to be seen as important. For example, subject age and the time of day of testing were now seen to systematically relate to different patterns of arousal in introverts and extraverts. Drive activation too was also conceptualized as having a curvilinear relationship to performance, with high activity helpful for simple but not more complex tasks. Moreover, arousal and activation were assigned equal status but were conceived as only partially independent. They were connected with two orthogonal dimensions, a potential contradiction that Eysenck was only able to partially resolve. Eysenck followed Ernst Gellhorn's separation of reticular and hypothalamic functions in emotion. The limbic system primed cortical arousal only during episodes of strong emotional activation. Nonetheless, it proved difficult to distinguish emotional and non-emotional arousal at a physiological level.[86]

The biological basis of personality had less of the lofty formalism of *The dynamics of anxiety and hysteria*. Far less emphasis was given to the importance of theory-led hypotheses, to bold Popperian predictions. Hullian equations and italicized postulates were nowhere to be seen. The biology of personality mark II was something altogether more cautious. It was an attempt

---

[85] Jeffrey Gray, interview, 25 July 2003.

[86] Reviewers chastised Eysenck for not being more up to date with the literature on the physiology of emotions. See Blackburn, R. (1970), Review of *The biological basis of personality* by Hans Eysenck, *British Journal of Social and Clinical Psychology* **9**, 398–9.

to explain the paradoxes and anomalies of more than a decade of conditioning and learning data, a transparent quest to find a better *post hoc* fit. Eysenck's conclusions were a short exercise in caution. 'Any theory in this perplexing field', he wrote, 'is almost by definition a weak theory. Its purpose is to guide research along promising and fruitful lines…'[87] Eysenck had subtly adjusted his intent. Developing complications in motivation, learning, and brain functionality had made the sweeping certainty of his first effort untenable. Linking what some saw as distinct 'interpretive systems' had to take these complications into account. Eysenck had to admit the connections he made between personality differences and brain structures were highly speculative. Commentators followed suit, suggesting the primary value of his biological model was 'heuristic', even if Eysenck indiscriminately mixed together data from 'different conceptual frameworks'.[88]

The biology of personality mark II was a quantum leap forward in terms of neurological sophistication, explicitly mapping personality differences to particular brain structures. Although it was written only a decade later, *The biological basis of personality* looks far more contemporary than its predecessor. With a little help from his friends, Eysenck had brought himself up to speed in neurology and blended a greater range of learning data from more complex tasks.

Much of the ideas and results Eysenck used to construct his new biological model had been around in some form when he wrote *The dynamics of anxiety and hysteria*. For example, the concepts of arousal and activation pre-dated World War II. While much of the key neuropsychological findings Eysenck wove in (e.g. those of Sokolov, Duffy, Hebb, Lindsley) came from Russia and research centres in the US, other sources of inspiration were far closer. Gordon Claridge had pointed to the importance of arousal in performance differences, albeit along the neurotic dimension, in the early 1960s.[89] Broadbent's former student at Cambridge, Dereck Corcoran, had linked arousal to introversion–extraversion in 1964, building on his Ph.D. work. Corcoran had proposed the characteristic inverted-U relationship in a *British Journal of Psychology* paper

---

[87] Eysenck (1967a), *The biological basis of personality*, p. 340.

[88] See Bannister, D. (1970), Review of *The biological basis of personality* by Hans Eysenck, *British Journal of Psychiatry* **116**, 103, and Smith, R.E. (1969), The other side of the coin. Review of *The biological basis of personality* by Hans Eysenck, *Contemporary Psychology* **14**, 628–30.

[89] Claridge, G. (1961), Arousal and inhibition as determinants of the performance of neurotics, *British Journal of Psychology* **52**, 53–63.

that appeared the following year.[90] Eysenck's debt to Gray was even more apparent. He duly cited Gray's meticulous analysis of Russian work, though not always accurately.[91] But, even though Gray admitted there was a trading of ideas between himself, Eysenck, and others, and even though Eysenck incorporated and extended Gray's suggestions in a grander theoretical design, Gray still felt miffed. He thought Eysenck should have given him more credit for sensing the potential of the arousal concept and for deepening the intellectual channel between East and West. It was, Gray remembered, the 'only time Hans ever played false to me'.[92]

Questions of priority and attribution were put aside when Gray subsequently developed his own biological model of personality from 1970 onward. Thus Gray was happy to refer to 'Eysenck's arousal theory' as just that. Gray's model related propensities to respond to reward and punishment to specific areas of the brain, and these were mapped on to two orthogonal psychometric dimensions of anxiety and impulsivity. It quickly came to be seen as the major theoretical rival to Eysenck's revised model. Yet the meaning of Gray's dimensions was still partly understood in reference to Eysenck's original grid. They were set at an oblique angle (30°) to Eysenck's neuroticism and introversion–extraversion dimensions. In turn, Eysenck would subsequently adjust aspects of this theory to account for reward/punishment phenomena that Gray had highlighted.

The reaction to *The Biological basis of personality* was less excitable than to *The dynamics of anxiety and hysteria*. It was also generally more positive. There were few roundhouse condemnations, and no rebels in the ranks. Reviewing *The biological basis of personality*, Australian psychiatrist Neil McConaghy likened Eysenck to a man rushing up a burning flight of stairs. 'As the foundations of the stairs behind him collapse, he is already being supported by a new

---

[90] See Corcoran, D.W.J. (1961), Individual differences in performance after loss of sleep, Ph.D. thesis, University of Cambridge; Corcoran, D.W.J. (1964), The relation between introversion and salivation, *The American Journal of Psychology* **77**, 298–300; Corcoran, D.W.J. (1965), Personality and the inverted-U relation, *British Journal of Psychology* **56**, 267–73.

[91] Eysenck freely cited Gray's book but tended to get the year wrong, listing it as 1964 at one point, 1965 at another. Eysenck was less generous with his referencing of Corcoran's work.

[92] Jeffrey Gray, interview, 25 July 2003.

and as yet unburnt section.' In a backhanded compliment, McConaghy wondered if was possible to be 'both a seminal person and a self-critical one'.[93]

Nevertheless, Eysenck had made sure he would not be hoist by his own petard again. In combining a number of interdependent concepts, Eysenck had increased the explanatory flexibility of his mark II model. Yet he had made it much harder, Claridge noted, to generate clear 'predictions either about static personality differences or about the underlying dynamics of the behavioural variations to which they give rise'.[94] Claridge gave a simple example of predicting reaction times to stimuli of varying intensity based on an individual's position on the personality dimensions. Any prediction would need to be mindful of the relationship between intensity and arousal (for that kind of individual), and the effect of anxiety-induced limbic activation (for that kind of individual). In short, Eysenck made his new formulation much harder to *falsify*, which could be one reason behind its ambiguous longevity.

Arousal theory was challenged strongly in the 1980s; physiological data tended to cast doubt on the inverted-U model and suggested that personality differences were greatly outweighed by stimulus effects. As Eysenck and son Michael commented in 1985: 'the imprecision of arousal theory often means that we are unsure whether a particular finding is consistent or inconsistent with H.J. Eysenck's theory of extraversion.'[95] However, Eysenck's theory still fares reasonably well when compared with other models, such as Gray's, but is still a long way short of explaining much of the complex situation–individual interactions in performance.[96]

Conditionability became a more multifaceted concept, with personality differences also greatly dependent on the type of task studied. In addition, Eysenck looked beyond the concept of inhibition to explain his performance test data. Reminiscence effects on the pursuit rotor were a case in point. Inhibition theory became complex and unwieldy but still failed to account for observed effects. Working with Chris Frith, Eysenck sketched a consolidation model that attempted to explain why performance boosts appeared to occur towards

---

[93] McConaghy, N. (1970), Review of *The biological basis of personality* by Hans Eysenck, *Australian and New Zealand Journal of Psychiatry* **4**, 113. McConaghy was by no means an opponent of Eysenck's outlook. At the time, he was a practitioner of behavioural techniques, including aversion therapy for homosexuals.

[94] Claridge (1986), Eysenck's contribution to the psychology of personality, p. 77.

[95] Eysenck, H.J. and Eysenck, M.W. (1985), *Personality and individual differences: a natural science approach*, Plenum Publishers, New York, p. 284.

[96] Matthews, G. and Gilliland, K. (1999), The personality theories of H.J. Eysenck and J.A. Gray: a comparative review, *Personality and Individual Differences* **26**, 583–626.

the end of the rest period rather than at the beginning.[97] Moving with the times, Eysenck conceded that learning was better thought of as task specific and more a matter of active cognitive strategies.

Dimensions of uncertainty

While the biological basis of Eysenck's personality dimensions was updated by the reformulations of mark II, the psychometric foundations were taken for granted, despite their age.

Eysenck's defence of his original two factors—based on the hysteric/dsythymic distinction—became more cursory as time went on. His last lengthy treatises on the basics of his dimensional approach tended to reproduce diagrams from his original 1944 factorial study. Eysenck otherwise did his best to wish away any problematic results generated in the meantime. Ex-student Peter Hildebrand was never less than forthright about the fundamental problems he had raised about Eysenck's dimensions in the mid-1950s, his outspokenness exacerbated by Eysenck's anti-psychoanalytic campaigning.[98] According to Alan Dabbs, this eventually led to a re-examination of Hildebrand's troublesome Ph.D. thesis.[99] The data was re-checked, essentially corroborating the original results.

Other researchers published similarly inconvenient results. The most notable of these was a 1973 study by another former colleague, Edgar Howarth, who had also worked with Cattell. Howarth got hold of Eysenck's original Mill Hill data and took the provocative step of re-analysing it.[100] His basic factoring did not differ much from Eysenck's original solution. However, when Howarth subjected it to oblique rotation to maximize loadings he came up with an interpretation sharply at odds with Eysenck's. He interpreted the largest factor as a cognitive one, in contrast to the temperamental interpretation of Eysenck's

---

[97] See Eysenck, H.J. and Frith, C.D. (1977), *Reminiscence, motivation and personality*, Plenum Publishing Corporation, New York, especially the epilogue where he considers and largely dismisses the single case methodology of B.F. Skinner.

[98] A write-up of Hildebrand's Ph.D. work had been published in 1958, but this was hardly the end of it. See Hildebrand, H.P. (1958), A factorial study of introversion–extraversion, *British Journal of Psychology* **49**, 1–11.

[99] What exactly occurred is unclear. I have only the testimony of Alan Dabbs to go on. At the time he was heavily involved with the British Psychological Society Division of Clinical Psychology. Alan R. Dabbs, interview, 13 October 2003.

[100] Howarth, E. (1973), A hierarchical oblique factor analysis of Eysenck's rating study of 700 neurotics, *Social Behaviour and Personality* **1**, 81–7.

neuroticism factor. In addition, Eysenck's second factor (i.e. introversion–extraversion) did not clearly emerge from the analyses. Eysenck tended to ignore studies such as this. Instead, he continued to collect a great range of data—across subgroups and cultures—that he claimed supported his original dimensional scheme. When he did mention dissenters like Howarth, he emphasized the positive. For example, in his co-authored 1985 book, Eysenck merely remarked (through gritted teeth?) that Howarth had found his original solution 'sound in regard to the mathematical aspect'.[101]

To cope with the drifting indeterminacy of factor interpretation, new versions of Eysenck's personality test took the load of both measuring and defining the meaning of his dimensions. The MPI obviously wouldn't do. However, revising the test would be a juggling act; it had to succeed in psychometric, substantive, and clinical terms. In the wake of Sigal, Star, and Franks's criticism of the MPI, Ralph McGuire and colleagues at the Southern General Hospital in Glasgow produced more damaging results in the early 1960s. The MPI measured their sample of hysterics as more introverted than normals and even dysthymics. The test might be a useful research measure of neuroticism, they concluded, but it had 'no value as a clinical tool'.[102] With help from his second wife, Sybil, Eysenck re-standardized the 1959 MPI as the Eysenck Personality Inventory (EPI) in 1964. However, psychoticism (P) was not included until the Eysenck Personality Questionnaire (EPQ) was developed in 1972. Meanwhile, Howarth and colleagues at the University of Alberta took aim at Eysenck's EPI, documenting the psychometric shortfalls and the substantive content drift of its scales. The E scale (introversion–extraversion) was particularly difficult to reconcile with Eysenck's dimensional scheme at a factorial level. With its origins in Guilford's scales that broke introversion–extraversion into various components, Howarth compared Eysenck's E scale with 'Humpty Dumpty (i.e. hard to put back together again).'[103] The EPI and

---

[101] Eysenck, H.J. and Eysenck, M.W. (1985), *Personality and individual differences: a natural science approach*, p. 53. Michael Eysenck confirmed that he and his father wrote various chapters of this book independently. This quote comes from a chapter that was assuredly from Eysenck senior.

[102] See p. 166 of McGuire, R.J. *et al.* (1963), The Maudsley personality inventory used with psychiatry patients, *British Journal of Psychology* **54**, 157–66. According to McGuire, Eysenck did not discuss or later address the problems he and his co-authors had highlighted, despite having the opportunity to do so. McGuire assumed this was probably because Eysenck already had the EPI on the way. Ralph McGuire, telephone interview, 11 August 2005.

[103] See p. 185 of Howarth, E. (1976), A psychometric investigation of Eysenck's personality inventory, *Journal of Personality Assessment* **40**, 173–85.

EPQ nevertheless remained widely used research tools in Britain during the 1970s and 1980s, though much less so in the US. They buttressed Eysenck's dimensional system, providing standardized measures for translating and comparing diverse studies of individual differences.

EPQ neuroticism was finally made up of traits like anxiety, depression, guilt, and tension. Extraversion was now characterized by sociability, as well as assertiveness. The psychiatric connotations of introversion–extraversion were otherwise downplayed, as the terms dysthymia and hysteria disappeared from psychiatric classifications. Conversely, psychoticism remained a work in progress, with Sybil Eysenck taking a prominent role in successive revisions of this scale. It was initially conceived as a dimension that, at the extreme, characterized schizophrenia and manic-depressive illness. Later versions complicated the picture, suggesting that intermediate levels of psychoticism were indicative of psychopathy, aggression, coldness (and perhaps genius). In the end, psychoticism became a strictly empirically defined dimension. Its measurement amongst normal populations, and any version of the continuity assumption this implied, remained a matter of debate. The psychoticism dimension traded off sensation-seeking and impulsivity with extraversion in certain situations. Of course, it would have helped to have a biological basis for psychoticism, but this remained relatively underdeveloped. Late in his life Eysenck linked psychoticism to brain serotonin levels, and even later to dopamine. Moreover, there was a persistent suggestion that neuroticism and introversion–extraversion were not independent amongst psychiatric samples. Introversion and neuroticism appeared to go together in many situations, leaving the extraverted-neurosis quadrant particularly uncertain. Eysenck's model predicted that extraverted neurotics (i.e. psychopaths and perhaps hysterics) should show high activation and low arousal. But the evidence suggested they showed the lowest activation and arousal levels of all.[104]

A stalling programme

Eysenck's 1967 revision essentially remained the theoretical endpoint in his attempt to provide a biological basis for his personality dimensions. His attention had already begun to shift to the practical implications of a tightly interlocking theoretical system. He went on to suggest that various forms of social distress were related to extreme positions on at least one of his three dimensions. For example, Eysenck had already claimed that introverts tended to have a more developed sense of morality—an implication more straightforwardly

---

[104] Claridge (1986), Eysenck's contribution to the psychology of personality.

drawn from his first biological model—as well as a greater capacity for academic achievement. His 1964 book *Crime and personality* made much of this.[105] Offenders should not be treated alike, Eysenck said; punishment and rehabilitation should take individual differences into account. As well as predicting that prisons would tend to be populated with extraverts, Eysenck suggested they should be given excitatory drugs to aid their moral reconditioning. The data on prisoner extraversion was quickly mired in debates about appropriate normative baselines and Eysenck's penal policy recommendations were greeted with scepticism.[106] Subsequent research manipulating the parameters of conditionability made unambiguous everyday life implications harder to draw out.

Eysenck's speculative practical pronouncements began to typecast him as a biological determinist with a tendency to make one inference too many. Yet Eysenck never maintained that personality was fully determined by biology. Instead, he viewed behaviour as the sum effect of genes and environment. Any given point in Eysenck's dimensional space represented accumulated life experience superimposed on genetic potential. This framework enabled Eysenck to explore the genetic basis of personality differences, and the way in which these differences determined conditioned learning. The explicitly interactive aspects of his second biological model have been taken up by some of his students' students, who champion what they see as Eysenck's prescient transcendence of the old nature–nurture opposition.[107] However, his core work on personality and its biological basis petered out, in part it seems, due to changing institutional circumstances.

The unique personal institutional basis for Eysenck's programmatic research did not last. Research student numbers, if not student quality, could not be maintained. While grant-backed research and a steadily increasing number of

---

[105] Eysenck, H.J. (1964b), *Crime and personality*, Routledge and Kegan Paul, London.

[106] For more on Eysenck's research on crime and how it was received by psychologists and sociologists see Rafter, N.H. (2006), H.J. Eysenck in Fagin's kitchen: the return to biological theory in 20th century criminology, *History of the Human Sciences* **19**, 37–56.

[107] In criminology, Peter Venables' student Adrian Raine focused on the paradoxical effects of conditioning and environment that Eysenck had implied but not emphasized. For example, poorly conditioned individuals would in most circumstances be expected to become less morally developed. However, if they were born into unsavoury law-breaking circumstances, this relative inability to learn might serve to protect them, enabling them to escape what would seem to be their criminal destiny. See Raine, A. (1997), Classical conditioning, arousal and crime: a biosocial perspective, in *The scientific study of human nature: tribute to Hans J. Eysenck at eighty* (ed. H. Nyborg), pp. 122–41, Elsevier, Oxford.

research appointments continued, student research output dropped markedly in the mid-1970s. The overall figures given by IoP annual reports suggest that the mid-1970s period was particularly lean time for both the psychology department in particular and the IoP in general.[108] Ph.D. output did pick up again in the early 1980s, however.

Changes in policy, budgeting, and/or staffing might well have been behind this downturn. Even so, a department with only a handful of research students was a far cry from the buzzing atmosphere of the 1950s. Perhaps the experimental conditioning and comparative research looked tired by the 1970s. Perhaps the ageing Professor began to look old-fashioned to research-oriented students looking for a high-flying career. There were now many more options for prospective students anyway. Professionally oriented postgraduate courses had sprung up all over Britain in the 1960s. They offered more bankable employment outcomes than research Ph.D.s. However, there is also anecdotal evidence that Eysenck's notoriety as a controversialist had begun to have negative impact. His intervention in the race and IQ controversy in the early 1970s ramped up his already high profile. Fairly or not, Eysenck became a hate figure amongst radical activists. The campus demonstrations they organized made sure that working with him carried a certain stigma, and this may well have put some would-be students off.[109] If this is correct, then it represented a strange twist of fate; the very thing that once worked so well for Eysenck, his reputation and his fame, came to work against him. Whatever the reason, the supply of cheap and committed intellectual labour suddenly dried up. Whether there was also a decline in the quality of students is difficult to substantiate. Apart from relatively high academic grades, there were few additional barriers to postgraduate student entry at the IoP. For example, Eysenck never did selection interviews. Fewer students might suggest the department might have been less choosy about who it took on, but this is still speculation. Eysenck must have found this turn of events hard to ignore and this may explain his relatively restrained response in the race and IQ wars.

The singular focus of the research undertaken too began to dissipate. As Irene Martin suggested, a coherent programme in personality research couldn't last forever.[110] The new IoP building and reorganization of the clinical programme in 1968 signalled the end of an era. The informal at home sessions—which were an important forum for intradepartmental discussion—had fallen

---

[108] For example, IoP annual reports cite only one psychology Ph.D. for the mid-1974 to mid-1975 period (i.e. David Nias), and none for the 1978–1979 period.

[109] IoP annual reports, 1972-1976; J. Philippe Rushton, interview, 3–4 July 2002.

[110] Martin (2001), Hans Eysenck at the Maudsley—the early years.

by the wayside. As a result, cooperation between the research and clinical sections became far less evident. The department became more pluralistic, with a greater mix of more senior researchers pursuing their own agendas. Eysenck's programme began to lose focus. Testing and modifying the core of his biological basis for personality appeared to stagnate and the ageing Professor became increasingly distracted with peripheral, off-beat areas. The IoP was never particularly well-equipped to do, for example, basic neuroanatomical and functional research. Eysenck could not count on being able to directly carry on with adjusting and extending this part of his biological model. Eysenck was challenged by the more modern neurobiological basis of Gray close to home—which he continued to agree to disagree with—and by radically different cognitive approaches elsewhere. With students scarcer and new research avenues more difficult to explore, Eysenck tended to reiterate claims that took his biological basis for personality as well-established. It also helps explain some of the more off-beat issues he took up late in his career, and why he was so keen to embrace people others would not.

Eysenck reprised his biological model in the mid-1980s without major modifications.[111] To do so was beyond his considerable resourcefulness. Even so, Eysenck had mapped the modern terrain of the biology of temperament. Others could only ignore it altogether, or attempt to respond to, query and adjust his work. Research on the biological basis of personality differences remained an exchange between a tight-knit and relatively privileged set of researchers—Gray, Claridge, Marvin Zuckerman, Robert Stelmack, Jan Strelau, Martin Seligman to some extent—those with the inclination and resources to have a go. All owed something directly to Eysenck, having been educated at the IoP or having visited Eysenck extensively. Only a few others have come at this speciality field more independently.[112]

In his memoirs, Eysenck half extolled and half bemoaned the fact that he left no 'Eysenckian school'. While those close to him trod in or around his footsteps, he generated no great army of followers exploring the biological basis of behaviour. This was partly a result of the way Eysenck organized things at the IoP. He left no school because he didn't recruit and train one. But Eysenck didn't persuade outsiders to come round to his point either—the stated aim of

---

[111] See Eysenck, H.J. and Eysenck, M.W. (1985), *Personality and individual differences: a natural science approach* and Eysenck, H.J. (1987a), Arousal and personality: the origins of a theory, in *Personality dimensions and arousal* (ed. J. Strelau and H.J. Eysenck), pp. 1–13, Plenum Press, New York.

[112] You could add J.A Brebner and C.R. Cloninger to this list of prominent researchers, and perhaps Robert Plomin, as well as a handful of younger American researchers.

his reconciliation project. If the data were as compelling as Eysenck claimed they were, this failure begs for some kind of explanation—especially since behavioural scientists are much keener to embrace neurophysiology now. There seemed to be a perverse duality lurking amidst critical responses to Eysenck's work. His results were deemed *too good* as well as not good enough. What was it about Eysenck that made his detractors impossible to satisfy?

It's all about trust, stupid
(…with apologies to Clinton political advisor James Carville)

A key selling point of Eysenck's psychological science was that it was objective. The measurement tools and experimental methods he favoured boasted of precision and reliability that made the data, and the reports constructed from them, trustworthy. Yet Eysenck's attempts to biologize personality inspired anything but trust. While this may seem banal, the interesting thing would be *why* Eysenck and his numbers were not trusted.

The bad reviews Eysenck's first biological model received initiated a persistently negative peer judgement set. Blindly universal scepticism is not the norm amongst scientists, as Robert Merton once claimed. Investigations of the way scientists talk amongst themselves have repeatedly suggested that they do not accord all results equal credulity. Instead, they look to a host of what might otherwise be called personal or social factors—the who, what, where, and how of the data—to help guide their judgements. While typically extraneous to public scientific discourse, these provenance elements feed into a central fund of knowledge in any given scientific community.

The criticisms of the 'IoP five'—Hamilton, and Storms and Sigal in particular—brought into the open the kind of informal communication that might normally take place between various researchers and groups. Their critiques amounted to a kind of cautionary red-flagging that would be hard to expunge. While this breaking of ranks could be attributed to a failure in personnel management on Eysenck's part, it was exacerbated by other aspects of his *modus operandi*.

Eysenck clearly concentrated his efforts in certain directions. His muscular output, the vast size of his public footprint, hid a relative neglect of other aspects of scientific activity. Craft skills, the tricks for making things work, are all crucial to scientific success—especially in complicated experimental work. This tacit knowledge is transferred by personal contact, by trial and error demonstration of apparatus, by simple informal explanations of techniques that transfer understandings of how to do things and what to expect. It is said to be caught, not taught. It helps create trust within a scientific community of shared scientific values. And it is a central ingredient for extending the domain of application and for enrolling like-minded researchers.[113]

It was an aspect of science that Eysenck did not give much attention to as head of department at the IoP. In running such a large operation, Eysenck might have found it hard to be highly involved in the nitty-gritty of experimental work. But even so, he deliberately styled himself as the big picture man, spreading himself across a broad array of concurrent investigations and projects.[114] He did not typically concern himself with practicalities and technical details. Many of those who observed him up close, such as Alan and Ann Clarke and Sidney Crown, remember Eysenck as a theoretician rather than a hands-on experimentalist. Latter-day collaborator Paul Barrett also recalled that, if you were regarded as competent, Eysenck would delegate completely. Barrett himself was charged with rigging-up the experimental equipment to investigate EEG measures of intelligence, and wrote 'the trust Hans put in me was absolute … I think Hans gave people two choices—either you "did the business" or you left.'[115] With his junior staff doing most of the mundane 'bench-top' work, Eysenck was content to get periodic progress reports. Instead, he made interpretation and global theorizing very much his domain, putting their results together in a grand framework. Thus the whispered accusation that Eysenck faked his data does not make much sense when you consider it is largely other people's numbers he was presenting. It was more the way he used their data that was the problem.

The 'IoP five' put Eysenck's representation of some of the procedural details of the research process on centre stage. They shone a spotlight on the adjustments to apparatus, the various matching criteria for conditioning groups that were debated, the decisions about which tests to use and report, and uncertainties about subject samples. These considerations mix trial-and-error with low-level inference, but are mostly left implicit in the write-up of experiments. It was for this reason that another Nobel Prize winner Peter Medawar labelled

---

[113] See, for example, Maynard, D.W. and Schaeffer, N.C. (2000), Toward a sociology of social scientific knowledge: survey research and ethnomethodology's asymmetric alternates, *Social Studies of Science* **30**, 323–70.

[114] Other roads could have been taken. Eysenck could have opted for something more specialized and narrow. He could have got more involved in the practicalities of particular topics. He certainly didn't *have* to spread himself so thinly or even publish so much.

[115] See p. 12 of Barrett, P. (2001), Hans Eysenck at the Institute of Psychiatry—the later years, *Personality and Individual Differences* **31**, 11–15.

the scientific paper 'a fraud'.[116] In 1963 Medawar argued that scientific papers systematically misrepresent the process of scientific research as something wholly planned and logical, omitting the blind alleys, silly mistakes, and dumb luck that is part of most successful research. Yet sociologists of science have since retorted that this misrepresentation has a functional power. Cleansing research reports of the messy and the particular helps make the knowledge they convey more mobile, transcendent, and universal.[117] Additional tacit craft elements necessary for other scientists to duplicate, use, and extend this knowledge are otherwise drawn from countless practicums and on-the-job experience. However, in the case of novel, cutting-edge procedures—experiments unreliable in unfamiliar hands—training, textbooks, and the relevant scientific papers often have to be supplemented with more informal instruction. Close attention to scientific practice has shown there is usually no substitute for face-to-face contact to make things work.

This, of course, was the rub in Eysenck's *modus operandi*. As the detached theoretician, he ignored or downplayed the importance of personal communication and tacit knowledge—both publicly *and* behind the scenes. His published work often glossed over crucial details of sampling and technique, even and especially when he was trying to do something new by merging disparate research traditions. Many of his publications, his books especially, tended to skim over methodological and procedural details.[118] His documentation and referencing habits left much to be desired. Those interested in pursuing his results, those familiar with (say) psychometric tests and factorial statistics but not eye-blink conditioning apparatus (or vice versa), would struggle to follow what he did. Instead, Eysenck put great store in the ostentatious rigour of his programme. Effectively this amounted to a bluff insistence that the logical, hypothetico-deductive research model displayed in his publications—especially *The dynamics of anxiety and hysteria*—represented the *actual* process he and his team had engaged in. It painted a target on his head for those with

---

[116] Medawar, P. (1991), Is the scientific paper a fraud? in Peter Medawar, *The threat and the glory: reflections on science and scientists*, pp. 228–33, Oxford University Press, Oxford (based on a BBC interview published in *The Listener*, 12 September, 1963).

[117] The seminal work in this area was that of Knorr-Cetina, K. (1981), *The manufacture of knowledge: an essay on the constructivist and contextual nature of science*, Pergamon Press, Oxford.

[118] To be fair, it must have seemed like a tedious task to Eysenck to reiterate such details to ensure each publication worked on a stand-alone basis. He tended to gloss over methodological details by referring to original publications, particularly when reviewing the research literature in his books. However, readers were often left chasing their tails trying to find adequate descriptions of what he did or where he got a particular idea from.

a grievance. But, as David Lykken noted, Eysenck's deductions appeared to 'follow rather than to lead the evidence'.[119] In lifting the lid on what 'really happened', Hamilton, Storms, and Sigal were calling Eysenck's bluff, as only insiders could.

While Eysenck was always careful to give formal attribution to researchers under his wing, some still felt uneasy. Eysenck's detachment was insensitive to the challenges they faced working within his programme—for example, to the difficulties inherent in comparing impure measures of his psychometric constructs, of obtaining suitable matching samples, of getting subjects to behave appropriately. He made their research look straightforward and theoretically consistent—annoying insofar as *they* knew it wasn't.[120] Moreover, Eysenck had also gained priority over the meaning of their work, its place, and importance.[121] Their methodological critique doubled as a challenge to Eysenck's power to represent their efforts. Hamilton, Sigal, and Storms were feeding a scientific star and apparently felt used in the process. Jeffrey Gray likewise felt a little short-changed. However, Gray coped by forging an independent reputation and did not air his grievances publicly. He was always mindful of how such comments would be seized upon by Eysenck's enemies. 'I didn't want people claiming Gray says "Eysenck stole his ideas"—well I never did say that!'[122]

For Eysenck, there was little more to the practice of science than its public performance. He never appeared to hoard his work. A fearful Darwin he was not. As far as I am aware, there were no unpublished notebooks, no hidden manuscripts, no half-written-up results. Not only did Eysenck put it all out

---

[119] Lykken (1959), Turbulent complication, Review of *The dynamics of anxiety and hysteria* by Hans Eysenck, p. 378.

[120] To get the flavour of some of their practical and theoretical criticism, they paid particular attention to Franks's eye-blink data. 'Eysenck discusses "conditionability" as though it had been demonstrated that conditionability is a general and unitary trait ... Eysenck would have been well advised to speak of eye-blink acquisition.' (p. 234) 'But were the groups [dysthymics and hysterics] in fact equated on sensitivity to tone? One of Franks' selection criteria was that no one would be retained in the sample who blinked to any of three presentations of the tone alone before conditioning began ... [However], it may be that more persons of one group than the other would blink at least once in another series of three tone presentations.' (p. 235). Storms and Sigal (1958), Eysenck's personality theory with special reference to *The dynamics of anxiety and hysteria*.

[121] For example, in his 1957 book *The dynamics of anxiety and hysteria*, Eysenck discussed Star's work on the problematic notion of conditioned inhibition without citation, since it was not yet published. See Eysenck (1957b), *The dynamics of anxiety and hysteria*, p. 254.

[122] Jeffrey Gray, interview, 25 July 2003.

there, he appeared to take most his intellectual cues from the public domain. Besides the in-house discussion groups, Eysenck's key ideas and inspiration came from already published material, some quite old. Pavlov and the other Russian researchers he drew on were prime examples of this. Eysenck was an avid consumer of the scientific literature. He was amazingly well versed in a wide array of obscure periodicals and he loved to be sent pre-publication copies of upcoming work in order to stay ahead of the game. However, he did not greatly depend on more informal information sources—the intellectual chat amongst his peers, the news and views from other research groups—that prominent researchers typically have available to them.[123]

Eysenck lived a life in the public eye, but at a remote level. Warmth and intimacy were not his forte; he was all cool formality. Eysenck had little time for glad-handing, explaining himself, or building personal networks. In this he was the polar opposite to, say, Donald Broadbent—an affable man who turned meet-and-greet into an art form. For his trouble, Broadbent came to be seen as an avuncular figure loved by his students and venerated throughout the discipline. He was one of the few Oxbridge experimentalists welcomed in IoP circles. In contrast, the introverted Eysenck did not care for such informalities. His uncompromising with-me-or-against-me outlook worked for and against him. It rallied the troops but made reaching out to potential allies more difficult. And, for a man trying to build intellectual bridges, he didn't travel as much as you might expect. He got out and about more toward the end of his career and during his retirement. People otherwise came to him, invariably those already sympathetic to his approach.

While Eysenck often talked to the media and gave many public lectures, his involvement with established professional associations tended to be limited and not particularly harmonious. In his younger days, Eysenck went to British Psychological Society (BPS) annual conferences regularly but was not involved at any deep organizational level. According to Sybil Eysenck, Eysenck found the increasing representation of social and educational psychology in the Society not to his liking, and eventually turned his back on it.[124] Eysenck also

---

[123] One notable occasion when Eysenck did so occurred when he adapted Martin Seligman's notion of response 'preparedness' in the early 1970s after Seligman had spent time at the IoP. Seligman also confirmed the odd kind of dissonance that surrounded Eysenck. In social situations, IoP colleagues felt most comfortable 'talking work' with the quiet but intimidating Eysenck, and he in turn was never much one for social chit-chat. However, they would talk endlessly *about* him behind his back. According to Seligman, the IoP cafeteria was full of 'toxic' gossip—most of it fanciful or erroneous. Martin Seligman, interview, 12 July 2002.

[124] Sybil Eysenck, interview, 25 July 2003.

resigned from the elite Experimental Psychology Group (later Society) in 1952 when Monte Shapiro's membership application was rejected. At the time he said he 'no longer wished to be a member of a group with such a light blue tinge'.[125] But protesting that this organization was dominated by Cambridge experimentalists was hardly likely to change it. It was a similar story with the medical section of the BPS a few years later. Unhappy with the section's Freudian orientation, he and his Maudsley colleagues attempted to engineer a wholesale shift in outlook. These moves were successfully resisted by the existing section leadership, and thereafter Eysenck appeared to lose faith in the idea of reform from within. (See the following chapter for more details.) Along with his uncompromising attitude, Eysenck's impatience with committee meant that he seldom took up a first-hand role guiding the discipline's professional development. To cite another example, in 1953 he quit the Committee of Professional Psychologists, the BPS body set up late in the war to guide the development of applied work. Not that this affected the direction of the Committee's work much, for he had never attended a meeting.[126] Eysenck found it easier to ignore or reject existing organizations and instead opted to set up new bodies for himself and likeminded colleagues. He helped found organizations like the International Society for the Study of Individual Differences and journals like *Personality and Individual Differences* and *Behaviour Research and Therapy*. While these initiatives directly reflected his intellectual priorities, they also tended to ensure he was always preaching to the converted.

Consensus anyone?

This was more than just a story of political antipathies and Dale Carnegie exercises. Making friends and influencing people is an important aspect of scientific consensus building.[127] As Jeffrey Gray dryly noted, 'Eysenck would not cross the road to mend a broken fence.'[128] Eysenck tended to treat

---

[125] Quoted from the resignation letter Eysenck wrote in 1952 Eysenck, H.J. (1953*f*), Letter of resignation, reproduced in the *Quarterly Journal of Experimental Psychology* **5**, 39, http://www.eps.ac.uk/society/eysenck.html.

[126] This point comes from the first of John Hall's recent articles on beginnings of clinical psychology in Britain, Hall, J. (2007*a*), The emergence of clinical psychology in Britain from 1943 to 1958, Part I: Core tasks and the professionalisation process, *History and Philosophy of Psychology* **9**(1), 29–55.

[127] This is a point that was most associated with the work of Bruno Latour; however, many other sociologists of science have demonstrated it in more specific case studies.

[128] Jeffrey Gray, interview, 25 July 2003.

disagreements as a competitive challenge. Criticism, if publicly voiced, was to be ruthlessly slapped down with a flurry of published replies and rejoinders. You cannot help but think that Eysenck was given little choice by some of his more hostile antagonists; he had to return fire. Even so, Eysenck did not appear to put much effort into behind the scenes conciliation. He didn't strive to find out what was behind critics' objections. It often suited him not to. Milton Rokeach's bid to recalculate some of Eysenck's data from his 1954 book, *The psychology of politics*, was a case in point. In private correspondence, Eysenck did little to clarify how he and his students had analysed the data.[129] Large discrepancies were the inevitable result, which Rokeach and his colleagues made much of. This pattern was repeated on several other occasions—for example, in his contribution to the debate over mental speed and EEG data in the 1980s and during the controversy over personality, smoking, and health in the early 1990s. For Eysenck, disagreement reflected his critics' failure to understand, rather than a weakness in his knowledge-claims or a failure to communicate them.

Similarly, Eysenck had a take-it-or-leave-it attitude to replication. His student-based research programme made it hard to do long-term follow-up. The high turnover of personnel and skills undermined any 'institutional research memory'. Each new student had to familiarize him or herself with the local facilities and the best ways of doing things. Uneven results were inevitable. It made it difficult to work through an area, to resolve anomalies and explain odd results—especially when news of non-replications arrived from other labs. Eysenck's perfunctory reporting practices didn't help either. There were often too many mistakes and too little detail for those wishing to check his work—the combined effect of Eysenck's dictation technique and detached standpoint, with a dash of strategic information withholding thrown in.[130] Moreover, he did not tend to go out of his way to visit other labs to show them how to do the eye-blink conditioning experiments they found unfamiliar or difficult. He did very little to combat the gulf of misapprehensions this

---

[129] It mostly boiled down to how 'don't know/can't decide' responses were dealt with. Eysenck was very vague about what procedure had been used when discussing the issue with Rokeach. This episode is recounted in detail in Chapter 7.

[130] As well as the conflagration with Milton Rokeach mentioned above, Eysenck's early 1990s exchange with Pelosi and Appleby was a good example of this point. See Pelosi, A.J. and Appleby, L. (1992), Psychological influences on cancer and ischaemic heart disease, *British Medical Journal* **304**, 1295–8; Eysenck, H.J. (1992*a*), Psychosocial factors, cancer and ischaemic heart disease, *British Medical Journal* **305**, 457–9; and, Pelosi, A.J. and Appleby, L. (1993), Personality and fatal diseases, *British Medical Journal* **306**, 1666–7.

perpetuated. Eysenck cited his preferred studies and otherwise stood on his honour and his word: either you believed him, or you didn't.[131]

In any case, Eysenck usually had only a second-hand knowledge of how his data was collected and occasionally appeared not to care—even when machines had malfunctioned, or his assistants and subjects departed from standardized protocols.[132] Colleagues tell how Eysenck ordered machines from catalogues because he liked the look of them, even though he had little initial idea of how he might use them.[133] He might have found it hard to practically re-educate those who disagreed or got divergent results even if he cared to. Thus his critics thought he avoided this because he had something to hide. I would argue that it was more the opposite, that he was treated with suspicion partly because he did not 'do' show-and-tell. And, although one might say he was trapped by his stiffly disengaged version of scientific practice, he was also able to play it to his advantage. It was his opponents who were personal and political he would say, liable to 'play the man and not the ball'. He was quite the proper scientist in this respect, almost to the point of caricature.

Distrust would become a key feature of peer appreciations of Eysenck, the prism through which many viewed his actions and output. His attraction to controversy in the latter part of his career would keep eyebrows permanently raised. For example, his involvement in the smoking and health debate and the conflict-of-interest accusations it raised made the trust issue overt. Yet one can detect higher-than-usual levels of cynicism *before* Eysenck weighed into divisive areas such as this. There was something more subtle in play as well. And that something was a matter of style, reflecting the diverse ways psychologists recognize themselves in normative images of scientific practice.[134] The ill

---

[131] The controversy over personality and politics was a good example of this, as John Ray later pointed out. See Ray, J. (1986), Eysenck on social attitudes: a historical critique, in *Hans Eysenck: consensus and controversy* (ed. S. Modgil and C. Modgil), pp. 155–73, Falmer Press, Philadelphia.

[132] Alan and Ann Clarke, interview, 25 April 2002; Sidney Crown, interview, 22 April 2002.

[133] Alan and Ann Clarke, interview, 25 April 2002.

[134] The normative images generated by specialist commentators on science (the philosophers, sociologists, ethnologists, ethicists, and so on) are not the only important influence on the way scientists represent their work. Broader public perceptions of what a scientist is and does, and what motivates him/her to do it, also play a role. Psychology's low epistemological profile and permeable borders have allowed such 'outside' ideas to circulate with remarkable freedom, making the discipline a compelling and challenging case study in the history of science. Highly literate psychological researchers like Eysenck have used any and all of these descriptive cum prescriptive ideas as methodological guides and rhetorical resources. Indeed, Eysenck moved with the times in this sense.

rumours were a function of the high-minded *and* high-handed way Eysenck chose to go about his work. It made him look both principled *and* slippery. However, his critics were seldom completely open about their accusations because they, like Eysenck, did not care to acknowledge the informal, micro-political aspects of building trust. Doing so might seem inappropriately personal and pernickety. It would make them look bad; it would make their science look bad; it would tend to strip away the allusions to universal truth that carry so much power.

Instead what Eysenck got was an intense but distorted kind of critical scrutiny. The higher Eysenck's star rose, the more 'intellectual capital' available to anyone who could confront, correct, or confound his ideas.[135] Those who attempted to do so could cloak themselves in scientific responsibility, arguing that Eysenck's widely deployed concepts and tools ought to be thoroughly checked. Hooking such a big fish was obviously hard to resist, and some— Edgar Howarth, for example—made a bit of a career of it. However, the *way* Eysenck was singled out went beyond this. Not many psychologists had their data re-analysed as often as he did; not many were subject to such vituperative attacks. There was an implicit assumption of probable cause in this kind of attention, a suspicion that there was a scandalous truffle lurking there in the data to be sniffed out.

So perhaps the most curious paradox of all was that, while Eysenck may have been a silent and withdrawn figure in private, his appetite for contentious, high-profile debates made his public persona very much part of the intellectual product. After a while it was as much about Hans Eysenck as it was about the ideas and the numbers. The ultimate value of his research came to be rather unhappily entwined with his personal credibility—the very pitfall his *über* version of objectivity was supposed to avoid.

One can detect elements of logical positivism and Popper in his earlier writings, Kuhn and Lakatos in his later work, and a lone-hero trope dominating his memoirs. His deployment of the history and philosophy of science was pragmatic rather than slavish; moreover, the very self-consciousness of this tactic tended to second-guess any attempt to analyse his work in these terms. My choice is to historicize this interplay by contrasting Eysenck's use of the philosophy of science with current sociological accounts. For an interesting commentary on related issues, see Feest, U. (2005), Operationism in psychology: what the debate is about, what the debate should be about, *Journal of the History of the Behavioral Sciences* **41**, 131–49.

[135] For a generalized statement of this kind of economic analysis of science see Bourdieu, P. (1975), The specificity of the scientific field and the social conditions of the progress of reason, *Social Science Information* **14**, 19–47.

Irreconcilable differences

Widespread emulation of Eysenck's biosocial research programme would have to wait. One can detect a degree of wariness in the underwhelming response to Eysenck's second biological model, despite its being markedly more successful in empirical terms. Crude as it seems, there arose some strong incentives to 'shaft Eysenck' rather than follow him. Despite a sneaking, imitative admiration, many fellow psychologists chose to do just that. Reconciling the experimental and the correlational would always be difficult, more so if such a project had his name on it. Eysenck's integrative vision was perceived as a hostile takeover, and not just because he was short on sympathy.

First and foremost, Eysenck wanted to bolster the ontological claims of correlational psychology. The experimental psychology of learning, motivation, and conditioning was pressed into service for his grand, psychometrically based plan. The crucial test for this ordering was which came first, which was the least dispensable, which was discarded or modified when non-supportive data came in. Eysenck was, in the first and last instance, a factorialist of mind. His mid-career position on the status of his dimensions reflected an astonishingly absolute realism. In his 1967 book he wrote, 'The position which will be argued in this chapter accepts as a fundamental reality the existence of two major independent dimensions of personality, E [extraversion] and N [neuroticism].'[136] As Peter Venables noted, such statements were bound to irritate readers, burdened 'under the weight of the orthogonal cross'.[137] Although Eysenck tempered his philosophical attitude in the latter part of his life, his personality dimensions were his articles of faith. While he cast around for experimental support, sifting through ideas, tests, and models readily, his three orthogonal dimensions remained a fixed ideal. The on-going quarrel he had with Peter Hildebrand plainly illustrated the non-negotiability of his dimensional trinity. The aim of uniting psychology's two disciplines was undermined by his realist inclinations. The causality of his factors always came

---

[136] This statement was contained in Eysenck's important chapter on activation, arousal, and emotion detailing the new neurobiological substrates for his dimensions. Eysenck (1967a), *The biological basis of personality*, p. 230.

[137] See p. 71 of Venables, P. (1970), Review of *The biological basis of personality* by Hans Eysenck, *Quarterly Journal of Experimental Psychology* 22, 70–1. The religious imagery comes direct from IoP sources. Apparently it was a kind of in-joke at the IoP in the 1960s and 1970s, especially amongst the clinical personnel, to make the sign of the cross whenever one mentioned Eysenck's dimensions. Alan R. Dabbs, telephone interview, 23 June 2005.

first when it came to the crunch. Eysenck would never satisfy representatives of either side because he was not attempting to unite them as equal partners.

Many experimentalists still saw Eysenck as just an individual differences personality theorist, and an uppity one at that. To them, his research programme was just as misdirected as those of his less vocal and less expansive London school compatriots. Not only did leading figures such as Oliver Zangwill distrust the heuristic power of factor analysis, the notion of individual variation was beside the point when investigating neuropsychological phenomena like brain lateralization or for constructing generalized models of models of memory and learning. Instead of explaining the phenomena itself, Eysenck's research simply mapped experimental indices on to what they saw as an artefactual descriptive scheme. Eysenck's blueprint for uniting experimental and correlational psychology looked more like a rapacious conquest, *their* intellectual resources put to *his* ends. Since Eysenck didn't ask the relevant questions, those who had staked careers on an experimental approach had little incentive to change tack. They had plenty of tractable problems to tackle already. They would not be forcibly converted; it seemed reconciliation would require a generational change of personnel at the very least.

Not all those with Oxbridge ties could be put in this blinkered, purist boat. But when Eysenck *did* manage to engage their attention it still ended badly. For example, Broadbent and his co-workers at Cambridge spent several unproductive years in the late 1950s and early 1960s exploring the effects of personality differences on cognitive performance. Patrick Rabbitt was one of Broadbent's students at the time, now a Manchester emeritus professor based at Oxford. Rabbitt recalled that negative results fed a growing disenchantment:

> During the 1950s and 1960s much of Broadbent's personal work, and even more the work of his post-doctoral and research students such as Colquhoun, Wilkinson, Corcoran, Fisher, Hockey, and Blake was on the interesting theme of modelling ways in which individual differences in personality and arousal, such as might possibly be captured by the introversion/extraversion dimension, affect vigilance, memory and information processing... They ... were not at all interested in challenging Eysenck's work or reputation. On the contrary, they would all have been very happy indeed to find his constructs reliable and his results replicable so that they could build upon them. They were simply frustrated that they found him problematic. Their interest in individual differences in human performance continued throughout their working lives, but they gradually took quite different directions—as did Broadbent's, though he continued work on introversion/extraversion even in his last years before retirement.[138]

---

[138] Pat Rabbitt, personal communication, 8 September 2003.

For Broadbent's group at least, it was a case of once bitten, twice shy. Disappointed with their investment in Eysenck's first model, they were less inclined to give him a second hearing. Ever imperious, Eysenck didn't do much to change their minds. Likewise, Eysenck's work on intelligence and speed of processing in the 1980s caught the attention of many of those working with generalized models of mental processes, including Rabbitt himself. But this work soon got bogged down in replication difficulties and debates over interpretation.[139]

To Rabbitt and his Cambridge colleagues, Eysenck was an extraordinarily talented and ambitious man who had a great deal of trouble 'accepting the limitations of his data'.[140] One cannot simply characterize such reservations as the inevitable expression of a London/Oxbridge cleavage set in stone since Darwin's time. Any kind of split has to be actively maintained or reinvented, and Eysenck was just the man to do it. He had been one of eight prominent psychologists invited to join the Experimental Psychology Group/Society in 1946, before he resigned in disgust six years later. As John Mollon confirmed, 'the antipathy was not a general one between Oxbridge and London. Every January, for example, the Experimental Psychology Society held its annual general meeting in London ... The suspicion was primarily between experimental psychology and Eysenck.'[141] But Eysenck's dominance at the IoP saw everyone he worked with tarred with the same brush. There was a deep stigma in Oxbridge circles attached to anyone who had anything to do with the Institute. Jeffrey Gray was virtually the only one who crossed over during the 1960s and 1970s. Gray himself recalled that Lindon Eaves was appointed at Oxford in 1979 over strong objections because he had worked with Eysenck. It was only more recently that IoP graduates Dick Passingham and Gordon Claridge got jobs there. Gray tried and failed several times to build the modest disciplinary consensus needed to get Eysenck elected an honorary fellow of the BPS. Although they deny it vehemently, it seemed it was mainly the blue-bloods from the north and west that stood in his way.[142]

Factorial methods might well have been in crisis just as Eysenck arrived on the scene, criticized by those at Cambridge in particular as misguided. But at

---

[139] This issue is discussed in greater detail in a later chapter on Eysenck's intelligence research.

[140] Pat Rabbitt, personal communication, 10 September 2003.

[141] John Mollon, personal communication, 24 June 2003.

[142] Jeffrey Gray recalled that, when canvassing votes, several prominent psychologists with Oxbridge links made it known they would oppose such a move vehemently. Jeffrey Gray, interview, 25 July 2003.

least there was *engagement*. Eysenck tended to put a stop to that. While often confrontational, he tended to insulate himself and his research group from those not sharing his point of view.[143] Intellectual exchange tended to occur only at a distance, as lofty one-way didacticism. For example, the first chapter in *The biological basis of personality* was a forthright challenge to experimental and correlational psychologists alike: ignore individual differences in personality at your peril! It was based on Eysenck's C.S. Myer lecture at the 1965 BPS annual conference. Such abrasive broadsides must have taxed the goodwill of his peers. As a rule, the more pervasive the implications of any claim, the greater the 'belief warrant' it must carry. Eysenck had a habit of making assertions that would oblige wholesale intellectual adjustments and costly re-directions of long-standing research. Building the trust required re-education and reassurance, and one way of achieving this would be to draw potential consumers within the knowledge-making process. What Eysenck needed to do, figuratively and perhaps literally, was to bring people closer to him at the IoP—to argue about and negotiate their differences. But it was not in his nature to do so.

Some British correlational psychologists remained off-side, but for different reasons. His mentor Cyril Burt, for example, saw Eysenck's grand programme as overly ambitious. A.R. Jonckheere suggested that Eysenck was a crude statistician who did not really want the factor analytic approach to succeed.[144] Eysenck's biological speculations created the impression that factorial methods were dubiously lightweight—that, by themselves, they were not enough.

Those who have been able to follow Eysenck and combine the experimental and the correlational are scattered across the UK, US, and Europe, mostly at the newer, middle-rung universities and research centres. In the US, intellectual divisions have tended to mirror institutional arrangements. The mid-century splitting of Allport's department at Harvard along social-correlational versus experimental-biological lines was a good example of this. Such entrenched, transnational divisions would be hard for any one person to overturn. Americans were also particularly wary of the political risks in connecting personality with biology in the immediate post-war era. Mainstream journals like the *Journal of Personality and Social Psychology* and the *Journal of Personality*

---

[143] Eysenck himself recounted the occasions he stood up to or annoyed Cambridge dons like Frederic Bartlett. See, Eysenck (1997*b*), *Rebel with a cause*, p. 111. Patrick Rabbitt noted that, when Eysenck perceived himself to be in hostile territory, he would take the attack up to his critics rather than tone things down. Pat Rabbitt, personal communication, 5 September 2003.

[144] A.R. Jonckheere, interview, 22 April 2002.

did begin to publish research on the genetics of personality. However, American personality researchers have by and large displayed an overriding conservatism, resisting biogenetic intrusions that might jolt them from their comfortable semantic-psychometric groove.[145] The sense of satisfaction surrounding the 'big five' model embodied this polite rejection of Eysenck's expansiveness, seeing in it more potential costs than benefits. Costa and McCrae perfectly encapsulated this New World outlook. In a reply to Claridge's critical 1986 summary of Eysenck's contribution they said:

> Few American psychologists would concur in Claridge's opinion that Eysenck's biological theories constitute the 'step which most drastically altered contemporary Western thought about personality.' Americans, noting all the perplexities in biological theory that Claridge points out, are more impressed with Eysenck's manifest success in describing and measuring personality.[146]

---

[145] As well as a fear of the biological, several other factors contributed to the old-school look of contemporary personality psychology. The field marked time for a large chunk of the post-war period as public criticism of personality assessment segued and overlapped with fundamental conceptual challenges from within the ranks. While personality testing came to be regarded as a potential infringement of individual and civil rights in the 1960s, some researchers also argued that it was compromised by test-taking response style. See Buchanan, R. (2002), On *not* 'giving psychology away': the MMPI and public controversy over testing in the 1960s, *History of Psychology* **5**, 284–309. Personality psychologists then had their hands full for much of the 1970s and early 1980s dealing with Walter Mischel's situationalist critique that suggested trait-driven consistency in behaviour was largely a myth. Eysenck conceded that the field was under siege in late 1970s, assailed by the forces of political correctness and behaviorism, as well as Mischel's 'doctrinaire and factually incorrect attacks'. P. 269 of Eysenck, H.J. and Eysenck, S. (1995), Editorial: report on the present state of *Personality and individual differences*, *Personality and Individual Differences* **19**, 269–73. By the mid-1980s, the concept of aggregation allowed everyone to save face and the person-versus-situation debate was closed down. See Epstein, S. and O'Brien, E.J. (1985), The person–situation debate in historical and current perspective, *Psychological Bulletin* **98**, 513–37. Proponents of the trait approach were able to regroup around some familiar themes. Eysenck played only a minor role in these US-centric controversies, however. For one statement attacking Mischel's position, see Eysenck, M. and Eysenck, H.J. (1980), Mischel and the concept of personality, *British Journal of Psychology* **71**, 191–204. Tim Rogers gives an excellent overview of these issues from a testing point of view. See Rogers, T.B. (1995), *The psychological testing enterprise: an introduction*, Brooks/Cole, Pacific Grove, California.

[146] See p. 86 of Costa, P.T. and McCrae, R.R. (1986), Major contributions to the psychology of personality, in *Hans Eysenck: consensus and controversy* (ed. S. Modgil and C. Modgil), pp. 63–72 and 86–7, Falmer Press, Philadelphia.

As Philip Corr concluded, the boundaries remained in place even as they were decried.[147] The divisions of Cronbach's two schools of psychology are still served by separate professional associations, journals, and degrees. Beyond the embodiment of Eysenck's select followers, the rapprochement that has occurred has taken place in a new, third space: the hybrid domain of cognitive neuroscience, with its new journals, associations, and institutional arrangements.

These developments had a way of adjusting the claims made on Eysenck's behalf. The enormous mind–body divide he leapt across has been filled by stepping-stone concepts that make his first model look especially naïve. Eysenck has come to be seen as a new-fashioned theorist of temperament, relating broad trait variations to primitive brain structures that develop relatively early in life and function at a basic level.[148] Operant learning could readily be related to reward and punishment sites at the base of the brain. But such 'old brain' substrates for classical associative learning have proved more elusive. Higher mental processes—the complexity of memory, learning, language, and consciousness—have long been implicated with the neocortex. They were barely addressed in Eysenck's grand scheme. Modelling such processes and mapping them to the 'new brain' with the aid of modern imaging technology became the bread and butter of cognitive neuroscience. And for the most part, this was not undertaken from an individual differences perspective.[149]

As the intellectual gatekeeper for the group, the strengths and weaknesses of IoP research during Eysenck's tenure were his strengths and weaknesses. There were several new areas he could not or would not countenance. Cognitive psychology and clinical neurology had little impact on his thinking. Higher mental processes, save for some intelligence work, were not extensively explored, and he was never tooled up for 'wet' neurological research. Eysenck was trying to

---

[147] Corr, P.J. (2000), Reflections on the scientific life of Hans Eysenck, *History and Philosophy of Psychology* **2**, 18–35.

[148] See, for example, Claridge (1997), Eysenck's contribution to understanding psychopathology. Eysenck's work, especially his experimental research, was largely restricted to adult subjects. The developmental implications of his final biological model have only recently begun to be related to a separate strand of bio-temperament research with children. See Strelau, J. and Zawadzki, B. (1997), Temperament and personality: Eysenck's three superfactors, in *The scientific study of human nature: tribute to Hans J. Eysenck at eighty* (ed. H. Nyborg), pp. 68–91, Elsevier, Oxford.

[149] One exception to this was Jeffrey Gray. Imaging individual personality differences was just the sort of research he was doing before his sad demise in 2004. More generally, time will tell whether the popularity of imaging research proves fertile or a technologically driven cul-de-sac.

merge fields unready to be merged. Reconciliation of the experimental and correlational could only be created by the willing migration of personnel and intellectual resources, not through prescriptive declarations or hostile takeovers. It develops via individual and institutional collaborations that promise to deliver more than the status quo. Now, under the vast interdisciplinary umbrella of the neurosciences, various researchers have come together from somewhat different methods and perspectives from those of Eysenck. How much credit can Hans Eysenck take for this kind of rapprochement? Some, certainly, even if it was not quite what he had in mind. Sad to say his general emphasis on the biogenetic basis of behaviour—so *de rigueur* nowadays—had an unfashionable, 'premature' quality that would make it destined to be undercredited.

Chapter 6

Clinical partisan

Just after World War II, Aubrey Lewis assigned Hans Eysenck the task of mapping out a clinical psychology programme for the planned Institute of Psychiatry. A clinical section was quickly set up to provide psychological services to the adjacent Maudsley Hospital, as well as perform research and train a new kind of specialist: the clinical psychologist. The results were momentous. The 'Maudsley clinical course', as it became known, would churn out a large proportion of the first generation of home-grown practitioners. Maudsley psychologists would also contribute some important research findings and play a key role in the development of a radically new form of treatment dubbed 'behaviour therapy'. While Eysenck never engaged in any clinical teaching or practice, he had a big hand in choosing staff and would put his spin on the theoretical and professional significance of their activities.

In his memoirs, Eysenck suggested he had an even more important historical role.[1] The way he told it, Lewis had effectively charged him with establishing clinical psychology in Britain—presumably because the Maudsley programme would serve as a model for clinical training and practice across the land. Eysenck further implied that Britain's clinical psychologists also had him to thank for freeing them from the shackles of medical dominance. According to Eysenck, Lewis's benevolence had its limits—particularly when it came to the issue of psychologists performing treatment. Eysenck had other ideas, rejecting a subservient role and asserting psychologists' right to use behaviour therapy when they became available. In a confrontation Eysenck characterized as short and decisive, he and his fellow psychologists scored a resounding professional victory over their medical counterparts. Behaviour therapy quickly became the technical hallmark of the independent psychological speciality Eysenck claimed he had launched.

But Eysenck's account was part of the process as much as a commentary on it. His memoirs left out the significant contributions of a range of people and much of the collective context. The Maudsley programme was never the only

---

[1] Eysenck, H.J. (1997*b*), *Rebel with a cause* [revised and expanded], Transaction Press, New Brunswick, New Jersey, Chapter 4.

game in town, even in the early years. While it was undeniably influential, other clinical programmes with contrasting approaches would also shape the British field. And as the field grew, some of the distinctive aspects of this Maudsley influence would inevitably wane. Moreover, the 'battle for behaviour therapy' was never quite as cut-and-dried as Eysenck's partisan tale suggested, especially within the Maudsley itself. Indeed, this episode was like a gestalt puzzle, flipping from one interpretative whole to another, depending on one's vantage point. Other Maudsley psychologists spoke of a complex and lengthy series of events, of a softly-softly approach rather than one of confrontation, of compromises rather than victories. Maudsley psychiatrists told a different story still. While they had to concede some strategic ground on the therapy issue, they hardly saw themselves as the losers.

Eysenck's broader professional influence was equally ambiguous. He was the most prominent voice in British clinical psychology for two decades after the war, his unstinting promotion of behavioural techniques accompanied by a scorched earth approach to Freudian therapeutics. But Eysenck could only hold the spotlight for so long. He otherwise had little appetite for the organizational nitty-gritty of profession-building; that was left to less heralded figures. Much to his chagrin, clinical psychology did not develop quite as he hoped. Eysenck's vision of systematic laboratory research feeding careful application ran hard up against the realities of professional practice. The 'applied learning theory' of behaviour therapy evolved as a mixed bag of cognitive–behavioural techniques geared to helping people help themselves. Clinical practitioners came to resemble adaptable highly trained artisans responding to the diverse demands of service, with academic researchers following as much as leading this evolution.

The beginning of the clinical course at the Maudsley

At the end of World War II, Lewis and Eysenck began planning for the clinical course. They effectively started with a clean slate. Other clinical psychology programmes would be independently initiated soon after at the Tavistock Clinic in London and the Crichton Royal Hospital in Dumfries. Lewis had no firm idea of what psychologists' role should be, other than to serve psychiatrists as part of a multidisciplinary approach to the problems of mental health. Psychiatrists could use the help. In the first decades of the twentieth century, the scientific achievements of this lowly speciality hardly stood comparison with the great strides made by mainstream medicine.[2]

---

[2] See for example Eliot Slater's vivid account of his time at Derby County Mental Hospital in Slater, E. (1971), Autobiographical sketch, in *Man, mind and heredity* (ed. J. Shields and

Psychoanalysis hadn't helped much. It had crept into Britain in the early 1900s, its radical iconoclasm popularized by a giddy mix of psychologists, psychiatrists, and the *literati*. The trauma of shell shock in the Great War boosted psychoanalytic modes of intervention. While many British psychiatrists became acquainted with analytic ideas in the aftermath of the war, this was seldom through the strict confines of their medical education. British psychiatrists never sought guild control over the practice of analysis as their counterparts did in the US. Thus psychoanalysis remained an abiding interest for a number of British academic psychologists as well, but not a therapeutic practice.[3]

The Maudsley had actually been a source of trenchant anti-psychoanalytic critique between the wars. Superintendent Edward Mapother was one of the most vocal opponents of Freudian methods, yet still covered his bases by maintaining a unit dedicated to dynamic psychotherapies. Much of the friction between the Maudsley and the more consistently psychoanalytic Tavistock Clinic amounted to personal enmities. However, this rivalry also invoked deeper intellectual questions about science and professional roles—particularly medical jurisdiction over the diagnosis and treatment of 'functional' neuroses. By the time Aubrey Lewis took over the re-opened Maudsley in 1946, the hospital employed several analytic psychiatrists. However, others at the hospital were actively hostile to this form of treatment, Lewis moderately so.[4]

I. Gottesman), pp. 1–23, Johns Hopkins Press, Baltimore. Slater described the comfortably numb life of an 'assistant medical officer'—one of administrative rituals and cushy routines that had little relation to improving patient welfare. Treatment regimes were limited at best, with chronic cases afforded only custodial care.

[3] The London Society of Psycho-therapeutics had been formed in 1901, the same year as the British Psychological Society. The London Psycho-Analytical Society was founded in 1913. The status and position of psychoanalysis within British psychology paralleled that of the discipline's occult doubles. Indeed, some psychoanalytic associations drew on a similar membership in their formative years. See Richards, G. (2000), Britain on the couch: the popularisation of psychoanalysis in Britain, 1918–1940, *Science in Context* **13**, 183–230, and Hinshelwood, R.D. (1995), Psychoanalysis in Britain: points of cultural access, 1893–1918, *International Journal of Psychoanalysis* **76**, 135–51. For more background on psychoanalysis and British psychology and psychiatry, see Hearnshaw, L. (1964), *Short history of British psychology, 1840–1940*, Methuen, London, and Berrios, G.E. and Freeman, H. (eds.) (1991), *150 years of British psychiatry, 1841–1991*, Vol. 1, Royal College of Psychiatrists, London.

[4] For some background on the anti-psychoanalytic movement in Britain, see Turner, T. (1996), James Crichton-Browne and the anti-psychoanalysts, in *150 years of British psychiatry, 1841–1991*, Vol. 2 (ed. H. Freeman and G.E. Berrios), pp. 144–55, Athlone Press, London.

Prior to World War II, psychology in Britain had little institutional overlap with psychiatry. Work with adult psychiatric patients was virtually unknown. The key sites for applied psychology were schools and child guidance clinics. The 'educational psychologists' working there had basic degree training at best, and often possessed some teaching experience. Most were women.

All that was about to change, as Lewis and Eysenck set about generating a template for a paramedical profession out of what was an essentially academic discipline. Lewis's multidisciplinary approach borrowed from the standard model in child guidance. It functioned according to a well-understood pecking order—but this would require constant policing. At the time, Lewis noted that most of the research departments and programmes at the new Institute of Psychiatry (IoP) would develop their own methods and concepts as separate sciences; experimental therapeutics was therefore best left to psychiatrists with on-the-job experience, backed by drug research. 'The alert clinician or the pharmacologist' could better initiate therapeutic investigations, Lewis argued, rather than full-time researchers assigned the task.[5] He clearly saw this area as exclusively medical, a sign of things to come.

Eysenck's initial views on the clinical course make fascinating reading. At one point he noted: 'As the people being trained as clinical psychologists will have to work under the heading of "educational" psychologists, it is imperative that our training should be recognised by those in power ... [by including] someone who was herself an educational psychologist...'[6] Eysenck went on to suggest that fellow émigré Hilde Himmelweit would make an excellent deputy as directory of the clinical course. She would be assisted to by Monte Shapiro, already working in the Maudsley Children's Department, and one other staff member.

Two things are striking. First, Eysenck accepted that clinical psychology would inevitably be seen as a kind of variant of psychologists' testing and counselling work undertaken in educational contexts. Second, he seemed to prefer Himmelweit to Shapiro. Perhaps he did not rate the inexperienced Shapiro highly, which would make for an interesting background element to the schism that developed between them in the 1950s. Nevertheless, Shapiro came to head the new clinical section. While it was not clear what sort of behind the scenes manoeuvring led to Shapiro's appointment, Eysenck later suggested that it was probably all for the best. According to Eysenck,

---

[5] AJL/JMT, 'Research', 11 July 1945, Aubrey Lewis Papers, Box 11, Bethlem Hospital Archives, p. 1.

[6] Hans J. Eysenck to Aubrey Lewis, c. 1946, Aubrey Lewis Papers, Box 11, Bethlem Hospital Archives, p. 1.

Himmelweit had 'Freudian tendencies' and, as a woman, might have been easily intimidated by patrician psychiatrists.[7] In any case, Himmelweit soon left the IoP for the London School of Economics, where she would head social psychology.

In 1947, Monte Shapiro assumed responsibility for the clinical course and for organizing psychological services for the joint hospital. On the face of it, Shapiro and Eysenck had much in common. They shared key intellectual sources, but each took very different lessons from them. Four years older than Eysenck, Shapiro was born in Germiston, South Africa to Jewish parents. He had taken a degree in psychology at Rhodes University in the 1930s and left for Cambridge to do graduate work with Oliver Zangwill.[8] Shapiro's studies were interrupted by the war, and he served in the Royal Air Force as a navigator—even though he apparently had limited spatial skills. He was shot down over the Netherlands in May 1943 and spent over a year as a prisoner of war.

During this period, Shapiro had the opportunity to practise psychotherapy on fellow prisoners. Shapiro read Pavlov's *Conditioned reflexes* and began to trial various conditioning principles, which would become the basis of his case-based approach at the Maudsley.[9] Demobbed near the end of the war, Shapiro met Desmond Furneaux at the 1945 British Psychological Society (BPS) conference in Exeter and, through him, got to know Eysenck. Even though Shapiro had no psychiatric experience or training in diagnostic testing, he obtained a half-time post in the children's department. He spent the rest of his time in the Maudsley library catching up on the formal clinical education he never had.

Clinical section staff initially had three roles: providing psychological services to the Maudsley and Bethlem hospitals; conducting research; and, teaching clinical trainees. Money for the academic appointments in the clinical section came from publicly funded hospital coffers for the service they provided, paid by the newly created National Health Service (NHS).

Officially, the clinical course began in mid-1947 with around a half dozen students completing the 12 month programme of instruction, although at the

---

[7] Hans J. Eysenck, interviewed by H.G. Gibson, 1 March 1979; Eysenck (1997b), *Rebel with a cause*, p. 141.

[8] Berger, M. (2000), Monte Shapiro [obituary], *The Independent*, 1 May 2000; Berger, M. and Yule, W. (1979), Retirement of an enthusiast—Dr. M.B. Shapiro, *Bethlem and Maudsley Gazette* winter 1979, 16–17.

[9] Payne, R.W. (2000), The beginnings of the clinical psychology programme at the Maudsley Hospital, 1947–1959, *Clinical Psychology Forum* **145**, 17–21.

time it led to no formal qualification.[10] Intakes grew to double that number by 1949 and the course expanded to 13 months in duration leading to the postgraduate Diploma in Clinical Psychology, analogous to the Diploma in Education. The following year the course was granted university recognition, the first in the country. It later became a one year M.Sc. course in 1966 and then two-year M.Phil. a decade later. Due to the labour-intensive nature of clinical supervision, staff–student ratios were kept very low, almost one-to-one. Clinical intakes hovered at around a dozen well into the 1970s. By 1949, Shapiro headed the clinical section, with seven other psychologists under him. The section always had more senior staff and fewer students than the bottom-heavy research section. Clinical section staff worked in different areas and, as a consequence, their activities were more diverse, and their recollections were more divergent, than those of the research section. Each was attached to a different part of the joint hospital, such as the children's department, adult outpatient department, Bethlem geriatric unit, and so on. Students were rotated through these departments and units, supervised by a psychology staff member who in turn worked under the direction of a psychiatrist.[11]

The clinical course attracted students from a variety of backgrounds, some of whom returned to their native countries after graduation. Most came specifically for the professional course; only a few did the clinical course and a research M.Sc. or Ph.D. as well. Bob Payne, Jack Ingham, and Irene Martin were three who did both in these early years. Payne and Martin were particularly important figures linking the research and clinical sections in the 1950s. Several students were employed as research assistants during the early years of the course, and subsequently joined the clinical staff. These included Aubrey Yates, Jimmy Inglis, and Reg Beech.[12] Gwynne Jones, who had graduated the clinical course with Bob Payne in 1950, returned from Wales to join the IoP clinical staff two years later.

---

[10] Behind the scenes, Cyril Burt had apparently tried to stymie their efforts and the course wasn't officially recognized by the University of London until 1950. There is also some confusion about when the course actually started. Official 50 year celebrations at the IoP put the start date at mid-1947. However, Shapiro suggested that the course started the year before. See Shapiro, M. (1955), Training of clinical psychologists at the Institute of Psychiatry, *Bulletin of the British Psychological Society* **26**, 15–20; Gibson, H.B. (Tony) (1981), *Hans Eysenck: the man and his work*, Peter Owen, London; and, Eysenck (1997b), *Rebel with a cause.*

[11] This information comes from the recollections of clinical staff, and various IoP annual reports.

[12] Payne (2000), The beginnings of the clinical psychology programme at the Maudsley Hospital, 1947–1959.

Preoccupied with his research, Eysenck left much of the initial direction and content of the clinical course to Shapiro. At first, it was not entirely clear what kind of role psychologists would take in the hospital. Most expectations revolved around psychologists' traditional expertise in psychological testing.[13] Maudsley psychiatrists saw these newly arrived psychologists as adjunct technicians, useful in clarifying and supporting their own diagnostic judgements. Such a view was consistent with NHS definitions. Psychiatrists would order particular tests to assess cognitive functioning and personality, the Rorschach a common choice for the latter task. The first battle Shapiro and his consultants faced was convincing psychiatrists that they were best qualified to judge which tests should be done and why. By standing firm on their status as independent experts, Shapiro emphasized tests that could be defended in empirical terms—a point Aubrey Lewis also strongly supported. Shapiro could not bring himself to teach and use the Rorschach, hiring Swiss projective expert Maryse Israel to handle demands for the technique in the late1940s.[14]

In his own image

Up until the late 1940s, Eysenck had taken a back seat in the affairs of the clinical section. In the middle of 1949, however, Lewis organized for Eysenck to spend six months as visiting professor at the University of Pennsylvania. Part of Eysenck's brief was to study the development of clinical psychology there. Nonetheless, he spent much of his time reading, writing, and promoting his point of view with guest speaking spots—and turning down the numerous professional offers that came his way. As far as clinical psychology went, his mind had already been made up. Prior to leaving and soon after his return, Eysenck wrote a series of articles delimiting the role of clinical psychologists.[15]

---

[13] Shapiro (1955), Training of clinical psychologists at the Institute of Psychiatry.

[14] Berger and Yule (1979), Retirement of an enthusiast—Dr. M.B. Shapiro.

[15] It appeared that the trip did little to change Eysenck's mind, given the consistency of his publications before and after. See Eysenck, H.J. (1949), Training in clinical psychology: an English point of view, *American Psychologist* **4**, 173–4; Eysenck, H.J. (1951*b*), Psychology Department, Institute of Psychiatry, Maudsley Hospital, University of London, *Acta Psychologica* **8**, 63–8; Eysenck, H.J. (1950*a*), Function and training of the clinical psychologist, *Journal of Mental Science* **96**, 710–25. The last of these papers explicitly incorporated his observations in the US, but these were used to reinforce points already made. See also Derksen, M. (2001), Science in the clinic: clinical psychology at the Maudsley, in *Psychology in Britain: historical essays and personal reflections* (ed. G.C. Bunn, A.D. Lovie, and G.D. Richards), pp. 267–89, BPS Books, Leicester.

He spoke not just to his local circumstances, but to present and prospective psychologists throughout Britain.

From high in his office on Denmark Hill, Eysenck gazed into the future of the clinical field and saw something resembling himself. The clinical psychologist was, above all, a detached scientist—one who applied rigorous intellectual skills and quantitative methods to the hitherto intractable problems of psychiatry. In these early writings, Eysenck argued that psychologists should be the ones to provide the research basis for clinical practices. They should respond only to the problems thrown up by their science, not to social need. Direct experience treating mental problems was unnecessary, and any requirement to this effect would result in a retreat from science. Eysenck saw US developments as a mistake, a gigantic one at that. The Americans had rushed at the opportunity to become service providers. Their scientist–practitioner model of training likewise put far too much emphasis on preparing students for such a role. It was craven imitation of a medical model capped off by an obsession with psychotherapy. For Eysenck, this was costing US psychologists both their science and their independence.

Eysenck wanted to distance clinical psychology from psychoanalysis. However, at this point he also wanted to appease the man in charge, Aubrey Lewis. He placed a strong prohibition on psychologists practising psychotherapy, which he defined as a fundamental change of the patient's outlook and behaviour by psychological means. Eysenck took particular issue with 1947 APA training recommendations (the 'Shakow report') that psychologists be trained in psychotherapy and receive some kind of therapeutic self-evaluation in the process. This kind of 'premature crystallization of a spurious orthodoxy' was inimical to the cultivation of the appropriate scientific mindset in the young trainee.[16] Psychologists were better off sticking to their historical strengths of research and diagnostic testing, Eysenck declared. This would result in a more efficient division of labour within the mental health care team. Besides, the frequent presence of organic complications gave medically trained psychiatrists what Eysenck thought of as an 'unanswerable' mandate over treatment.

Even then, though, there was a 'chink of light' in this prohibition.[17] Eysenck suggested there were certain 'borderline areas which are clearly the psychologist's

---

[16] Eysenck was quoting D.C. Wright on this point, whom he took to be an authority on psychoanalysis. Eysenck (1950a), Function and training of the clinical psychologist, p. 720.

[17] This phrase comes from Jack Rachman's insightful piece, Rachman, S. (1981), H.J. Eysenck's contribution to behaviour therapy, in *Dimensions of personality: papers in honour of H.J. Eysenck* (ed. R. Lynn), pp. 315–30, Pergamon Press, Oxford.

prerogative ... [including] the treatment of educational difficulties, such as reading defects, speech impairments, and other difficulties.'[18] Psychologists could engage in a kind of re-training akin that done by educational psychologists. Eysenck was rationalizing what was already beginning to occur in children's department at the Maudsley. But he otherwise appeared to deplore the *sub rosa* practice of psychotherapy on children by poorly trained psychologists.

Eysenck's vision for clinical psychology was that of an independent but complementary science. While psychiatrists retained ultimate responsibility for patient care, psychologists would provide the basic research in psychopathology and diagnostics to facilitate psychiatrists' treatment. Moreover, psychologists would be the ones best placed to evaluate the results.

Stage one: attacking analytic psychiatry

Eysenck always tried to lead by example. In the mid-1940s, he had already outlined a dimensional approach to personality as an alternative to psychiatric taxonomy. Now he framed the professional activities for clinical psychologists as independent but complementary to psychiatry. Eysenck then went on to demonstrate the critical contribution psychologists could make arbitrating on psychiatric practices in the clinic. When he first arrived at Mill Hill, Eysenck computed the reliability of diagnoses given by the psychiatrists there, but was prevented by the medical hierarchy of the hospital from publishing the embarrassingly low figure he obtained. The 1949 US trip apparently gave him another idea though, one that would create an even bigger fuss.

Eysenck again targeted the 1947 APA report that served as a whipping boy for his 'English viewpoint'. He wondered aloud whether its positive recommendation for training in psychotherapy could be empirically justified. Eysenck had those forms of therapeutic intervention identified with psychoanalysis squarely in his sights. He recalled that 'in New York alone, I was told that there were 100 different types of psychoanalysis that people were using.'[19] However, Eysenck saw a conspiratorial absence of evaluation. 'Where was the evidence that psychotherapy worked?' he asked.

The timing of this question was historically apposite, to say the least. It was the golden age of psychotherapy in the US, a cultural high watermark for psychoanalysis in particular. The practice of psychotherapy had helped bring

---

[18] Eysenck (1950*a*), Function and training of the clinical psychologist, p. 721.

[19] See p. 424 of Feltham, C. (1996), Psychotherapy's staunchest critic: an interview with Hans Eysenck, *British Journal of Guidance and Counseling* **24**, 423–36.

American psychiatry out of asylums in the interwar years, sold as a potent means to self-fulfilment. It promised bloodless, civil encounters, and made consultative, office-based practice fashionable. American psychologists thereby found themselves in a dilemma. Psychoanalysis was the antithesis of their experimental science. Yet psychiatrists, the professional leaders in the mental health field (and their chief rivals outside academia), had made so much of it and made it their own. Moreover, psychiatrists had attempted to claim a monopoly over the practice of all forms of psychotherapy, a situation their psychological counterparts could not afford to leave unchallenged.

Psychoanalysis had already spawned many 'talk therapies' by mid-century. Some of the more notable variants, such as Carl Rogers' client-centred approach, had been developed by psychologists. Since the 1920s, American behaviourists and learning theorists had also suggested a range of practical therapeutic techniques. These scattered ideas of John B. Watson, Mary Cover Jones, Knight Dunlap, the Mowrers, and Andrew Salter remained undeveloped. American psychologists remained fixated with psychoanalysis—half wanting it, half rejecting it, but prevented from truly having it.[20]

It would be easy to claim that professional priorities forbade sober assessment, that analytic psychiatrists especially did not want to question the efficacy of their work. While half true, it also ignores the assumptions and uncertainties of the period. Psychoanalysis had always sat uneasily with materialist medicine that emphasized discrete syndromes and hitherto undiscovered organic aetiology. Psychoanalysis had developed as a kind of participatory healing, insular and transactional. It was a process, a journey that might not ever end. Outcome measures made little sense in a discourse where deterioration was interpreted as a 'negative therapeutic reaction'. In any case, psychotherapeutic practitioners did not appear to be plagued by professional self-doubt. Personal experience, case studies reports, and anecdotal evidence had given them an overwhelming impression that psychotherapy helped patients. If more formal evaluation seemed unnecessary it might also have seemed premature, for there was no consensus on whether empirical comparisons were feasible or appropriate. Moreover, diverting resources in order to carry out lengthy evaluation studies did not appear to be a desirable option in the face of pressing therapeutic needs.

Compelled to go where others feared to tread, Eysenck said he quickly 'collected the papers and did my study'. It was not 'unique or very original',

---

[20] For a general history of behavioural therapies see Kazdin, AE. (1978), *History of behaviour modification: experimental foundations of contemporary research*, University Park Press, Baltimore.

he allowed.[21] Indeed, it involved no new data. It was based on a review of around two dozen tentative evaluation studies and hospital reports, including several from the Maudsley. Other researchers had cautiously raised many of the same points. However, the uncompromising way Eysenck put it all together produced a thunderclap that reverberated through the clinics, consulting rooms, and colleges of the Western world.

Shock, horror: psychotherapy doesn't work!?

> [The figures] fail to prove that psychotherapy, Freudian or otherwise, facilitates the recovery of neurotic patients (p. 322 of Eysenck, H.J. (1952e), The effects of psychotherapy: an evaluation, *Journal of Consulting Psychology* **16**, 319–24)

The first shock was to break ranks and emphatically state a need for evaluation *now*. The second shock was to collate all the major studies and present them in the form of a standardized, quasi-experimental group comparison in an area where naturalistic *in vivo* studies had appeared to most service providers as, well, more natural.[22] The third shock was to highlight a new baseline for comparison. Eysenck suggested that many patients got better without the benefit of any form of psychotherapy, a phenomenon he termed 'spontaneous remission'. Others—such as Carney Landis at the New York State Psychiatric Institute—had pointed to this phenomenon. But no one had emphasized it with such telling effect. Two criteria were considered. Following Landis's psychiatric research, Eysenck ascertained the percentage of neurotic patients discharged annually from various US hospitals as recovered or improved. Alternatively, he collated the percentage of disability claimants recovered or improved over a *two*-year period from Denker's study of disability insurance claimants. Both sets of figures appeared to agree quite well, suggesting a spontaneous remission rate for neurosis of just over 70%. In a show of magnanimity, Eysenck settled on a slightly lower figure of two-thirds (i.e. 66%). Therapeutic efficacy now had to be measured against a much higher zero-point than many had imagined: it had to improve on the high rate at which neurotics were expected to get better with no treatment at all. It gave Eysenck the means to undermine therapists' positive personal experience.

---

[21] Feltham (1996), Psychotherapy's staunchest critic: an interview with Hans Eysenck, pp. 423–4.

[22] For an account of the tension between competing approaches to evaluation in the US, see Rosner, R. (2005), Psychotherapy research and the National Institute of Mental Health, 1948–1980, in *Psychology and the National Institute of Mental Health: a historical analysis of science, practice and policy* (ed. W.E. Pickren and S.F. Schneider), pp. 113–50, American Psychological Association, Washington, DC.

Finally, the fourth shock was to shift the burden of evidence on to the therapist. Aware that existing evaluation studies were full of holes, Eysenck did not risk trying to confirm a negative. Instead, he argued that psychotherapy should be *proven to work* before it was used and before students were trained in it, *not* the other way round. It was a logical master stroke, a sly evocation of the Hippocratic principle. It laid a trap that would claim many a hapless critic.[23] Eysenck argued that all psychiatric treatments should be held to the same standard—never mind that the practice of psychotherapy seemed relatively innocuous compared with some of the more radical somatic treatments (e.g. psychosurgery and various shock therapies) being pioneered at this time.[24] He treated the necessity of demonstrable efficacy as though it was observed in practice as much as it was in principle.[25] In effect, he dared psychotherapists to say this wasn't so, dared them to argue for widespread application before proper evaluation.

Protected by a cherished oath, Eysenck loaded the dice against psychotherapy. For example, Eysenck assumed that his control group cases had not received *any* kind of psychotherapy whilst in a psychiatric hospital or in the care of a doctor unless specifically stated.[26] He also classified deaths and drop-outs,

---

[23] For example, the Menninger Foundation's Lester Luborsky appeared to fall into this trap. Luborsky argued Eysenck's paper was based on poor data that invalidated Eysenck's conclusions. Luborsky argued that indeterminate data should not preclude students being trained in psychotherapy. Luborsky, L. (1954), A note on Eysenck's article: 'The effects of psychotherapy: an evaluation', *British Journal of Psychology* **45**, 129–31. Eysenck's straightforward reply highlighted Luborsky's misunderstanding of his approach. Eysenck pointed out that, even if one disputed the comparison he had set up, the non-proven verdict would still hold true. Eysenck, H.J. (1954c), A reply to Luborsky's note, *British Journal of Psychology* **45**, 132–3.

[24] Then and now, psychotherapy was seen as more appropriate for less severe cases. Chronic cases languishing in the back wards of public institutions were seen as beyond the reach of talk therapy. Class also played an undeniable role. Psychoanalysis in particular was pitched to the well-educated and highly verbal, and lent itself to consultative private practice. Even so, the widespread mid-century adoption of more radical somatic forms of treatment tended to defy such class generalizations.

[25] Luborsky, for one, articulated the opposite view, suggesting that 'if medical doctors had followed such advice their entire science would not have developed'. Luborsky (1954), A note on Eysenck's article: 'The effects of psychotherapy: an evaluation', p. 129. However, Eysenck's cautionary approach was easier to sell to a broad audience, even if it simply collapsed together the potential risks and benefits of all forms of treatment.

[26] Eysenck later qualified this assumption considerably. See Eysenck, H.J. (1994a), The outcome problem in psychotherapy: what have we learned? *Behaviour Research and Therapy* **32**, 477–95.

along with 'slightly improved' and 'not improved', as therapeutic failures. His spontaneous remission estimate bluntly equated cure with discharge or discontinuance of care. Moreover, his 66% figure was based on data from the overcrowded US state asylums and the insurance industry, both contexts where there was a strong incentive to mark up improvement and move patients on. However, any methodological flaws in the evaluation studies he reviewed merely added weight to his argument, Eysenck argued. If his paper was inconclusive, it was because he was faithfully reporting studies that were themselves problematic. His doubting point remained: psychotherapists still had to prove themselves.

Eysenck's results were striking. Eclectic psychotherapists were apparently performing poorly, their cure rate of 64% on a par with that of spontaneous remission. However, psychoanalytic practitioners were doing even worse. Their 44% cure rate saw them damningly outperformed by a strategy of doing nothing at all. Eysenck had produced remarkable numbers suggesting that the analyst's couch was actually detrimental for a significant number of its occupants.[27]

Eysenck's article was published in the American-based *Journal of Consulting Psychology*, as well as in abbreviated forms in the local *Quarterly Bulletin of the British Psychological Society* and *Proceedings of the Royal Society of Medicine*. [28] The impact was dramatic. Eysenck's conclusions caught the approving attention of some academic psychologists; they had their skepticism confirmed.[29] However, this reaction was overwhelmed by the almost hysterical condemnation from those with a stake in psychotherapy, a kaleidoscope of psychologists

---

[27] It could be argued that Eysenck's indictment of the negative effects of psychoanalysis anticipated the condemnations of latter-day anti-Freudians. The fate of other therapeutic innovations also made Eysenck look prescient, though slightly off the mark. The psychosurgical procedures and shock treatments widely adopted in the 1940s and 1950s are *now* cited as salutary examples of reckless experimentation prior to cool evaluation, a black stain on the history of the profession. See Pressman, J.D. (1998), *Last resort: psychosurgery and the limits of medicine*, Cambridge University Press, Cambridge. A less well-known example from earlier in the twentieth century comes from Scull, A. (2005), *Madhouse: a tragic tale of megalomania and modern medicine*, Yale University Press, New Haven, Connecticut.

[28] Eysenck, H.J. (1952e), The effects of psychotherapy: an evaluation, *Journal of Consulting Psychology* **16**, 319–24; Eysenck, H.J. (1952d), The effects of psychotherapy, *Quarterly Bulletin of the British Psychological Society* **3**, 41; Eysenck, H.J. (1952c), The effects of psychotherapy, *Proceedings of the Royal Society of Medicine* **45**, 447.

[29] While my evidence for this point is mostly anecdotal, the University of Miami's Edwin Erwin gave a latter-day articulation of this viewpoint in 1980. See Erwin, E. (1980). Psychoanalytic therapy: the Eysenck argument, *American Psychologist* **35**, 435–43.

and psychiatrists who saw it as a crude and outrageously slanted attack. Yet they were caught in a bind. While they were convinced Eysenck was mistaken, they lacked good counterevidence. Otherwise, harping on the biases of Eysenck's study would only give it more attention than it deserved. Thus, in the 1953 *Annual Review*, Nevitt Sanford suggested it would be best to ignore Eysenck.[30] But turning the other cheek would soon become impossible.

This one short article gave Eysenck his biggest, most direct impact across the Atlantic. 'The effects of psychotherapy' became Eysenck's most cited work.[31] Joseph Zubin—who had worked with Landis in New York—confided to Eysenck that he and his American colleagues would probably 'get the sack' if they said something like this publicly. But Zubin also warned that such confrontational tactics might well be counterproductive, undermining working relations with psychiatrists across the board.[32] Unperturbed, Eysenck refused to back off.

Many read (or misread) Eysenck as arguing that psychotherapy did not work.[33] Eysenck was careful not to make such a claim—apart from mischievously suggesting there may be 'an inverse correlation between recovery and psychotherapy'.[34] But his conclusions said one thing; his comparative numbers said another. Many critics instead attacked Eysenck's baseline of comparison and tried to establish spontaneous remission rates considerably lower than

---

[30] As a left-leaning social psychologist with Freudian sympathies, Sanford represented almost everything Eysenck was beginning to define himself as against. See Sanford, N. (1953), Psychotherapy, *Annual Review of Psychology* **4**, 317–42.

[31] I can't be absolutely conclusive on this point, but a sample of citations make this claim appear assured. From 1961 to 1980, 'The effects of psychotherapy' was cited over 275 times. *Current Contents* (1980), This week's citation classic, *Current Contents* **46** (11 August 1980), 275. It was one of four citation classics identified by J. Philippe Rushton, along with a 1968 paper on the P scale (cited more than 100 times from 1968 to 1986), the 1967 book *The biological basis of personality* (cited more than 855 times from 1967 to 1987), and the 1975 manual for the EPQ (cited more than 770 times from 1975 to 1990). Rushton, J.P. (2001), A scientometic appreciation of H.J. Eysenck's contribution to psychology, *Personality and Individual Differences* **3**, 17–39.

[32] Eysenck (1997b), *Rebel with a cause*, p. 128; Sybil Eysenck, interview, 13 September 2004.

[33] See for example, Luborsky (1954), A note on Eysenck's article: 'The effects of psychotherapy: an evaluation'; Strupp, H. (1963), The outcome problem in psychotherapy revisited, *Psychotherapy: Theory Research and Practice* **1**, 1–13; and, Eysenck, H.J. (1964d), The outcome problem in psychotherapy: a reply, *Psychotherapy: Theory Research and Practice* **1**, 97–100. For a summary of this criticism see Smith, M. et al. (1980), *The benefits of psychotherapy*, Johns Hopkins Press, Baltimore.

[34] Eysenck (1952e), The effects of psychotherapy: an evaluation, p. 322.

his provocative two-thirds dictum.[35] The exchanges were quite heated and illustrated a clash of cultures between the rule of science and the demands of the clinic. It was an especially rude awakening for practitioners who were not well-versed in statistical evaluation and unused to having their work assessed by outsiders. Underneath lurked historic tensions between participatory understandings and detached calculation—one side exemplified by in-depth, discursive case studies, the other represented by standardized group comparisons and quantitative statistics.[36]

No one could ignore the new terms of debate that Eysenck had helped impose. While psychoanalytic diehards began to flounder, the less aligned took on board the need to demonstrate efficacy with hard numbers.[37] More than just creating shockwaves, Eysenck's article helped create a whole new field—not because those with an interest in psychotherapy wanted to emulate Eysenck so much as they wanted to prove him wrong.[38] Psychotherapy evaluation research took off in the US, in part supported by the newly formed National Institute of Mental Health. A community of scholars emerged—mostly psychologically trained—where no identifiable subgroup had existed before.[39] The practical and methodological issues of study design, criteria, and case

---

[35] Rosenzweig, S. (1954), A transvaluation of psychotherapy—a reply to Hans Eysenck, *Journal of Abnormal and Social Psychology* **127**, 330–43; Bergin, A. (1963), The effects of psychotherapy: negative results revisited, *Journal of Counseling Psychology* **10**, 244–50; Bergin, A. and Garfield, S. (eds.) (1971), *Handbook of psychotherapy and behaviour change*, Wiley, New York.

[36] For an example of the resistance to the rule of hard numbers see Hall, S.B. (1955), Psychotherapy: misapprehensions and realities, *British Journal of Medical Psychology* **26**, 295–9.

[37] The writings of Paul Meehl also helped reinforce this shift in the mid-1950s. Meehl's book (Meehl, P.E. (1954), *Clinical versus statistical prediction: a theoretical analysis and a review of the evidence*, University of Minnesota Press, Minneapolis) made a similar impact in the US to that of Eysenck's article. Meehl argued that the available evidence pointed to the superiority of mechanical actuarial methods over intuitive expert judgement in almost all cases, prompting a similar outcry from many of the same people Eysenck had outraged.

[38] For example, Gene Glass said Eysenck's writings motivated him to develop meta-analysis to get a grip on the effects lurking in the diverse range of evaluation studies available. 'Looking back on it, I can almost credit Eysenck with the invention of meta-analysis by anti-thesis.' Gene V. Glass, "Meta-Analysis at 25," January 2000, p.5, http://glass.ed.asu.edu./gene/papers/meta25.html.

[39] Even Hans Strupp, one of Eysenck's many sparring partners on this issue, gave him credit for getting the research ball rolling. See Strupp, H. and Howard, I.I. (1992), A brief history of psychotherapy research, in *History of psychotherapy: a century of change* (ed. D.K. Freedheim), pp. 309–34, American Psychological Association, Washington, DC.

comparability—not to mention the role of suggestion and placebo effects—occupy investigators to this day.

Broader developments would only help Eysenck's cause. The widespread promulgation of psychotherapy in the US in the post-war years introduced a raft of new stakeholders, all with an interest in whether it worked. By the early 1960s, various levels of American government wanted to play a greater role in regulating psychotherapeutics. Insurance companies wanted to calculate premiums and limit reimbursements. And a more curious and informed public wanted reasons to believe the assurances of its mostly medically trained practitioners—despite the doubts expressed by *other* kinds of experts like Eysenck. These stakeholders helped transform the technical content of the debate, for their influence reached inside the scientific beltway to change or create new forms of specialized knowledge. Each pressed for accountability in a manner that favoured the use of impersonal numbers and robust, translatable yardsticks. This trend would peak in 1980 with the arrival of DSM III—touted as reliable, accurate, and easy-to-use—a triumph of standardization that de-skilled and democratized diagnostic practices.[40]

Eysenck revisited the issue in 1960 and in 1965 and essentially came to the same conclusions.[41] By the early 1970s, Eysenck claimed his negative conclusion was all but proven: dynamic psychotherapy did not work.[42] Eysenck's sceptical stance had a long-term cost, however. His 70% spontaneous remission rate remained a tremendous hurdle for any psychological therapy to surmount, including the new kind of technique he began to advocate in the late 1950s. Any failure to do better than this was a potential point of criticism. Any failure to carry out thorough empirical evaluations in the first place left him open to a charge of hypocrisy.

---

[40] See Buchanan, R. (2003), Legislative warriors: American psychiatrists, psychologists and competing claims over psychotherapy in the 1950s, *Journal of the History of the Behavioral Sciences* **39**, 225–49, and Mayes, R. and Horwitz, A.V. (2005), DSM-III and the revolution in the classification of mental illness, *Journal of the History of the Behavioral Sciences* **41**, 249–67.

[41] His original 1952 data was re-analysed by European researchers who suggested his calculations were faulty and biased. See Duhrssen, S. and Jorswieck, E. (1963), Zur Korrektur von Eysenck's Berichterstattung über psychoanalytische Behandlungsergebnisse, *Acta Psychotherapeutica* **10**, 329–42, and Eysenck, H.J. (1964*f*), The effects of psychotherapy reconsidered, *Acta Psychotherapeutica* **12**, 38–44.

[42] See Eysenck's comments on his original 1952 article reprinted in Eysenck, H.J. and Wilson, G.D. (eds.) (1973), *The experimental study of Freudian theories*, Methuen, London, p. 374.

Reformation the Maudsley way

For Eysenck, 'The effects of psychotherapy' marked the beginning of a long anti-Freudian offensive. He would attack the scientific credibility of Freudian theory and practice to his dying days. Eysenck began to dominate as the public spokesman for the Maudsley brand of clinical psychology, a programme he deliberately contrasted with the interdisciplinary psychodynamics of the Tavistock.[43] Emboldened, Eysenck next attempted to impress the 'Maudsley way' on the organizational structure of British psychology. With his representatives Shapiro and Gwynne Jones leading the charge, Eysenck sought to reform the medical section of the British Psychological Society (BPS). Formed in 1920, the section had been a professional home for medically trained practitioners with an interest in psychological therapeutics. What united many section members was their interest in psychoanalysis, an interest not readily catered for by existing medical bodies. A substantial proportion of the medical section were London-based medical practitioners. Many were also European émigrés; a few were psychiatrists or lay practitioners. Despite an influx of psychologists into the section after World War II, the Harley Street influence and psychoanalytic focus remained. Those at the Maudsley saw the section as a Tavistock front, an organizational anachronism that was holding back the development of scientific clinical psychology.[44] Time to storm this Freudian citadel.

In 1952, Eysenck complained in print that the section's flagship periodical, the *British Journal of Medical Psychology*, was increasingly filled with 'speculative' papers on 'idiographic, psychoanalytic, and other "dynamic" topics'.[45] He and Maudsley psychiatrist Wilhelm Mayer-Gross later ruffled feathers by

---

[43] The approach taken at the Crichton Royal lay somewhere in the middle of the intellectual–ideological poles represented by the 'Tavvy' and the Maudsley. The key figure in the early development of the Crichton Royal course—'contemplative and humanistic' in style—was noted test developer John Raven. The quotation comes from Richards, B. (1983), Clinical psychology, the individual and the welfare state, Ph.D. thesis, North East London Polytechnic, p. 137.

[44] The section did in fact have a broader base than this. Nevertheless, Maudsley psychiatrist Eliot Slater remembered the political divisions of the interwar years as quite sharply drawn. 'If the enemy [the Tavistock group] had captured the Medical Section of the British Psychological Society, we had a firm grip on the Psychiatric Section of the Royal Society of Medicine. Neither group paid much attention to the effete and recumbent Royal Medico-Psychological Association, which was regarded as just a club for medical superintendents!' Slater (1971), Autobiographical sketch, p. 17.

[45] Eysenck, H.J. (1952b), Letter to the editor, *Quarterly Bulletin of the British Psychological Society* 3, 97–8.

having these concerns taken to the BPS council, bypassing the section's leadership. Characterized as a 'small group of clinical psychologists well known for their intolerance of dynamic viewpoints', Maudsley psychologists pushed for a broader intellectual agenda.[46] Section stalwarts—like secretary Alfred B.J. Plaut, the Tavistock's Michael Balint, and psychiatrist T.F. (Tom) Main—saw this as a hostile takeover, a dire attempt to exorcise the section's traditional psychoanalytic outlook and sever its medical links. They did everything in their power to resist these psychologists pounding at the gates, whose numbers were swelling by the year. Two extraordinary general meetings were held, late in 1955 and early in 1956, to deal with the issue. At the first of these meetings in November 1955, Eysenck argued that in a developing field, all ideas, papers, and treatments should be judged on their scientific merit, not according to their allegiance to any particular approach. He was up against his intellectual antithesis, a school of thought that argued that the 'therapeutic relationship admits the psychotherapist a point of observation about human behaviour inaccessible to any other observer or experimenter, and that the findings made from this point are of cardinal importance to the general body of psychology…'[47]

Eysenck had relatively strong backing from BPS psychologists generally. He also had some support within the medical section membership, though it was not clear how much. Fearing a take-over, the section's leadership held over the election of new members and engineered a return of the section's ruling committee by post in January 1956. After the second extraordinary meeting in February 1956, the Maudsley group was offered its own subdivision within the section. Apparently, this smacked of tokenism. Shapiro and Jones rejected the offer, preferring their vision of a 'unified, single and liberalised' section.[48] Later that year a record number of membership applications were dealt with.

---

[46] 'Elections to membership: report of the Medical Section Committee', Agendum: business for BPS Council Meeting, 1 December 1956, BPS Medical Section, Secretary's Correspondence 1955–1956, History of Psychology Centre, BPS, London, p. 1.

[47] This was part of a paragraph the medical section committee moved to add to its charter rules in 1955 at the height of the squabble. 'Notes on historical and current trends within the Medical Section of the British Psychological Society', 12 July 1955, BPS Medical Section, Secretary's correspondence 1954–1955, History of Psychology Centre, BPS, London, p. 3.

[48] A. Plaut to H. Gwynne Jones, c. May 1956, BPS Medical Section, Secretary's Correspondence 1955–1956, History of Psychology Centre, BPS, London, p. 1. Plaut was quoting Shapiro and Jones's demands, adding that while he 'greatly respected the idea of a "unified, single and liberalized section" …this does not offer us a practical solution of out present impasse.' The differences in approach were simply too great for this to work, Plaut concluded.

These 22 applications were subject to stricter than usual scrutiny, Main admitted, given that routine admission had led to 'unsympathetic' members like Gwynne Jones, Jimmy Inglis, and Hans Brengelmann being accepted in the past. Seven applicants were not even put to the ballot because they had not made 'a sufficient case for membership', including Sybil Eysenck, Reg Beech, and Cyril Franks.[49] Despite this, 10 out of the 15 applications were still voted down—an extraordinary event given the section had vetoed only nine applications over the past decade. Amidst accusations of vote rigging and legal wrangling, the BPS council moved to suspend further membership elections on the 3rd of November 1956.[50] Eventually things settled down to business as usual. Despite impassioned arguments for a more scientific, psychological organization, the 'Maudsley coup' was stymied by the existing leadership—who apparently harboured one or two regrets over the 'irregular procedures' used to do it.[51] The section remained a medical-analytic preserve well into the 1960s, effectively declaring itself independent from the BPS and discouraging attempts to professionalize the clinical field. One of Eysenck's young ex-students, Joseph Sandler, cut his combative teeth here, having been quickly elevated to the section's executive echelons. Eysenck's scorched-earth approach to psychoanalysis would help clarify Sandler's allegiances for the rest of his career.

The episode can't have done much to persuade Eysenck of the value of working for change from within, nor of the intellectual openness of psychoanalysts. Blocked on this occasion, he did not engage with psychoanalysts in a direct organizational manner again. Instead, Eysenck shifted his attack to a print propaganda war where he was far more comfortable. Eysenck's support for Shapiro in his dealings with the medical section followed his protest over Shapiro's rejection by the Experimental Psychology Society a few years earlier. The sum of these disheartening experiences appeared to make Eysenck shift tack, more determined to create his own associations and forums rather than having to accept the compromises of established ones. However, he and Shapiro had begun to fall out on other scores, in a way that was both cause and effect for Eysenck's second-stage strategy.

---

[49] 'Elections to membership: report of the Medical Section Committee', Agendum: business for BPS Council Meeting, 1 December 1956, BPS Medical Section, Secretary's correspondence 1955–1956, History of Psychology Centre, BPS, London, p. 2.

[50] Pilgrim, D. and Treacher, A. (1992), *Clinical psychology observed*, Tavistock, London, pp. 15–16.

[51] Michael Balint to Alfred Plaut, 28 November 1955, BPS Medical Section, Secretary's correspondence 1954–1955, History of Psychology Centre, BPS, London, p. 1.

Fall-out from the Eysenck–Shapiro split

Hans Eysenck and Monte Shapiro had a quietly fraught relationship. Personal events appeared to strain their relationship. A committed member of the Communist Party, Shapiro had first brought the like-minded Sybil Rostal to the Maudsley in the late 1940s. By all accounts, Shapiro was a very moral man, almost rigidly so.[52] He was not entirely at ease with the budding relationship between the still-married Eysenck and Ms Rostal, the glamorous young research assistant. When Eysenck went entirely public with this relationship after returning from the US in January 1950—where Sybil had joined him as his secretary—he shocked his more conservative colleagues. The idea of divorce was still unusual, even scandalous. Then and now, the roles cast in this kind of drama have a way of manipulating perceptions and dividing loyalties. Margaret Eysenck did not want a divorce and many thought of her as a wronged woman.[53] Shapiro tried to remain neutral and resented any attempt to enlist him to one side or the other.[54] He and his wife had been great friends with Margaret Eysenck too. While all this hardly led to an immediate breakdown in relations between Eysenck and Shapiro, it probably exacerbated fundamental intellectual differences that only widened over the years.

While Shapiro and Eysenck could complement each other's knowledge in some ways, they were never likely to see eye-to-eye in the practice of their science, for Eysenck was not a practitioner at all. He was, as Jack Rachman put it, very much the 'theoretical clinician'. He never ventured into the hospital or talked with patients. Shapiro, in contrast, was much more hands-on. He had to be. A lack of formal training had forced him to learn on the job. Moreover, Shapiro's service role put him in the firing line, with his fellow psychologists under pressure to give opinions and recommendations. Shapiro's response was to be patient and careful, to never go beyond the data of a particular case.

---

[52] Shapiro's strong moral convictions might seem a little old-fashioned nowadays, perhaps touchingly so. He would never, for example, use a hospital phone for anything that could be construed as personal use. Even calling home to say he would be late meant using a pay phone down the road. Shapiro left the Communist Party as a matter of principle after the Hungarian uprising in the mid-1950s. Alan and Ann Clarke, interview, 25 April 2002; David Shapiro, interview, 14 October 2003.

[53] Monte had married Alan Clarke's sister Jean; both Alan and Ann Clarke worked in the social psychiatry Unit at the IoP.

[54] For example, when Eysenck suggested Shapiro provide a surety for the new flat he and Sybil had taken residence in after returning from the US, Shapiro saw it as tantamount to asking for his blessing for their union and was greatly offended. William Yule, interview, 22 July 2003.

Key observations should be followed up, odd results needed to be investigated, and all claims needed to be substantiated. While Shapiro bowed to standardized testing as the bread and butter of applied practice, he questioned the reliability and validity of tests psychiatrists asked his fellow psychologists to do. Moreover, routine interpretation might gloss over important aspects of patient self-presentations. Shapiro argued that objective diagnostic devices could be adapted to a rigorous but flexible approach, enabling the clinician to zero-in on a particular patient's problems. To this end, Shapiro developed his own diagnostic device, the personal questionnaire, that allowed the clinician to assess patient's mood changes and condition at successive stages of the clinical process.[55]

Hans Eysenck was more inclined to speculate and was given to quick generalizations. While he was hardly 'anti-test' either, he came to dismiss those measures that did not fit into his notion of a detached, applied science. *Dimensions of personality* had contained an attempt to put the Rorschach in a standardized format, and several of Eysenck's students used the test in their research in the early 1950s. By 1955, professional circumstances had changed and Eysenck took a more aggressive stance coordinating his clinical colleagues' attacks on the Rorschach. Maryse Israel and Bob Payne were encouraged to broadcast their doubts about the test at the 1955 BPS conference in Durham, England.[56] Eysenck himself did not attack projective tests like the Rorschach—in English at least—until 1958.[57] By that stage, Maudsley psychologists had begun to challenge psychiatrists on the therapy issue and Eysenck had developed his own standardized personality test, the MPI.

---

[55] Berger and Yule (1979), Retirement of an enthusiast—Dr. M.B. Shapiro, p.17.

[56] Eysenck had apparently challenged those working in the hospital, including Maryse Israel, to blindly discriminate normal and neurotic Rorschach protocols. They could not do so. Gibson (1981), *Hans Eysenck: the man and his work*, p. 85.

[57] Their initial critiques of projective tests were buried in French journals. See, for example, Eysenck, H.J. (1955d), La validité des techniques projectives: une introduction, *Revue de Psychologie Appliquée* 5, 231–3, and Payne, R.W (1955), L'utilité du test de Rorschach en psychologie clinique, *Revue de Psychologie Appliquée* 3, 255–64. Three years later the professional Rubicon had been well and truly crossed and Eysenck published similar general conclusions in English. See Eysenck, H.J. (1958b), Personality tests: 1950–1955, in *Recent Progress in Psychiatry*, Vol. 3 (ed. G.W.T.H. Flemming), pp. 118–59, Churchill, London. Eysenck got much more attention the following year with his trenchant review of the Rorschach in Buros's influential *Yearbook*. See Eysenck, H.J. (1959c), The Rorschach test, in *The fifth mental measurement yearbook* (ed. Oscar Buros), pp. 276–8, Gryphon Press, Highland Park, New Jersey.

Both Shapiro and Eysenck's standpoints served to undermine the ancillary role of clinical psychologists—but for subtly different reasons. As Bob Payne recalled, it soon became apparent to any budding clinical trainee in the early 1950s that Monte and Hans agreed on very little:

> Hans was mainly interested in those characteristics, 'traits' or 'types' that patients had in common. He seemed to regard the specific actions of an individual patient as being due to 'error' in the context of psychological theory and therefore largely unpredictable.[58]

With his London school training, Eysenck had a nomothetic mind set. His theories were couched in terms of group membership and continuous distributions. Shapiro, in contrast, focused on the individual. His more idiographic frame aimed to bring the patient's problematic behaviour 'under experimental control'.[59] Shapiro sought to map this process within a systematic protocol, the grammar of his practice always first-person singular. He got far closer to his subject matter than Eysenck ever did.

Bob Payne said he often felt Shapiro's approach owed something to his background in experimental psychology at Cambridge and South Africa. But it was also a response to the position Shapiro occupied as head of clinical section, charged with service delivery and teaching. It was also the kind of perspective that was easily grafted on to medical and administrative categories, the basis of the first therapeutic endeavours at the Maudsley. In this, Shapiro had a certain kind of priority.

Stage two: behavioural science on the couch

Psychiatrists initially saw psychologists and their tests as useful means to confirm their diagnostic judgements. Inspired by Shapiro's example, Jones, Yates, Inglis, and Payne began to press the single-case hypothetical deductive model in the early 1950s that would 'yield implications for disposal and treatment'.[60] At that stage, they were not allowed to follow up these implications in any way, and psychiatrists seldom acted on them. If they did, it was for the wrong reasons—a process Vic Meyer later described as 'a comedy'.[61] Maudsley psychologists soon

---

[58] Payne (2000), The beginnings of the clinical psychology programme at the Maudsley Hospital, 1947–1959, p. 18.

[59] Ibid., p. 19.

[60] See p. 7 of Jones, G. (1984), Behaviour therapy—an autobiographic view, *Behavioural Psychotherapy* 12, 7–16.

[61] Victor Meyer, quoted in Derksen (2001), Science in the clinic: clinical psychology at the Maudsley, p. 278.

learned to resist being restricted to mere 'labelling'.[62] According to Bob Payne, the psychologist–psychiatrist exchange soon came to be framed in terms relevant to treatment, prognosis, or some other practical issue.

Gradually, Maudsley psychologists began to trade on a new basis. Their clinical services were rewarded by permission to use patients in training courses and with greater latitude in the children's department.[63] Maudsley psychologists were able carry out some novel remedial measures on cognitive and educational problems of juvenile patients. As it was in the US before the war, the treatment of children became the Trojan horse for clinical psychologists' professional expansion.[64]

In the winter of 1954–55, Shapiro and Jones were discussing a case involving involuntary incontinence with a young psychiatric registrar, P.M. Middleton, in the staff cafeteria. Fatefully, Middleton suggested using some form of conditioning. Jones worked in the children's department and had been trying persuade the senior psychiatrist to allow them to use the Mowrers' bell-and-pad method for bed-wetting. With an electrical mechanism triggering an alarm when urination begins, the method trains the patient to either wake up anticipating urination or, more commonly, to sleep through the night without an accident.[65] However, getting the go-ahead to treat a young adult—the famous case of the 'peeing ballerina'—was a professional–bureaucratic breakthrough. Shapiro did the literature review while Jones undertook the treatment.[66] A training programme was devised to re-condition the abnormal bladder response and then eliminate the accompanying anxiety. The case was presented in April 1955 at the BPS conference in Durham and published the following year, taking precedence as the first recorded use in the UK of a new form of treatment soon to be given the appellation 'behaviour therapy'.[67]

---

[62] Payne (2000), The beginnings of the clinical psychology programme at the Maudsley Hospital, 1947–1959, p. 20.

[63] Derksen, M. (2000), Clinical psychology and the psychological clinic: the early years of clinical psychology at the Maudsley, *History and Philosophy of Psychology* 2, 1–17.

[64] Buchanan (2003), Legislative warriors: American psychiatrists, psychologists and competing claims over psychotherapy in the 1950s.

[65] Mowrer, O.H. and Mowrer, W.M. (1938), Enuresis: a method for its study and treatment, *American Journal of Orthopsychiatry* 8, 436–59. What made the Mowrer method significant was the learning theory used to conceptualize the problem and its solution: the specific bladder 'stimulus' was failing to elicit an appropriate 'response'. Thus, it was a matter of conditioning a more appropriate waking or suppression response.

[66] Berger and Yule (1979), Retirement of an enthusiast—Dr. M.B. Shapiro, p. 17

[67] Jones, G. (1956), The application of conditioning and learning techniques to the treatment of a psychiatric patient, *Journal of Abnormal and Social Psychology* 52, 414–19.

Eysenck took note of these developments at the hospital. Behind the scenes, he had begun to soften his stance on the prohibition of therapy. In his stunningly popular 1953 book *The uses and abuses of psychology*, Eysenck observed that, in the US, many psychiatrists 'rightly frowned upon' psychologists carrying out treatment.[68] But in equally frowning upon the unproven claims of psychoanalysis he introduced practical alternatives 'recently derived' from learning theory—the bell-and-pad method for treating bed-wetting, along with aversive conditioning to combat alcoholism. Eysenck clearly regarded these as promising techniques, but more work was needed by research-oriented psychologists. At this stage, Eysenck was still not publicly advocating that psychologists deliver such treatment, let alone orient their professional identity on such practices.

On a broader stage, other clinical courses were starting around the country and professional opportunities for clinical psychologists in the NHS were being consolidated. Although official reports initially defined psychologists as 'medical auxiliaries', this did not necessarily reflect the way day-to-day practice was developing at the Maudsley and elsewhere. Maudsley psychologists had already rejected the role of diagnostic ancillary. But what would go in its place? The practice of treatment presented itself as a clear and perhaps necessary development. Led by Shapiro, those in the clinical section had already embarked on treatment delivery. They had managed to do so largely without reproach because they were able to disguise it as research. Shapiro's contribution to this was more important than Eysenck ever allowed. Shapiro's cautious approach represented a professional way-point, a subtle shift in professional boundaries. His case-study emphasis downplayed therapeutic content at the same time as servicing professional and administrative requirements. Cast in the language of experimental psychology, it did not strike Maudsley psychiatrists as a rival form of healing. Bob Payne's publications in the late 1950s illustrated the point. Although strongly involved in the development of Eysenck's notion of behaviour therapy in the late 1950s, Payne's diagnostic reviews and case study reports of the period were framed in Shapiro's terms. Not coincidentally, Payne worked in the professorial unit under the watchful eye of Sir Aubrey Lewis.[69]

---

[68] See p. 14 of Eysenck, H.J. (1953*b*), *Uses and abuses of psychology*, Penguin Books, London.

[69] See, for example, Payne, R.W. (1957), Experimental method in clinical psychological practice, *Journal of Mental Science* **103**, 189–96, and Payne, R.W. and Jones, H.G. (1957), Statistics for the investigation of individual cases, *Journal of Clinical Psychology* **13**, 115–21.

1955 and all that

The year 1955 was a watershed for Eysenck, as it was for Shapiro. By then, the newly anointed professor of psychology had become well aware of Joseph Wolpe's work and its implications. And here the plot thickens considerably. Wolpe was another South African, a psychiatrist not comfortable amongst his disapproving or uninterested medical colleagues at home.

Eysenck saw in Wolpe a man after his own heart. Wolpe trained and practised as a psychiatrist, but possessed a broad scientific education and research inclinations. During his military service, Wolpe had employed dynamic therapy with limited success. In his doctoral research just after the war, Wolpe looked at research on experimentally induced neurosis that Pavlov had helped inspire. Taking his experimental cues from Gantt, Liddell, and Masserman and adopting the behavioural language of Hull, Wolpe re-conceptualized neurotic anxiety as a conditioned response. His results led him to question the prevalent belief that conflict was an essential component of neurotic disturbance. Persistent exposure to an aversive situation alone was enough to induce such behaviour—in cats at least. Moreover, what could be conditioned could also be *de-conditioned*. The key to this was the observation that neurotic anxiety tended to generalize, extending to other situations or objects according to their resemblance to the original scenario. Wolpe trialled various treatment techniques on his human clients in the early 1950s, mindful of Mary Cover Jones's work on childhood fears. He eventually settled on a basic strategy of gradual familiarization. Whilst remaining in a state of pleasant relaxation, the patient was confronted by succession of stimuli that progressively approximated the key fear-inducing stimuli. The technique broke the maladaptive stimulus–response connection with the introduction of an alternative response (e.g. relaxation) that inhibited anxiety. The phobia or fear was systematically de-sensitized, a process Wolpe soon embedded in learning theory as 'reciprocal inhibition'.[70] Here was a new set of techniques straight out of the laboratory and readily applicable to clinical practice. Eysenck read Wolpe's 1954 article and set to work.

According to Jack Rachman, neither Eysenck nor Wolpe could recall exactly 'when and how their association first began'.[71] Their initial contact may have

---

[70] Wolpe, J. (1954), Reciprocal inhibition as the main basis of psychotherapeutic effects, *Archives of Neurology and Psychiatry* **72**, 205–26. See also Jack Rachman's obituary for his mentor, Rachman, S. (2000), Joseph Wolpe (1915–1997), *American Psychologist* **55**, 431–2.

[71] Rachman (1981), H.J. Eysenck's contribution to behaviour therapy, p. 318.

dated back to the early 1950s. However, by 1955 they were corresponding enthusiastically. They first met in London in 1956 when Wolpe stopped off en route to a year-long sabbatical at Stanford. Eysenck recalled that he had 'immediately recognised and welcomed a redoubtable ally in the battle against pretence and make-believe that was Freudian "dynamic" psychology'.[72] As Eysenck told it, both were outsiders battling against an intolerant but misguided orthodoxy. To underline the synchronicity, Wolpe had published a paper critical of the therapeutic claims of dynamic psychotherapy in 1952, the same year Eysenck opened fire.[73]

Eysenck had concerns both large and small that made Wolpe's arrival all the more welcome. Ordinarily, Eysenck was not one to pine for an ally. He was perfectly happy and more than capable of fighting the good fight on his own. Unpopular causes were becoming a speciality. But Wolpe gave him several things he didn't have. For one thing, Wolpe gave Eysenck an alternative line into the clinic, a proxy interface with patients that didn't involve those already working with or against him at the Maudsley. Eysenck now had a show-piece alliance with medicine that fostered credibility amongst psychiatrists and blunted any attack based on traditional demarcations of expertise. Most importantly, Wolpe gave Eysenck a set of techniques compatible with learning theory and a quantitative, nomothetic approach. Wolpe gave Eysenck a big big stick just as Maudsley psychologists sought to redraw a line in the sand.

If Maudsley psychologists were going to do therapy, Eysenck wanted it to be done his way—as an intellectually coherent set of techniques he would later characterize as 'applied learning theory'. The implicit contrast was hard to miss. Shapiro had been informally outlining a distinctive, if not rival approach for several years and had just begun the process of systematizing it in published articles. This was not without its inherent difficulties, however, for Shapiro was a cautious, equivocating voice.

In many respects, Shapiro hardly differed from Eysenck. Like Eysenck, Shapiro put a heavy stress on the 'objective contribution' psychologists could make to psychiatry.[74] Clinicians should not neglect close observation and

---

[72] See p. 685 of Eysenck, H.J. (1996), Review of *Joseph Wolpe* by Roger Poppen, *Behaviour Research and Therapy* **34**, 685–6.

[73] Wolpe, J. (1952), Objective psychotherapy of the neuroses, *South African Medical Journal* **26**, 825–9. See also Salter, A. (1952), *The case against psychoanalysis*, Holt, New York. Neither of these two attacks had half the impact Eysenck's did, however.

[74] Shapiro, M. (1957), Experimental method in the psychological description of the individual psychiatric patient, *International Journal of Social Psychiatry* **3**, 89–102. Shapiro took issue with those in the BPS he saw as exemplifying this retreat from science.

ignore hunches; rather, they should try to make them as rigorous as possible and test the hypotheses so generated. They should not be afraid of adopting a single-case framework using the patient as their 'own experimental control'. In contrast to Eysenck, Shapiro argued that each case should be treated as a new challenge. While some patients might exhibit similar behaviour, it always remained to be seen whether the same kind of diagnostic judgements and recommendations applied. Shapiro made no blanket assumptions as to the applicability of a specific treatment regime to a class of patients and, up to 1957, made few general treatment recommendations *per se*.

For Eysenck, reciprocal inhibition was manna from heaven. It came from someone who was a one-man exemplar of the research-feeding-into-practice model that Eysenck had tried to institutionalize at the IoP. It was as if Wolpe was a Maudsley insider, and Eysenck would try hard to make him one in the ensuing years. Reciprocal inhibition was a generalized treatment that could be applied to a great swag of neurotic patients where anxiety was a central component—especially overconditioned neurotic introverts. It could be demonstrated and evaluated accordingly. And it was the kind of treatment that could be easily accommodated in Eysenck's vision of the detached clinical scientist, and lent itself to the boundary work he was engaged in.

Eysenck made sure his clinical colleagues were familiar with Wolpe. Gwynne Jones said he had been unaware of the South African's work until Eysenck placed a reprint under his nose as he was writing up his first case in July 1955. Jones subsequently cited Wolpe's success with 'similar methods' to back up his claims. This cross-fertilization was deliberately cultivated by at-home sessions held regularly at this time. Shapiro was not a regular attendee.[75] Although Shapiro was better placed to influence his students and colleagues at a local level, a head-to-head battle for hearts and minds of British clinicians more generally would inevitably leave him chagrined, spluttering in Eysenck's wake. Few could match the speed and power of Eysenck's output—least of all Shapiro, a man with a pronounced tendency for procrastination. Eysenck had a stunning capacity to draw diverse threads together into a generalized theoretical

---

He argued that qualitative data was not to be neglected but should be made quantitative and thus testable. All it took was ingenuity and time. See also Shapiro, M. (1951), An experimental approach to diagnostic testing, *Journal of Mental Science* **97**, 748–64, and Shapiro, M. and Nelson, E.H. (1955), An investigation of an abnormality of cognitive function in a co-operative young psychotic: an example of the application of the experimental method to the single case, *Journal of Clinical Psychology* **11**, 344–51.

[75] Irene Martin, quoted in Eysenck (1997*b*), *Rebel with a cause*, p. 156.

whole, appropriating certain elements and marginalizing all that remained as unimportant, irrelevant, or wrong.

By 1957, Eysenck was praising Wolpe's approach as the 'only *rational* theory of treatment available' [orig. italics].[76] But Eysenck still had concerns about its generality, hoping similar techniques could be developed to tackle the under-conditioned psychopaths and hysterics. No matter. That year Eysenck suddenly began to go public with his advocacy, as his colleagues began to publish more cases of successful treatment with this new kind of 'behaviour therapy'. Around this time the name itself was coined at one of the at-home sessions, a name that deliberately divorced it from the analytic connotations of psychotherapy. The concluding chapter of his 1957 book *The dynamics of anxiety and hysteria* devoted much space to this new scientific approach to therapy, with Wolpe's work particularly prominent. Eysenck also published a similar piece in *Medical World* making sure psychiatrists knew about it too. Behaviour therapy was different from traditional psychotherapeutic methods. It was less talk, more a planned course of remedial training. According to Eysenck, it demanded no reflexive understanding from the therapist, no sense of empathy. Instead, it was based on the opposite, the kind of remote, neutral scientific attitude that Eysenck embodied. The gun was loaded.

Provocation and backlash

The full impact of this new therapeutic approach was yet to be felt, however, for Eysenck still hedged ever so slightly over the implications. For example, how did behaviour therapy stack up against existing regimes in theoretical and practical terms, and who was best placed to research and deliver this treatment? The flashpoint, from Eysenck's point of view, came the following year, when he was invited to present at the Royal Medico-Psychological Association of Psychiatrists (RPMA). Eysenck claimed he chose the forum and the timing carefully, banking on having enough material to cope with the inevitable backlash. Even so, Eysenck still said he was shocked at its vehemence.

Just what went on at this meeting on the 3 July 1958 is hard to say. Eyewitness testimony is difficult to come by, apart from that Eysenck supplied. We also have his presentation in article form. Even in this format, Eysenck was at his provocative best. He defined this new 'behaviour therapy' in terms of the way it differed from dynamic psychotherapy, waving a red rag at psychoanalysts by making the contrasts as stark as possible. According to Eysenck, behaviour

---

[76] Hans J. Eysenck to Joseph Wolpe, quoted in Rachman (1981), H.J. Eysenck's contribution to behaviour therapy, p. 319.

therapy dealt with the here-and-now rather than the past and it focused on overt problems rather than underlying conflicts. As an applied science rather than a clinical art, behaviour therapy was a logically consistent, targeted technique summed up by Eysenck's pithy slogan *'Get rid of the symptom and you have eliminated the neurosis'* [original italics].[77] Symptom relapse or substitution would not be a problem. Since it was derived from learning principles, Eysenck made it clear that the development and delivery of behaviour therapy belonged to psychologists. It therefore gave them a significant mandate over all disorders it could be applied to. The diagnosis and treatment of neurosis were now the 'special competence' of psychological scientists, Eysenck argued, even if such patients might remain under general medical care. Neurotic symptoms were a result of inappropriate or insufficient conditioning, learning processes that were themselves normal. Psychiatrists could still have the treatment of all those suffering from organic impairments, including many forms of psychosis, all to themselves.

After Eysenck spoke and Gwynne Jones presented case study material in support, the floor was opened to discussion. The already tense atmosphere became overtly hostile. According to Eysenck, most questions were an ill-disguised excuse for rabid denunciation. Some in the audience rushed down the aisles shouting and the chair had to appeal for calm several times. For Eysenck, this episode illustrated how highly educated people can be highly irrational, especially psychoanalysts. 'I have seldom seen an audience behave in such an infantile manner.'[78] But then again, much was at stake. He was never invited to the RMPA again.

There may also have been an element of payback in Eysenck's upfront performance. Aubrey Lewis had given Wolpe a frosty reception a year earlier when Wolpe had presented a talk on reciprocal inhibition at the Maudsley. While Wolpe was a less than engaging performer, Lewis was even less impressed with Wolpe's very positive results. Lewis got up and declared that a cure rate of 90% was simply unbelievable. It could only be explained, Lewis said, by the fact that Wolpe himself had done the evaluation. Apparently little discussion followed.[79]

---

[77] See p. 65 of Eysenck, H.J. (1959d), Learning theory and behaviour therapy, *Journal of Mental Science* **105**, 61–75. This paper, the first conscious use of the term 'behaviour therapy' by the Maudsley group, was also reprinted in Eysenck, H.J. (ed.) (1960a), *Behaviour therapy and the neuroses*, Pergamon Press, Oxford.

[78] Eysenck (1997b), *Rebel with a cause*, p. 146.

[79] Wolpe apparently gave two talks at the IoP in the 1956–57 period—one on the way to, and one coming back from, the US. Rachman precisely dated one talk, 'Relationships between experimental neuroses and psychotherapy', that Wolpe gave 31 July 1956.

Eysenck's RMPA presentation was taken as a flagrant challenge. According to Eysenck, many Maudsley psychiatrists refused to speak to him in the aftermath, and some demanded his resignation. His talk also provoked a more organized response from the IoP leadership, headed by Lewis. But, even though Eysenck's upstart psychologists were clearly subordinate to him, Lewis had difficulty defining and enforcing what they were *not* supposed to do. Lewis reportedly used the committee of management of the joint hospitals to tighten rules on psychologists performing treatment. However, Lewis was never assured of complete support from the psychiatrists heading the various departments and units within the hospital that were not directly under his control. Many of these psychiatrists—Linford Rees, Denis Leigh, and Denis Hill in particular— were sympathetic to the psychologists' cause. They would still freely hand over certain types of patients—such as phobics, obsessive-compulsives, and alcoholics—because they did not wish to treat them or had little idea how to do so. To make sure this alliance could not continue to function in some form, with clandestine referrals for treatment and doctored case notes, Lewis would have had to put a ban on all patient access for psychologists or disband the clinical section of the psychology department altogether. Eysenck suggested that this is what Lewis actually tried to do. But by 1955, psychology had been made an independent department within the Institute. Moreover, the services the clinical section provided could not easily be replaced or dispensed with.[80]

Having privileged science in a clinical–medical context, and having empowered an outsider to define this, Lewis had to give ground. Eysenck had made a dual appeal to science and patient care, highlighting the empirically rigorous claims of behaviour therapy and contrasting it with the subjective, insular conceits he attributed to analytic psychotherapy. If you took Eysenck's point of view—and many did—withholding behaviour therapy was scientifically and ethically indefensible. Conversely, Lewis framed the issue in terms of professional encroachment, as a threat to psychiatrists' control over treatment that went with their ultimate responsibility for patient welfare. Nonetheless, he too conceded that behaviour therapy might well have its merits. Thus Lewis attached psychiatrist George Laverty to the psychology department to oversee

---

This was given on the way to Stanford. Rachman (1981), H.J. Eysenck's contribution to behaviour therapy, p. 318. The witness statement comes from Payne, who remembered Wolpe's controversial talk as occurring 'about 1957'. Payne (2000), The beginnings of the clinical psychology programme at the Maudsley Hospital, 1947–1959, p. 20.

[80] According to Maudsley psychiatrist Isaac Marks, there was a widely held belief amongst his medical colleagues that Eysenck was even more forthright with them once he had got his chair. Isaac Marks, telephone interview, 5 October 2005.

its activities and from 1962 encouraged two young psychiatrists, Isaac Marks and Michael Gelder, to study and practise behaviour therapy.[81] A medically directed course in behavioural techniques was also instigated with Maudsley nurses. Marks and Gelder would become important figures in the development of behaviour therapy as a broad interdisciplinary church, as opposed to the tight psychological speciality Eysenck had in mind.

The clear-cut victory Eysenck claimed was more equivocal from other points of view. Eysenck's clinical colleagues were not allowed to go beyond what they were already doing at the Maudsley. They continued to do various forms of therapy on a range of neurotics and other recalcitrant cases. They were simply able to do so more openly. However, psychiatrists and other paramedicals were quickly able to appropriate their techniques. Meanwhile, Eysenck tried in vain to bring Wolpe to the IoP. In 1957, Eysenck first attempted to secure money for a research position funded from private foundations. Unsuccessful, Eysenck then tried to get Wolpe on the faculty of the Maudsley, but this avenue was comprehensively blocked (by Lewis presumably). Finally in the early 1960s, Eysenck and his clinical section colleagues drew up an ambitious plan to found an independent Institute of Behaviour Therapy, with Wolpe as director. They were never able to secure the private financial backing, however, and the plan came to nought. By this time Wolpe had settled in the US and would remain there for the rest of his career.

As Jack Rachman pointed out, the failure to bring Wolpe to the Maudsley probably helped the international development of behaviour therapy by spreading its institutional base, ensuring there were key sites on both sides of the Atlantic.[82] However, Eysenck's repeated attempts to recruit his key psychiatric ally implicitly acknowledged that the 'turf war' within the hospital was far from over.[83] Maudsley psychologists still depended on psychiatrists for patient access and permission to do treatment. Eysenck had little appetite for administrative work and seldom personally attended joint meetings of the hospital hierarchy where any professional policy changes could be addressed. Thus his

---

[81] Jones, G. (1984), Behaviour therapy—an autobiographical view. Michael Gelder recalled that he became interested in behaviour therapy in 1962 'when I shared in the care of a phobic patient with Victor Meyer, a member of the Psychology Department who was using the new methods. Lewis encouraged me to find out more about behavioural treatments, and to try to devise a way of evaluating them.' Quoted in *Current Contents* (1983), This week's citation classic, *Current Contents* **46** (14 November 1983), 22. With the backing of Lewis, Isaac Marks joined Gelder in 1963 in a Medical Resarch Council funded project on behaviour therapy, which included controlled randomized trials.

[82] Rachman (1981), H.J. Eysenck's contribution to behaviour therapy.

[83] The phrase comes from Isaac Marks, telephone interview, 5 October 2005.

psychologists did not finally win the right to take full responsibility for their own patients until the late 1970s—well after many practitioners outside the NHS system, and even a few within, had been enjoying such privileges.[84] The dominance of psychiatry within the joint hospital made sure there was no sudden change in professional arrangements.

The start of something big

While Eysenck's colleagues made limited professional progress within the Maudsley through the 1960s, it was a different story outside. It was an exciting time for clinical psychologists in Britain generally, with the behavioural evangelists of Denmark Hill particularly convinced they were at the birth of something special. Encouraged by Eysenck, Maudsley psychologists published a string of studies drawn from their work. Jones's pioneering case study on incontinence had already become well-known. Vic Meyer recounted two case studies of phobia alleviated by 'systematic de-sensitization' (as the original stepwise version of Wolpe's reciprocal inhibition soon came to be known). In addition, Aubrey Yates detailed the treatment of tics by generating a conditioned inhibition response by repeatedly practising the tic, and he gave extensive theoretical elaborations in the process.[85] Cyril Franks reviewed the literature on aversive treatments for alcoholism.[86]

By 1960, Eysenck was able to bring all this work together into one edited volume—*Behaviour therapy and the neuroses*.[87] Drawing on a history represented by classic papers by Watson and Rayner, and Mary Cover Jones, this landmark compendium brought two main research groups together: those who had worked with Eysenck and those who had worked with Wolpe. Eysenck's general theoretical introduction provided the overarching rationale, lending coherence to a diverse set of practices across the English-speaking world. Suddenly a burgeoning international movement had appeared.

---

[84] William Yule, interview, 22 July 2003; Desai, M. (1969), The function of clinical psychologists in relation to treatment, *Bulletin of the British Psychological Society* **22**, 197–9.

[85] Meyer, V. (1957), The treatment of two phobic patients on the basis of learning principles, *Journal of Abnormal and Social Psychology* **55**, 261–6. Yates put Knight Dunlap's notion of negative practice in a Hullian learning framework. Yates, A.J. (1958), The application of learning theory to the treatment of tics, *Journal of Abnormal and Social Psychology* **56**, 175–82.

[86] Franks, C. (1958), Alcohol, alcoholics and conditioning: a review of the literature and theoretical considerations, *Journal of Mental Science* **104**, 14–33.

[87] Eysenck, H.J. (ed.) (1960a), *Behaviour therapy and the neuroses*, Pergamon Press, Oxford.

Fig. 6.1 Eysenck in his office at the IoP on Denmark Hill.
Source: Centre for the History of Psychology, Staffordshire University.

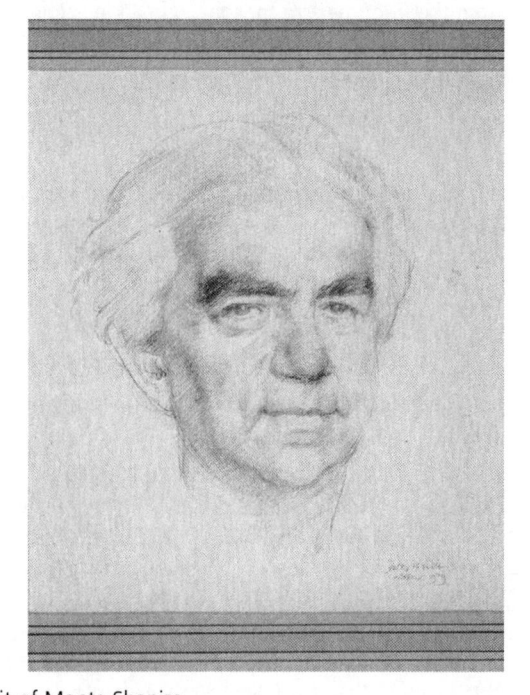

Fig. 6.2 Portrait of Monte Shapiro.
Source: Centre for the History of Psychology, Staffordshire University.

Fig. 6.3 Eysenck amongst the psychiatrists—a formal dinner at the inaugural meeting of the European Association for (Consultation–Liaison Psychiatry and) Psychosomatics, held at the Maudsley, 22–23 May 1955. A young Hans Eysenck is seated on the left at the near table, second from the far end. Aubrey Lewis is standing in the centre at the top table, third from the left, flanked by Alfred Meyer on his right, and D.L. Davies and possibly Denis Leigh on his left. Michael Shepherd is seated at the near side of the far table, second from the front, next to the lone woman, possibly Frieda Goldman-Eisler.
Source: Bethlem Royal Hospital Archives and Museum.

With Eysenck as chief publicist, extensive press coverage was assured. Donald Kendrick told of heady times, of media doorstop interviews and press write-ups of a paper he presented on cat phobia at the 1960 BPS conference in Hull.[88] Here was a new, scientific cure for mental problems that had bedevilled civilized societies through the ages.[89]

Clinical psychologists in Britain had stolen an intellectual march on their medical colleagues. They could make a claim for the treatment of functional

---

[88] Freeman, H. and Kendrick, D. (1960), A case of cat phobia: treatment by a method derived from experimental psychology, *British Journal of Medicine* **11**, 497–502; Donald Kendrick, interview, 13 October 2003.

[89] There was plenty of press coverage for Eysenck and this new treatment method in the early 1960s. For a flavour of this see Maddox, J. (1961), The influence of Eysenck, *The Guardian*, 18 April 1961; *Oxford Mail* (1963), Freud ousted by Pavlov: professor, *Oxford Mail*, 29 August 1963; and, McIlraith, S. (1965), Revolutionary new treatment for sexual aberration, *People*, 5 May 1965.

neuroses with a mode of therapy grounded within a long history of psychological theorizing. The early 1960s came to be seen as an era dominated by the positivistic scientific promise of behaviour therapy. In Eysenck's 1964 edited collection—subtitled *Readings in modern methods of treatment of mental disorders derived from learning theory*—he welcomed another important group into the fold.[90] The Americans had finally made good on the practical vision of their behaviourist forebears. The Skinnerian operant conditioning techniques they favoured could be applied to the management of psychotic patient populations as well, which appeared to broaden the reach of the behavioural coalition advantageously. With the help of Robert Maxwell, Eysenck also started the key journal *Behaviour Research and Therapy* the year before. The *BRAT* acronym summed up the spirit of youthful intellectual rebelliousness. The behavioural approach to psychotherapy began to gather pace across the English-speaking world in the mid-1960s.

A key ingredient of this Eysenck-inspired consensus was a sense of common opposition. It was, Irene Martin recalled, like marching as to war. Eysenck made sure his clinical foot-soldiers—Jones and Payne especially in the late 1950s, along with Jimmy Inglis, Russ Willett, and Martin herself—knew exactly who the enemy was: 'psychoanalysts, dangerous through their wealth and influence, psychiatrists through their dominance, unscientific psychologists'.[91] They very nearly managed to banish Freud from British psychology altogether, with few clinical psychologists describing their work in analytic terms in this period. Only the Tavistock, with a mission geared to the development and deployment of psychodynamic treatment within the NHS and beyond, remained a significant exception. However, with comparatively larger student numbers, the Maudsley dominated clinical training in the immediate post-war period.[92] It had produced almost 200 graduates by the mid-1960s. The founding 1966 membership of the Division of Clinical Psychology was only 159, although there were probably at least 200 more psychologists working in some sort of clinical capacity in Britain at the time. The Maudsley course thereby supplied a large proportion of the available qualified personnel for this

---

[90] Eysenck, H.J. (ed.) (1964c), *Behaviour therapy and the neuroses*, 2nd edn, Pergamon Press, Oxford.

[91] Irene Martin, quoted in Eysenck (1997b), *Rebel with a cause*, p. 155.

[92] In comparison, the Tavistock course was always two years long, and took in less than half the students the Maudsley course did. Sutherland, J.D. (1951), The Tavistock Clinic and the Tavistock Institute of Human Relations, *Quarterly Bulletin of the British Psychological Society* 2, 105–11.

relatively small but growing speciality.[93] This proportion would inevitably decline as clinical courses sprouted across Britain in the late 1960s and 1970s. Nevertheless, several ex-Maudsley staffers and graduates would come to head some of these programmes and carry on aspects of the Maudsley tradition.[94]

While Eysenck failed to bring Wolpe to the Maudsley, he landed one of Wolpe's most capable students. After graduating from Wolpe's alma mater, the University of Witswatersrand in Johannesburg, Stanley J. ('Jack') Rachman joined the IoP in 1959 to do Ph.D. research on de-sensitization. Eysenck subsequently got Rachman on to the clinical faculty 'very much against Monte Shapiro's wishes'. Eysenck said Shapiro probably saw Rachman 'as a threat, and not unreasonably so'.[95] Rachman helped maintain a bridge between the research and clinical sections in the late 1960s when it looked like these sections might go completely separate ways. Rachman was quickly promoted. He became Eysenck's right-hand man in the clinic—a productive researcher and skilled practitioner, as well as long-time *BRAT* editor. Rachman succeeded Shapiro as head of the clinical section in 1974 and gained a professorship in 1976. Shapiro retired in 1979, still a reader.[96]

---

[93] I have based this figure on course intake numbers, along with Shapiro's mid-1950s snapshot. Shapiro (1955), Training of clinical psychologists at the Institute of Psychiatry. According to Shapiro, the clinical course produced 57 graduates up to 1954. Only ten had gone abroad since completing the course. The rest were employed in the British mental health field as clinical and educational psychologists, and as researchers and teachers. Estimates of professional numbers of clinical psychologists in Britain were derived from Pilgrim and Treacher (1992), *Clinical psychology observed*. These figures are very rough, being highly dependent on professional definitions.

[94] In the 1960s and 1970s, IoP staff and graduates began to fan out across the country, including the north of England, and rose to leading positions in various academic psychology departments or psychiatric units. Gwynne Jones and Alan Dabbs took senior positions at Leeds; Donald Kendrick did likewise at Hull. Reg Beech went to the University of Manchester, while Victor Meyer became a founding member of the department of psychiatry at the Middlesex Hospital Medical School. Others went further afield. Aubrey Yates migrated to Australia, Bob Payne went back to Canada, as did Rachman somewhat later on. For more commentary on these various affiliations and migration patterns, see Hall, J. (2007*b*), The emergence of clinical psychology in Britain from 1943 to 1958, Part II: Practice and research traditions, *History and Philosophy of Psychology* 9(2), 1–33.

[95] Eysenck (1997*b*), *Rebel with a cause*, p. 149.

[96] This information was gathered from various, IoP annual reports and William Yule, interview, 22 July 2003. Shapiro's clinical colleagues gave him an armchair as a retirement gift—the chair he never got but deserved.

In 1965, Eysenck and Rachman published *The causes and cures of neurosis*, covering the theory, research, and practice of behavioural methods.[97] It was an attempt to relate Eysenck's biology of individual differences with explanations of the origins of various neurotic disorders, and to provide a rationale for their use. Teasingly, it contained a brief outline of Eysenck's second biological model in the making. Neurotic behaviour was 'learned behaviour', with anxiety-related Pavlovian conditioning the central mechanism.[98] Two main classes of neurotics were considered: those whose maladaptive behaviour was the result of surplus conditioning (e.g. dysthymics); and, those whose behaviour stemmed from a lack of appropriate conditioning (e.g. hysterics and psychopaths). Many dysthymic disorders (phobias, obsessive compulsive behaviour, anxiety states) would be expected to show a greater tendency for spontaneous remission, thereby explaining a phenomenon Eysenck had put great store in. While this was often observed, Eysenck and Rachman argued that the reinforcing effects of successful avoidance served to preserve symptoms in some cases—still leaving the behaviour therapist with a job to do. The final chapter once again pummelled the therapeutic claims of traditional psychotherapy at the same time as tackling the vexed question of why particular kinds of disorders exhibited 'relapse'. At last it seemed Eysenck's vision of a fully integrated science and practice of clinical psychology had arrived.

Out damn Freud!

In Eysenck's partisan hands, behaviour therapy conveyed an explicit challenge to other therapeutic approaches. The differences were black and white. On one side were the good and the rational, the scientific and the empirical; on the other, the bad and the irrational, the pseudoscientific and the mythical. Behaviour therapy was not just an additional or complementary set of techniques but a total rival approach. Most, if not all, of those deemed suitable for traditional talk therapy could be better treated by behavioural methods. And, by defining behaviour therapy in terms of what it *was not* as much as what it *was*, Eysenck invited an aggressive counter-attack.

Eysenck repeatedly baited psychoanalysts in print and in person, taking the fight to a wider audience in the early 1960s. He published several lowbrow pieces lambasting Freud's deluded heirs. They were little better than charlatans,

---

[97] Eysenck, H.J. and Rachman, S. (1965), *Causes and cures of neurosis*, Routledge and Kegan Paul, London.

[98] Ibid., p. 3.

Eysenck claimed, foisting baseless fictions on the gullible masses.[99] According to Eysenck, Freud was no scientist, neither insightful nor honest. Psychoanalytic psychology was thus fundamentally flawed. Its components could not easily be made testable, and were usually found wanting when they were. Repeated therapeutic failure was the only possible result, Eysenck concluded. This kind of critique was hardly unheard of, even at the time, and it would tend become a little passé over the years as analysts scaled back their scientific pretensions. Eysenck's anti-Freudian polemics leant heavily on the writings of others—Karl Popper, Frank Cioffi, Henri Ellenberger, and Frank Sulloway, to name a few—and this gave his book-length treatments of the issue in the 1970s and 1980s a scrapbook or compendium feel. Nevertheless, Eysenck had the public's ear and he had been a central player in the design and delivery of mental health services in Britain; this meant psychoanalysts had to take him seriously.

By 1966, Eysenck was ready to pronounce psychoanalysis moribund, if not dead.[100] Perhaps this was wish-fulfilment; it certainly was premature, since he ended up giving the Freudian empire its last rites for the next two decades.[101] An incident at the 1963 BPS conference in Reading was also indicative of the way Eysenck played stray analysts for fools. Wolpe, Eysenck, and a number of others were presenting papers in a joint symposium on behaviour therapy.[102] At one point, a member of the audience with a thick European accent (taken as a sure sign of psychoanalytic inclinations) asked for a clarification. The unfortunate gent began with the phrase: 'Professor Eysenck, I did not understand when you said …' Former clinical section head Bill Yule recalled that Eysenck's response was devastatingly curt. Eysenck got up, walked slowly to the podium, and said with his slow Germanic inflection: 'Yes—you did not understand'. And then he sat down again.[103] When it came to psychoanalysis, peaceful coexistence was not an option.

Nonetheless, the response from those sympathetic to psychoanalysis was less vociferous than one might expect. As one commentator later put it, they tended to be patronisingly dismissive of these new breed of psychologists, but rarely to

---

[99] Eysenck, H.J. (1960b), What's the truth about psychoanalysis? *Reader's Digest*, January 1960, 38–43; Eysenck, H.J. (1961), Psychoanalysis—myth or science? *Inquiry* 4, 1–15.

[100] Eysenck, H.J. (1966c), Psychoanalysis—a necrology, *The Twentieth Century* 2, 15–17.

[101] Eysenck, H.J. (1971f), The decline and fall of the Freudian empire, *Penthouse* 6, 28–30, 70, 84, and 86; Eysenck, H.J. (1985c), *Decline and fall of the Freudian empire*, Viking, London.

[102] Eysenck presented on 'Behaviour therapy or psychotherapy' and Wolpe on 'Behaviour therapy in complex neurotic states' at the 1963 BPS Annual Conference in Hull.

[103] William Yule, interview, 22 July 2003.

their faces.[104] Those committed to the analytic approach began to avoid engaging Eysenck and his allies on the terms he had set out—that of laboratory-led, quantitatively assessed science—because they saw them as both inappropriate and inherently unfavourable. It was a fight they knew they couldn't win. Instead, the leadership of British psychoanalysis sought to tar behaviour therapy as a collection of naïve techniques wielded by those with little patient empathy or therapeutic aptitude. Edward Glover had given Wolpe's 1958 book on reciprocal inhibition a tepid review, characterizing it 'as nostalgic as the scent of a late Victorian wardrobe'.[105] It was a throw-back to the kind of morally misguided attempts to change 'bad habits' like masturbation through punitive physical means. The apparent successes of these techniques, Glover said, could otherwise be attributed to the powerful suggestion effect of an authority figure.

However, Glover and his analytic colleagues still harboured anxieties about the demonstrable effectiveness of their work, given that behavioural therapists were now beginning to publish promising numbers on successful therapeutic outcomes. They wanted to reject the focus on symptoms and overt behaviour but had trouble articulating clear alternatives.[106] 'Self-discovery' and 'personal insight' hardly seemed enough. Eysenck had intimidated them; analysts' comebacks usually came with a tinge of defensiveness. They habitually cordoned off the sanctity of the therapist–patient interaction as a counter to externally imposed outcome criteria. For example, Thomas Freeman was at pains to avoid discussing 'therapeutic results' because 'any serious attempt to treat psychoneurotic patients ... will obtain its quota of recoveries, improvements and failures'.[107] Instead, Freeman accused behaviour therapists of ignoring the patient's inner thoughts and feelings underlying their participation in therapy. Behaviourists might obtain some success with their suggestion-driven techniques, but they had no recourse once the patient's attitude to treatment shifted in a negative direction.

---

[104] Berry, D. (1985), In a mind field, *New Statesman* 6 September 1985.

[105] Although at great pains to dismiss Wolpe's approach, Glover's lengthy review suggested that he took it as a very serious threat. P. 68 of Glover, E. (1959), Review of *Psychotherapy by reciprocal inhibition* by Joseph Wolpe, *British Journal of Medical Psychology* 32, 68–74.

[106] Crown, S. (1968), Criteria for the measurement of outcome in psychotherapy, *British Journal of Medical Psychology* 41, 31–7. See also the anxieties over a lack of good empirical studies of psychoanalysis expressed by Glover himself. Glover, E. (1955), *The technique of psychoanalysis*, International Universities Press, New York, pp. 376–7.

[107] See p. 53 of Freeman, T. (1968), A psychoanalytic critique of behaviour therapy, *British Journal of Medical Psychology* 41, 53–9.

Other analysts also conceded that behaviour therapy might have its uses. For instance, New York analyst Bernard Weitzman suggested that it would be unwise for his peers to ignore systematic de-sensitization in particular, given its effectiveness with simple monosymptomatic phobias and anxiety.[108] However, support for other techniques was less clear-cut, and it was still uncertain whether behavioural approaches could handle more complex, existential complaints. Ex-IoP staffer Sidney Crown was another to highlight the limitations of behaviour therapy, having crossed the floor to train as a psychiatrist and dynamic psychotherapist. In 1965, he described *The causes and cures of neurosis* as 'slipshod arrogant and naïve'. The 'immoderate claims of Eysenck and Rachman do it [behaviour therapy] a disservice.'[109] In sum, Crown concluded, behaviour therapy fell short of its own exacting theoretical and empirical standards.

These skirmishes began to peter out in the late 1960s, however, as psychoanalysts and behaviourists came to operate in quite different professional spaces. Eysenck had managed to lever his Freudian opponents out of the way of his behavioural juggernaut. One of the stranger effects of Eysenck's campaign was to drive psychoanalysts closer to the mainstream of British medicine. Many of the medical analysts of the BPS medical section eventually migrated to the safety of their parent body, now reconciled to their presence. From 1971, they were accommodated within the psychotherapy division of the newly formed Royal College of Psychiatry, the re-badged RMPA. Certainly this was against the contemporary trend toward *de-medicalization* of psychoanalysis in the US. Medical control of the *British Journal of Medical Psychology* didn't change, however. British psychoanalysts remained insulated within the ranks of psychiatrists, their practice protected by medical dominance in most interdisciplinary contexts. They simply learnt to ignore Eysenck's attacks. As Sidney Crown said, we knew he 'couldn't hurt us'.[110]

Psychologists had in the meantime shored up their own institutions. Since the end of the war the BPS Committee of Professional Psychologists (Mental Health) had engaged in the somewhat torturous process of professionalization, liaising with representatives from various levels of government and the NHS on qualifications, practices, salary structures, and recognition. With the medical analysts all but gone, clinical psychologists were even able to establish

---

[108] Weitzman, B. (1967), Behaviour therapy and psychotherapy, *Psychological Review* 74, 300–17.

[109] See p. 1235 of Crown, S. (1965), Review of *Causes and cures of neurosis* by Hans J. Eysenck and Stanley Rachman, *British Journal of Psychiatry* 111, 1234–5.

[110] Sidney Crown, interview, 22 April 2002.

a BPS division to call their own in 1966. True to form, Eysenck played no role in this thankless committee work.[111] He was the ideas man; liaising with interested parties, drawing up regulatory guidelines, and administrative implementation were for others. Maudsley-trained clinicians—such as Alan Dabbs—would play a significant role in the future direction and workings of the new division of clinical psychology. The BPS had also launched the *British Journal of Social and Clinical Psychology* in 1962 as an alternative to the *British Journal of Medical Psychology*. Completing an exorcism of sorts, the Tavistock clinical course wound down in the 1970s, bowing to pressure to broaden and standardize in line with the rest of the discipline.[112]

By the early 1970s, the fight with the 'alien forces' of psychoanalysis was all but over. In 1972 Perry London called for an 'end to ideology' in behaviour therapy, declaring that the borders were secure.[113] Behaviour therapy, or the more American 'behaviour modification', had cemented its place on both sides of the Atlantic. In 1969, there were two English language journals specifically devoted to behavioural therapies. A decade later there were 21, rivalling a similar number of psychoanalytic periodicals. Of course, many articles on behavioural theory and techniques were published in more general forums, over 4000 all told by 1971.[114] The influential multidisciplinary group the Association for the Advancement of Behaviour Therapy (AABT) was formed in 1966 in the US, with British and European bodies soon following. The 'polemical era' in clinical psychology had ended—and with it, Eysenck's lengthy moment in sun.[115] Much of what followed reflected his declining influence. The behavioural movement he kick-started, became sprawling, heterogeneous enterprises, far too big for any one person to direct. However, it is

---

[111] Historian John Hall details at length how this committe work was mainly done by small group of hard-working but less vocal figures—women such as May Davidson, Grace Rawlings, and Lucy Fildes—the 'educational' and child guidance psychologists who migrated to the clinical field after the war. See Hall, J. (2007a), The emergence of clinical psychology in Britain from 1943 to 1958, Part I: Core tasks and the professionalisation process, *History and Philosophy of Psychology* **9**(1), 29–55.

[112] Pilgrim and Treacher (1992), *Clinical psychology observed*.

[113] London, P. (1972), The end of ideology in behaviour modification, *American Psychologist* **27**, 913–20.

[114] Data cited in London, P. (1983), Science, culture, and psychotherapy: the state of the art, in *Perspectives on behavior therapy in the eighties* (ed. M. Rosenbaum, C. Franks, and Y. Jaffe), pp. 17–32, Springer, New York, p. 19.

[115] See pp. 4–5 of Franks, C. and Rosenbaum, M. (1983), Behavior therapy: overview and personal reflections, in *Perspectives on behavior therapy in the eighties* (ed. M. Rosenbaum, C. Franks, and Y. Jaffe), pp. 3–16, Springer, New York.

worth exploring why Eysenck was ignored. Not only does this encompass some interesting turns in the history of clinical psychology, it illustrates the limitations of Eysenck's faith in high science as a guiding principle in human affairs.

The cognitive challenge

Just when Eysenck had achieved a kind of integration of theory and practice in the mid-1960s, aspects of it slowly began to unravel. Defined by an opposition that ceased to engage him, Eysenck could only pull together *his* version of the behavioural approach for so long. Behaviour therapy was now a bandwagon, a success story that others wanted to have a say in. Eysenck's purist model thus faced two related challenges from somewhat unexpected directions. The first encompassed a critique of its theoretical underpinnings, highlighted by the fall from grace of behaviourism and the rise of the new cognitivism. The second challenge came in the form of a broad political attack on the social interventions the behavioural turn made possible and encouraged.

As early as 1965, American psychologists Louis Breger and James McGaugh opened up a new front by attacking the foundations of behaviour therapy.[116] Eysenck's claim that it was 'derived from modern learning theory' was misleading, they argued, since there was not one learning theory but many. While these theories often contradicted each other, some had already been discredited. The heated theoretical disputes of Hull and Tolman had become passé, overtaken by enthusiasm for radical operant models. However, Skinner himself had been under attack. In 1959 a young, relatively obscure linguist named Noam Chomsky had suggested that empirical learning approaches could not possibly account for the acquisition of complex capacities and skills like language.[117] While Chomsky's intervention hardly represented the rout of behaviourism, it was clear from the early 1960s onwards that mentalist concepts started to be seen as both allowable and necessary. Sniffing this intellectual wind, Breger and McGaugh claimed behaviour theory ignored inner events. The S-R terms behaviourists habitually recited were at best allegorical in the therapeutic situation, better represented as S-O-R, where 'O' represented the acquisition of information as a 'central process'. Enter the notion of mediating cognitive mechanisms.

---

[116] Breger, L. and McGaugh, J. (1965), Critique and reformation of 'learning theory' approaches to psychotherapy and neurosis, *Psychological Bulletin* **63**, 338–58.

[117] Chomsky, N. (1959), Review of *Verbal behavior* by B.F. Skinner, *Language* **35**, 26–58.

In reply, Rachman and Eysenck acknowledged that learning theory was far from perfect; however, it was the best theory available, generating successful practical results.[118] Breger and McGaugh's reformulations were too vague to be helpful, Rachman and Eysenck concluded, and simply unnecessary. Having the last word, Breger and McGaugh pointed to the gap between the laboratory and the consulting room. 'The prior existence of the techniques as well as the great *dissimilarity* between what goes on in behaviour therapy and in most learning experiments indicates that the relationship between theory and practice is *non-specific*.'[119]

By 1970, the hubris that marked Eysenck's initial promotion phase had almost vanished.[120] In his paper 'Behaviour therapy and its critics', Eysenck insisted on the link between theory and practice, but recommended that critics go easy on this fledgling science. Drawing on Kuhn, he argued that the strict behavioural approach constituted a new paradigm worth persisting with. Theoretical confusions and partiality of reach should not see it prematurely buried. Again he doubted that cognitive elements were necessary, sanguinely adding that 'if these work better, then they will no doubt supersede the less complex and inadequate [behavioural] ones'.[121] Otherwise, theoretical simplicity was a virtue, not a fault.

Eysenck was probably at his most defensive when it came to the question of controlled comparison trials. Aware of the fall he had set himself up for, Eysenck pleaded that all forms of psychotherapy should be judged according to the same standard. Over the years, the lofty break-even point for spontaneous remission that Eysenck had 'unwisely' committed himself to would give critics ample opportunity to attack.[122] However, in citing Marks and Gelder's

---

[118] Rachman, S. and Eysenck, H.J. (1966), Reply to a 'critique and reformation' of behaviour therapy, *Psychological Bulletin* **65**, 165–9.

[119] See p. 171 of Breger, L. and McGaugh, J. (1966), Learning theory and behaviour therapy: reply to Rachman and Eysenck, *Psychological Bulletin* **65**, 170–3.

[120] Eysenck summarized these intradisciplinary challenges in Eysenck, H.J. (1970*d*), Behaviour therapy and its critics, *Journal of Behaviour Therapy and Experimental Psychiatry* **1**, 5–15.

[121] Ibid., pp. 7–8.

[122] The quoted word comes from Crown (1965), Review of *Causes and cures of neurosis* by Hans J. Eysenck and Stanley Rachman, p. 1235. In 1977, Monte's son David Shapiro suggested a double standard was in play. With his wife Diana, David Shapiro argued that Eysenck and Rachman were not as tough on their own brand of therapy as they should be, as they were when other forms of therapy were evaluted. See Shapiro, D. and Shapiro, D. (1970), The 'double standard' in evaluations of psychotherapies, *Bulletin of the British Psychological Society* **30**, 209–10.

1968 survey of the evaluation literature for support, he inadvertently drew attention to an embarrassing problem at home. Eysenck's clinical crew had found it hard to put their money where their spokesman's mouth had been. While they continued to perform and publish case work, their research ambitions had been largely stymied by Lewis.[123] For example, Eysenck and Rachman's 1965 opus *The causes and cures of neurosis* contained an extensive evaluative chapter on the results of behaviour therapy, one of the first of its kind. Much of the encouraging data they reported came from Wolpe and his students; none of this data came from Eysenck's fellow psychologists. The most notable Maudsley research in this period, including group comparison trials, was initiated by Lewis's psychiatrists Gelder and Marks. They had access to funds and patients that their psychological counterparts did not—much to Eysenck's chagrin. He would later complain bitterly of being denied research funds by 'embattled analysts on grant-giving bodies' and generally being treated 'as an outcast and a pariah'.[124] Maudsley psychologists formed research partnerships in these larger studies—doing much of the actual therapy—but the psychiatrist held sway through to the late 1970s.[125] It was the kind of alliance Eysenck accepted in practice, despite maintaining that practitioners in behaviour therapy should be learning theory specialists. It certainly made it much harder for Maudsley psychologists to lead by example.

In a series of publications in the late 1960s, Marks and Gelder compared behavioural techniques with brief psychotherapy. In one key study with simple phobia cases, de-sensitization appeared to work better and faster.[126] Marks and Gelder interpreted their work as suggesting various approaches might complement each other. While Wolpe complained furiously from afar that Marks and Gelder were underselling his techniques, Eysenck kept closer

---

[123] In 1976, Eysenck put together a set of case studies. Eysenck, H.J. (ed.) (1976*b*), *Case studies in behaviour therapy*, Routledge and Kegan Paul, London. Case study methods were hardly the kind of studies Eysenck valued, suggesting that he was scratching for material. One is tempted to assume that this volume served as a substitute for the kind of bigger research programme he would have liked to undertake. Moreover, critics pointed out that many of these case studies were far from ideal demonstrations of the efficacy of behavioural methods. See Blackman, D. (1976), Be on your 'best' behaviour, *Times Higher Education Supplement*, 22 October 1976.

[124] Hans Eysenck, quoted in *Current Contents* (1980), This week's citation classic, p. 22.

[125] For example, Marks undertook a long collaboration with Jack Rachman in the mid-1970s using exposure therapies with obsessive-compulsive patients.

[126] Gelder, M., Marks, I. and Wolff, H.H. (1967), De-sensitization and psychotherapy in the treatment of phobic states: a controlled inquiry, *British Journal of Psychiatry* **113**, 53–73.

counsel, knowing on which side his bread was buttered.[127] Marks and Gelder were otherwise seen as honest brokers by psychiatrists and psychologists not committed to the distinctions Eysenck campaigned on. Gelder later recalled that this was the beginning of a 'more constructive exchange of ideas between psychotherapists and behaviourists'.[128] Across the water, a task force set up by the American Psychiatric Association reported in 1973 that behaviour therapy had much to offer.[129] Even analytic psychiatrists were able to bring themselves to advocate behavioural techniques for certain classes of patients. In hindsight, Isaac Marks suggested that what was needed was 'a more nuanced approach'—more nuanced than Eysenck's, that is.[130]

The importance of theory

Eysenck thought of theory and technique as indivisible. What defined behavioural techniques was their origin in laboratory experimentation. Maintaining the connection was paramount. Behaviour therapy was not mere technique and it was not compatible with other theoretical orientations, especially dynamic ones. Mowrer's 'neurotic paradox' needed explaining if the behavioural theory were to survive. Why did people persist with behaviour that was clearly maladaptive and poorly received? In the late 1960s, Eysenck had already begun to outline a theory of incubation as a more sophisticated version of the simple Pavlovian learning framework he said he had borrowed from Watson.[131] Simple conditioning was insufficient to explain the origin, strength, and persistence of fears and phobias—now the home turf of behaviour therapy. They often appeared without an initial high trauma, and tended to escalate without apparent reinforcement. Explaining why fears and phobias did not spontaneously remit would also ensure behaviour therapy took the credit as a means of combating them.

Characteristically, Eysenck's way of modelling this problem came direct from the laboratory. Eysenck came to think of anxiety and phobic fears as special cases of conditioned learning—which he came to label as B-type

---

[127] See Wolpe, J. (1965), Letter to the editor, *British Medical Journal* **250** (19 June 1965), 1609.

[128] Michael Gelder, quoted in *Current Contents* (1983), This week's citation classic, p. 22.

[129] American Psychiatric Association Task Force on Behavior Therapy (1973), *Behavior therapy in psychiatry*, report 5, American Psychiatric Association, Washington, DC.

[130] Isaac Marks, telephone interview, 5 October 2005.

[131] Eysenck, H.J. (1967c), Single-trial conditioning, neurosis and the Napalkov phenomenon, *Behaviour Research and Therapy* **5**, 63–5; Eysenck, H.J. (1968b), A theory of the incubation of anxiety–fear responses, *Behaviour Research and Therapy* **6**, 309–21.

Pavlovian conditioning. The key idea was that, as a response, anxiety was also a drive (as was sex). Contact with the conditioned stimulus alone was enough because the anxiety accompanying it was virtually identical to that produced by the unconditioned stimulus. Anxiety-reducing behaviour thus had reinforcement properties. It could create a positive feedback-loop under certain conditions. Eysenck suggested that when the unconditioned stimulus was strong, short exposures to a conditioned stimulus would generate an escalating (i.e. phobic) response because it was reinforced by the subsequent reduction in anxiety. Conversely, long exposures that prevented such reinforcement—akin to 'flooding'—would tend to result in habituated extinction. This phenomenon was best exemplified by research with dogs and rats—especially Napalkov's isolated and relatively obscure study demonstrating the escalating reactions of dogs to the threat of pistol shots.[132]

Confusing as it was, Eysenck's notion of incubation was designed to save his version of learning *theory*.[133] But it seemed the effectiveness of behaviour therapy as *practical technique* was another thing all together. For example, de-sensitization was explainable in terms of four different models of learning theory alone, but it was still judged to work.[134] While Eysenck would struggle to maintain the link between theory and practice, others within the movement sought to dissolve it. Another of Wolpe's protégés, Arnold Lazarus, exemplified the problem-solving eclecticism of case work. Like many others, the US-based Lazarus would cease to describe himself as a behaviour therapist *per se*. Theory and origins were irrelevant to the question of whether a particular technique worked with a particular patient, he argued. Many practising

---

[132] Napalkov, A.V. (1963), Information process of the brain, in *Progress in brain research*, Vol. 2 (ed. N. Weiner and J.P. Schade), pp. 182–6, Elsevier, Amsterdam.

[133] Eysenck's explanations of his incubation model were less than clear. For example, in one 1976 publication describing the work of Napalkov, he appeared to get the behavioural labels UCR and UCS mixed up. See Eysenck, H.J. (1976c), Behaviour therapy—dogma or applied science? in *Theoretical and experimental foundations of behaviour therapy* (ed. M.P. Feldman and A. Broadhurst), pp. 333–63, Wiley, London, p. 342. Perhaps he could be forgiven for this general lack of clarity. Napalkov provided only the most cursory description of his procedures in his crucial 1963 study, and the strong effects he reported were never easy to replicate. See Levis, D.J. and Malloy, P.F. (1982), Research in infrahuman and human conditioning, in *Contemporary behaviour therapy* (ed. G.T. Wilson and C.M. Franks), pp. 65–118, Guilford Press, New York and London, p. 85. Moreover, Eysenck was not able to clearly illustrate his model with case examples from the human clinical field.

[134] Eysenck, H.J. and Beech, H.R. (1971), Counter conditioning and related methods, in *Handbook of psychotherapy and behaviour change* (ed. A. Bergin and S. Garfield), pp. 543–611, Wiley, New York.

clinicians were especially keen to distance themselves from the negative connotations of behaviourism, especially since critics tended to conflate the two. In the US, behaviour therapy was less case-driven than in the UK. Those working an operant paradigm often sought to mould responses of large patient populations. On top of this, leading radical behaviourists had made a daring pitch at social engineering. In making the leap from the confines of the laboratory to all of life, B.F. Skinner's *Walden two* and *Beyond freedom and dignity* helped generate a broad critical swell that would wash up in Britain in the early 1970s.

Correctional politics

Boosted by Eysenck's advocacy, behaviour therapy was the most visible treatment modality of British clinical psychology. Behaviour therapy had technocratic efficiency on its side. It was a goal-directed correction of strictly defined problems. As such, it appealed to government and business, as well as the insurance industry. However, the enthusiastic reception amongst these groups appeared to rather prove the point when psychologists and psychiatrists were being castigated as the 'servants of power', as disciplinarians as much as healers.

Success obviously had a price. Anti-psychiatrists of various political stripes, the left-wing mental health journal *Red Rat*, and sundry social commentators characterized behaviour therapy as narrow, mechanistic, and authoritarian, and their critiques moved through media spaces devoted to wider public consumption. Aversion therapy got particularly bad publicity. Anthony Burgess's 1962 novel *A clockwork orange* and Stanley Kubrick's 1971 film version of it created a huge cultural splash. Aversive treatments were portrayed as cruel and de-humanizing, producing only the most dubious of cures. Aversion had been used to tackle alcoholism, various compulsions, and even writer's cramp. However, the use of electric shock or nausea-inducing drugs to correct sexual 'deviances' like homosexuality—itself under siege as a pathological category—made sure advocates were singled out.

Eysenck did his best to ignore the hubbub. However, he could not resist engaging his new nemesis of the period, R.D. Laing. 'Ronnie' to his patients, the fiery Glaswegian was the empathetically engaged *yin* to Eysenck's coolly detached *yang*. Although he counted himself as some kind of analyst, Laing was hardly mainstream Freudian. His Marxist-influenced critique of psychiatric practices was perfectly in tune with the anti-authority mood of the period. Laing labelled aversion therapy 'torture'—even if it was, statistically speaking, successful. So were all forms of punishment when judiciously applied, he argued. In a 1970 issue of the rationalist periodical *Question*, Eysenck counter-attacked.[135] Eysenck playfully suggested that Laing's conception of schizophrenia as a rebellion against oppressive circumstances was so absurd it 'must have been dreamt up by a very clever man'. Laing and his collaborator, David Cooper, were simply using patients for political grandstanding, Eysenck wrote. If their fanciful ideas were ever followed, they would only 'increase the misery of mentally-ill patients'.[136]

Eysenck also had a lengthy print exchange with Alejandro Portes in 1971. Portes attacked behaviour therapy as narrow and limited, succeeding primarily through its technocratic appeal to practical action. Eysenck once again defended the notion of partiality as a necessary stage in scientific progress, and dismissed Portes's humanistic concerns as irrelevant to the question of empirical merit.[137]

Critical broadsides from those Eysenck saw as badly informed outsiders seemed to warrant much less attention, however. Eysenck and Isaac Marks had at least one run-in with well-known gay activist Peter Tatchell.[138] The scene was a symposium of the London Medical Group in November 1972. Gatecrashing this professional forum, Tatchell made it clear he thought aversion therapy was a brutal treatment often coercively undertaken, a form of repression geared to reactionary sexual norms. Claiming the discomfort was no worse than a 'trip to the dentist', Eysenck said there was 'no ethical principle involved in aversion therapy that is not involved in any psychological treatment'. For his part, Marks described the portrayal in *A clockwork orange* as exaggerated and misleading, having pioneered such treatments himself. Even without patient consent, Marks suggested, such treatments might be invoked in 'society's best interests'. Tatchell's outraged response saw him thrown out.

---

[135] Eysenck, H.J. (1970c), The ethics of psychotherapy, *Question* 3, 3–12. The newspapers lapped it up, of course, for it made good copy. See also Atticus (1970), The row which is splitting psychology in two, *The Sunday Times*, 25 January 1970.

[136] Hans J. Eysenck, quoted in Atticus (1970), The row which is splitting psychology in two. In his autobiography, Eysenck recalled that he also joined in a public debate with David Cooper at University College around this time, describing it as a surreal experience, 'but not untypical of the times'. Eysenck (1997b), *Rebel with a cause*, p. 190.

[137] See Portes, A. (1971a), On the emergence of behaviour therapy in modern society, *Journal of Consulting and Clinical Psychology* **36**, 303–13; Eysenck, H.J. (1971e), Behaviour therapy as a scientific discipline, *Journal of Consulting and Clinical Psychology* **36**, 314–19; and, Portes, A. (1971b), Behaviour therapy and critical speculation, *Journal of Consulting and Clinical Psychology* **36**, 320–4.

[138] See materials at http://www.petertatchell.net/psychiatry/psychiatry%20index.htm.

If these kinds of attacks worried Eysenck, it was not particularly evident in his published output.[139] On record as opposing the use of behaviour therapy without patient consent, Eysenck bemoaned the public focus on aversion as misleading. It represented, he maintained, only 1% of behavioural treatment.[140] The 'purely imaginary' portrayal of aversion therapy in *A clockwork orange* would be 'abhorrent' as well as 'useless'.[141]

It wasn't quite enough, though, to turn around a growing image problem. The psychoanalytic put-down, that behaviour therapy was merely punitive correction refashioned, finally began to bite when it came from sources unencumbered by a self-serving oracular mystique. The image of rigorously impersonal science suddenly became a liability. As a rule, advocates of behavioural interventions were far more ready to concede the technical rather than the political point. Yet they still felt the pressure to soften the public presentation of their work, and this created a backdraft for theory change. Gwynne Jones would write later that, while the process of aversion could produce a curious bond between patient and therapist, 'one felt degraded by the procedures and often doubted whether the end justified the means'.[142] It was not terribly effective either, Jones happily noted, since it induced only temporary suppression that allowed alternative response patterns to be built in. The various theoretical models for aversion techniques—including Eysenck's—were still said to be unsatisfactory.[143]

---

[139] As late as 1982, Eysenck was still discussing the 'extinction and neutralisation of homosexual tendencies'. Eysenck, H.J. (1982a), Neobehaviouristic (S-R) theory, in *Contemporary behaviour therapy* (ed. G.T. Wilson and C.M. Franks), pp. 205–76, Guilford Press, New York and London, p. 260. Eysenck also co-authored a book on media standards in 1978 in which he came across as cautiously in favour of censorship. He suggested that the proliferation of images of sex and violence could trigger antisocial behaviour in some people and should therefore be regulated more tightly. This moral conservatism was couched in empirical scientific terms, of course. See Eysenck, H.J. and Nias, D. (1978), *Sex, violence and the media*, Maurice Temple Smith, London.

[140] Eysenck, H.J. (1965a), Letter to editor, *Times Educational Supplement*, 10 December 1965.

[141] This explicit condemnation was contained in Eysenck, H.J. (1973e), *The inequality of man*, Maurice Temple Smith, London, p. 201. It backed up his earlier attempt to sell the behavioural approach in psychiatry and education with the paperback, Eysenck, H.J. (1972a), *Psychology is about people*, Penguin, London. See also Eysenck, H.J. (1977d), *Crime and personality*, 2nd edn, Routledge and Kegan Paul, London, p. 174.

[142] See p. 11 of Jones, G. (1984), Behaviour therapy—an autobiographic view, *Behavioural Psychotherapy* **12**, 7–16.

[143] Rachman (1981), H.J. Eysenck's contribution to behaviour therapy, p. 327.

The image problem had been exacerbated by the expansion of mental health services. Outside insulated hospital settings, psychotherapy services needed to be sold on public faith. The shift toward consultative services and private practice in psychiatry and clinical psychology had been far more rapid in the US, with its fractured and largely private health system. So, not coincidentally, it was Americans who led the way in re-packaging their therapeutic services for broader consumption. As APA President George Miller's famous phrase of 1969 put it, psychology needed to be 'given away'. People should be helped to change their lives, the goal being to build competencies rather than correct deficits.

While applied behaviourism crested as an academic discourse, the *techniques* themselves endured—incorporated in community-based services that were sold on the basis of self-empowerment rather than social control.[144] Reaching out to the public also helped pave the way for more explicitly cognitive approaches. Couched in accessible everyday language, cognitive regimes blurred the distinction between service-providing expert and lay consumer. Not only could these techniques readily express and enable self-directed change, their addition to the behavioural arsenal solved some sticky technical problems. For example, cognitive models explained vicariously transmitted fears and obsessional thinking in a simple and direct manner that conditioning theory did not. The Americans started to dominate the international scene by the mid-1970s. They brought an elephant of theoretical diversity into the consulting room that only Eysenck seemed to want to ignore. The rise of Albert Ellis and Aaron Beck in particular amounted to rival systems that Eysenck was increasingly reluctant to acknowledge, except in critical terms.

From high science to low technique

Eysenck's carefully nurtured baby had grown up and left home. Many of the central figures in the tight-knit group that developed behaviour therapy said farewell to the Maudsley for greener professional pastures. Yates, Inglis, and Payne left in the late 1950s, Jones and Meyer in the early 1960s. Each put their own spin on what they had learnt.

Eysenck's vision of the detached scientist remotely applying results of high science in the clinic was only partially realized. For a start, it did not map

---

[144] For a fine account of how behavioural techniques were 'given away', see Baistow, K. (2001), Behavioral approaches and the cultivation of competence, in *Psychology in Britain: historical essays and personal reflections* (ed. G.C. Bunn, A.D. Lovie, and G.D. Richards) pp. 309–29, BPS Books, Leicester.

easily on to the structure of British health care. The NHS, the key employer of clinical psychologists in Britain, was unlikely to ever fully embrace psychologists as principally scientists, and successive reviews implicitly rejected it.[145] Moreover, Eysenck was never able to fully impose *his* integration of theory feeding into clinical practice even in his own backyard. Maudsley insiders—Rachman included—openly acknowledged that Eysenck's dimensional model of individual differences did not translate easily into clinical practice. It was not therapeutically and administratively useful to know where on Eysenck's grid a patient was located. While those in the clinical section may have treated Eysenck's dimensions with (sometimes mock) reverence, they found them unhelpful in case work. Eysenck's dimensions and trait breakdowns were neither sensitive nor broad enough to capture the range of patient symptomology. The clinical section rarely used his standardized tests—the MPI, EPI, and EPQ. Moreover, the quasi-deterministic implications of Eysenck's biosocial model sat uneasily with changing patient presentations and the assumption of plasticity in the therapeutic process. It was not as if Shapiro's intensively idiographic approach fared any better, however, even though it was explicitly derived from case work. By the time Shapiro had fully elaborated a coherent alternative to behaviour therapy, it became clear it was far too demanding and time-consuming to be practical.[146]

As British clinical psychology diversified in the 1970s, work with the handicapped, children, and the aged expanded, as did private practice. Pushed to respond to a greater variety of service needs, the strict Maudsley formula of rigorous diagnostics and behaviour therapy seemed nowhere near enough. Surveys showed many more practitioners were using behavioural techniques than had been taught them—illustrating the success of Eysenck's promotional campaign—but practitioners tended to describe their orientation as 'eclectic'.[147] Led by the Americans, clinicians on both sides of the Atlantic would increasingly position themselves as partners in the process of helping people help

---

[145] Pilgrim and Treacher (1992), *Clinical psychology observed*.

[146] See Hamilton, V. (1964), Techniques and methods in psychological assessment: a critical appraisal, *Bulletin of the British Psychological Society* **17**, 27–36; Barbrack, C. and Franks, C. (1986), Contemporary behaviour therapy and the unique contribution of H.J. Eysenck: anachronistic or visionary? in *Hans Eysenck: consensus and controversy* (ed. Sohan Modgil and Celia Modgil), pp. 233–46, Falmer Press, Philadelphia; and, Gibson (1981), *Hans Eysenck: the man and his work*, pp. 159–61.

[147] Pilgrim and Treacher (1992), *Clinical psychology observed*.

themselves. By 1975, the AABT promoted behaviour therapy for the alleviation of suffering *and* the enhancement of functioning. It involved:

> environmental change and social interaction rather than the direct alteration of bodily processes by biological procedures. The aim is primarily educational. The techniques facilitate self-control…a contractual agreement usually negotiated, in which mutually agreeable goals are specified… guided by generally accepted ethical principles.[148]

While psychologists were the driving force in the development of behavioural approaches, the movement was always a mixed bag when it came to practice and professional representation. These easily taught techniques were 'given away' to nurses, teachers, social workers, parents, and the public. Like the AABT, the British Association for Behavioural Psychotherapy was an interdisciplinary group: only a third were practising clinical psychologists, along with a number of academics. Maudsley alumni were prominent in its formation in 1972, with Meyer inaugural president, Jones chairman, and Marks vice-chairman.

Behaviour therapy came to encompass an ever-expanding range of tools and modes of intervention. Throughout the 1970s, 'the cognitive trend became increasingly apparent'.[149] Joining de-sensitization, flooding, and aversion were biofeedback, behavioural medicine and behavioural pharmacology, group and community therapy, along with self-control procedures. High science had become low technique.

Eysenck was a bystander in these shifts in service philosophy and delivery modality, and a disapproving one at that. He would complain of the relative lack of interest of psychiatrists in behaviour therapy, of the 'mish-mash of theories' advocated by the likes of Lazarus, of the retreat from basic research, and a neglect of animal work.[150] Therapeutic ethics were primarily a matter of conducting the most empirically effective style of treatment.[151] It was not that Eysenck didn't move with the times—he did, but only slowly. For example, while burying the Freudian empire yet again in 1985, Eysenck warned behavioural

---

[148] Franks, C. and Wilson, G.T. (1975), *Annual review of behaviour therapy: theory and practice*, Vol. 3, University Park Press, Baltimore, p. 1.

[149] Franks and Rosenbaum (1983), Behavior therapy: overview and personal reflections, p. 5.

[150] For his critical evaluation of Lazarus's approach, see Eysenck, H.J. (1970*b*), A mish-mash of theories, *Journal of Psychiatry* **9**, 140–6; for other complaints, see Eysenck, H.J. (1988*b*), Preface and behaviour therapy, in *Theoretical foundations of behaviour therapy* (ed. H.J. Eysenck and I. Martin), pp. vii–ix and 3–35, Plenum Press, New York.

[151] Eysenck (1994*a*), The outcome problem in psychotherapy: what have we learned?

practitioners not to overdo the impersonal aspects of their work.[152] He was well aware that psychoanalytic practitioners had managed to survive partly by selling themselves as a more humane alternative to the 'mechanical' behaviourists. Moreover, as Donald Kendrick noted, it had become obvious that the personal attributes of the therapist affected therapeutic outcomes. These effects could, with some intellectual work, be incorporated within a behavioural framework.[153] Even so, this amounted to quite shift for a man who in the late 1960s seriously imagined that behaviour therapy could be automated, most efficiently delivered to needy patients by computer.[154] As a 'theoretical clinician', practical issues were not Eysenck's strong suit. For example, Isaac Marks observed that Eysenck often looked uncomfortable dealing with specific questions about particular kinds of patients and treatments, and would defer to those with clinical experience. Just as well, because former Maudsley clinical crew members started to bridle at his continued assumption of the spokesman role as if there were complete agreement as to what behavioural therapy was, as if he too were a practitioner.[155]

Last shouts

Unhappily reconciled to being a bit player, Eysenck shifted attention to future directions and his place in history. His parting shots attempted to shore up the coherence of behavioural models and articulate his opposition to particular new trends. In the mid-1970s, Eysenck augmented his incubation model with Seligman's notion of preparedness.[156] Phobic fears were not randomly distributed. We are 'prepared' by our evolutionary history to be fearful of things like heights, snakes and spiders, and small and large spaces because such fears had survival value. Anxious individuals, usually high on neuroticism, had an overweaning fear of particular objects and situations incubated by the random happenstance of life. Even so, Eysenck found it difficult to cross the divide between animal work and clinical practice. When Eysenck's incubation model was criticized by Bersh in 1980 as imprecise and circular, Eysenck admitted it

---

[152] Eysenck (1985c), *Decline and fall of the Freudian empire*.

[153] Donald Kendrick, interview, 13 October 2003. According to Kendrick, Eysenck had Jones re-formulate these personal effects in behavioural terms.

[154] See Davies, C. (1968), What is behaviour therapy? *The Listener*, 15 February 1968.

[155] Victor Meyer, interviewed by Maarten Derksen, 26 April 1999.

[156] Eysenck, H.J. (1976f), The learning theory model of neurosis: a new approach, *Behaviour Research and Therapy* **14**, 251–61.

was still in the development stages and more research was needed.[157] It was a hope unrealized. Very little research designed to test or support his incubation model was done with human subjects—aside from that by Rachman at the Maudsley. The notion of preparedness generated more attention. However, research by Seligman, the Uppsala school, and others in the late 1970s and early 1980s generated mixed results. By 1982, Eysenck had virtually signed off in terms of theory development and retirement loomed.

In the late 1980s, Eysenck increasingly emphasized the importance of genetic factors, taking his inspiration from his collaborations with Lindon Eaves and David Fulker.[158] In Eysenck's last essays on clinical topics, Ronald Grossarth-Maticek's researches made their first, slightly jarring appearances. Eysenck had become preoccupied with Grossarth-Maticek's 'creative novation therapy' and the role it could play in altering the course of physical disease. Oddly enough, Grossarth-Maticek's approach was distinctly cognitive in content, although its roots lay in psychoanalysis. But that, as they say, was not the half of it.[159]

Eysenck maintained a hard and fast distinction between behaviour therapy and verbal forms of 'psychotherapy' to the end. In his mind, talking therapies were always associated with the sins of analysis. Newer cognitive approaches were either nonsensical or behaviour therapy re-labelled. 'To add "cognitive" to behaviour therapy is either an oxymoron or a redundancy. I prefer the latter.'[160] Behaviour therapy had always taken cognitive factors into account, Eysenck claimed, in effect conceding their importance. Pavlov had anticipated this with his second signalling system, a system still subject to the laws of conditioning. When introducing concepts like 'conflict' and 'frustrative non-reward' in his later writings he was quick to add that they could be defined in 'operational terms'.[161] In a grudging 1983 acknowledgement that he had lost the interest of the new clinical generation, Eysenck admitted that cognitive approaches were increasingly fashionable because they seemed to address

---

[157] See Bersh, P.J. (1980), Eysenck's theory of incubation: a critical analysis, *Behaviour Research and Therapy* 18, 13–17; Eysenck, H.J. (1983a), The theory of incubation: a reply to Bersh, *Behaviour Research and Therapy* 21, 303–5; and, Bersh, P.J. (1983), The theory of incubation: comments on Eysenck's reply, *Behaviour Research and Therapy* 21, 307–8.

[158] Eysenck, H.J. (1988c), The role of heredity, environment, and 'preparedness' in the genesis of neurosis, in *Theoretical foundations of behaviour therapy*, (ed. H.J. Eysenck and I. Martin), pp. 379–402, Plenum Press, New York.

[159] See Chapter 9 for a lot more on Eysenck's collaboration with Grossarth-Maticek.

[160] Eysenck (1994a), The outcome problem in psychotherapy: what have we learned?, p. 491.

[161] Eysenck (1982a), Neobehaviouristic (S-R) theory, p. 216.

more distinctively human attributes. They were more 'acceptable to people who still fail to appreciate the importance of evolutionary concepts in the explanation of human behaviour'.[162] Some people 'ought to know better', he chided.[163] Cognitive approaches functioned as *de facto* behaviour therapy in an ideational realm—often amounting to, Eysenck suggested, a kind of imagined de-sensitization. The success of other forms of therapy could likewise be explained by their accidental or surreptitious incorporation of behavioural principles. They were simply less successful because they did not match specific symptoms with the most effective re-conditioning strategy. The neglect of animal research was also a mistake, Eysenck added, for such work generated new and symptomatically specific kinds of treatment that were hard to develop with human subjects.

For Eysenck, it was all about the theory, the link to the laboratory, and the credit. Behaviour therapy was 'a science, not a dogma', Eysenck argued in 1976—taking the extraordinary step of turning to perhaps *the* philosopher of science of the day, Imre Lakatos, to back him up.[164] Quoting Lakatos's reply to a private letter, Eysenck argued that the behavioural approach was a *progressive research programme*. Psychoanalysis and its variants clearly were not. Adding Kuhn for good measure, Eysenck proclaimed that a *scientific revolution* had occurred, and that revolution was behavioural not cognitive. A few key components of the cognitive perspective could be incorporated within the theoretical fold. Otherwise, these poorly formalized, disparate, and contradictory ideas should be rejected, for they lacked evidential support. They were not 'with the programme', and Eysenck predicted that the march of science would see them left behind. Eysenck reiterated these warnings throughout the 1980s and early 1990s, demanding to be heard as a contemporary leader. By then, though, far fewer people were listening.

Yesterday's man?

> ...having failed to keep pace with the times, Eysenck exists far behind the mainstream of current thought and evidence in the burgeoning field of behaviour therapy. (Lazarus, A. (1986), On sterile paradigms and the realities of clinical practice, in *Hans*

---

[162] Eysenck, H.J. (1983e), Classical conditioning and extinction, in *Perspectives on behavior therapy in the eighties* (ed. M. Rosenbaum, C. Franks, and Y. Jaffe), pp. 77-98, Springer, New York, p. 90.

[163] Eysenck was referring to those trained in behaviour therapy—such as his former student Cyril Franks. Eysenck, H.J. (1979d), Special review: behaviour therapy and the philosophers, *Behaviour Research and Therapy* 17, 511–14.

[164] Eysenck (1976c), Behaviour therapy—dogma or applied science?

Eysenck: consensus and controversy (ed. S. Modgil and C. Modgil), pp. 247–57 and 260–1, Falmer Press, Philadelphia, p. 261)

No one could reasonably be expected to remain in the vanguard of such a rapidly developing field forever. For all his anti-psychiatric posturing, Eysenck remained a curious kind of medical empiricist. Up against the Freudians mid-century, Eysenck looked radical. The unease experimental psychologists and biologically oriented psychiatrists felt toward psychoanalysis helped him generate significant traction. But when it came to the tribulations of the psyche, Eysenck was very much a traditionalist. He repeatedly evoked a deficit model that defined pathology as the absence of normality, its touchstone a materialist tradition dating back to Kraepelin.[165] With his bluff mix of the biological and the social, Eysenck transposed a strict quantitative yardstick over psychiatric categories that were taken to be relatively non-negotiable. His seemingly neutral evaluation of psychotherapy hid endless debate over what the problems were and what defined success. For those not enamoured with analysis, behaviour therapy was readily acceptable as an alternative because—in Eysenck's hands at least—it was laboratory-based and biologically grounded. It targeted observable behaviour, invoking a symptom-only version of medical healing that had a straightforward administrative appeal.

Eysenck had formulated his ideas in opposition. But what happened when that opposition retreated or disappeared? As the Freudian ascendancy passed and the neo-Kraepelin diagnostics of DSM III crystallized, as pharmacology and genetic research flourished, Eysenck's conceptual language made less and less sense to his thriving clinical progeny. Eysenck's alignment with materialist medicine became more problematic when it became more dominant within psychiatry. His post-1960s persistence with terms like 'symptom', 'remission', 'relapse', and 'cure' began to look quaintly old-fashioned, as though behavioural disorders were contracted like a disease. His preoccupation with the Pavlovian origins of neurosis was a side-bar to those dealing with the 'worried well' rather than the 'ill', to those helping their clients with 'life issues' in all their varied, native forms.[166] Ironically, *nouveau* cognitive-behaviourists had taken on some of the goals of psychoanalysis but not the content. For clinical

---

[165] Even as late as 1982, Eysenck still maintained that behaviour therapy essentially aimed at 'curing' emotional disorders by means of classical conditioning. Conversely, the management of personality disorders through positive operant methods fell outside its ambit, and fell into the realm of behaviour modification. Eysenck (1982*a*), Neobehaviouristic (S-R) theory, pp. 206–207.

[166] Lazarus, A. (1986), On sterile paradigms and the realities of clinical practice, in *Hans Eysenck: consensus and controversy* (ed. S. Modgil and C. Modgil), pp. 247–57 and 260–1,

psychologists cut loose from medical imperatives, Eysenck was not helpful any more. His was the language of a rival profession, not theirs. His boundary rhetoric had been emptied of its relevance and function. The 'end of ideology' in behaviour therapy made him look like yesterday's man.

Warm feedback

Eysenck likewise retained his commitment to a classical, feed-forward model of research and practice. 'Indeed, like Kraepelin', Derksen commented, 'Eysenck defined clinical psychology as a laboratory in a clinic.'[167] Theory and research drove practical application, not the other way round. But to make the science work in this way, the laboratory had to be extended to the domain of application. In this case, Eysenck matched behavioural techniques to restricted classes of patients and problems, and subtly re-defined medical notions of success. These techniques could then best be applied in a remote, standardized manner. For Eysenck, the clinic should be a cool and rational place—much like the laboratory—a place devoid of politics and gender, kindness and cruelty. Since he did not seek to understand through empathy, practitioners need not either. They did not have to 'feel' what it was like to be a dysthymic neurotic in order to treat one.[168]

However, taking this kind of behavioural science to the people demanded a remodelling of the way mental health services were conceptualized, regulated, and funded. Eysenck and his behavioural evangelists just didn't have the power and the reach to impose laboratory-style clinics across the board. Specific work contexts could hardly be made to order—witness the aborted Institute of Behaviour Therapy. There was too much input from emerging or rival professional groups, too many stakeholders, too much political feedback. Diverse professional settings inevitably placed very different demands on clinical practitioners. 'Applied learning theory' alone was not sufficient. Practitioners informally combined or derived alternative techniques to suit, cherry-picking

---

Falmer Press, Philadelphia. In this exchange, Lazarus characterized his colleagues' opinion of Eysenck's recent clnical writings as a 'time-capsule back to 1963' (p. 261).

[167] Derksen (2001), Science in the clinic: clinical psychology at the Maudsley, p. 274.

[168] For Eysenck, this followed directly from his detached mode of knowledge gathering. The best way to learn about such conditions was to read the literature and conduct the necessary long-term empirical research, preferably along experimental-correlational lines. As Gordon Claridge pointed out, R.D. Laing he certainly was not. See Claridge, G. (1997), Eysenck's contribution to understanding psychopathology, in *The scientific study of human nature: tribute to Hans J. Eysenck at eighty* (ed. H. Nyborg), pp. 364–88, Pergamon, Oxford.

from the research literature where necessary. Clinical practice became a pluralistic affair, 'multi-theoretical' rather than 'atheoretical'. Key IoP alumni came to see emotionally detached therapy as undesirable and even Eysenck appeared to agree in the end. In turn, these practical developments had a way of informing the basic science. Several theorists systematized their approach from their consulting work, generating considerable academic research interest in the process. The Americans Ellis and Beck were notable examples of this, despite the fact that Beck was a psychiatrist by training.

In one respect, this was a situation of Eysenck's own making. At the peak of his audience and influence in the late 1950s, circumstances prompted Eysenck to suggest that clinical psychologist *already had* a scientifically consistent set of practical tools to achieve professional independence. Eysenck was so successful in getting this message across that he helped undermine the careful, research-first approach he otherwise advocated. His last missives on clinical topics could be read as a belated attempt to repair the unintended effects.[169]

Carving up the spoils

Eysenck's importance in the development of clinical psychology in Britain was a function of circumstances. He found himself in a key institutional position, perhaps *the* key position, at a pivotal time in history. Up to 1945, clinical psychology barely existed in the British Isles. Eysenck *and* his colleagues would change all that, training a large significant portion of the start-up manpower and framing much of the technical content for this new applied speciality. For over two decades after the war, the Maudsley crew largely had the floor. Collectively, they made history, and with that went a tremendous urge to squabble over priority rights. In print, Eysenck gave Shapiro's contribution little more than lip-service.[170] In response perhaps, Bob Payne and Aubrey Yates felt that Shapiro had been written out of the story and sought to redress this.

---

[169] If this meant taking on board the constructivist philosophy of science, Eysenck was prepared to do it. Not only did he make the incommensurability thesis of Kuhn work for him, he began to emphasize the theory-ladenness of facts to bolster his notion of the indivisibility of theory and practice. Eysenck (1994*a*), The outcome problem in psychotherapy: what have we learned?, p. 479.

[170] When interviewed in the late 1970s, Eysenck attributed Shapiro's intellectual limitations and stubbornness to wartime brain injuries. He was dismissive of Shapiro as a researcher and impatient over his inability to decisively commit thoughts to paper. Hans J. Eysenck, interviewed by H.G. Gibson, 1 March 1979.

Shapiro in turn championed Gwynne Jones's pioneering efforts, a debt of thanks echoed by Eysenck on at least one occasion.[171]

Then there was the 'name thing'. Both Skinner and Lazarus had independently used the description 'behaviour therapy' in print prior to Eysenck's 1959 article—as Eysenck acknowledged in 1970.[172] However, in a 1980 APA Convention dialogue with Eysenck, Skinner conceded that his group's 1953 experimental study had no therapeutic aspect; their use of the term was a misleading ruse to attract funding.[173] In 1958, Lazarus had also applied the term to Wolpe's techniques of reciprocal inhibition.[174] However, Eysenck and his crew made a conscious decision at that time to apply it to all forms of behavioural treatment, the better to emphasize their programmatic intent. This allowed Eysenck to claim that he had inaugurated 'the term "behaviour therapy" in its current meaning.'[175] Shapiro hated it, for it conveyed to him a kind of slavish, formulaic approach. Instead, Shapiro had an intense flirtation with the principles of Rogerian therapy in the mid-1960s, underlining his differences with Eysenck even more. Yet history did not quite come down on Shapiro's or Eysenck's side.

Eysenck never quite got rid of the Freudians either. They never entirely left Denmark Hill, with Joseph Sandler and Peter Hildebrand returning to teach at the IoP in the 1970s. Gallingly one might think, Sandler also became professor of psychoanalysis at Eysenck's alma mater University College in 1984, and later established a new analytic teaching and research unit there. However, Eysenck said he just thought this 'extremely funny', contrasting his response with Wolpe's immense disappointment with his former student Arnold Lazarus.[176]

---

[171] Eysenck, H.J. (1985a), H. Gwynne Jones, 1918–1985—an appreciation, *Behavioural Psychotherapy* **13**, 171–3.

[172] Eysenck (1970d), Behaviour therapy and its critics.

[173] I am indebted to Frank Farley for sharing a tape of this APA dialogue. For Skinner's original research report, see Skinner, B.F. *et al.* (1953), *Studies in behavior therapy*, status report no. 1, Naval Research Contract N5 ori-7662. Much of the impetus for this work, essentially the experimental study of psychotics' behaviour in room-size Skinner boxes, came from Skinner's co-worker Ogden Lindsley. For an historical account of this research see Rutherford, A. (2003), Skinner boxes for psychotics: operant conditioning at Metropolitan State Hospital, *The Behavior Analyst* **26**, 267–79.

[174] Lazarus, A. (1958), New methods in psychotherapy: a case study, *South African Medical Journal* **33**, 660–3.

[175] Eysenck (1983e), Classical conditioning and extinction, p. 77.

[176] See p. 686 of Eysenck, H.J. (1996), Review of *Joseph Wolpe* by Roger Poppen, *Behaviour Research and Therapy* **34**, 685–6.

In his autobiography, Eysenck pointed to the diverse roots of behaviour therapy. Not only did he give little credit to the important professional–therapeutic advances achieved by Shapiro, he gave surprisingly little attention to Wolpe's contribution. Instead, Eysenck said he had been greatly impressed by the graded task method of émigré 'active psychoanalyst' Alexander Herzberg, whom he had become acquainted with during the war.[177] Eysenck devoted more space to describing Herzberg's obscure work in London than he did to Wolpe's—as if to buttress his contention that he had already formulated his theory of behaviour therapy by the early 1940s. The only thing that stopped him being more upfront about it at the time, he said, was the implacable opposition of Lewis to lay therapy.

Eysenck claimed that behaviour therapy perfectly fitted his vision for clinical psychology as an applied psychological science. It could be easily bolted on as an extension of his research endeavours without contradiction. He thus claimed he had long entertained some kind of therapeutic role for psychologists, and downplayed the contradiction by ceasing to refer to his earlier ban. Either he feared the repercussions—as he said—or he simply changed his mind. In hindsight, it looked more a case of the latter. Changing professional circumstances and new treatment possibilities forced his hand. If he had always envisaged psychologists as therapists, such a strong prohibition would seem unwise and inexplicable, as would his very tough stand on the efficacy of existing methods. Both laid him open to later attack.

Eysenck played a crucial part in the development of clinical psychology in Britain—just not quite the one he claimed. At a professional level he was more vanguard advocate than founder, a role that came with a definite use-by date. Likewise, at a technical level he was more facilitator than hands-on innovator, behaviour therapy's 'midwife'. Nevertheless, his contribution to this new form of psychological treatment, indirect as it was, was probably the most concrete impact of his career.[178] While critics might dismiss his dimensional schemes and bio-behavioural models as so many theoretical castles in the air, behaviour therapy made a direct connection with the world outside the laboratories of the IoP. It changed many people's lives. For this, they can look to Hans Eysenck—promoter, patron, and point-man for the behavioural cure.

---

[177] Eysenck (1997b), *Rebel with a cause*, pp. 132–4.

[178] Eysenck's (cognitive) behavioural progeny saw him as a foundational figure, and were especially keen to acknowledge this after he was gone. See Sanavio, E. (ed.) (1999), *Behaviour therapy and cognitive behaviour therapy today: essays in honour of Hans J. Eysenck*, Elsevier, Oxford. The book was a compilation of papers from the 1997 Congress of the European Association of Behavioural and Cognitive Therapies, held a month after Eysenck passed away.

Chapter 7

Mr. Controversial: the psychology of politics

Hans Eysenck was obviously no ordinary psychologist. As well as being a compulsive controversialist he was one—perhaps *the*—pre-eminent popularizer of psychological science in the twentieth century. These two aspects of Eysenck's career ran together. When he wrote accessible texts on psychology and its relationship with crime, sex, or astrology, he sought to sell a demystified version of his discipline to lay audiences. Yet these popular works had much in common with the more technical efforts he directed at his peers. They were not just an adjunct to his scientific output; they were an intrinsic part of the way he played the game. Through them, Eysenck was able to ascend to a lofty stratosphere populated by a select few, becoming a public intellectual whose opinions and life were the stuff of mass consumption. It gave him an unparalleled audience reach, not to mention fame and fortune, but it wedded him to a role that was inherently volatile and contradictory.

Many of the controversial topics mentioned above could each be given a full-length treatment—as I do for smoking, personality, and cancer in the penultimate chapter. Here, in successive chapters, I want to focus on two areas of Eysenck's output that had high- and low-brow components: the psychology of politics and the biogenetic basis of intellectual differences. His against-the-grain model of political attitudes was developed in the first half of his career. His later pronouncements on race differences in intelligence attracted media attention and political protests that would forever cast him as some kind of 'IQ warrior'—ironic given that much of his later research on intellectual differences attempted to go beyond psychometric testing.

But context was everything. In the aftermath of World War II, psychologists did double duty. Already in the habit of investigating the ill effects of war, they felt an overriding obligation to help heal the psychological wounds and prevent such conflicts in the future. However, not all would come to the party, least of all Eysenck. Almost inadvertently at first, Eysenck became a dissenting voice. This kind of socially engaged science was not for him. By articulating a symmetric left–right view of authoritarianism in the mid-1950s, Eysenck made sure the liberal intelligentsia could not claim him as one of their own.

Instead, they attacked him for all they were worth, illustrating the destructively adversarial side of 'normal' science—especially when Hans Eysenck was involved.

Political attitudes and post-war renewal

During his post-Ph.D. spell with the Air Raid Precaution Service in 1941, J.C. Flügel had given Eysenck some sensitive data from a pre-war survey of various political and social groups. According to Eysenck, Flügel had undertaken not to publish this research. Eysenck did not feel so honour-bound; he canvassed the literature and did the analyses, submitting a manuscript to the *Journal of Social Psychology* in July 1942. However, the journal's editorial office apparently delayed its publication until 1944.[1] Perhaps some of the content was potentially inflammatory in a time of war, for it assessed the attitudes of various groups agitating for social change. Nothing indicates that Eysenck revised the manuscript in this period.[2]

Eysenck's first 'General social attitudes' study built on the factor analytic techniques he used in his doctoral work on aesthetics. It contained his now familiar dimensional cross as a determinative structure, like that of his landmark study of neurotic soldiers published the same year. If anything, Eysenck's political attitudes work predated his initial two-factor model of personality—a reversal on the usual story of his intellectual progression. The tedious analyses were performed sometime in the late 1941 to early 1942 period, and the write-up was one of the first things Eysenck did after joining the staff at Mill Hill in June 1942.

Eysenck re-analysed attitude studies by Thurstone, Carlson, and Ferguson, using Burt's summation and centroid techniques. He extracted two comparable, unrotated factors from all three studies and asked several expert judges to assess the content of high loading items. The first factor contrasted radicals versus conservatives, already a familiar finding in the field. The second was a more novel theoretical versus practical dichotomy. With reasoning more implicit than apparent, Eysenck suggested this factor could be connected with temperamental differences: 'The *practical* attitude is that of [William] James's "tough-minded" man, of the extravert; the *theoretical* attitude is that of the "tender-minded" introvert.'[3] Eysenck gave no theoretical justification for

---

[1] Eysenck, H.J. (1944*b*), General social attitudes, *Journal of Social Psychology* **19**, 207–27.

[2] Eysenck did not cite any reference in this 1944 paper, including his own work post-dating 1941.

[3] Eysenck (1944*b*), General social attitudes, p. 214. In making this connection to personality, Eysenck was taking a lead from pre-war American work, for example, Allport, F.H.

equating these concepts; perhaps it simply appeared self-evident. Alternatively, this logical leap might be explained as a post-hoc rationale generated by his subsequent analysis of Flügel's unorthodox groups. Either way this was a significant point, for Eysenck would later associate tough-mindedness with authoritarianism. When Eysenck superimposed his two-factor structure on the attitude data from these groups, he claimed the pattern obtained could be interpreted in these terms: radical versus conservative, practical/tough-minded versus theoretical/tender-minded. Nevertheless, Eysenck admitted to being selective in which items he included.[4] This was the basic template he set in stone in 1942, the basis for his future research on political attitudes. As with his personality work, Eysenck partitioned a relatively well-worn set of results in a new way. At the time, he seemed most interested in challenging American attitude research, suggesting that: 'More attention to manner in which various attitudes are inter-related and structured' was needed, rather than simple atomistic description.[5]

After returning to the Maudsley after the war, Eysenck used his research students to pilot new questionnaires. A 1947 factor analytic study reprised his not-quite-orthogonal dimensions of radicalism versus conservatism (R) factor and tough- versus tender-minded (T) and he quickly developed a set of scales to measure them.[6] Eysenck stressed that T was not an ideological dimension but a projection of personality differences on to the social attitudes field.

However, at this point Eysenck's research would begin to connect with highly sensitive attitudinal research taking place across the Atlantic, and his priorities and emphasis would change accordingly.

---

and Hartman, D.A. (1925), The measurement and motivation of atypical opinion in a certain group, *American Political Science Review* **14**, 735–60.

[4] As well as being selective, Eysenck needed considerable conceptual flexibility to make both sets of factor interpretations consistent. For example, he suggested that 'abstemiousness', 'non-smoking', and 'vegetarianism' could be equated with a theoretical/tender-minded outlook. Eysenck (1944*b*), General social attitudes, p. 222.

[5] Eysenck (1944*b*), General social attitudes, p. 224. Unacknowledged but lurking in the wings was a rival European approach that fused empirical questionnaire techniques with psychoanalytic and Marxist sociological insights. Continental researchers sought globalized explanations of ideological commitment by detecting how various political ideas and actions went together. Even if Eysenck was only dimly aware of the Europeans at the time—for he does not refer to them—he straddled both Old and New World traditions.

[6] Eysenck, H.J. (1947*c*), Primary social attitudes. 1. The organization and measurement of social attitudes, *International Journal of Opinion and Attitude Research* **1**, 49–84.

Up to the Depression, studies of race differences had a relatively central place in American psychology, less so in Europe.[7] However, heavy criticism and some notable changes of heart saw explicit investigations of race-related topics began to disappear in the lead up to World War II. Instead, a curious turnaround occurred: if race differences were largely illusionary then any belief in such differences amounted to irrational, possibly pathological prejudice.[8] The rise of Nazi-inspired anti-Semitism lent American research a greater sense of urgency. In 1944, the American Jewish Committee sponsored a series of 'Studies in prejudice', edited by Max Horkheimer and Samuel Flowerman and undertaken by a diverse collective of social scientists at Berkeley and Stanford, and in New York. Several leading figures in this research were pre-war émigrés from Germany. They included the part-Jewish Theodor Adorno, who with Horkheimer, had set up Frankfurt Institute in exile at Columbia. Of the various papers and books that emerged, by far the most influential was *The authoritarian personality*, a two-volume opus published in 1950.[9] Authored by Adorno, Else Frenkel-Brunswik, Daniel Levinson, and Nevitt Sanford, *The authoritarian personality* combined questionnaire measures with projective tests and clinical interviews to build up a psychoanalytic interpretation of fascism. The authoritarian was an extreme and rigid type who could be readily exploited by ruthless demagogues. Adorno *et al.* even developed a handy scale for measuring 'potentiality for fascism', otherwise known as the 'California F scale'. Their results suggested plenty of racial antipathy lay beneath the surface of American life, anti-Semitism by no means the end of it.[10] *The authoritarian personality* highlighted the link between personality and ideology, framing irrational prejudice as individual psychopathology.[11] This was not a particularly

---

[7] US racial research had focused mainly (but not exclusively) on differences between those of African versus European descent, extending from intelligence testing to personality and psychopathology, social efficiency, and fatigue. See Richards, G. (1997), *'Race', racism and psychology: towards a reflexive history*, Routledge, London, Chapter 4.

[8] As Graham Richards argued, controversy in racial psychology was partly the means through which New World psychologists became aware of 'racism' (a term not used until 1936, according to Richards) as a problem. See Richards (1997), *'Race', racism and psychology: towards a reflexive history*, p. 112.

[9] Adorno, T.W. *et al.* (1950), *The authoritarian personality*, Harper and Row, New York.

[10] See Herman, E. (1995), *The romance of American psychology: political culture in the age of experts*, University of California Press, San Francisco. As Herman observed, the ill-feeling generated by internment of Japanese Americans and deadly race riots during the war made this dislocating effects of racism seem all too real.

[11] Ackerman, N.W. and Jahoda, M. (1950), *Anti-semitism and emotional disorder: a psychoanalytic interpretation*, Harper and Row, New York helped ram home the idea of the anti-Semitic bigot as a sick, conflicted individual.

new idea, even in the US.[12] However, it found a receptive audience amongst social psychologists anxious to ensure domestic social harmony, and it provided a suitably damning explanation of Nazi horror.

The chilly ideological climate of the Cold War in the early 1950s upped the ante considerably. By equating psychopathology with one end of the political spectrum, *The authoritarian personality* was open to the charge of one-sidedness. Caught up in the debate about the threat of authoritarian communism abroad, some of those involved with 'Studies in prejudice' were eyed with suspicion, as foreigners with an agenda.[13] Marxist theorists Adorno and Horkheimer quickly returned to Germany, while other contributors were careful to downplay the overt political content of their work. In 1954, Else Frenkel-Brunswik introduced the more psychological notion of rigid cognitive style, an idea taken up at length by another young researcher associated with the project, Milton Rokeach.[14]

Although *The authoritarian personality* spawned a huge amount of follow-up work, it struck many American psychologists as empirically unconvincing as well as ideologically myopic.[15] Continental critical theory did not readily mix with the positivistic orientation of US attitudinal research.[16] In a follow-up volume edited by Richard Christie and Marie Jahoda, *The authoritarian personality* was picked apart from various methodological and political perspectives. Herbert Hyman and Paul Sheatsley pilloried the naïve survey designs and

---

[12] For example, it had been anticipated in the pre-war work of Harold Lasswell.

[13] The communist-hunters of the era cracked down on the very freedoms of association and expression they claimed to be defending, for they were the means by which enemy doctrines could enter and undermine American democracy. Just as *The authoritarian personality* was published, Sanford and Levinson became embroiled in an infamous oath-signing controversy at Stanford University. See Roiser, M. and Willig, C. (2002), The strange death of the authoritarian personality: 50 years of psychological and political debate, *History of the Human Sciences* **15**, 71–96.

[14] See Frenkel-Brunswik, E. (1954), Further explorations by a contributor to *Authoritarian personality*, in *Studies in the scope and method of the authoritarian personality* (ed. R. Christie and M. Jahoda), pp. 226–75, Free Press, Glencoe, Illinois.

[15] *The authoritarian personality* spawned over 64 studies in the five years after its publication. Richards (1997), 'Race', racism and psychology: towards a reflexive history, p. 232.

[16] Gordon Allport, for example, responded with a broadly eclectic approach to the problem of prejudice, drawing on trait psychology, theories of social group processes, and developmental psychology. See Allport, G (1954), *The nature of prejudice*, Addison-Wesley, Reading, Massachusetts.

inadequate sampling used.[17] The scales presumed much of what they were deemed to discover, and had already been shown to be vulnerable to acquiescence.[18] Conversely, political scientist Edward Shils claimed far too little attention had been paid to the opposite end of the spectrum, to varieties of leftist ideology.[19] Fascism and bolshevism had much in common, Shils suggested. However, Shils lacked the data to back up his assertion of the symmetry of the political extremes. Enter Hans Eysenck.

Correcting the consensus

Eysenck had already begun to respond by gearing his research to American concerns. And they had to take note, for Eysenck had something they didn't. In Britain, groups such as Oswald Mosley's ultra-nationalist Union Movement and the Communist Party were legal and operated openly. Eysenck and his team were able to test groups openly identifying themselves as communists or as fascists, rather than the university students and 'fellow-travellers' that made up much of US researchers' samples. In 1951, Eysenck published a study of various political persuasions and classes. These included middle and working class supporters of the main political parties: Conservative, Liberal, and Labour. (Labour party members were dubbed 'socialists' by Eysenck). A sample of almost 150 middle and working class Communist Party members was gathered, as well as a handful of fascists identified as followers of Mosley. Eysenck's tough-mindedness dimension functioned as the crucial concept, linking to and cutting across political attitudes research across the Atlantic. Lurking in the background was the 'Budenz–Bentley syndrome'—the idea that the extremes of left and right were so similar that switching from one to the other was surprisingly easy.

Eysenck's notion of tough-mindedness allowed him to differentiate varieties of left- *and* right-wing commitment. In a survey published in 1951, his T scale did not discriminate the main political parties. However, 'communists and fascists were ... found to be tough-minded in comparison with conservatives,

---

[17] Hyman, H. and Sheatsley, P. (1954), *The authoritarian personality*: a methodological critique, in *Studies in the scope and method of the authoritarian personality* (ed. R. Christie and M. Jahoda), pp. 50–122, Free Press, Glencoe, Illinois.

[18] See Cohn, T.S. (1953), The relation of the F-scale response to a tendency to answer positively, *American Psychologist* **8**, 335.

[19] Shils, E.A. (1954), Authoritarianism, right and left, in *Studies in the scope and method of the authoritarian personality* (ed. R. Christie and M. Jahoda), pp. 24–49, Free Press, Glencoe, Illinois.

liberals and socialists', a result he duplicated with overseas samples.[20] Eysenck restricted himself to comparing his tough-minded conservative with the authoritarian personality of Adorno *et al.* Nonetheless, his capacity to connect the extremes was hampered by his small fascist sample.[21]

One of Eysenck's Ph.D. students, Thelma Coulter, would rectify the problem. Infiltrating both fascist and communist organizations, Coulter gathered data from 43 active members of each extreme group, plus a control group of around 83 soldiers.[22] Coulter tested them on measures of R and T that had been revised by another of Eysenck's Ph.D. students, D. Melvin.[23] She also gave them the F scale, thus measuring the 'fascist potential' of real fascists. It gave Eysenck the jump on his American colleagues, and it made him hard to disregard.

Although Eysenck published several more works in the 1970s and 1980s, his 1954 book, *The psychology of politics* remained his most significant work in the area.[24] In it, Eysenck brought together all his social survey research to date. The first three chapters dealt with the basics of demographic and class sampling, opinion polling, and attitude testing. The remaining chapters fleshed out the personality and experimental correlates of his four-quadrant attitudinal space. The last chapter contained a highly speculative treatment of the relationship of beliefs to behaviour, connecting political attitudes with Hullian habit conditioning.

Eysenck reiterated his contention that the extremes of left and right were mirror images of each other: communists were tough-minded radicals and fascists tough-minded conservatives. Presenting Coulter's T scale results as a graphic scatter-plot, most of her fascist sample tested as very tough-minded. Communist T scores lay between that of the fascists and the low mean score of

---

[20] For his overseas study, see Eysenck, H.J. (1953e), Primary social attitudes: a comparison of attitude patterns in England, Germany and Sweden, *Journal of Abnormal and Social Psychology* **48**, 563–8. The quote comes from p. 563.

[21] Eysenck, H.J. (1951a), Primary social attitudes as related to social class and political party, *British Journal of Sociology* **2**, 198–209. In this study, fascist and communists were lumped together because the fascist sample was so small. If anything, class differences in tough-mindedness seemed the most striking feature of Eysenck's data.

[22] The exact number in Coulter's control group was not clear; Eysenck's accounts cited two figures, 83 and 86.

[23] Coulter, T.T. (1953), An experimental and statistical study of the relationship of prejudice and certain personality variables, Ph.D. thesis, University of London; Melvin, D. (1955), An experimental and statistical study of two primary social attitudes, Ph.D. thesis, University of London.

[24] Eysenck, H.J. (1954a), *The psychology of politics*, Routledge and Kegan Paul, London.

her soldier control group, a result Eysenck said was in line with expectations. Furthermore, Coulter's mean scores on the F scale showed a similar kind of ordering. Eysenck was thus able to take the fateful step of equating tough-mindedness with authoritarianism. He then argued that Adorno *et al.*'s F scale was a 'measure of tough-mindedness and that [was] not restricted to the measurement of *Conservative* authoritarianism' [original italics].[25] While fascists and communists espoused distinct political ideals, they could share a deep authoritarian streak. It balanced out the political picture, Eysenck felt, explaining the 'same but different' paradox he had witnessed in the volatile politics of pre-war Germany.

The backlash

According to Eysenck, *The psychology of politics* first brought upon him the wrath of the left. 'Severe criticisms of the book were made, particularly by the London School of Economics (which was traditionally very left-wing) and in America by M. Rokeach and R. Christie.'[26] The London School of Economics (LSE) attack that Eysenck mentioned was merely a short critique published by a student member of the Communist Society. Eysenck also neglected to mention that fascist groups apparently objected as well.[27] Understandably, neither liked being lumped with the other. In bracketing Rokeach and Christie with left-wing political groups, Eysenck implied they had similar axes to grind. While there was a grain of truth to this, it wasn't quite that simple.

Milton Rokeach and Richard Christie were two social psychologists who had cut their teeth on *The authoritarian personality* project. Rokeach had been a student of Nevitt Sanford and Else Frenkel-Brunswik at Berkeley in the late 1940s, relating prejudice to mental rigidity, before getting a job at Michigan State.[28] He met Eysenck when in England early in 1954, though no doubt Rokeach was

---

[25] Ibid., p. 153.

[26] Eysenck, H.J. (1990*b*), *Rebel with a cause*, W.H. Allen, London, pp. 85-6. The bracketed comment on the politics of the LSE is Eysenck's, not the author's.

[27] See Gibson, H.B. (Tony) (1981), *Hans Eysenck: the man and his work*, Peter Owen, London, pp. 224–7. The LSE criticism had come from a member of the student communists, Colin Sweet, who wrote a negative review for the student newspapers. Sweet, C. (1955), Mr. Eysenck and the state of social science, *Clare Market Review* **50**, 34–9. Nevertheless, Eysenck had visited the LSE in 1955 with no protesters to greet him. Gibson claimed fascist groups also took exception to Eysenck's work but did not cite any evidence of this.

[28] Rokeach, M. (1948), Generalised mental rigidity as a factor in ethnocentrism, *Journal of Abnormal and Social Psychology* **43**, 259–78.

aware of Eysenck's work before then. Rokeach buttonholed Eysenck at the British Psychological Society (BPS) conference in April with questions relating to Eysenck's 1951 article. Rokeach was particularly perplexed with Eysenck's T scale. Items in the scale seemed to work in separate directions, and in an unbalanced fashion. In subsequent correspondence, Rokeach asked Eysenck to explain the rationale and scoring of the scale, making it clear he had been unable to reconstruct the scale means for the various political groups based on the item endorsement rates Eysenck had published.[29] The items were of the five point Likert type, from 'strongly agree', to 'agree on the whole', to 'don't know/can't answer' (marked as '?'), to 'disagree on the whole', to 'strongly disagree'. Eysenck duly explained that the T scale was an indirect measure of tough-mindedness, necessarily saturated with the political content of his R scale. 'None of [T scale] items are unidimensional, so that the scale as a whole has to be constructed by cancelling out the contribution of the other [R] factor.'[30] In other words, Eysenck suggested one can't simply measure tough-mindedness in isolation; one has to measure it in terms of being tough-minded about a range of (R scale) sociopolitical issues. As to the scoring strategy, he told Rokeach:

> I think it could only be a chance effect if your calculations should agree with the T-scores as given because, if I remember rightly, "?" answers are sometimes scored with a "Yes", and sometimes with a "No" reply. Consequently it would be impossible to reconstruct the scores with any degree of exactitude.[31]

That was all he would say. It is hard to know what to make of this rather vague reply. Given Eysenck's detached *modus operandi,* he might not have been able to remember the scoring details because he wasn't particularly familiar with them in the first place. But assuming he kept the data, it appeared he could not be bothered to go back to check or he preferred not to tell.[32] While Eysenck generously enclosed the just-completed proofs of *The psychology of politics,* this was not quite the answer Rokeach was *ostensibly* looking for. Richard Christie,

---

[29] Milton Rokeach to Hans J. Eysenck, 12 April 1954, Milton Rokeach Papers, Archives of the History of American Psychology.

[30] Hans J. Eysenck to Milton Rokeach, 21 April 1954, Milton Rokeach Papers, Archives of the History of American Psychology, p. 1.

[31] Ibid., pp. 1–2.

[32] Even allowing for the fact he was corresponding with someone he must have suspected to be a hostile critic, Eysenck's reply seemed strangely uncooperative. While Eysenck was probably just being strategically defensive, other more negative interpretations are possible, i.e. that he had used some kind of procedure that would be hard to defend and was thus loath to make explicit.

in touch with Rokeach, also had his doubts about Eysenck's data and had undertaken to review *The psychology of politics* for the *American Journal of Psychology*.

So began a notorious confrontation. Rokeach and Christie launched a blistering, multipronged attack that, in principle, hinged on the reality of left-wing authoritarianism and the adequacy of Eysenck's factor-analytically derived measures. However, much more was at issue, as we shall see. It is a compelling case study in scholarly conduct dictated by the explicit and implicit norms of scientific debate.

Attacking Eysenck

Rokeach and Christie both prepared lengthy critiques—Rokeach on Eysenck's earlier work and Christie on newer material contained in *The psychology of politics*. Christie went first with his short review of *The psychology of politics* in 1955. It was scathing. Christie suggested Eysenck must have been attempting 'some kind of practical joke'.[33] He dismissed the book's substantive content in just a few lines, claiming 'these pages are grounded upon samples which are at best atypical ... upon the use of scales which do not measure what they are purported to measure, and upon a misleading analysis of results.'[34] He spent the rest of the review highlighting mistakes, misprints, and omissions. *The psychology of politics* was certainly uneven even by Eysenck's standards, parts of it giving the impression of being put together in great haste. Christie pointed to tables that didn't quite tally, and to Eysenck's 'annoying' referencing habits that made it hard to track down his sources.[35] Eysenck had made numerous references to his students' work. One of these, Melvin's crucial 1955 dissertation, was cited inconsistently.[36] Christie's request for it was met with 'we have no such thesis on file' from IoP library staff, and Christie let his readers know. As Eysenck's biographer Gibson commented, this disclosure carried

---

[33] See p. 702 of Christie, R. (1955), Review of *The psychology of politics* by Hans Eysenck, *American Journal of Psychology* **68**, 702–4.

[34] Ibid., p. 703.

[35] For instance, Eysenck discussed studies by Guilford and Hildebrand in the text of *The psychology of politics*, but didn't cite them in his bibliography. Other important researchers were mentioned in passing without an indication of which of their studies he was talking about. Overlooked by Christie was an even stranger omission. Eysenck didn't cite *The authoritarian personality* in *The psychology of politics* at all, even though it was a central counterpoint to the arguments he made in the middle sections of the book.

[36] Eysenck cited Melvin's 1955 thesis as 'Melvin (1954)' in the text of *The psychology of politics*, and gave the full reference with the year 1953 in his bibliography.

a degree of innuendo, an (unfounded) suggestion that something suspicious was going on.[37]

After consultation between Brewster Smith and Wayne Dennis, respective editors of the *Journal of Abnormal and Social Psychology* and *Psychological Bulletin*, both Rokeach's and Christie's subsequent critiques were funnelled to *Psychological Bulletin* and published in 1956. Rokeach and Christie worked closely together to minimize overlap and produce a 'coordinated one–two punch'.[38]

Teaming up with Charles Hanley, Rokeach mounted a detailed critique of Eysenck's earlier papers, focusing on Eysenck's 1951 article in particular.[39] Rokeach and Hanley suggested Eysenck's T scale measured at least two different things. For example, communists were the most tough-minded on six of the 14 items (tending to be anti-religious and morally liberal), but the *least* tough-minded on the remaining eight (tending to be egalitarian, anti-racist pacifists). Rokeach and Hanley sought an explanation for this by recalculating the means for all the groups, leaving out the small fascist sample. This altered the picture significantly, with several of the groups appearing much less tough-minded—especially the communists.[40] Rokeach and Hanley leaped on this point, concluding that Eysenck's contention that 'communists are more tough-minded than conservatives, liberals and socialists, [was] not supported by his published data.'[41]

---

[37] See Gibson (1981), *Hans Eysenck: the man and his work*, p. 225. Christie seemed to be implying that Eysenck had dreamt this study up. Such a suggestion was totally unfounded; it was merely the result of the confusion brought about by Eysenck's careless citation practices and the ambiguity of doctoral completion dates in the British system. Christie also suggested Eysenck was in the habit of plagiarizing the words of others. However, the example Christie cited concerned Eysenck's discussion of another researcher's ideas, i.e. Gardner Lindzey's scapegoat theory. Eysenck made it clear he was talking about another researcher's viewpoint, not his own. This was not plagiarism, just a failure in formal attribution. The source of his information was not cited properly and a portion of the text was in need of quotation marks.

[38] Richard Christie to Milton Rokeach, 18 April 1955, Milton Rokeach Papers, Archives of the History of American Psychology, p. 1.

[39] Rokeach, M. and Hanley, C. (1956), Eysenck's tender-mindedness dimension: a critique, *Psychological Bulletin* **53**, 169–76.

[40] Rokeach and Hanley's recalculation changed the overall orderings of some of the groups too, with conservative and socialist voters appearing to be more tough-minded than communists amongst working class groups. The ordering remained the same for middle-class groups, but communists were now only marginally the most tough-minded.

[41] Rokeach, M. and Hanley, C. (1956), Eysenck's tender-mindedness dimension: a critique, p. 173.

Although they didn't explicitly say so in their initial critique, Rokeach and Hanley's recalculation had been guided by information contained in *The psychology of politics*.[42] Eysenck had reproduced in an appendix the scoring rules Melvin had used for the revised R and T scales.[43] Melvin had scored the mid-point 'don't know/can't answer' option as a zero, effectively making it a tough-minded response on a scale scored in the *tender-minded* direction. The information Eysenck had privately given Rokeach over a year earlier was *potentially* consistent with Melvin's rules, but his 'sometimes yes and sometimes no' reply left plenty of room for other possibilities. Finally, Rokeach and Hanley questioned Eysenck's interpretation of the T factor based on his factor solution. Since items used on the T scale tended to cluster in the middle of each of the four quadrants, a 45° rotation led to near zero loadings on one of the two factors. The T scale item content in this rotated solution could readily be interpreted in terms of 'religiosity' and 'humanitarianism'—in line with the findings of previous researchers they said.

Eysenck, of course, was given the right of reply by *Psychological Bulletin* editor Wayne Dennis. Interpreting Rokeach and Hanley's article as an attack on *The psychology of politics* generally, Eysenck defended his R and T factor solution in the context of his developing theory of personality.[44] His tough-versus tender-minded dimension exhibited important correlations with personality variables and experimental data, vindicating its placement over other possible solutions. Eysenck cited additional studies—some too recent to include in *The psychology of politics*—to back up his notion of left-wing authoritarianism. However, he was completely unmoved by Rokeach and Hanley's re-computations. The usual checks had been done, Eysenck wrote. In any case, obtaining equivalent figures from the data he had published was unlikely and irrelevant, given that such re-computations would 'leave out part of the data'.[45] Was he suggesting his published item endorsement rates did not include data generated by the 'don't know/can't answer' option? It is hard to say. Rokeach and Hanley had made it clear they assumed Eysenck always scored this option

---

[42] They did make it clear they were drawing on *The psychology of politics* in their subsequent reply to Eysenck's reply, however. See Hanley, C. and Rokeach, M. (1956), Care and carelessness in psychology, *Psychological Bulletin* 53, 183–6.

[43] Eysenck (1954a), *The psychology of politics*, p. 276.

[44] Eysenck, H.J. (1956c), The psychology of politics: a reply, *Psychological Bulletin* 53, 177–82.

[45] Ibid., p. 178. This suggested that his original scoring rules differed to that of Melvin and that his published item endorsement rates did not necessarily include scores generated by the 'don't know/can't answer' option.

in the tough-minded direction, *à la* Melvin. In a footnote in his reply, Eysenck said this was 'quite untrue'.[46] He had experimented with several scoring systems, Melvin's being the latest and best. Nonetheless, Eysenck saw no value in telling *Psychological Bulletin* readers what he *had* done. In effect, he played a game of 20 questions with his critics, putting the onus on them to guess the correct answer if they felt obliged to make such checks. Eysenck otherwise implied that the level of detail they were demanding was unreasonable. Beyond that, he simply stood on his word: either you believed his figures or you didn't.

Commenting on Eysenck's draft reply in April 1955, Rokeach thought it evasive, with a bit of 'red baiting' thrown in for good measure.[47] Rokeach counselled Dennis not to allow Eysenck to change his rejoinder in proof after he had seen their final reply, for it exhibited more sleight-of-hand in referencing. Dennis agreed, having also been pressured by Rokeach to accept Christie's publication *in toto*, regardless of whether its technical complication inhibited reader understanding of the broader issues involved.[48]

Rokeach and Hanley's reply to Eysenck, mockingly titled 'Care and carelessness in psychology', argued that Eysenck's dual loadings of the T scale were not handled in a consistent fashion. Rival factorial solutions should be directly compared by pitting their correlates with variables of interest against each other. The questions regarding re-computed means remained. They wondered how much store should be placed in data so affected by 'don't know/can't answer' response scoring procedures. And they again complained of Eysenck's misleading referencing.[49] Eysenck's reply had also mentioned two additional 'independent' studies with results even more striking than his 1951 study. These new studies were actually IoP dissertations he helped supervise—Thelma Coulter's 1953 effort and a more recent one by 'Nigneiwitzky' (or 'Nigniewitzky').[50]

---

[46] Ibid.

[47] Milton Rokeach to Richard Christie, 14 April 1955, Milton Rokeach Papers, Archives of the History of American Psychology, p. 1.

[48] See Milton Rokeach to Wayne Dennis, 23 December 1955, and Wayne Dennis to Milton Rokeach, 17 January 1956, Milton Rokeach Papers, Archives of the History of American Psychology.

[49] Hanley and Rokeach (1956), Care and carelessness in psychology. Hanley and Rokeach pointed out that Eysenck did not cite in his original article a 1941 paper by Ferguson that Eysenck implied he had in his reply to them.

[50] Eysenck did not give the full reference for this 'Nigniewitzky/Nigneiwitzky' study in his reply to Rokeach and Hanley, and had trouble spelling his student's name correctly. To be fair, though, Eysenck wasn't the only one with spelling difficulties. The 1954–55 IoP Annual Report listed this work as Nigniewitzky, R.D. (1955), A statistical study of rigidity

However, such studies needed to be evaluated on their own merits, Rokeach and Haley argued, they could not be used to support the consistency and computational correctness of his earlier data.

Then it was Christie's turn again. While working as a research fellow at the Centre for Advanced Study in the Behavioral Sciences at Columbia, Christie prepared a longer but more conventional attack on *The psychology of politics*.[51] He took aim at Eysenck's sampling methods, as well as his unbalanced scales and scoring procedures, which could produce strange results. For instance, the Melvin's T scale scoring procedure meant those selecting the 'don't know/can't answer' option were automatically cast as very tough-minded. Echoing Rokeach and Hanley, Christie reconstructed some of Eysenck's 1951 data to demonstrate how Eysenck had rounded the averages of class means up or down in line with his expectations.

Perhaps Christie's most substantive points centred on his analysis of Coulter's data. He did a rough calculation of the means of her groups from the scatter-plot reproduced in *The psychology of politics*. Communists scored significantly more *tender-minded* than fascists (11.02 versus 7.85, with the control group estimated at 14.2), making it difficult to understand why Eysenck had lumped them together as tough-minded. The contrast with the control group was the key. In *The psychology of politics* Eysenck had merely said that 'with few exceptions both Fascists and Communists have more "tough-minded" scores than the average of the soldier [control] group.'[52] Coulter's F scale data exhibited a similar pattern. Her communist's average F-scale score lay well below that of fascists (94 versus 159), and somewhat higher than the control group (75). When he presented these numbers, Eysenck commented that 'the communists scored significantly more "Fascist-minded" than did the soldiers.'[53] But four pages earlier he had suggested 'Communists make almost as high scores on this scale as Fascists.'[54] The item endorsement averages were even more illuminating. With a theoretical neutral point of 4.0 on a scale of 1.0 to 7.0, Coulter's communist's item average of 3.13 indicated a general tendency

---

as a personality variable, MA thesis, University of London, while the 1955-56 Annual Report cited the doctoral dissertation Nigniewitzky, R.D. (1956), A statistical and experimental study of rigidity in relation to personality and social attitudes, Ph.D. thesis, University of London.

[51] Christie, R. (1956a), Eysenck's treatment of the personality of Communists, *Psychological Bulletin* 53, 411–30.

[52] Eysenck (1954a), *The psychology of politics*, p. 142.

[53] Ibid., p.153.

[54] Ibid., p. 149.

to reject items. American students typically scored in the 3.0 to 4.0 range, Christie wrote.[55] The item average for Coulter's fascists was an all-time high 5.30, while the control group's average of 2.50 was one of the lowest on record. For Christie, this data highlighted the *dissimilarity* of communists and fascists, not to mention the 'fascinatingly aberrant' nature of Eysenck's control group sample. In sharp contrast to Eysenck, Christie argued that the F scale was *not* a good measure of left-wing authoritarianism. But given that real fascists scored very highly on it, Coulter's data suggested it actually measured what it was designed to.

For his part, Eysenck seemed relatively unconcerned by the psychometric problems Christie pointed to, highlighting aspects of his unbalanced scales that made left-wing and right-wing similarities less likely.[56] Tellingly, he flatly ignored Christie's critique of Coulter's data. Staying one step ahead, Eysenck again relied on in-house research for support, particularly the 'striking' results of 'Nigniewitzky's' doctoral study of French political groups.[57] He otherwise complained that his critics ignored the framework of his bio-behaviourist theory, an aspect of Eysenck's work that Christie privately dismissed out of hand.[58]

The last word in this episode went to Christie.[59] Having finally got hold of both Coulter's and Melvin's theses, he concluded that Eysenck had never clarified the tough-minded concept, and had never really tried. Eysenck's T scale did not measure it anyway. It might have been reliable, as Eysenck had maintained, but it was not valid across different political groups. According to Christie, communist subjects tended to respond to T scale items according to the item's 'R' component, not the 'T.' Christie finished on an exasperated note:

> Eysenck's responses to these critical points which he takes note of invariably evade the specific issue. Reliance is placed upon an extensive citation of the research of others. Those that are available do not support his position but indicate the cogency of the criticism.[60]

---

[55] Christie (1956a), Eysenck's treatment of the personality of Communists, p. 425.

[56] Eysenck, H.J. (1956f), The psychology of politics and the personality similarities between fascists and communists, *Psychological Bulletin* **53**, 431–8.

[57] Ibid., p. 436. This time he gave a full citation for 'Nigniewitzky's' Ph.D. work.

[58] Richard Christie to Milton Rokeach, 27 April, 1955, Milton Rokeach Papers, Archives of the History of American Psychology, p. 1.

[59] Christie, R. (1956b), Some abuses of psychology, *Psychological Bulletin* **53**, 439–51. Like Hanley and Rokeach, Christie pointedly parodied Eysenck's popular book titles.

[60] Ibid., p. 450.

An analysis not a verdict

Eysenck's methodological vagueness in print and in private could be read as an insulating strategy—a sensible one his defenders might say. Likewise, his heavy reliance on in-house work kept his critics at bay, for it was hard to dispute something one had not seen. For example, in his reply to Rokeach and Hanley, Eysenck wrote that 'the readers interested in the construction of the scales, and the problems encountered, are referred to a separate publication by Melvin.'[61] However, Melvin's thesis would remain unpublished and hard to get. Likewise, Eysenck's write-up of Coulter's work did not make it into print until 1972, well after her death in a road accident. Short of visiting the IoP library, student theses could only be obtained as microfilm copies and would take up to five months to arrive for those overseas. The interested reader would have to be very interested indeed! As it happened, these Americans were and they let their irritation about the way Eysenck leant on unpublished work show. They never apologized for any misapprehensions they might have created about the existence of these theses once they did emerge. Rokeach and Hanley had ruled this cite-a-thesis tactic out of order. With more time on his hands, Christie was a little more tolerant.[62]

Defensive as these strategies might have been, they also had a counter-attacking aspect. When Eysenck vetted a draft of Rokeach and Hanley's paper he told Rokeach he was 'rather surprised at some of your views'. Nevertheless, he did not recommend modifications 'as the points I might have suggested have already been covered in conversation and correspondence'.[63] Eysenck could have cleared up the uncertainty over scoring procedures, but once again chose not to. Instead he covered himself in ambiguity by implying that he had already done so, making sure Rokeach and Hanley would go into print half-cocked. In addition, Eysenck did little to clarify the honest confusion about Melvin's thesis, merely thanking his critics for alerting him to minor mistakes and misprints. In turn, this invited Eysenck's sympathizers to think of Rokeach, Hanley, and Christie as nit-picking zealots.

---

[61] Eysenck (1956c), The psychology of politics: a reply, p. 180. Like Christie, Hanley and Rokeach criticized Eysenck's confusion over the date of Melvin's thesis, also pointing out that it was not actually a 'publication' insofar as it had not been published.

[62] Rokeach and Hanley were never able to get hold of Melvin's thesis before going into print, but they did obtain Coulter's. Christie had Coulter for his initial write-up and Melvin for his subsequent reply, but never got his hands on Nigniewitzky's work.

[63] Hans J. Eysenck to Milton Rokeach, 25 March 1955, Milton Rokeach Papers, Archives of the History of American Psychology, p. 1.

The aggressive control of information went both ways. Rokeach and Hanley's paper did not allude to their 'unhelpful' correspondence about scoring procedures with Eysenck because they liked it like that. Apart from the awkwardness of bringing private communication into the public domain, they were only too happy to press on with their recalculations, confident the differences they were finding would embarrass Eysenck. Likewise they were only too happy to see Eysenck make mistakes and use misleading references, going to some lengths to ensure they remained in his published script. At the same time, Hanley and Rokeach were able to claim that Eysenck could have turned over his raw data in order to resolve the disagreement but had made 'no effort to do this'.[64]

In the end, Christie conceded that there was never enough detail in Eysenck's publications to decisively adjudicate 'where, if anywhere, the truth lies…'[65] Was anyone looking? In this war of words, truth was not so much the first casualty, more missing-in-action. The pursuit of truth—or at least the resolution of differences—was clearly not what Rokeach and Christie were after. They spread their charges scatter-gun style, distracting from genuinely productive points of agreement. They didn't try to close the distance between them, and Eysenck was only too happy to widen the gap. It was as if they had set against him because, quite frankly, they did. They saw Eysenck as a 'charlatan' and they wanted to take him down.[66] Under the guise of cooperative exchange they played by a set of adversarial rules, the kind of rules Eysenck favoured and evoked.

Passive-aggressive to a T

One wonders what it was about Eysenck that brought out the inner hatchet-man—and they were almost always men—in his peers.[67] The psychology of politics debate was a relatively early example of what became a recurring pattern of controversy for Eysenck, only just preceded by the outrage generated by

---

[64] Hanley and Rokeach (1956), Care and carelessness in psychology, p. 184.

[65] Christie (1956b), Some abuses of psychology, p. 450.

[66] In sharing the opportunity to attack Eysenck, Rokeach wrote to Christie to assure him: 'We think that there is certainly enough charlatanism in Eysenck's work to justify two critiques.' Milton Rokeach to Richard Christie, 14 April 1955, Milton Rokeach Papers, Archives of the History of American Psychology, p. 1.

[67] Short exchanges with psychometrician Jane Loevinger and philosopher Kathleen Nott were two of the few exceptions to this rule. See Eysenck's responses, Eysenck, H.J. (1956b), Diagnosis and measurement: a reply to Loevinger, *Psychological Reports* 2, 117–18 and Eysenck, H.J. (1964e), Philosophers and behaviourists: a reply to Kathleen Nott, *Encounter* 23, 53–5.

his 1952 psychotherapy paper. No doubt that pugnacious paper had primed Americans to Eysenck's style, even if it had engaged somewhat different audience. Nevitt Sanford, for one, was an influential figure who straddled both the social and clinical fields. Sanford can't have been neutral about Eysenck. As already noted in the previous chapter, Sanford hated Eysenck's therapy paper and Sanford was Rokeach's supervisor at Berkeley. Eysenck had visited California in 1954–55 and had got to know Else Frenkel-Brunswik, ensuring that everyone there knew of his political attitudes work and how it cut across theirs.

Without dismissing his genuine capacity to nurture those he took under his wing, Eysenck tended to heighten the sense of combat with those who did not accept significant aspects of his work. Not only was his style that of science at a distance, it was science as a *competitive game* characterized by a debunk-and-destroy ethos rather than caring-and-sharing. Moreover, his competitiveness had a mannered, passive-aggressive quality cleverly adapted to a double standard in scientific mores. While Eysenck's fellow psychologists might have been loath to acknowledge the importance of social ties and personal interaction for building consensus and trust, they also downplayed the destructive, cut-throat side of their business.[68] A need to succeed predicated on the failure of others was an impulse no one likes to own up to.[69] But make no mistake, discrediting other scientists' work is professionally rewarding, especially if they are high-flying rivals. It is generally seen as a dirty but necessary part of scientific life, however, nothing to be proud of nor examined too closely. Judgements about who and what to go after evoke uncomfortable considerations of character and motivation that go beyond the sanctities of pure data. In contrast, scientists have always tended to represent their profession as an overwhelmingly civil pursuit—thereby shoring up the 'useful illusion' that truth was not in the

---

[68] A focus on winning overlooks a landscape often littered with losers—the ideas, researchers, and programmes that are actively killed off rather than just fall by the wayside. Most major philosophical models of science acknowledge the effects of ruthless competition but have seldom highlighted it. For example, Lakatos suggested unsuccessful research programmes generally just fade away. Recent constructive accounts have recognized that it is important to account for failure as well as success, but they have still tended to emphasize the positively constructive rather than the negatively destructive.

[69] While I spoke to many of Eysenck's most trenchant critics, few were willing to admit they disliked Eysenck or wanted to take him down a peg. His first biographer, Tony Gibson, also mentioned how reluctant Eysenck's opponents were in identifying themselves as such.

telling, that 'the results of science depend not on argument but on nature herself'.[70]

Eysenck milked this double standard for all it was worth.[71] He chose his interventions carefully, and he flushed out opponents accordingly. Far-reaching claims were typically couched in qualified, matter-of-fact tones. His psychotherapy paper could be presented as exhibit A here, its seemingly neutral empiricism a red rag to Freudian bulls. Many of Eysenck's publications were constructed in this way, written with an eye to provoking responses that fell within a certain range. He put out over 40 replies or rejoinders to criticism, as well as responding to opposing viewpoints in the course of his other publications. He drew fire in order to return it. More importantly, it allowed Eysenck to avoid being seen to be the one initiating critical exchanges. It gave him the opportunity to play the innocent party time and time again, attacked for merely speaking unpopular but empirically supported truths.

The tactical possibilities afforded by the polite niceties of scientific debate were never far from his mind. For example, he was fond of reminding his interlocutors that character assassination just would not do, that arguments should be *ad rem* rather than *ad hominen*.[72] While such injunctions could themselves be construed as *ad hominen*—implying his critics were willing to stoop to below-the-belt tactics—it was usually enough to make them back off. Courteous in print and in private, he played by the rules in part to exploit them.

The discomfort Eysenck generated was often palpable, for he created a dilemma grounded on fear. Did one take him on at his own game, or dispute its legitimacy? The problem with the first option was he was so good at it, so politely merciless toward those who chose to have a go.[73] Those unwilling to

---

[70] Quote comes from Gross, A.G. (1990), *The rhetoric of science*, Harvard University Press, Cambridge, Massachusetts, p. 32. I am also drawing on Soyland, A.J. (1994), *Psychology as metaphor*, Sage, London, especially Chapter 7.

[71] For an excellent account of life under Eysenck, see Jensen, A.R. (1997a), Eysenck as teacher and mentor, in *The scientific study of human nature: tribute to Hans J. Eysenck at eighty* (ed. H. Nyborg), pp. 543–59, Elsevier, Oxford. Even if he sounded a little awed by the Professor, Jensen illustrated Eysenck's capacity to nurture his students and gave an insightful analysis of outsider versus insider perceptions.

[72] See Gibson (1981), *Hans Eysenck: the man and his work*, pp. 22 and 116.

[73] In contrast to the combative fireworks he displayed when debating (say) the merits of psychotherapy or his first biological model of personality, Eysenck gave more sympathetic criticism less attention, if not more respect. See, for example, his response to Bersh's analysis of his incubation hypothesis. Eysenck did not regard Bersh as 'a hostile critic' (p. 305). Without the necessary competitive challenge, Eysenck's reply contains little of his usual rhetorical flourishes. In fact, much of it was in point form, like a rough first

risk it—and those who wished they hadn't—were forced to fall back on the inherently submissive second option, clinging to a principle that left Eysenck free to ride roughshod over their work.

Eysenck brought a big bag of rhetorical tricks to any intellectual joust. One ploy Eysenck admitted to using, one he especially favoured in public presentations, was 'seeding' questions by leaving out the most obvious negative study or counterargument. As Cyril Franks and Ann and Alan Clarke attested, Eysenck would memorize the details of this potential objection 'chapter and verse', along with a well thought out response.[74] When it was inevitably raised, Eysenck could then graciously but ruthlessly dispatch it. Such commanding displays had an intimidating effect; it also led some people to believe that Eysenck planted stooges at his public lectures to show off his debating skills. In a way he did—it was just that the hapless interrogators weren't aware they were being set up.[75]

Other Eysenck moves were both obvious and sly. Always on top of the relevant literature, Eysenck was not above using the time-honored academic gambit of referring to material that particular audiences or opponents were unlikely to be familiar with.[76] Since he spoke several languages, this worked particularly well. Obscure researchers and foreign language articles (especially those in German) could be used to silence all but the most cosmopolitan journal-addicts. As illustrious protégé Jeffrey Gray wryly commented, 'you had to be up early to catch Hans out'.[77] Statistics and higher mathematics could likewise be used to intimidate the innumerate, especially philosophers and

---

draft. Eysenck, H.J. (1983a), The theory of incubation: a reply to Bersh, *Behaviour Research and Therapy* 21, 303–5.

[74] Cyril Franks, telephone interview, 13 July 2002; Alan and Ann Clarke, interview, 25 April 2002. Late in his life Eysenck acknowledged his use of this strategy. He told *The Guardian*'s Joan Freeman: 'I prepare everything in meticulous detail, and leave carefully organised gaps in what I am saying. Since the questions always refer to what I've missed out, my answers are already prepared.' Freeman, J. (1997), The pugnacious psychologist [obituary for Hans Eysenck], *The Guardian*, 8 September 1997.

[75] Jensen gave an observer's account of this kind of spectacle in Jensen (1997a), Eysenck as teacher and mentor, pp. 549–50 as did Gibson in Gibson (1981), *Hans Eysenck: the man and his work*, p. 185.

[76] In his autobiography, Eysenck gave a run-down on the rules he observed when debating, which could be paraphrased as: be concise, make sure you're an expert in the area, do your homework, concentrate on what you think are the most important points, and force your opponent to address them. Eysenck, H.J. (1997b), *Rebel with a cause* [revised and expanded], Transaction Press, New Brunswick, New Jersey, p. 76.

[77] Jeffrey Gray, interview, 25 July 2003.

psychiatrists, while the history and philosophy of science could be used to disarm empirically oriented psychologists.[78] According to latter-day research associate Paul Barrett, Eysenck could 'inflame others to the point of apoplexy by his impassive style of argument ... arguing with Hans was like arguing with a logic machine with a large working memory and a huge long-term storage capacity.'[79]

To IoP insiders and supporters, it was all in a good cause. However, the anecdotes outsiders and opponents tell were often less charitable. For instance, psychologist and child-care guru Penelope Leach recalled how she was taken aback by Eysenck's maverick advocacy when she first encountered him. Leach attended an Eysenck lecture on personality and politics in the 1960s while a graduate student at the LSE:

> My PhD concerned authoritarianism so I had some knowledge of that literature. Eysenck made reference to a paper and body of evidence that I felt I should have heard of but hadn't, so, greatly daring, I asked for the reference. He said 'ask me at the end' so I waited to speak to him. When I finally got to ask my question he said: 'Oh there isn't an actual paper; I was just making a point.'[80]

Leach was shocked at the time. As a wide-eyed young researcher, she looked up to senior figures like Eysenck. Older and more cynical, she said she would be less surprised today.[81]

Almost inevitably, those whose work had been contradicted or reinterpreted would respond. Psychoanalysts were a favorite target, with Eysenck constantly baiting them into playing to his strength in quantitative analysis. Likewise, Eysenck would rail against the one-sidedness of correlational and/or experimental psychology, positively begging a representative of either school to hit back. In that case, he generally got a little less than he hoped. In other cases—such as the psychology of politics and race and IQ—he got a little more.

---

[78] Eysenck admitted to using these tactics to deal with Aubrey Lewis and Cyril Burt, for example. See Eysenck (1997b), *Rebel with a cause*, p. 117. He also recounted humiliating the philosopher C.E.M. Joad in a debate on aesthetics by shifting the argument to his strengths in experimental psychology and mathematics. See Eysenck (1997b), *Rebel with a cause*, pp. 75–7.

[79] See p. 14 of Barrett, P. (2001), Hans Eysenck at the Institute of Psychiatry—the later years, *Personality and Individual Differences* 31, 11–15.

[80] Penelope Leach, personal communication, 8 October 2003.

[81] Ibid.

Playing the man

Rokeach and Christie's exchange with Eysenck illustrated just how ugly things could get—even allowing for the fact that such critique/reply scenarios are inherently adversarial. While the vindictiveness behind the scenes may not surprise most scientists, it was certainly at odds with the way they present their activities to the outside world. It hardly embodied the beneficent ideal of cooperative knowledge-making. Rather, this was science as a blood sport, capped only by libel laws and a sense it might reflect badly on all concerned.[82]

Eysenck's guardedness walked a tightrope of sanctioned appearance and underlying purpose. For instance, he felt obliged to cover his refusal to respond to requests for information with gracious ambiguity, and claimed that it was an 'overwhelming honor' to have one's writings subjected to such scrutiny—all appeals to the appearance of openness and cooperation.[83] Rokeach and Christie went through these same motions, dressing up their vendetta as a laudatory attempt to reconcile different viewpoints. All went in hard for debating match points. If anything, though, it was Eysenck's critics that took it to the limit, attempting to make every aspect of Eysenck's work look as bad as possible. Thus Eysenck's data were deemed weak and misleadingly presented, his responses to criticism slammed as selective and evasive. Worse, his critics implied he made up studies, plagiarized the words of others, massaged data to suit hypotheses, and just couldn't add up. By pointing a finger at Eysenck's cagey strategies, they tried to make him look like he was *obviously* hiding something. Eysenck's responses seemed restrained in comparison.

Rokeach and Christie's synchronized assault on Eysenck embodied some of the very aspects of Eysenck's partisan style they found reprehensible. As Christie admitted to Rokeach, it would colour their reputations as less-than-gentle critics from that moment on.[84] They also risked counter-attacks from Eysenck, his supporters, and anyone else who thought their critiques just too nasty and aggressive. Could Rokeach and Christie's work withstand such scrutiny? Probably not, according to some of Eysenck's sympathizers and

---

[82] Looking back on the debate, John Ray said Eysenck could well have sued his critics if he had been so inclined. However, this was not as common a practice as it is today. See Ray, J. (1986), Eysenck on social attitudes: a historical critique, in *Hans Eysenck: consensus and controversy* (ed. S. Modgil and C. Modgil), pp. 155–73, Falmer Press, Philadelphia.

[83] Eysenck (1956*f*), The psychology of politics and the personality similarities between fascists and communists, p. 431.

[84] Richard Christie to Milton Rokeach, 8 April, 1955, Milton Rokeach Papers, Archives of the History of American Psychology, p. 1.

Eysenck himself.[85] Luckily for them—and here was the rub for Hans—they were much less likely to attract such attention.

Rokeach and Christie wanted to 'shaft Eysenck'.[86] As Rokeach wrote Christie: 'The major goal we had set for ourself [sic] was to discredit Eysenck's reputation as a scientific investigator because we had come to the conclusion that he is not.'[87] They wanted to de-legitimize the man as much as his work. And they were not alone. Rokeach received several congratulatory letters from those irritated by Eysenck's work. For example, Albert Kurtz at the University of Florida told Rokeach he thought he and Hanley had 'slaughtered' Eysenck.[88] At Western Reserve, Jan Bruell read their papers with 'great satisfaction', having been annoyed at the way Eysenck presented data in another unrelated paper.[89] Like the critique soon to be directed at Eysenck's first biological model of personality by IoP insiders, the Rokeach–Christie attack divided opinion for years to come. It left Eysenck's supporters feeling that he was unfairly singled out and held to a higher standard. Conversely, it provided his detractors with ample evidence that there was something shonky about Hans.

Eysenck developed a penchant for the adversarial side of science because it worked so well for him. By his own admission, he loved the cut-and-thrust of debate—though only when it was played by the kind of civil rules governing

---

[85] For example, see Eysenck's reply to Christie condemning his critic's sampling and armchair theorizing. John Ray also pointed out these Americans tended to base their conclusions on samples largely restricted to undergraduates. Ray (1986), Eysenck on social attitudes: a historical critique.

[86] In his correspondence with Rokeach, Christie was in the habit of signing off with a rallying line of 'shaft Eysenck', or 'down with Eysenck', which Rokeach echoed. See, for example, Richard Christie to Milton Rokeach, 27 April 1955, Milton Rokeach Papers, Archives of the History of American Psychology, p. 1.

[87] Milton Rokeach to Richard Christie, 14 April 1955, Milton Rokeach Papers, Archives of the History of American Psychology, p. 1

[88] Albert K. Kurtz to Milton Rokeach, 28 May 1956, Milton Rokeach Papers, Archives of the History of American Psychology, p. 1.

[89] Jan H. Bruell to Milton Rokeach, 7 April 1956, Milton Rokeach Papers, Archives of the History of American Psychology, p. 1. Rokeach's reply is worth noting, if only for the level of zeal it suggested. Rokeach encouraged Bruell to keep complaining to officials, suggesting that Eysenck should be 'continually exposed, to the point where he [Eysenck] is asked to resign from the APA.' Eysenck was an American Psychological Association (APA) fellow at the time. Milton Rokeach to Jan H. Bruell, no date, c. April 1956, Milton Rokeach Papers, Archives of the History of American Psychology, p. 1.

scientific journals, seminars, and conferences.[90] Well-served by sharp argumentative skills and a cool and calculating outlook, such confrontations almost always saw him emerge a winner. There was much more than rhetorical value to the countless victories he achieved. They protected the various aspects of the Eysenck edifice from being seen to be discredited or refuted, and warned off repeat attacks. They were public displays of competence and status that thrilled his sympathizers, and helped generate attention and rally support in a research community full of potential collaborators and prospective students.

Yet Eysenck's appetite for the fight would sow the wind and reap the whirlwind. Outside his coterie, he came across as hypercompetitive, as an arrogant know-it-all who didn't know when to back off. And given that such adversarial exchanges tend to *penalize* rather than reward openness, a reputation for caginess would follow him. When he got the better of them, those left battered and bruised resented it for years, complaining that Eysenck's tactics fell outside the spirit of communal science. Conversely, on the few occasions he did not have it his own way, a sense of *Schadenfreude* abounded, the unresolved doubts only adding to the distrust generated by his aloof praxis.

The politics of disdain

The mid-1950s marked the beginning of some kind of fight for a liberal intelligentsia buffeted by 'red scare' politics. Rokeach and Christie saw themselves as defending everything the 'Studies in prejudice' project stood for: inclusiveness and fairness, tolerance and internationalism.[91] Both Rokeach and Christie were not completely adverse to commonalities between the political extremes, one of Eysenck's central contentions. What they took exception to, however, was the idea that the left could be tarred with the same kind of authoritarian brush so carefully applied as a pathology of the right—which they took as a slur that was both empirically unjustified and politically insensitive. Christie suggested that Eysenck's data clearly refuted the idea that communists and fascists were mirror images of each other. If anything, the orthogonal component of Eysenck's T scale reflected a moral ruthlessness common to the extremes of left and the right, those who believed ends justified means.[92]

---

[90] Eysenck (1997b), *Rebel with a cause*, p. 274. Conversely, Eysenck voiced distaste for what he called 'no holds barred political street-fighting', the kind of 'external' popular controversy he nonetheless courted.

[91] Roiser and Willig (2002), The strange death of the authoritarian personality: 50 years of psychological and political debate. See also Stone, W.F. et al. (eds.) (1993), *Strength and weakness: the authoritarian personality today*, Springer Verlag, New York.

[92] Christie (1956a), Eysenck's treatment of the personality of communists, pp. 428–9.

In this vein, some of Christie's later work examined Machiavellian strands in political behaviour. Rokeach, in contrast, was busy developing a more psychological model of political engagement. By associating authoritarianism with 'dogmatic cognitive style' he made it easier to think of authoritarianism as pan-ideological. Even so, Rokeach would argue that dogmatic thinkers tended to be right-wing.[93]

To Eysenck, *The authoritarian personality* and the F scale had been etched in popular consciousness in a way that necessitated a corrective. *The psychology of politics* adapted some of the ideas of Arthur Koestler, another maverick figure whose 'premature' anti-communism had put him at odds with the liberal left.[94] Nevertheless, it did not come across as a labour of love, not just because of the limited care and attention with which it was written. Harking back to his Berlin boyhood where roving gangs made life far too interesting, it conveyed Eysenck's disdain for political extremism of all sorts, as well as a degree of dyslexia in reading their subtleties.[95] To Eysenck, the hard left and the hard right were all much the same and should rightly be condemned. Both extremes were tough-minded, and both were associated with a conception of extraversion that he linked with higher rates of criminality and acting-out behaviour.[96] Eysenck cast a plague on both their houses: communist and fascist groups were full of impulsive, aggressive, immoral types.

The popular Pelican series

According to Eysenck, his popular books were simply an exercise in pedagogy and demystification, nothing more. However, they also gave him a handy source of extra income and a means to ascend to the status of public intellectual. Because of them, Eysenck was able to bring his political attitudes work to

---

[93] Rokeach, M. (1960), *The open and closed mind*, Basic Books, New York.

[94] See Eysenck (1954a), *The psychology of politics*, p. 132. Always mindful of what Koestler had to say, Eysenck subsequently fell out with him over Eysenck's negative review of Koestler's book *The act of creation*. Eysenck, H.J. (1964a), Review of *The act of creation* by Arthur Koestler, *New Scientist*, 18 June 1964. See Koestler's reply in the correspondence section, 2 July 1964. Nevertheless, it probably helped get Eysenck interested in the topic of genius and creativity, which he would take up later in his life.

[95] As Jensen attested, politics was never a favorite topic of discussion for Eysenck. Psychological research was the thing, and those around him would keep to this preference. Jensen (1997a), Eysenck as teacher and mentor.

[96] Eysenck, H.J. (1964b), *Crime and personality*, Routledge and Kegan Paul, London. See also Gudjonsson, G. (1997), Crime and personality, in *The scientific study of human nature: tribute to Hans J. Eysenck at eighty* (ed. H. Nyborg), pp. 142–64, Elsevier, Oxford.

a much broader audience than his critics could ever hope for. His popular 1957 paperback *Sense and nonsense in psychology* would give him the opportunity to wave away all their previous criticisms.

The search for a broad audience had started early for Eysenck. His first book, *Dimensions of personality*, was a nod in this direction. *The structure of human personality* in 1953 was more so, even though both were principally aimed at fellow psychologists. Many of his books were gift wrapped as primers. Even some of his most intellectually significant works utilized an accessible textbook model to reach beyond narrow subdisciplinary boundaries. Refusing to be pinned down to any one speciality, their selling point was the integrative 'big picture'. His simplifications were justified insofar as they allowed all to understand. *The psychology of politics* was a good example of this. It juxtaposed opinion and attitude research, with factor analyses of personality and behavioural learning theory, each leaning on the other for support. *The psychology of politics*, Eysenck wrote, was intended to be 'intelligible to the layman', obliging his peers to evaluate it in these terms. Yet in it, Eysenck presented summaries of masses of data and it served as his major scientific statement in the area. It brought together a range of material, much of which had never seen the light of day before. There was no other more scholarly book-length treatment of this topic in the Eysenck catalogue.[97]

It was never clear with Eysenck where the 'serious' ended and the 'popular' began. Up to the early 1950s, he had been doing a bit of the latter under the guise of the former. However, when the opportunity arose to unambiguously chase lay audiences, Eysenck took it up with enthusiasm and aplomb. The head of Penguin Books, Allen Lane, had approached Aubrey Lewis for ideas about bringing the human sciences to the masses. Lewis recommended Eysenck. His first Penguin paperback under the Pelican imprint appeared in 1953. *Uses and abuses of psychology* was a chatty and cheerful rephrasing of his lectures to psychiatric trainees and it sold spectacularly well. It included chapters on intelligence testing, abnormal behaviour, techniques and effects of psychotherapy, and personality and attitude testing, and more. The follow-up *Sense and nonsense in psychology* covered more borderline but crowd-pleasing topics like hypnosis, dreams, lie-detectors, telepathy, and clairvoyance, as well as social attitudes.[98]

---

[97] The only other book Eysenck published in the area was Eysenck, H.J. and Wilson, G. (eds.) (1978a), *The psychological basis of ideology*, University Park Press, Baltimore. It contained a collection of Eysenck's articles and those of a number of other researchers, as well as a jointly written introduction and conclusion.

[98] Eysenck, H.J. (1957a), *Sense and nonsense in psychology*, Penguin Books, London.

Penguin paperbacks had revolutionized publishing in Britain in the late 1930s. With their extraordinarily low cost—equivalent to that of a packet of cigarettes—they sold in unprecedented numbers. Good marketing was backed by innovative retailing via vending machines. Of course, a strong line-up of expert authors writing on topical subjects helped too. Eysenck was not particularly well-known in the early 1950s but still managed to connect with a public receptive to this promising young science. Word-of-mouth was probably a factor, but so too was Eysenck's clear and direct style. He had a way of flattering the sensible reader by taking them into his confidence—the better to poke fun at the unjustified pretensions of some of his fellow psychologists whilst instructing the reader on what they should believe in and why. Eysenck guided the reader through the do's and don'ts of psychological research, drawing from personal experience and the general literature to illustrate his points—of how, say, a well-designed study may contradict common sense, or why projective tests had seduced his colleagues despite demonstrations they didn't work.

With few competing titles, Eysenck's Pelican paperbacks came to represent much of what the British public understood about psychology in the 1950s and 1960s. Initially at least, Eysenck's peers welcomed the popular arm of his writings. While these books helped sell Eysenck and his point of view, they also helped promote psychology as a discipline. Popularization helps create cultural space for any given scientific discipline, enrolling sympathetic patrons in government and industry, as well promoting it to a lay audience of potential consumers. Popularization helps give identity and purpose to any given research community, feeding images of what they are supposed to be doing and why. Most prosaically, it helps recruit new workers. Eysenck's popular books turned countless would-be psychologists on to the discipline generally, as well as attracting students to the IoP.[99] Eysenck recalled how many people told him his Pelican books were the first taste of psychology, piquing their interest as youngsters in career-moulding ways. Many psychologists I have met in the course of this project—some associated with Eysenck, others not—told me the same thing.[100]

Popularization also has its costs, as we shall see. It is not simply looked down upon by scientists, as is commonly suggested (even by Eysenck himself). Nevertheless, it is a process that is still policed by media and literary gatekeepers

---

[99] I am drawing on the ideas of science studies scholars such as Brian Wynne, amongst others. See, for example Wynne, B. (1991), Knowledge in context, *Science Technology and Human Values* **16**, 11–121; Wynne, B., Wilsdon, J., and Stilgoe, J. (2005), *The public value of science*, Demos, London.

[100] See also Jensen, A.R. (1997*a*), Eysenck as teacher and mentor.

as well as the public, and by scientist themselves. Popularizers have to be seen as sufficiently eminent and fair-minded to rightfully shop their science to the people. Those deemed too lowly risk being dismissed as bumptious or deluded.[101] Those not representing their discipline in a balanced and inclusive fashion invite accusations of bias. Nonetheless, these scholarly prerequisites do not alone guarantee ultimate success, for those not adept at contextualizing their science in an engaging manner are simply boring—ignored by the media and unread by the masses.

The debunking strategy of Eysenck's Pelican books played well to his lay readership and, up to a point, to his peers. While critical notices were generally good, reviewers noted a certain slant to the way he presented particular intellectual traditions.[102] Part of the fun of reading these books, noted one IoP colleague in the *Bethlem and Maudsley Hospital Gazette*, was to wonder who would be next to 'get it in the neck'.[103] Eysenck had a way with an insult that entertained everyone except those he targeted. For example, when concluding his summary of recent political attitudes research in *Sense and nonsense* Eysenck told readers that:

> unfortunately experimental social science is not welcomed very much in academic quarters, where the quiet somnolence of the reading-room and the dead and forgotten writings of past nonentities are considered much more soothing that the fresh air of empirical investigation, and the intoxicating flood of empirical data.[104]

In this chapter, Eysenck gave a run-down of his political research, just as other chapters summarized some of his other academic titles and papers. It culminated with a presentation of the means for various political groups plotted against his R and T axes, including Coulter's communists and fascists.

---

[101] The downstream metaphor, suggesting a largely one way flow of information and intellectual capital, comes from Hilgartner, S. (1990), The dominant view of popularization: conceptual problems, political uses, *Social Studies of Science* **20**, 519–39.

[102] Eysenck's paperbacks were widely translated and favourably reviewed for the most part. See *Lancet* (1954), Review of *Uses and abuses of psychology* by Hans Eysenck, *The Lancet* **263**, 348, and *Times Literary Supplement* (1954), Review of *Uses and abuses of psychology* by Hans Eysenck, *The Times Literary Supplement* 19 February 1954. However, reviewers often expressed reservations about the particular hobby-horses Eysenck chose to ride. For instance, Maurice Richardson in Richardson, M. (1954), Review of *Uses and abuses of psychology* by Hans Eysenck, *The New Statesman and Nation*, 23 January 1954 suggested Eysenck's attack on psychoanalysis was 'so passionate that one cannot help cocking a clinical eyebrow'.

[103] See p. 121 of GMC (1954), Dr. Eysenck's table talk, Review of *Uses and abuses of psychology* by Hans Eysenck, *Bethlem and Maudsley Hospital Gazette*, **1**, 121–2.

[104] Eysenck (1957a), *Sense and nonsense in psychology*, p. 302.

While Eysenck did not give the numerical mean values here, nor had he in *The psychology of politics*, their placement on the diagram was at odds with the figures worked out by Christie. Communists and fascists did indeed appear to be similar in terms of tough-mindedness, with T score means of around 9.6 and 8.3, respectively (remember the higher the score, the more *tender*-minded).[105] Moreover, Eysenck made no allusion to the kind of questions Rokeach and Christie had raised about this work. Obviously, he did not accept their criticisms, or the alternative figures they had calculated. He also tempted curious readers to test themselves on his R and T social attitudes inventory, providing the scoring direction key and scoring rules so they could categorize their political dimensionality.[106] It was an inspired little move. Many other dedicated 'test yourself' titles would follow.[107]

Popular presentation was a format that suited Eysenck. With nitty-gritty details deemed distracting and unnecessary, technicalities and critical disagreements could be justifiably glossed over. Thus he could quickly and safely dictate from memory. With nary a reference in sight, it was difficult to source, let alone dispute, much of what he was saying. The reader simply had to invest in his authority, to take it from him. Eysenck's treatment of the psychology of politics in *Sense and nonsense* also illustrated his strategy of moving on, unbowed and on-message. For most his career he rarely looked back—except to even up scores—and he had a habit of simply leaving behind the nagging worries critics had raised.[108]

---

[105] Ibid., p.303.

[106] Eysenck also gave an example of a personality inventory, a forerunner of his MPI/EPI/EPQ. See Eysenck (1957a), *Sense and nonsense in psychology*, pp. 195–6.

[107] Eysenck, H.J. (1966b), *Check your own IQ*, Penguin Books, London; Eysenck, H.J. and Wilson, G.D. (1975), *Know your own personality*, Maurice Temple Smith, London; Eysenck, H.J. and Sargent, C. (1983), *Know your own PSI-Q*, Multimedia Publications, London; Eysenck, H.J. and Evans, D. (1994), *Test your IQ*, Thorsons/Harper-Collins, London.

[108] Both colleagues and opponents noted Eysenck's tendency to neglect criticism and ignore past negative results. Liam Hudson made this point in Hudson, L. (1971), Science and popularisation, Review of *Race, intelligence and education* by Hans Eysenck, *New Society*, 1 July 1971, 29–30. Former student Richard Passingham also alluded to this when discussing Eysenck's research on criminality and the extraversion hypothesis. Richard Passingham, interview, 24 July 2003.

Changing research priorities, changing context

Subsequent research was not entirely kind to Eysenck's T scale. At UCL, Robert Green and Barrie Stacey took a long handle to it.[109] They noted the conceptual transformations that James's concept had undergone in Eysenck's hands. Eysenck's original T scale and Melvin's revision actually had very few items in common, with the revised version more a measure of moralizing religiosity. Such a scale would fail to measure those widely regarded as tough-minded authoritarians. For example, the religious zeal of Spanish Inquisition organizer Tomás de Torquemada would cast him as tender-minded on this scale. The T scale score clusters would only confirm that, if fascists and communists had anything in common, it was a tendency to reject all dogmas other than their own. Apart from recommending a name change for Eysenck's scale, Green and Stacey suggested a purer, more psychological measure was needed to do justice to William James's original idea.

The search for left-wing authoritarianism would quietly continue. However, the heat went out of the issue in the early 1960s as the perceived threat of fascism dwindled. In his widely read textbook of 1965, Roger Brown felt communists and fascists had not been shown to be similar, and described Eysenck's conclusions as 'truly extraordinary'.[110] With his attention elsewhere, Eysenck shelved the topic for the time being. Perhaps this also reflected a dearth of students willing to take it on, especially those with the commitment and resourcefulness of Thelma Coulter. However, the race and IQ wars of the 1970s would resurrect Eysenck's interest in the psychology of politics in a big way.

---

[109] Green, R.T. and Stacey, B.G. (1964), Was Torquemada tenderminded? *Acta Psychologica* **22**, 250–71. See also Green, R.T. and Stacey, B.G. (1962), The T concept, *Nature* **196**, 94.

[110] Brown, R. (1965), *Social psychology*, Collier-Macmillan, London, p. 542.

Chapter 8

Mr. Controversial: race and IQ

On an ordinary day, the 8th of May 1973, Hans Eysenck was delivering a talk at the London School of Economics (LSE). For his trouble, he was denounced, punched, and kicked. He was repeatedly assaulted by those in the crowd who had come not to praise him, but to bury him.

So it had come to this. The most renowned British psychologist of his generation had become such a hate figure amongst radical activists that they thought outright violence was justified, even obligatory. Never one to run with the crowd, Eysenck stayed true to his version of biologically grounded psychological variation, a framework that had now fallen out favour. Eysenck was already seen as an enemy of the left, but one little book published in 1971 came to dominate his wider reputation. He became more than just a public intellectual; he became a martyr.

When in 1969 Arthur Jensen reignited the debate on race differences in IQ, Eysenck backed him up with a quick and cavalier treatment of the issue. Eysenck's book *Race, intelligence and education* was greeted with outrage. The reaction from many of his peers conveyed a deep sense of betrayal. Eysenck had been wildly successful in selling psychology to the public and they had grudgingly accepted him as a spokesperson for their discipline. After 1971, however, all bets were off.

One does not have to dig deep to explain why Eysenck's contribution to the race and IQ debate stirred the pot. It was a dynamite topic already, touching on pervasive social problems that were by no means solved. Far more elusive questions revolve around his motivations. What, other than loyalty and an attraction to controversy, made Eysenck write that book? Why did he take such a seemingly unnecessary risk? These are difficult questions to answer in the absence of personal archival material. The limited correspondence with peers that has survived sheds only a smattering of insights. The recollections of those close to him are not particularly illuminating either, for Eysenck was a man who played his cards close to his chest. Nevertheless, such questions deserve answers, however incomplete and speculative.

I have already suggested there was an element of fatalism to his actions. With his level of acceptance in established circles already beginning to plateau, he had nothing to lose by publicly supporting Jensen. Moreover, in jumping in at the deep end of that debate, Eysenck gave the distinct impression of a man driven to settle scores dating back to his dust-up over the authoritarian personality. Stranger still would be the suggestion that it was all intentionally strategic. After all, marginality has its uses. It provides the freedom to engage in heterodoxy of all sorts, unencumbered by a need to protect a conventional reputation.[1] But was the rebel outsider pose he subsequently cloaked himself in a prior motivation or a *post hoc* rationalization? Any sneaking suspicion that Eysenck deliberately took himself out of the post-war mainstream has to be balanced with an entirely opposite proposition: that the race and IQ book was one step too far for this inveterate controversialist, that he got more than he bargained for, more than he was prepared to admit.

Personality, intelligence, and genetics

If Eysenck's 1950s work on political attitudes represented a defiantly middle-ground position, his emphasis on the biogenetic basis for variation in personality and intelligence would put him firmly on the conservative side of the political divide. Up to 1940, the distinctions implied by Francis Galton's 'convenient jingle of words'—nature and nurture—had woven their way through the discipline without necessarily stratifying it politically.[2] The environmental emphasis of John Watson's behaviourism coexisted with the hereditarian psychometrics of Cyril Burt. The differences were more philosophical and methodological, rather than ideological. Despite championing the primacy of innate intelligence, Burt was a left-leaning educational reformist, while Galton's biometric protégé, Karl Pearson, considered himself a socialist. The eugenic movement Galton founded was supported by a cross-section of the political spectrum in Britain prior to 1940, although there was already a distinctly right-wing racialist element amongst their American counterparts. Even if its reformist ideals were expressed in language that now makes us wince—all condescension and dire warnings—some genuinely good intentions were involved.

---

[1] For a discussion of some of these issues see Bos, J. *et al.* (2005), Strategic self-marginalization: the history of psychoanalysis, *Journal of the History of the Behavioural Sciences* **41**, 207–24.

[2] Francis Galton, quoted in Fancher, R. (2003), The concept of race in the life and thought of Francis Galton, in *Defining difference: race and racism in the history of psychology* (ed. A.S. Winston), pp. 49–75, American Psychiatric Association, Washington, DC, p. 70. See also Richards, G. (1997), *'Race', racism and psychology: towards a reflexive history*, Routledge, London.

Eysenck was very much a product of this old-school political mix. As a fellow of the British Eugenics Society since 1947, Eysenck was acculturated by the standard tales of a better, scientifically engineered society. The society enabled Eysenck to link up with like-minded colleagues in other fields and was a valuable source of research funds. Just after the war, Eysenck sounded like Galton's long-lost German heir. In 1947 he looked forward to the day when Galton's eugenic utopia of 'Kantsaywhere' could be fully implemented—where everyone was assigned a status and reproductive role reflecting their measured talent and character. It was, Eysenck suggested, a classically noble vision, an update of Plato's *Republic* based on modern psychological science. Most of the necessary know-how was already available; what was lacking was the will and wisdom to use it.[3] In a follow-up piece Eysenck warned that time was not on our side.[4] Citing Burt's research on intelligence, Eysenck suggested the average IQ in Britain was plummeting because the poorly endowed were out-breeding the more able. While civilization might not survive such a catastrophe, he stopped short of spelling out the kind of steps necessary to remedy it. While Galton wasn't shy about such specifics, half a century later Eysenck felt obliged to leave the reader to fill in the blanks.

World War II changed everything, investing the nature–nurture dichotomy with unavoidable ideological implications. Biogenetic accounts of human behaviour had come to be seen as almost inherently right-wing, with environmental theories cast as left-wing, if anything.[5] The lessons repeatedly drawn from the Holocaust were enough to taint eugenics as the science of genocide. In turn, the disastrous effects of the Lysenko affair in the Soviet Union would put the scientific left in an uncomfortable position, arming their conservative opponents with a cautionary tale of political interference in science.

By the late 1960s, the history of science had two competing moral tropes characterizing the proper relationship of science and politics. It was a situation Hans Eysenck both exploited and attempted to transcend. He attacked 'environmentalism' with those same cautionary tales of Soviet excesses, but was

---

[3] Eysenck, H.J. (1947*b*), The measurement of socially valuable qualities, *Eugenics Review* **39**, 103–7.

[4] Eysenck, H.J. (1948), Some recent studies of intelligence, *Eugenics Review* **40**, 21–2.

[5] See Winston, A.S. *et al.* (2003), Constructing difference: heredity, intelligence, and race in textbooks, 1930–1970, in *Defining difference: race and racism in the history of psychology* (ed. A.S. Winston), pp. 199–229, American Psychological Association, Washington, DC. In their study of textbook representations of the race differences in intelligence, Winston *et al.* point out that 'it was not the horrors of World War II that produced such a change [i.e. a sense of taboo], but perhaps later societal events …' (p. 221).

able to trump the 'Nazi card' by drawing on his own experiences in pre-war Germany. Eysenck would always maintain that science and politics were, or should be, quite distinct realms. Certain that his brand of rational empiricism was the neutral point in a universe of emotive ideology, he constantly pointed to the corrosive effects that politics had on other people and in others places to back up this distinction. Eysenck's hybrid theorizing straddled both nature and nurture. He would thus suffer the slings and arrows aimed at both extremes. However, his willingness to make biology the first and principal cause of individual differences would set him apart—as a doggedly unfashionable realist to his supporters, a hereditarian fundamentalist to his opponents.

Eysenck had always wanted to do for personality what Burt had done for intelligence. The genetics of personality was mooted as a key research project in his research reports at Mill Hill during the war.[6] This was always going to challenge Eysenck both logistically and technically. He would thus draw heavily on the data and expertise of others, cheerfully admitting he was the junior partner in most of these collaborations.[7]

In the lead up to World War II, a small number of researchers in the US had investigated the inheritance of personality via the twin study method. At the Maudsley, Eliot Slater had already begun a programme of research on genetic factors in psychiatric disturbance. However, it wasn't until the late 1940s that Eysenck began to investigate the genetic basis for his dimensions, supported by the Eugenics Society. Teaming up with gregarious American research student Donald Prell, the general goal was to estimate the heritability of neuroticism by combining factor analysis with a twin study model. Without a standardized measure—Eysenck's MPI was still several years away—Prell re-isolated Eysenck's neurotic dimension with a battery of tests.[8] Using 25 identical and fraternal twin pairs located via the London birth registry, they came up with an astonishing heritability estimate (h^2) of 0.81. Given that this figure was higher

---

[6] Hans J. Eysenck, 'Programme of research, psychological department', Aubrey Lewis Papers, Bethlem Hospital Archives, p. 1.

[7] Eysenck, H.J. (1990b), *Rebel with a cause*, W.H. Allen, London, p. 200. Even so, Eysenck would chastise his fellow psychologists for their ignorance in matters biological and genetic.

[8] These included performance measures like the body sway test, self-report inventories, and Rorschach scores. Intra-class correlations for the pairs of twins yielded figures of 0.851 for identical twins and 0.217 for fraternal twins. It was a huge differential, suggesting very high levels of heritability. Eysenck, H.J. and Prell, D.B. (1951), The inheritance of neuroticism: an experimental study, *Journal of Mental Science* **97**, 441–65.

than for any single test, Eysenck and Prell went so far as to suggest this factor constituted a 'biological unit which is inherited as a whole'.[9]

It must be remembered that the scientific meaning of the term heritability is more specific than that of popular understanding. It partitions variance only, providing an index of how much phenotypic (i.e. observed) variation can be attributed to genotypic factors.[10] It is a way of understanding the sources of the differences within a particular population, not how much an individual's attributes are due to their genes or their environment. So what was at issue in the nature–nurture wars was how much the range in the measured intelligence could be attributed to genetics. No sensible social scientist denied that the capacity to think intelligently *per se* was part of the basic biogenetic blueprint for human beings. Moreover, heritability is a population-specific measure greatly affected by environmental homogeneity. The more homogeneous the environment, the more genetic factors would be expected to account for observed variation.

On the face of it, Eysenck and Prell's 1951 paper was stunning. It suggested that over 80% of the variation in neuroticism scores could be attributed to genetic factors, higher than Burt and others had proposed for intelligence differences. Not only did it account for family resemblances in temperament, it suggested that a new understanding of the aetiology of neurotic complaints was just around the corner. If only it could be replicated.

Eysenck and Prell had put forward their results with 'great hesitation', given the almost embarrassingly high estimate obtained.[11] Not surprisingly perhaps, other students working with Eysenck had trouble matching Prell's results. For instance, D.B. Blewett had trouble identifying the same neuroticism factor with the tests Prell had used. The factor Blewett did manage to isolate produced low heritability estimates.[12] Nonetheless, Eysenck was able to salvage a paper from this negative data, illustrating his capacity to spin gold from empirical hay. By re-analysing Blewett's twin data and that of another doctoral student, Eysenck found a *third* factor resembling his introversion–extraversion dimension.

---

[9] Ibid., p. 461.

[10] The usual illustrations of the concept cite simple attributes like the *number* of eyes, toes, or fingers. While everyone understands that these features are inherited in a biogenetic sense, they have a *heritability* close to zero. Variations in the number of eyes, toes, or fingers are typically the result of accidents or gestational deformities—all usually thought of as environmental causes.

[11] Eysenck and Prell (1951), The inheritance of neuroticism: an experimental study, p. 462.

[12] See Blewett, D.B. (1953), An experimental study of the inheritance of neuroticism, Ph.D. thesis, University of London.

The heritability of this factor for 26 pairs of fraternal and 26 pairs of identical twins was duly ascertained as 0.62.[13] However, such a figure did not carry 'very much meaning', Eysenck allowed, for it was based in part on a very awkward result. The intra-class correlation for fraternal twins was calculated at *negative* 0.33—implying that these non-identical siblings were systematically *dissimilar*. It was a surprising finding, one Eysenck could only attribute to 'chance variation'.[14]

Eysenck had several additional reasons at the time to be cautious about specific heritability estimates. This work was based on a simple twin study model—like most heritability research of the era. Such a model assumed additive genetic variance, random mating, no gene–environment correlation or interaction, and equal resemblance of identical and fraternal twin environments. Moreover, Eysenck's limited samples made it impossible to separately analyse twins reared together and twins reared apart. Obtaining better samples that included scarce identical twins reared apart was a coup in itself. They were a lauded (and then much disputed) feature of Burt's studies on the inheritance of intelligence.[15] Specific reservations had also been raised about the assumptions of the Holzinger formula they used to calculate heritabilities, assumptions that Eysenck acknowledged were 'unrealistic'.[16]

While Eysenck's early work on the genetics of personality was applauded by his colleagues in the Eugenics Society, wider critical reaction was more subdued. It wasn't until 1958 that Eysenck and Prell's original article attracted substantial critical notice.[17] Karon and Saunders questioned whether the assumptions of their model held true, and pointed to some odd characteristics

---

[13] Eysenck, H.J. (1956a), The inheritance of extraversion–introversion, *Acta Psychologica* 12, 95–110.

[14] Ibid., p. 109. According to Arthur Jensen, statistical naïvety was the most likely explanation for this 'impossible' finding. Arthur Jensen, personal communication, 23 January 2008.

[15] Eysenck had only limited twin data to work with. Burt was of no help at this time, nor when Eysenck came back to this topic in the early 1970s. See Burt's famously evasive letter to Eysenck, no date *c.* July 1971, Cyril Burt Papers, 191/18/17, Special Collections and Archives, University of Liverpool.

[16] Eysenck (1956a), The inheritance of extraversion–introversion, p. 23.

[17] Karon, B.P., and Saunders, D.R. (1958), Some implication of the Eysenck–Prell study of 'The inheritance of neuroticism': a critique, *Journal of Mental Science* 104, 350–8. See Eysenck's rebuttal, Eysenck, H.J. (1959e), The inheritance of neuroticism: a reply, *Journal of Mental Science* 105, 76–80.

of this study (e.g. the variances were generally larger for identical rather than fraternal twins) that fuelled lingering doubts about this work.[18]

The small samples Eysenck was able to obtain at this time placed great constraints on the generality of his results, given that all such heritability estimates are subject to sampling error.[19] Few would follow Eysenck's lead, especially his idiosyncratic combination of factor analysis with the twin paradigm.[20] The only other similar programme in this period was that of Burt's former student, the equally nativistic Raymond Cattell in the US. Designs and sample sizes varied and heritability estimates ranged from the modestly to highly positive (generally in the 0.3 to 0.6 range). None was as impressive as the Prell data; over the years it came to look like a high-end statistical aberration.

In hindsight, these initial investigations look primitive. Testing complex models would also have to wait. According to Arthur Jensen, Eysenck was still trying to learn quantitative genetics at the time, getting to grips with the subject by reading C.D. (Cyril) Darlington's 1953 book *The facts of life*.[21] But if Eysenck was something of a neophyte, many of the students were, in comparison, babes in the woods. Nevertheless, Eysenck would later cite this research with pride as a pioneering attempt to pin the tail on the donkey and put a number on the heritability of personality.

During this period Eysenck and his team found another entry point into the genetics of personality in the IoP animal laboratory in Bethlem. This time he had the good sense to enlist collaborators well-versed in the developing new field of behavioural genetics. Before moving back to the US, his doctorate incomplete, Prell had introduced Peter Broadhurst to Eysenck. Broadhurst was amongst the last of Burt's undergraduate students at University College. At Eysenck's suggestion, Broadhurst took a scholarship to study animal psychology at Stanford. By the time he got back in 1952, Roger Russell had departed and the lab was run by J.G.L. Williams. After several less than fruitful

---

[18] Prell's stint at the IoP evoked mixed memories and opinions. While some remember him as a very gregarious student researcher, others expressed considerable reservations about his competence as an experimenter. For example, Ann and Alan Clarke did not put much faith in the data he obtained with the body sway test. Alan and Ann Clarke, interview, 25 April 2002.

[19] See Loehlin, J. (1986), H.J. Eysenck and behaviour genetics: a critical view, in *Hans Eysenck: consensus and controversy* (ed. S. Modgil and C. Modgil), pp. 49–57, Falmer Press, Philadelphia.

[20] Later in the 1960s, Eliot Slater's student James Shields used a more sophisticated twin study design that included identical twins reared apart. See Shields, J. (1962), *Monozygotic twins: brought up apart and brought up together*, Oxford University Press, Oxford.

[21] Arthur Jensen, personal communication, 23 January 2008.

years, Broadhurst took over the lab and was able to focus on breeding high and low 'emotionally reactive' strains of rats by crossing inbred strains.[22]

Broadhurst's rats were used to explore the genetic basis for the behavioural differences displayed by the 'neurotic' strains. Daunted by the technicalities of the task, Eysenck and Broadhurst looked for help. They took the data to C.D. Darlington, and Eysenck would strike up a long-time friendship with the eminent Oxford geneticist. Like Eysenck, Darlington was a member of the British Eugenics Society and an arch polemicist, authoring a number of books linking biological science with cultural history. Darlington suggested biometric analysis and put Eysenck and Broadhurst in touch with John L. Jinks at the University of Birmingham. By teasing out complex genetic pathways, the results revealed partial dominance in the polygenetic expression of rodent emotionality traits, with moderate to high heritabilities.[23] Broadhurst's pedigree rats were also used to investigate endocrinal and sex differences, and the effects of psychopharmacology and maternal fostering. After another stint at Stanford in the early 1960s, Broadhurst took up a chair at the University of Birmingham. It cemented a lasting connection with those at Birmingham, leading to collaborations with Lindon Eaves and David Fulker in the 1970s. Fulker eventually joined Eysenck at the IoP reinvigorating the languishing animal lab where Broadhurst had left off.

Discouraged by the resources that research in quantitative genetics demanded, Eysenck turned his back on the area in the early 1960s. But he would return with reinforcements and more sophisticated techniques in the late 1970s. In the meantime, the nature and nurture of intellectual differences had to be put to rights.

Intelligence, race, and social policy

Prior to the late 1960s, Eysenck had said little about intelligence, and had certainly not done any significant research on the subject. Former student Arthur Jensen has suggested that Eysenck's interest in the topic of intelligence arose late in his career partly because his mentor Burt had carved out the area so assiduously. When he did turn to it, Eysenck did it from both the pure and applied ends, readily taking on Burt's mantle of disciplinary spokesman and policy advisor. Within the intelligence field, his role was like that he played in the clinical domain—as much proxy promotion as first-hand research.

---

[22] Peter L. Broadhurst, interviewed by H.B. Gibson, 30 January 1979.

[23] Broadhurst, P.L. (1959), Application of biometric genetics to behaviour in rats, *Nature* **184**, 1517–18.

In contrast to most of his colleagues, Jeff Gray suggested, Eysenck saw himself as representing a distinct approach to psychology.[24] With that went a certain power and responsibility. As a torch bearer for his fusion of experimental individual differences, Eysenck took it upon himself to set the methodological rules for intelligence research, to champion particular approaches, and dictate future directions. Eysenck would also cash in his intellectual chips in high-profile spats over race and IQ, the heritability of intelligence, and educational policy. If these were debates the discipline had to have, then he had to have a say in them.

Just before the war, Eysenck had weighed into the debate on general versus specific mental abilities. His very first research paper argued Thorndike's primary mental abilities could be treated as a mathematical subdivision of Spearman's g.[25] However, it was an awkward experience, and he left the field alone for several decades. In his popular 1953 book *The uses and abuses of psychology*, Eysenck pointed to the influence of non-cognitive factors like persistence, and bemoaned the neglect of the notion of speed of processing as a central component of g. Prior to late 1960s, the only other notable statement Eysenck had otherwise made was to hive off intelligence as a separate but interacting construct from that of personality.[26] However, in 1967 Eysenck published his second and final biological model of personality, along with what Ian Deary called his 'theoretical manifesto' on intelligence.

Eysenck had been invited to contribute to a symposium on new aspects of intelligence assessment at the 1966 British Psychological Society (BPS) conference at Swansea. After summarizing the existing achievements of the field, he set out what turned out to be an influential road map for the future.[27] The psychometric approach of Spearman, Binet, Thurstone, and Guilford had stagnated, he said. Intelligence testing was a technology divorced from theory, and factor analysis alone would not move it forward. The time was ripe for a more experimental approach that incorporated the results of laboratory learning studies. As an example, Eysenck picked up the somewhat overlooked work of his friend and IoP colleague Desmond Furneaux. Furneaux had isolated performance constants related to item difficulty and speed of processing.

---

[24] Jeffrey Gray, interview, 25 July 2003.

[25] Eysenck, H.J. (1939*b*), Critical notice of 'Primary mental abilities' by L.L. Thurstone, *British Journal of Educational Psychology* **9**, 270–5.

[26] See for example, Eysenck, H.J. and White, P.O. (1963), Personality and the measurement of intelligence, *British Journal of Educational Psychology* **33**, 197–202.

[27] Eysenck, H.J. (1967*b*), Intelligence assessment: a theoretical and experimental approach, *British Journal of Educational Psychology* **37**, 81–98.

This was the way to go, Eysenck said. Individual differences in intellectual functioning should be parsed according to highly measurable component processes. Not only that, personality differences had to be taken into account. Integration of Cronbach's two disciplines was again the guiding light—intelligence testing had to come back into the academic fold.

Another one-time IoP colleague had also done just as Eysenck recommended. In 1964, American researcher Arthur Jensen produced results suggesting that extraverts attacked intellectual tasks far more quickly than introverts. Coming from a clinical background, Jensen had read Eysenck's *Uses and abuses of psychology* while at Columbia. He was much impressed, enough to take a National Institute of Mental Health post-doctoral appointment with Eysenck from 1956 to 1958. One of the last pieces of research Jensen did before leaving the US was a study on the authoritarian personality, correlating the F scale and other such measures with psychopathology patterns of the MMPI.[28] Jensen shifted his attention to educational research when he returned to the US.

Jensen's stint with Eysenck appeared to thoroughly ground him in a London school perspective. He remained in close contact with Eysenck for the rest of their joint careers, returning to London in the mid-1960s on a lengthy sabbatical and on many other occasions. Early in 1969, this hitherto obscure psychologist provoked a huge controversy in his homeland with an article in the 1969 winter edition of the *Harvard Educational Review*. Jensen had been invited to address the issue of race differences in intelligence in light of compensatory education programmes. Jensen's 'How much can we boost IQ and scholastic achievement', thrust him into the national media spotlight and re-opened an old debate.[29]

The American scientific community had by and large attempted to bury race as a scientific category, though not without some counterproductive effects.[30] For example, successive UNESCO-backed statements in the 1950s and 1960s

---

[28] Jensen, A.R. (1957), Authoritarian attitudes and personality maladjustment, *Journal of Abnormal and Social Psychology* **54**, 303–11.

[29] Jensen, A.R. (1969), How much can we boost IQ and scholastic achievement? *Harvard Educational Review* **39**, 1–123.

[30] The best known of these was Ashley Montagu's *Man's most dangerous myth: the fallacy of race*, which went through multiple editions. Montagu was a former student of Franz Boas, and had a hand in various UNESCO statements that rejected the notion of race differences in intelligence. Such proclamations helped create the idea of a taboo, however, that some pro-difference researchers pointed to in arguing the issue was not closed. Nevertheless, US psychology textbooks covered the issue of Black and White IQ differences all through the 1950s and 1960s, though often as a cautionary tale. See Winston *et al.*

tended to equivocate between the moral and the scientific, arguing that race research was wrong in principle as well as being (or perhaps because it was) scientifically indefensible. In the US, victories by the civil rights movement likewise conflated empirical data and moral suasion.[31] Race differences research had been relegated to the margins within psychology, although it was still a popular dissertation topic in the American south. Perhaps the most well-known work of the immediate post-war period was Audrey Shuey's lengthy compilation—first published in 1958 and revised and expanded in 1966—which incorporated much of this student thesis material.[32] However, compensatory educational programmes of the 1960s turned US schools into social science laboratories, imperfectly realized attempts to redress historic inequalities wrought by poverty and minority status. They would provide new psychometric grounds to contest old divisions of nature versus nurture.

By the late 1960s, a climate of radical dissent and conservative counterreaction strained the social fabric of American life. Jensen's article became a focal point for these opposing forces, a clarion call to both left and right to man the barricades. Jensen could have been ignored as, for example, the racialist writings of ex-American Psychological Association (APA) President Henry Garrett had generally been. Even Nobel laureate William Shockley had not achieved the necessary critical mass, despite some outrageous publicity-seeking proposals. Jensen, however, was different. His article was hard to miss, appearing in a prestigious, high-profile journal as target piece with seven respondents set to critique it. Moreover, Jensen could not be dismissed as a racist crackpot. His work had the appearance of sober science, a data-driven change of heart from a researcher previously favouring environmental explanations of race differences.[33] Jensen's starting assumption, that compensatory education had already failed, set up the gist of his paper. What followed was his explanation why.

(2003), Constructing difference: heredity, intelligence, and race in textbooks, 1930–1970.

[31] See Tucker, W.H. (1994), *The science and politics of racial research*, University of Illinois Press, Urbana, Chapter 4.

[32] Shuey, A.M. (1966), *The testing of Negro intelligence*, 2nd edn, Social Science Press, New York.

[33] The anger and indignation the article sparked surprised the editors and Jensen himself, with the cycle of attack and defence running on and on. The *Harvard Educational Review* editorial office played a big role in generating this firestorm, sending out pre-publication copies of Jensen's article and critical responses to media outlets. Under pressure, they later tried to restrict the article's circulation and hedged over whether they had actually asked Jensen to address class and race issues in the first place. Jensen also did his part, releasing pre-publication copies of his article and responding with alacrity to almost every

Taking his cues from the outspoken Shockley, Jensen evoked the 'simplest possible hypothesis': low African American IQs could be traced in part to genetic factors.[34] Since environmental explanations of minority underachievement were the rationale for compensatory initiatives of the period, Jensen argued, they had to be assessed in a dispassionate manner. His pessimistic verdict undercut the optimism surrounding the new federal Head Start and Title I programmes, threatening alliances that helped get such programmes up and running. And it appeared just as the more conservative Nixon administration took office. It was a match thrown in a tinder box.

Jensen was taken to task by many social scientists, media commentators, and student groups, as well as some notable geneticists. Nonetheless, he also drew some support—especially from pro-heredity psychometricians.[35] For a time the backlash threatened to get out of hand, with anonymous death threats and a 'fire Jensen' campaign instigated by the Students for a Democratic Society (SDS) at Berkeley. University security had to screen his mail and shadow his movements for years. While the media heat soon tapered off, these skirmishes furnished a much longer technical debate with a treacherous, high-stakes flavour. Moral pleadings intersected with scientific argument in a manner usual for academic discourse.

Meanwhile in Britain

Without a history of slavery or high levels of non-European immigration, race had never been the hot-button issue in Britain that it always was in the US.[36]

---

   critique and challenge. Jensen gave a revealing account of the controversy as he saw it in a lengthy preface to Jensen, A.R. (1972), *Genetics and education*, Methuen, London.

[34] Jensen had come in contact with Shockley during a one year stint at the Centre for the Advanced Study in the Behavioral Sciences at Stanford in 1966–67. Tucker suggested Shockley's influence was crucial to Jensen's change of heart. See Tucker (1994), *The science and politics of racial research*, pp.196–7. When I interviewed Jensen, it was clear he remained fascinated by the brilliant but abrasive Nobel Laureate. Arthur Jensen, interview, 1 July 2002.

[35] Not many of his colleagues were willing to support Jensen publicly, however. Several prominent professional bodies stepped in to condemn Jensen's position, including the Society for the Psychological Study of Social Issues and the American Anthropological Association. While published opinion ran largely counter to Jensen, he did claim to receive a lot of private support from colleagues who were otherwise unwilling to openly endorse his views.

[36] As Graham Richards argued, race and IQ has been very much an 'American thing', especially as a topic for empirical research. Richards, G. (2003), 'It's an American thing': the 'race' and intelligence controversy from a British perspective, in *Defining difference: race*

With a couple of key exceptions, British psychologists had not concerned themselves with questions of racial hierarchy and had led attempts to combat anti-Semitism. However, higher levels of immigration from the West Indies and the subcontinent in the 1960s and 1970s would disrupt this rosy picture. In 1968 Conservative British MP Enoch Powell—a political rebel Eysenck seemed to identify with—had sparked intense debate about immigration levels with his notorious 'rivers of blood' speech. Underground racist elements began to resurface in Britain, given a visible political home by the foundation of the right-wing National Front.

As you would expect, Eysenck followed Jensen's travails closely. Jensen's conclusions had not taken him by surprise, he recalled. After reading the 1966 edition of Shuey's *The testing of Negro intelligence*, Eysenck said he had reached a similar verdict but had chosen to keep it to himself.[37] 'The Negroes, or so it seemed to me, were having enough problems without my adding another one.'[38] But Eysenck changed his mind with the appearance of Jensen's *Harvard Educational Review* monograph. It had released a 'genie' that could not be put 'back in the bottle'.[39] Jensen was a friend and a careful scholar, Eysenck added, implying that he deserved support. He was very much the loyalist.[40] But the issue came to him in other ways, as we shall see. For a start, he could count on being courted by the media to comment on topics such as this, having become one of Britain's most notable and quotable psychologists. Even though the race and IQ debate had an American tone, it evoked parallel social and educational concerns at home that made it a good story.

Quick off the mark, the British journal *New Scientist* set up a discussion in the 1 May 1969 issue that linked Jensen's work with recent pronouncements by political leaders and the local school committee's treatment of immigrant children.[41] With an eye to provocation perhaps, leading hereditarians Burt and Eysenck were invited to comment, along with the more environmentally inclined Sidney Irvine. Both Burt and Eysenck were cautiously supportive of

---

and racism in the history of psychology (ed. A.S. Winston), pp. 137–70, American Psychological Association, Washington, DC.

[37] See Richards' discussion of Shuey's work in Richards (1997), *'Race', racism and psychology: towards a reflexive history*, pp. 245–8.

[38] See p. 80 of Eysenck, H.J. (1972b), The dangers of the new zealots, *Encounter* **39**, 79–91.

[39] Ibid.

[40] Jensen recalled that Eysenck did feel a strong obligation to 'stick up for me', especially since few others would do so publicly. Arthur Jensen, interview, 1 July 2002.

[41] See *New Scientist* (1969), Intelligence and 'race' [editorial], *New Scientist* 1 May 1969, 219.

Jensen's position. Burt, now in his eighties, doubted there were large innate racial differences in intelligence.[42] Eysenck, however, gave far more credence to the possibility and to the educational implications that might follow—such as matching schooling to relative abilities. He concluded with a plea that 'research will precede action'.[43] Some readers faulted Burt and Eysenck for bending over backwards to massage unpleasant facts.[44] Conversely, leading geneticists Charlotte Auerbach and G.H. Beale, both fellows of the Royal Society (FRS), took umbrage at the very notion of biological definitions of race. Moreover, 'a correlation between genes for skin colour and genes for intelligence is unlikely in the extreme.'[45] Other critics warned that old arguments for selective schooling now had a new scientific basis that could be used to justify racial segregation—as Jensen's work apparently was in the American south.[46]

Sweeping educational reforms taking place in Britain at the time provided an additional, overlapping invitation into the race and IQ debate. The Wilson Labour government had set about overhauling the tripartite system of secondary education that had been put in place at the end of World War II, a system in which Burt had played an important legitimating role. The notion of innate and enduring intellectual aptitudes was a cornerstone for selective streaming into grammar (for the academic elite), technical (for the scientifically and mechanically minded), and secondary modern (for the rest, the majority) schools, based on the 11-plus examinations. While the left had initially seen the tripartite system as a means of overcoming class-bound destiny, by the mid-1960s many saw it as producing precisely the opposite, including Education Secretary Anthony Crosland. Reform centred on the introduction of general comprehensive schools with much later subject specialization. Amidst fears this would mean a lowering of standards, a series of 'black papers' (so-called because government policy documents were called 'white papers') was published in *Critical Quarterly*.

For defenders of the *status quo*, the promise of 'grammar schools for all' was a deceptive ruse. Eysenck joined with a number of other psychologists like Burt and Richard Lynn, in the second black paper of October 1969. Eysenck stressed

---

[42] Burt, C. (1969), Intelligence and heredity, *New Scientist* 1 May 1969, 226–8.

[43] See p. 229 of Eysenck, H.J. (1969a), A critique of Jensen, *New Scientist*, 1 May 1969, 228–9.

[44] Meakin, C. (1969), Letter to the editor, *New Scientist*, 22 May, 1969, 429.

[45] See p. 491 of Auerbach, C. and Beale, G.H. (1969), Letter to the editor, *New Scientist*, 29 May 1969, 491–2.

[46] Simon, B. (1969), Intelligence does not depend on race, class or colour, *Morning Star*, 25 August 1969.

the meritocratic value of standardized intelligence testing, especially its capacity to identify talent irrespective of background.[47] To Eysenck, blinkered ideology rather than impartial science was dictating educational practices. Scientists should be allowed to investigate any problem, no matter how sensitive, he argued, for the consequences could not be foreseen.

Already the die was cast. Eysenck was implicated in the race and IQ debate by the involvement of a close colleague. He had already outlined a position and taken to the policy advice trail blazed by his mentor Cyril Burt. With his four children from his second marriage beginning to embark on their secondary schooling, educational practices had become understandably dear to Eysenck's heart. By the end of a turbulent decade, he had firmly established himself as a personal opponent of 'environmentalists'. They represented the 'new dogmatism', he claimed in a *Times Educational Supplement* piece in December 1969, and he condemned the educational practices flowing from it.[48] Compensatory measures—such as more money for facilities and teacher salaries in the poorer regions of the country—would be ineffective or even counterproductive, Eysenck argued. Moreover, constitutional psychological differences were being glossed over. They should be seen to be at the heart of such debates, with those well-versed in their reality best qualified to contribute to the debate. Instead, his London school perspective was being increasingly dismissed as conservative elitism dressed up as science. Too much of what he held dear was at stake to leave race and IQ alone.[49]

---

[47] Eysenck, H.J. (1969d), The rise of the mediocracy, in *The crisis in education: Black paper No. 2*, pp. 34–40, Critical Quarterly Society, London. Burt claimed there was evidence of a marked decline in educational attainment that comprehensives would only increase. However, Lynn explicitly linked the idea of class and talent, claiming poverty was correlated with low innate intelligence and poor family upbringing. For Lynn, the solution was to return to the pre-tripartite era of private fee paying grammar schools with scholarships for the poor. See *Times Educational Supplement* (1969), Black paper two savages the progressives, *The Times Educational Supplement*, 16 October 1969.

[48] Eysenck, H.J. (1969c), Environment—the new dogmatism, *The Times Educational Supplement*, 12 December 1969, 4.

[49] See, for example, Bird, R. (1969), Letter to the editor, *New Scientist* 15 May 1969, 376 and Eysenck's somewhat condescending reply, Eysenck, H.J. (1969b), Letter to the editor, *New Scientist*, 29 May 1969, 490. Already Eysenck was of the opinion that the 'balance of probability was now firmly against' environmental explanations of 'black–white' differences in IQ.

That little book

Through 1970, Eysenck kept relatively quiet on race and IQ.[50] In the interim, Jensen visited again, along with I.I. Gottesman, discussing the topic at one of Eysenck's at-home sessions. Jensen took the trouble to defend his position at a Cambridge meeting arranged by the British Society for Social Responsibility in Science. As well, a conference on 'Human differences and social issues' was organized at the Maudsley by conservative Berkeley political scientist A.J. Gregor in August 1970, featuring Jensen, Cattell, and C.D. Darlington.[51] This was the calm before the storm. All that changed with the appearance of *Race, intelligence and education* in the middle of 1971. Anyone closely following Eysenck's train of thought would not be totally surprised at the position he adopted. What was more unexpected was the form this position statement took. *Race, intelligence and education* was a light and breezy paperback on a subject that was getting more touchy and technical by the day.

Late in 1970, Eysenck had asked Jensen whether he was planning to write up his *Harvard Educational Review* article as a book. Jensen told him he did not, and six months later Eysenck's book appeared.[52] According to Jensen, the idea was to harness Eysenck's capacity to connect with a wide audience. Many of Eysenck's colleagues saw the book as unnecessary, but they were missing the point. It was not an attempt to change their opinions, for Eysenck had no new data to report and few novel insights. The intention was to go for the hearts and minds of the general public. Eysenck felt the issue had been misrepresented, that this 'mainstream science' was not being put sympathetically and accurately. Clearly, Eysenck knew he would be in for a roasting from some of his peers anyway. As he showed his biographer Tony Gibson a pre-publication copy he reportedly remarked: 'Now for the crucifixion'.[53] Like Jensen, Eysenck would adopt what he saw as a rational position on race differences, cannily cautious yet still provocative. If left-wing 'environmentalists' in psychology, sociology, and education did not hate him already, they would now. And they would hate him for reasons *he* thought he could turn to his advantage.

---

[50] He did make one addition to the debate on Shuey's work, however. See Eysenck, H.J. and Castle, M. (1970), More on *Testing negro intelligence*, *The Humanist*, March/April 1970, 34.

[51] Gregor had links with segregationalist right in the US and the journal *Mankind Quarterly*. Like Eysenck, Gregor suggested there was a certain symmetry to the political extremes, notably arguing that fascism was essentially a left-wing philosophy. See Tucker (1994), *The science and politics of racial research*, pp. 171–9.

[52] Arthur Jensen, interview, 1 July 2002.

[53] Gibson, H.B. (Tony) (1981), *Hans Eysenck: the man and his work*, Peter Owen, London, p. 234.

They would hate him, he thought, because he would expose them as pretenders and ideologues. They would hate him because he dared question the consensus enforced by this collusion of non-experts, a consensus at odds with the opinion of the truly expert, a consensus inimical to his version of open scientific inquiry.

Eysenck's book caused considerable ructions even before it saw the light of day. Eysenck had lined up Temple Smith as his British publisher, presumably because Penguin didn't want it. Temple Smith also published the magazine *New Society*. Eysenck's book was one of a series on pressing social issues overseen by *New Society* editor Paul Barker. However, his staff refused to have anything to do with the book and wrote to *The Sunday Times* to dissociate themselves from it.[54] The UK edition carried a similar disclaimer inside the dust jacket. Maurice Temple Smith stood by his man, however, responding with a sympathetic interpretation of Eysenck's conclusions. Temple Smith claimed the book 'neither says nor implies the black races are inferior to the white. It is concerned with a particular subgroup in the United States, descended from the victims of a slave system … If part of the damage done to them is genetic, then the blame lies squarely with the whites.'[55] On firmer ground, Temple Smith added that Eysenck's book specifically denied that the data supported segregated education.

The dissent within the publishing house resulted in an extremely shoddy-looking production for the UK market. Titled *Race, intelligence and education*, it was poorly laid out, unevenly printed, and included a host of uncaptioned and possibly inflammatory illustrations of slave-trading and school classrooms, reinforcing its image as a scarlet document.[56] Given the more innocuous title of *The IQ argument*, the US edition was more professional-looking, with the illustrations of the UK edition removed. To ward off damaging misappropriation, an extra preface was added with an italicized section emphasizing Eysenck's opposition to segregation. Eysenck had been sensitized by the way Jensen—an avowed anti-segregationist—had been portrayed by his critics as the very opposite. The preface to the US edition also conveyed Eysenck's rationale for race research. Like Shockley had done previously, Eysenck suggested he and his fellow scientists were really trying to help the 'black man'. Factual research as much as goodwill would help him 'climb out of his

---

[54] Ball, D. et al. (1971), Protest by *New Society* staff [letter to the editor], *The Sunday Times* 27 June 1971. This letter was also run in ads in the US promoting the American edition. See, for example, *The New York Times*, 8 October 1971, p. 41.

[55] Temple Smith, M. (1971), Letter to the editor, *The Times*, 30 June 1971.

[56] Eysenck, H.J. (1971b), *Race, intelligence and education*, Maurice Temple Smith, London.

disadvantaged status'.[57] Like Jensen, Eysenck was making a direct riposte to affirmative educational measures being trialled in the US and the UK.

While the UK edition went through several printings, the book remained hard to get in the US, with local bookstores unwilling to stock it for fear of reprisals from student groups like the SDS. Outwardly confident that the public would see it his way if they could read only him, Eysenck was particularly irked by this kowtowing to intimidation.[58] Any hopes he had for a bestseller in this huge market were comprehensively dashed.

Eysenck's book cheerfully echoed Jensen's more scholarly monograph, and was barely any longer. Philip Vernon half-jokingly suggested Jensen should sue the publisher for plagiarism, such was Eysenck's willingness to quote large slabs of the American's work.[59] Jensen, though, was grateful for the support. The book followed the style of Eysenck's previous Penguin paperbacks, with little or no attribution and only the briefest of bibliographies. A more scholarly monograph was hardly an option anyway, for he had very little relevant data of his own to contribute and relatively little experience in race differences research.

As with the psychology of politics, Eysenck's intervention in the race and IQ debate betrayed his outsider status. Hardly familiar with the history and social nuances of race politics, Eysenck mixed erudite exposition with wild speculation and inappropriate light-hearted asides. Leading Yale researcher Sandra Scarr-Salapatek described *The IQ argument* as 'maddeningly inconsistent ... with something in this book to insult everyone except WASPs and Jews.'[60] Some of Eysenck's colleagues would claim much of this offence was inadvertent, a product of Eysenck's limited experience in his self-imposed ivory tower on Denmark Hill. Even Jensen thought Eysenck's colorful language a little ill-advised, a trap Eysenck would have avoided if he had a more sophisticated appreciation of US race relations. The school-ground rhymes, the American admitted, could well have been left out.[61]

Eysenck began the book with an anecdote about his boxing career, how he scoffed at a trainer's injunction to 'watch it, these niggers have heads made

---

[57] Eysenck, H.J. (1971c), *The IQ argument*, The Library Press, New York, p. iii.

[58] For a published version of these events, see Eysenck (1972b), The dangers of the new zealots, p. 88.

[59] Arthur Jensen, interview, 1 July 1970.

[60] See p. 1224 of Scarr-Salapatek, S. (1971), Unknowns in the IQ equation, *Science* **174**, 1223–8.

[61] Arthur Jensen, interview, I July 2002. See, for example, Eysenck's use of the 'Miss Starkey' limerick on p. 113.

of iron'.[62] At the time, Eysenck put this down to prejudice, but soon found his trainer was correct. What then is the reader to make of this story? For Eysenck, the moral was that any proposition, no matter how racist it appeared, should be carefully investigated before it could be dismissed. But, in adding to an already overheated debate, Eysenck's anecdote carried a potentially hurtful implication—that the suggestion those of African descent were inherently dull might well be true. And why use the term 'nigger', even in quote marks? Eysenck had a knack for being calculatedly offensive, but this was going too far. Years later he appeared to acknowledge his mistake in his autobiography. Tip-toeing around sociopolitical sensitivities, he rehashed his boxing anecdote but did not use the 'n-word'.[63]

Eysenck's main focus was on explaining racial differences in intelligence, as measured largely by IQ tests. Given the apparently high heritability of IQ differences in Whites, Eysenck reasoned, they are probably highly heritable amongst those of African descent as well. There was scant evidence to support this presumption, however, since little relevant research had been done within Black communities. Ploughing on regardless, Eysenck went on to suggest that the stubborn differential in the distribution of Black and White IQ scores probably reflected genetic differences between these populations. However, he still noted that the evidence was inconclusive. But, as Sandra Scarr commented, a quick reading of the book would tend to suggest otherwise.[64]

Much hinged on both the validity and relevance of within-group measures of the heritability of IQ. For Eysenck, evidence of high heritability amongst Whites came from five sources: classic twin study data; orphanage data, where genetically dissimilar children were raised in similar environments; foster family data comparing the IQ of children with their biological and adoptive parents; the phenomenon of regression to the mean; and, the comparison of selectively bred bright and dull rats. However, critic after critic noted that the heritability of within-group differences *could not and should not* be extrapolated to between-group differences.[65] In fact, there was no logical relationship

---

[62] Eysenck (1971c), *The IQ argument*, p. 1.

[63] See Eysenck's re-telling of this anecdote in Eysenck, H.J. (1997b), *Rebel with a cause* [revised and expanded], Transaction Press, New Brunswick, New Jersey, p. 52.

[64] Scarr-Salapatek, S. (1971), Unknowns in the IQ equation, *Science* **174**, 1223–8.

[65] Perhaps the most notable critic to raise this point was Richard C. Lewontin with his famous corn seed example, originally published as Lewontin, R.C. (1970), Race and intelligence, *Science and Public Affairs, Bulletin of Atomic Scientists*, March 1970, 2–8. This piece was re-published, along with a number of other critiques of Eysenck and Jensen, in Montagu, A. (ed.) (1975), *Race and IQ*, Oxford University Press, New York.

between the two; high within-group heritabilities did not necessarily make a genetic explanation for between-group differences more likely.[66] Leading geneticists such as L.L. Cavalli-Sforza and W.F. Bodmer concluded that the race differences question was essentiality irresolvable. There was no ethical way to design the necessary human experiments. Obtaining good within-group estimates of heritability was hard enough, requiring samples of a controlled range of environments and the assurance of limited gene–environment interactions. Most geneticists saw intelligence as a far too complex, multicausal trait, and had their doubts about self-ascription definitions of race being employed.[67] But like Jensen, Eysenck took within-group heritability as a strong pointer, 'a necessary but not sufficient condition', and attempted to infer the rest from other evidence.[68] He glossed over the problems of confounded interactions and any problems associated with the notion of intelligence as a measurably distinct, acultural attribute.

In a short chapter on race (but a longer treatment than Jensen had attempted), Eysenck borrowed heavily from Gottesman to explain race as a multilevel concept. For Eysenck, the term 'race' implied any recognizable group defined by differing gene pools, despite overlap and mixing. He went on to suggest that 'American Negroes' were still racially distinct, and were much closer genetically to native West Africans than to Whites. This gave him the opportunity to speculate on population movements and selective pressures to moderate the impression he was dismissing all those of African descent as inferior. Eysenck suggested African Americans were a non-random subset of native African

---

> Lewontin invoked the concept of variation in seed plot environment to illustrate the paradoxical relationship between within-group heritability and between-group differences. For example, two plots of corn could be grown in nutrient-rich versus nutrient-poor soil. The differences in growth would be largely environmental, even though heritability within each plot might be very high. Amongst human populations, height can be cited as a similar example. This distinctly heritable attribute displays significant differences between different ethnic groups and/or regions. But are these differences genetic? Maybe, but pronounced increases in height have occurred over the past few generations, more in some groups and regions than others. These increases are probably due to environmental factors like better nutrition and less disease.

[66] As Sussex geneticist James H. Sang commented, this well-known stumbling block had been 'pointed out by Lancelot Hogben in 1933'. Sang, J.H. (1971), Letter to the editor, *New Society*, 1 July 1971, 30. It is a logical fallacy that can be traced back to the previous century according to some writers.

[67] Cavalli-Sforza, L.L. and Bodmer, W.F. (1971), *The genetics of human populations*, Freeman, San Francisco, pp. 753–804.

[68] Eysenck (1971c), *The IQ argument*, pp.112–13.

populations whose forebears were selected for dullness by slave traders. But he left the door open as to whether there were 'genetic differences in ability between Negroes in general and whites'.[69] Likewise, Eysenck attributed other ethnic and national differentials in measured intelligence to similar mechanisms—explaining low Irish IQs, for example, by suggesting their best and brightest were the first to emigrate.[70] This kind of rough and ready Darwinism was both uncharacteristic and unfortunate, winning him few friends.[71] According to *The Sunday Times*, Eysenck would only say that he had read about such possibilities 'somewhere'.[72] One potential source was not hard to trace. A respondent to his 1969 *New Scientist* piece had proposed just this kind of selective effect in the history of slavery, chastising Eysenck for overlooking it.[73]

The rationale and the social consequences

For all the drama it was to cause, *Race, intelligence and education/The IQ argument* had very little new to say. It made many wonder why Eysenck wrote it in the first place. As Michael Eysenck asked, why would his father write a whole book on such a delicate question just to say he didn't have an answer?[74] Ostensibly, it was a bid to convince the public that, stripped of the political posturing, this was a legitimate question, one that could be tackled as an extension of mainstream psychological science. Perhaps too he thought it was one in which the issues were so sensitive that it was necessary to prepare the ground for the practical steps that might flow from some unwelcome facts.

Yet the book came across as more illogical than illiberal, with Eysenck equivocal and inconsistent about social policy implications. The real problem was dullness, Eysenck argued, not race.[75] Positive discrimination based on race was wrong in principle, just as compensatory programmes that flew in the face of

---

[69] Ibid., p. 43.

[70] Naturally this did not go down too well with the Irish. See Akass, J. (1971), Come outside, stupid, and say that again, *The Sun*, 18 June 1971.

[71] See also Sang, J.H. (1971), Letter to the editor.

[72] Page, B. *et al.* (1971), The fallibility of H.J. Eysenck, *The Sunday Times*, 20 June 1971.

[73] See Bowen, R. (1969), Letter to the editor, *New Scientist*, 15 May 1969, 376. One can be sure Eysenck read Bowen's letter, for Eysenck took note of his views in his reply to various correspondents in Eysenck (1969*b*), Letter to the editor.

[74] Michael Eysenck, interview, 28 October 2001.

[75] Eysenck would still tend to merge the two issues with a clunk, however. For example, in responding to Eliot Slater's 1973 suggestion that American Negroes should regard IQ as 'the bagatelle it is', Eysenck wrote: 'If the mean IQ of the inhabitants of this island should sink to that characteristic of American Negroes, then not only would our living standards

scientific research were bound to fail. Since relatively high heritability estimates still left open the possibility of producing significant change through environmental modification, it remained unclear how a demonstration of African genetic inferiority would help them.[76] Even so, Eysenck claimed it should be enough to send educators 'back to the drawing board'.[77] He concluded that, in his personal *opinion*, proof of the genetic hypothesis would mean we should focus on 'countervailing environmental pressures which would as far as possible redress the balance and bring the Negro up to white standards'.[78] But this did not sit comfortably with all that had preceded it. If instead the Black–White IQ differential was shown to be purely environmental in origin, similar recommendations would surely apply. Moreover, Eysenck left these compensatory possibilities—and much of the research into them—relatively unexplored. He tended to dismiss the environmental hypothesis as 'too vague to be properly tested at all'.[79]

Despite the focus of his book, Eysenck stressed that IQ was hardly the only measure of a person's worth. However, he seemed oblivious to the damage that labelling one group as intellectually inferior might do. Eysenck still liked to think of himself an anti-racist. To him, a racist was 'one who views other races with hatred, distrust and dislike; one who wishes to subordinate them and keep them in an inferior position'.[80] Eysenck professed his 'admiration for their many outstanding qualities' of non-White groups and 'a deep sympathy for their suffering'.[81] But that should not blind us to evidence of genetic differences 'favouring one race (or ethnic subgroup) as against another'.[82] To modern ears his tone sounds faintly patronizing, like that of a Victorian era British scientist praising the contribution that various dominion groups could

---

sink dramatically, widespread famine would kill off many millions…' Eysenck, H.J. (1973g), Letter to the editor, *New Statesman*, 27 April 1973, 616.

[76] Even so, Sandra Scarr maintained that ascertaining heritability could still tell something about how to intervene, even if it was not clear *a priori* what this should be. See Scarr-Salapatek, S. (1971), Unknowns in the IQ equation, *Science* **174**, 1223–8 and Richards (1997), *'Race', racism and psychology: towards a reflexive history*, pp. 266–8.

[77] Eysenck (1971c), *The IQ argument*, p. 143.

[78] Ibid., p. 139.

[79] Ibid., p. 114.

[80] Ibid., p. 5.

[81] Ibid.

[82] Ibid.

make to sustain Empire.[83] When the book was launched he said he 'found it difficult to believe' that extremists—diehard segregationists, for example—would gain solace from his words.[84] Their ideologies were based on irrational prejudice rather than 'facts'. Science was of a different ilk entirely, and scientists should be left to perform their research free of political and religious interference. Nazi racial science and Soviet genetics were testament to this, Eysenck reminded his readers.

The reaction

Eysenck's book made by far the biggest impact in Britain. Much to Eysenck's annoyance, the Americans largely ignored him. The controversy over 'Jensenism' had already peaked in the US media and, wary of the repercussions, key media outlets like *The New York Times Book Review* did not see fit to review his book. Nonetheless, several scientific journals did him the honour, with the respected *Science* carrying Sandra Scarr's harsh assessment.[85] While open to the possibility of genetically based racial differences, Scarr found the evidence Eysenck presented fairly unconvincing.

Back home it was a different story as the 'crucifixion' Eysenck expected got underway. *The Guardian* and *The Times Educational Supplement* both carried negative reviews from various experts.[86] More outrageous, from Eysenck's point of view, was a lengthy commentary in *The Sunday Times* written by a trio of journalists.[87] They judged Eysenck's book as both unbalanced and blunderingly insensitive, spicing the piece with critical 'grabs' from other scientists. The cultural neutrality of intelligence tests and the high heritability of the differences they measured were far less settled than Eysenck would have us believe, they cautioned. It was a pseudo-question in any case, they argued, backing this

---

[83] Thankfully Eysenck's writing was free of the overtly derogatory racial descriptions peppering the work of such scientists—including that of his hero Francis Galton. See Fancher (2003), The concept of race in the life and thought of Francis Galton.

[84] Hans J. Eysenck quoted in Page *et al.* (1971), The fallibility of H.J. Eysenck.

[85] Scarr-Salapatek (1971), Unknowns in the IQ equation. See also the review from York College biologist Amedeo D'Adamo. D'Adamo, A. (1971), Review of *The IQ argument* by Hans Eysenck, *Commonwealth*, 8 October 1971.

[86] See Nisbett, J. (1971), Review of *Race, intelligence and education* by Hans Eysenck, *The Times Educational Supplement*, 18 June 1971, and Coard, B. (1971), Review of *Race, intelligence and education* by Hans Eysenck, *The Times Educational Supplement*, 18 June 1971. See also Halsey, A.H. (1971), Discriminations, Review of *Race, intelligence and education* by Hans Eysenck, *The Guardian*, 17 June 1971.

[87] Page *et al.* (1971), The fallibility of H.J. Eysenck.

up with an opinion from psychologist Donald Hebb. Behavioural geneticist Jerry Hirsch was quoted as suggesting the heritability statistic was more useful for 'breeding cattle' than in formulating social policy. A sympathetic review from Eysenck's friend C.D. Darlington subsequently appeared in the same newspaper, while Peter Evans penned a supportive piece in *The Times*.[88] Some readers also applauded Eysenck's stance and the potential light he might shed on this murky topic. Nonetheless, a chorus of negative reviews was double-tracked by readers complaining of the social damage Eysenck's book might do.[89]

In a letter to the editor, Eysenck damned *The Sunday Times* article as 'tittle-tattle' that misrepresented his own writings as well as general expert opinion.[90] Having Hebb's words used against him especially bothered Eysenck, for he had gotten to know Hebb on sabbatical in California in the mid-1950s and counted him a sympathetic colleague. Much of the popular press would follow these 'non-experts', Eysenck complained, with their 'curious outpourings of fancy' allowed to overshadow the judgements of those well-informed in the technical details.[91] Eysenck was being selective: he had in mind the favourable appraisals of Darlington and Cyril Burt rather than the far more critical opinions of, say, Scarr or Burt's old nemesis, Oxford sociologist A.H. Halsey.[92]

According to Eysenck, this was trial by journalism, an additional cross popularizers had to bear. Yet he had written a book expressively for the lay person on a cross-disciplinary issue where fact and value, science and social policy were widely seen to be intimately intertwined. Many of his peers were much less willing than Eysenck to separate these spheres. Lionel Penrose, emeritus professor of eugenics at UCL, said Eysenck had been led astray by the idea that

---

[88] See Darlington, C.D. (1971), Review of *Race, intelligence and education* by Hans Eysenck, *The Sunday Times*, 27 June 1971, 29 and Evans, P. (1971), Review of *Race, intelligence and education* by Hans Eysenck, *The Times*, 16 June 1971.

[89] See for example Mungo, C.J. (1971), Letter to the editor, *The Times*, 28 June 1971. As editor of *Enquiry* (organ of the Trinidad and Tobago Society), Mungo wrote: 'one must conclude these august gentlemen [Eysenck and Jensen], for some undisclosed reason, are doing their best to undermine this new confidence in himself which the black man has begun to acquire…' See also Wober, M. (1971), Race and intelligence, *Transition* **40**, December 1971, 17–26, and Hudson, L. (1971), Science and popularisation, Review of *Race, intelligence and education* by Hans Eysenck, *New Society*, 1 July 1971, 29–30.

[90] Eysenck, H.J. (1971*a*), Letter to the editor, *The Sunday Times*, 27 June 1971.

[91] Eysenck (1972*b*), The dangers of the new zealots, p. 85.

[92] In a publicity blurb for the American edition, Burt was quoted as describing Eysenck's book as 'sound' and 'impartial,' a 'lucid and readable account of the available evidence.' See *The New York Times*, 8 October 1971, p. 41.

intelligence was a 'precisely measurable characteristic like stature'.[93] Instead, Penrose suggested it was an inevitably value-laden concept reflecting the goals of the society in which it arises. Several other critics accused Eysenck of perfidy, of not meeting the standards required for such a delicate subject. 'In matters of race', A.H. Halsey cautioned, 'the scientist must address the layman with the highest possible standards of cautious precision, recognising the task as a political act with responsibilities not only to science but also to society.'[94] Across the Atlantic, Sandra Scarr dubbed Eysenck's 'racial thesis' both careless and dangerously provocative. Whilst recognizing the scientific legitimacy of the questions being explored, she famously likened Eysenck's book to shouting 'FIRE! ... I think' in a crowded theatre.[95]

The critical reception that greeted Eysenck's book illustrated the perils and paradoxes of the scientist as public intellectual. Peer-granted intellectual capital is a prerequisite to play that can be parlayed downstream into a wider social authority. However, such capital is seldom recouped in the process. It is simply spent. The public intellectual can enhance his or her social authority with savvy and sensible presentations, but must maintain credibility within the scientific community or lose risk losing the lot.[96] Moreover, the public intellectual is assumed to be an insider representing the narrow specialist field in which they achieved eminence but, as a commentator on its relevance to wider society, is required to stand outside it. The sense of disappointment and betrayal voiced by Eysenck's peers can be understood in terms of these contradictory demands. Eysenck was *not* being excoriated for 'selling out'—the charge levelled at those deemed too eager (and too good) at pleasing the masses by debasing their science. After all, this was hardly a crowd-pleasing topic. Despite Eysenck's attempts at levity, this little paperback was not likely to make him enviably popular. Rather, critics of Eysenck's book faulted him for being too *narrow*, too much the inside advocate. He represented the issue as a purely scientific one, one that could be adjudicated by the very psychological methods he was expert in. Eysenck was not so much overspending but *misspending* his intellectual booty. He was socially irresponsible—as Scarr-Salpatek, Hudson, and Halsey suggested. Or he was not up to speed in related

---

[93] Penrose, L. (1971), Negro intelligence, *The Friend*, 10 September 1971.

[94] Halsey (1971), Discriminations, Review of *Race, intelligence and education* by Hans Eysenck. Halsey was an implacable opponent of those advocating policies based on hereditarian positions, and was a notable sparring partner for Cyril Burt.

[95] Scarr-Salapatek, S. (1971), Unknowns in the IQ equation, p. 1228.

[96] I am drawing on Collini, S. (2006), *Absent minds: intellectuals in Britain*, Oxford University Press, Oxford, as well as the work of Pierre Bourdieu and other scholars.

fields like anthropology, sociology, genetics, and neurobiology—as Halsey, Penrose, Amedeo D'Adamo, and Mallory Wober suggested. Or he was simply one-sided in his survey of the psychological evidence—as nearly all the critics suggested.

While reactions from Eysenck's peers and the press were immediate, public political protests took a little longer to ferment. At first, Eysenck did not shy away from publicly debating the race and IQ issue. An appearance on the BBC TV's *Controversy* programme in August 1971 no doubt helped identify Eysenck with the issue in the public mind. The three geneticists were included as discussants, including Penrose, and a trio of social scientists (including Halsey, I suspect). According to Eysenck it was the latter group that flunked the test, failing to conform to the rules as he saw them. These social scientists gave his book strong moral readings rather than debating the question within the confines of the empirical evidence.[97] On the whole it was an unproductive experience for Eysenck, with the set-up for the programme conveying the impression of an intellectually isolated scientist flailed by critics on all sides. The taping of the programme was also disrupted by a vitriolic student protest.

If Eysenck felt played by the media, it was partly his own doing. His unashamedly low-brow book added little to the technical discourse and ensured he would get more attention in popular rather than scientific forums. Nonetheless, Eysenck initially struggled to come to terms with the way the game was played there.[98] As the American storm over 'Jensenism' had shown, race and IQ brought the popular media to the scientists like bees to a honey pot. But it also brought the scientists to the media. Whilst co-opted to serve the short-term ends of various electronic and print outlets, Eysenck and his opponents used the popular media to respond directly to each other. Such downstream exchanges feed off and, to a lesser extent, feed back into what is traditionally thought of as 'internal' scientific discourse. However, popular debates obey a looser and more awkward set of rules than scientific ones. Political implications and social consequences are seen to matter as much as the presentation of evidence according to conventions dictated by scientists. The competition is as much for *moral* as intellectual capital, the kind of capital underpinning a scientist's broader social authority.

Scientists squabbling in the popular media can thus be seen to perform a discursive dance with one eye on public appreciations of what a good and

---

[97] *Controversy*, BBC2 TV broadcast, 23 August 1971.

[98] His peers saw Eysenck as the consummate media performer, even as they pilloried him. See, for example, Hudson (1971), Science and popularisation, Review of *Race, intelligence and education* by Hans Eysenck.

responsible scientist should be and do. They are not just vying to be seen as correct in the traditional scientific sense. They are competing for the right to present to the public the correct, if not the orthodox, opinion of the discipline they represent, as well as spell out its social ramifications. Moreover, scientists taking their arguments downstream almost invariably succumb to the expectations and demands of such forums. The adage that the press loves a fight applies to sober and well-mannered intellectuals as much as it does to those in politics, entertainment and sport. With time frames shorter, the arguments become more concise and the language more florid. Name-calling is par for the course—'he called me a WHAT!?'—in the mad scramble for high ground. It all tends to increase intellectual differences, with polite disagreements becoming unbridgeable chasms.[99] So, while Eysenck was famously thick-skinned, he was clearly annoyed to be described as (for example) a 'master of the persuasive art' and a 'naïve controversialist'.[100]

The trip downstream opened the doors to non-specialists and lay commentators. It allowed for multiple inputs and an escalating babble that was hard for any one person to rein in—Hans Eysenck included. As his message echoed through the public domain, it was amplified and altered by the journalists, commentators, and political activists who joined the throng. His position was stripped of the nuances and qualifications he gave it, and loaded with others he did not approve of. Poorly informed scientific expositions mixed with dire political commentary, and the results were often less than enlightening. For example, the popular but erroneous belief that heritable meant unchangeable, or changeable only via eugenic or genetic engineering, wove its way through the debate—variously deployed to up the ante in social policy implications or demonstrate the technical ineptitude of those who subscribed to it.

For many of Eysenck and Jensen's critics, suggesting that low IQs of African Americans were anything but a function of culturally loaded tests and a history of deprivation was a dangerous act of scientific myth-making.[101] Armed with the idea that science was a cultural product inseparable from its social

---

[99] One finds this point amply illustrated in Segerstråle, U. (2001), *Defenders of the truth: the sociobiology debate*, Oxford University Press, Oxford. Many protagonists in the sociobiology debate ceased to be on speaking terms once the fight went downstream.

[100] See Hudson (1971), Science and popularisation, Review of *Race, intelligence and education* by Hans Eysenck, and Clare, A.W. (1973), Eysenck the controversialist, *The Spectator*, 21 April 1973. See also Eysenck's reply to Clare: Eysenck, H.J. (1973*d*), Letter to the editor, *The Spectator*, 5 May 1973.

[101] See for example the various essays collected in Montagu (ed.) (1975), *Race and IQ*, especially Gould, S.J. (1975), Racist arguments and IQ, in *Race and IQ* (ed. A. Montagu), pp. 145–50, Oxford University Press, New York.

function, anti-hereditarian critics saw genetic explanations for African American underperformance as a bid to naturalize racial inequalities. It was enough for leftist political groups to condemn Eysenck as a racist and a fascist. Hereditarian scientists, in contrast, saw the issue as one of intellectual autonomy, and were variously amazed, angered, or aghast over the politicization of their work. In these volatile times they struggled to find a voice in the din. Nevertheless, a fight-back of sorts began in 1972 when a petition supporting Jensen signed by 50 like-minded scientists was circulated in the US.[102] The petition compared those giving credence to genetic influences on intelligence with persecuted scientific heroes like Galileo, Darwin, and Einstein. It was also signed by six British luminaries—Eysenck, C.D. Darlington, Philip Vernon, Eliot Slater, and Nobel prize-winners Francis Crick and John Kendrew. As the protests against Eysenck mounted, he and his supporters would take on the now familiar role of noble victimhood. Like Jensen, Eysenck would be lionized by the political right for calling it as he saw it, a courageous race realist braving repressive leftist orthodoxy.

As 1972 wore on, Eysenck remained embittered by the response his book had generated. He saw the media as set against him, distorting the debate and his place within it, just as it had with Jensen.[103] Eysenck leapt at the chance to shift back upstream where he felt more comfortable—to debate intelligence differences with Steven Rose and Leon Kamin, for example—backed up by more research from his own labs. As a more immediate corrective, Eysenck spelt out the social consequences as he saw them more fully in 1973 in *The inequality of man*.[104] It came hot on the heels of his attempt to give applied behavioural science a human face in his last Pelican paperback, *Psychology is about people*.[105] It was also more technical and slightly narrower in scope, steering clear of race for the most part. The main theme was the limits imposed by psychological and biological differences. Such differences did not translate

---

[102] The 'Resolution on scientific freedom,' signed by 50 eminent scientists, was published in the APA's flagship journal (amongst other places): Page, E.B. (1972), Behavior and heredity, *American Psychologist* **27**, 660–1. See also the response in Proshansky, H.M. (1973), Behavior and heredity: statement by the Society for the Psychological Study of Social Issues, *American Psychologist* **28**, 620–1, and Tucker's commentary in Tucker (1994), *The science and politics of racial research*, pp. 276–7.

[103] Eysenck listed the various underhand tactics he saw the press as using to discredit his book and destroy his reputation in Eysenck (1972b), The dangers of the new zealots—including ignoring the book's central arguments, (mis)quoting authority figures for critical purposes, and rebutting arguments that were never made.

[104] Eysenck, H.J. (1973e), *The inequality of man*, Maurice Temple Smith, London.

[105] Eysenck, H.J. (1972a), *Psychology is about people*, Penguin, London.

directly in action. However, social inequality—if not political inequality—was their inevitable consequence. We must acknowledge that people weren't equal in this sense, Eysenck argued, the better to structure opportunities and resources in education and industry, in mental health and the penal system. A good and efficient society must revise the socialist dictum to read: 'to each according to their ability ...' Everyone would then be happy to assume their proper, scientifically determined station in life—even if it was the bottom rung—because it reflected largely innate capacities and propensities. *The inequality of man* thus updated Galton for the late twentieth century. More than that, it was an attempt to offset the specifically racial connotations of his position on IQ and social policy and put debate back on a more technical footing.

The LSE incident

Meanwhile, a farcical series of events would play into Eysenck's hands. By the spring of 1973 he had already endured political protests from radical left groups when speaking at Birmingham and Leicester universities. Activists had first tried to break up one of his lectures at Birmingham in November 1971, and again in February 1973, on the latter occasion subjecting him to heckling and ink bombs. At Leicester, they resorted to condemning him for 'not having the courage' to address racial issues. While rancorous enough, these protests garnered comparatively little media coverage, partly because Eysenck had still managed to complete his presentations. However, they did alert news agencies to a growing story that would come to a head on 8 May 1973. The headlines would continue for months.

There was clearly trouble brewing at the LSE that day. A string of disruptions and denunciations threatened to cancel Eysenck's talk before it began. When he was finally given the chance to speak, Eysenck mounted the podium, with 'a gentle, perhaps even ironic smile on his lips'.[106] The topic was individual differences in brain processes, but Eysenck barely got a word out before a young woman protester rushed at him and a general melee ensued. Some members of the audience, including LSE graduate student Philippe Rushton, did their best to protect him. Eysenck was quickly escorted from the auditorium and bundled into a taxi. He had not been seriously injured, apart from a cut he sustained when his glasses were broken. Shaken but not stirred, he returned to the Maudsley and immediately got back to work. Despite the

---

[106] The quote comes from J. Philippe Rushton's eye-witness account in Eysenck, H.J. (1991f), Introduction: science and racism, in *Race, intelligence and bias in academe* (ed. R. Pearson), pp. 1–55, Scott-Townsend, Washington, DC, p. 35.

protestors openly bragging of their 'revolutionary action', they would not be held accountable. When Eysenck preferred not to press charges, the police let the matter drop.

The incident severely embarrassed and divided students at the LSE. While no one wanted to defend Eysenck on the charge of racism, many had grave misgivings about the means used to oppose him. The LSE Student Union passed a motion affirming 'free expression' as the best way to deal with such issues, but also reserved the right to muzzle those with 'declared fascist leanings'.[107] The LSE Social Science Society, which had invited Eysenck in the first place, also put out a statement regretting what took place, whilst agreeing to disagree with his views.[108] Eysenck took strong exception to their implication that he was nonetheless a racist, protesting in the *New Statesman* that he had 'admitted a higher proportion of coloured students to my department than any other psychology department in this country (as far as I know)'.[109]

While the LSE was renowned for its student radicalism, the impetus came from elsewhere. The protest was actually organized as a show of strength by members of the Communist Party of England (Marxist–Leninist) (CPE-ML). The CPE-ML was an isolated offshoot of an obscure Canadian Maoist group called the Internationalists, founded by Indian-born Hardial Bains in 1963. The anti-revisionist CPE-ML should not to be confused with the major hard left parties operating at the time—especially the old pro-Soviet Communist Party of Great Britain and the pro-China, trade union-led Communist Party of Britain (Marxist–Leninist). One reason the CPE-ML singled Eysenck out was that a significant proportion of their membership was non-White, elite foreign students from Africa and Asia. Race was just the kind of rallying issue this tiny party needed to demonstrate its power and relevance. According to contemporary investigative reports, the party's key organizers came from outside London. They were members of the Birmingham Student Movement, a 30 strong Maoist group formed as a local affiliate of the CPE-ML.[110] Drawing on the

---

[107] Blair, J. (1973), The assault on Eysenck at the LSE by Jon Blair, an eye-witness, *The Listener*, 17 May 1973.

[108] London School of Economics Social Science Society (1973), Letter to the editor, *New Statesman*, 11 May 1973, 693.

[109] Eysenck, H.J. (1973a), Letter to the editor, *New Statesman* 18 May 1973, 732.

[110] Chaytor, R. (1973), Revolutionaries on the campus, *Birmingham Evening Star*, 16 July 1973. In the murky world of British radical politics, the CPE-ML was variously rumoured to be a CIA front, and/or involved in other violent incidents, and/or linked to the very fascist groups they professed to oppose. In later rejecting Maoism, the CPE-ML was reorganized as the Revolutionary Communist Party of Britain (Marxist–Leninist).

tactics they had used on their home turf, leaflets distributed before Eysenck's LSE talk were headed by the iniquitous, counterproductive slogan: 'Fascist Eysenck has no right to speak.'

In the short term, the CPE-ML did Eysenck an enormous favour: they turned him into a martyr. Thereafter, media coverage swung dramatically his way. Eysenck was transformed as a libertarian icon for freedom of expression. Now he had all the moral capital he needed, the beleaguered scientist standing up for truth. The newspapers were full of sympathy. They bemoaned the 'bully boys on campus' and called for their punishment. Even tabloid 'red tops' that had published hostile pieces on his book—such as the *Daily Mirror*—came down hard on Eysenck's side.[111] Liberal academics felt obliged to condemn the violence, but many could only manage to do so in muted tones.[112]

Mastering the media spin cycle

In the aftermath of the events of 8th May, the media called incessantly for interviews, and several major newspapers did features. While the CPE-ML attempted to use Eysenck as a visibility-raising target, Eysenck would use their actions to turn the issue into one of political interference in the conduct and reporting of science. One of the most effective ways of doing so would be to draw on his past. Eysenck had already introduced this personal element to the debate prior to the appearance of his race and IQ book. Responding to his original 1969 'critique' of Jensen, one *New Scientist* reader had suggested that bad racial science and bad politics (e.g. Nazism, Powellism) tend to go together. Eysenck replied that he had 'left Hitler's Germany in protest and saw his family suffer at his hands ... Such remarks have no place in a scientific controversy.'[113]

It remained very much a fringe party, however. See Bagnall, N. and Taylor, P. (1973), Student rowdyism: is there a link? *Sunday Telegraph*, 24 June 1973.

[111] For tabloid editorial commentary, see *Daily Mirror* (1973), Throw them out [editorial], *Daily Mirror*, 10 May 1973, as well as *Sunday Express* (1973), Bring them to justice [editorial], *Sunday Express*, 13 May 1973. Left-of-centre papers like *The Guardian* also editorialized against the attack; see *Guardian* (1973), People who don't want to know [editorial], *The Guardian*, 10 May 1973. See also Izbicki, J. (1973), Charges urged over Eysenck attack at LSE, *Daily Telegraph*, 10 May 1973.

[112] See for example, Meredith, P. (1973), Eye synch, *The Guardian*, 26 May 1973 and National Union of Students president Digby Jacks' comments to *The Guardian*, 26 May 1973 deploring such tactics because he felt they would 'only achieve greater credibility for Eysenck's views...'

[113] Eysenck (1969b), Letter to the editor.

Separation of the rational and empirical from the personal and political was a rule Eysenck held dear, even if he retained the luxury of defining when it should be observed. Eysenck would exploit his German youth time after time whilst ignoring the complications and contradictions of doing so. As the political demonstrations began to escalate late in 1972, Eysenck warned of 'The dangers of the new zealots.'[114] Eysenck fondly remembered the 'old left' of his youth, contrasting their 'belief in rational argument' with the 'unreason, intolerance, and veneration of force' of those protesting against him.[115] 'The psychology of the fascists under whose rule I grew up in Germany ... has been taken over holus-bolus by the scattered troops of the 'New Left'.'[116]

In the lead up to the LSE incident, Eysenck also penned an opinion piece in the conservative *Daily Mail* that took aim at the National Union of Students (NUS). The NUS had drawn up a blacklist of 'racist' speakers, and had placed him at the top. Pressed to comment, Eysenck penned an article with the strident headline: 'I defied Hitler and now I defy these students'.[117] Was it a calculated provocation? If so, it worked. The CPE-ML activists at Birmingham who had talked of halting Eysenck's future engagements made good their threat. Now Eysenck had the floor that went with widespread public empathy. Two days after the LSE attack, the *Daily Mail* did an interview-based feature that pressed home the point. 'They call ME a racialist', it thundered, 'but these 'students' are the new Hitlers.' Comparing the protesters' tactics with those of the Third Reich, Eysenck concluded 'they appear nothing but left-wing fascists'.[118] Other media outlets would take up Eysenck's free speech cause and follow his cues.[119]

---

[114] Eysenck (1972*b*), The dangers of the new zealots. Much of material in article was republished in Eysenck (1991*f*), Introduction: science and racism, and in his account of this episode in his biography.

[115] Eysenck (1972*b*), The dangers of the new zealots, pp. 79 and 89.

[116] Ibid., p. 89.

[117] Eysenck, H.J. (1973*b*), I defied Hitler and now I defy these students, *Daily Mail*, c. early May 1973.

[118] Mather, I. (1973), They call ME a racialist ... But these 'students' are the new Hitlers, Interview with Hans J. Eysenck, *Daily Mail*, 10 May 1973.

[119] See, for example, Watt, D. (1973), Bad year for the universities, *The Spectator*, 18 August 1973, and Parry, M. (1973), Bully boys on the campus, *Yorkshire Post*, 11 July 1973. In spite and perhaps because of the protests, Eysenck's viewpoint got more airplay than it might have otherwise, a counterproductive result he liked to remind his readers of. See Eysenck (1991*f*), Introduction: science and racism, p. 37.

Eysenck's use of personal material in public image-making discourses gave him valuable moral traction that conflated the validity of his opinion with his right to state it. By drawing on his German upbringing to illustrate the folly of extremism, he turned politicized interpretations of his work to his advantage. 'Many critics of my book have said that in order to judge a person's views you must know his political allegiance ... Like many others', Eysenck wrote, 'I made the journey from Marxism to Socialism, and finally to liberalism.'[120] Eysenck drew on his own political history to testify to the apolitical neutrality of his views, never mind the paradoxes of doing so.

To be fair, Eysenck was drawn into this kind of defence by slippages in popular perceptions. Here was a pillar of the scientific establishment, who still spoke with a pronounced German accent, making much of racial differences. Throwing a certain kind of mud was easy. Unwittingly or not, the anti-racist left played on ethnic stereotypes when they tagged Eysenck a 'Hitlerite fascist'.[121] Such overwrought put-downs called out for a like response. Thus Eysenck wrote of being a 'strong critic of Hitler' who had 'gone into exile to avoid living in a fascist state. To even suggest that I *was* a Fascist is absurd.' [italics added][122] Eysenck's use of the past tense was crucial, converting a political critique of his work into an accusation that depended on his biographical record that almost he alone was privy to. It was a line not easily disputed. To my knowledge, no one ever did so publicly. In popular forums, the credibility of the source officially and overtly matters, as Eysenck had clearly grasped. Upstream it was a different story. None of Eysenck's scientific peers suggested he was any more authoritative on racial differences because he said he had 'defied Hitler'. But in any case, examination of the historical record makes this claim questionable. Urged by his mother and her de-facto partner, Eysenck did a very understandable thing all those years ago. He finished his final year of schooling and he got out—by necessity as much as in protest. 'Sensible self-preservation' would be one way of describing it; 'defiance' was a bit of a stretch.[123] His immodest assertions could be construed as a little insulting to those who stayed on and tried to resist as much as possible.

---

[120] Eysenck (1972*b*), The dangers of the new zealots, p. 79.

[121] This phrase comes from Tony Gibson's vitriolic article defending Eysenck, Gibson, T. (1973*a*), Professor Eysenck's nose, *Anarchist Weekly* **34**, 19 May 1973. Gibson labelled Eysenck's critics in academia and left-wing political circles as 'scum'.

[122] Eysenck (1973*b*), I defied Hitler and now I defy these students.

[123] Ibid. To be fair, Eysenck might not have written that particular headline, even if the article carried his name. It was a reasonable summation of his views, however.

Amidst the accusations and counter-accusations over the LSE incident, Eysenck did not have it all his own way. Many of those who sympathized with him joined the dots to presume Eysenck was a Jewish émigré, lulled by the easy historical comparisons he was encouraging. For example, the 10th May *Mail* article labelled him a Jew.[124] Writing at the time, Eysenck's future biographer Tony Gibson did likewise, describing 'the beating up of Jewish professor' as 'quite like old times again—old times in Munich, Nuremberg, and Berlin where an academic's right to speak was measured by the length of his nose'.[125]

Eysenck himself seemed to avoid the 'Jewish question' when he could, but would deny his mixed heritage when specifically asked. For instance, in an interview appearing in a 1977 edition of the right-wing National Party's journal *Beacon*, Eysenck had simply said: 'no', he was not Jewish, before adding, 'if I wanted to go to university I would have to join the Nazi party, or the storm troopers, and this just wasn't on, so I decided I'd better leave'.[126] In private correspondence with friends he did likewise.[127] He gave John Baker, author of the 1974 book *Race* that created a storm of its own, a similar impression.[128]

Eysenck had his own reasons for sticking by denials he had made when he first arrived in England as a youth. Added to that now was the moral capital he had been able to accrue in promoting himself a principled opponent of Nazi fascism. Admitting to his partial Jewishness would potentially weaken this line in special pleading. It explains why he continued to be so sensitive about it, why he waited so long to reveal his mixed heritage.

---

[124] Mather (1973), They call ME a racialist ... But these 'students' are the new Hitlers, Interview with Hans J. Eysenck.

[125] Gibson (1973a), Professor Eysenck's nose. This article generated considerable debate in anarchist circles. See correspondence from Jerry Westfall, *Anarchist Weekly*, 26 May 1973, and Andrew Williams, *Anarchist Weekly*, 16 June 1973, as well as Gibson's replies in Gibson, T. (1973b), Letter to the editor, *Anarchist Weekly*, 2 June 1973, and Gibson, T. (1973c), Letter to the editor *Anarchist Weekly*, 30 June 1973. Gibson was more circumspect about these issues in his authorized biography a decade later, however.

[126] *Beacon* (1977), Interview with Hans Eysenck, *Beacon*, February 1977, 8.

[127] Hans J. Eysenck to Ann Clarke, 19 November 1979. Copy kindly supplied by Ann and Alan Clarke.

[128] Baker was still not sure about Eysenck's background, however, and bent over backwards to assure Eysenck he was not 'anti-Jewish'. John R. Baker to Hans J. Eysenck, 24 January 1977, Section E Anthropology and Race, CSAC 114.5.86/E.99, Papers and Correspondence of John Randal Baker, Special Collections and Western Manuscripts, Bodleian Library, Oxford University.

Deferential restraint

There is a suggestion that Eysenck displayed a degree of restraint on the race and IQ issue, though it was restraint of a particular variety. In the wake of the criticism of his little book, Eysenck did not counter-attack in his more usual ways. Where was the flurry of rejoinders in the journals, as Phil Rushton pointed out, the courteous but cutting replies that we might normally expect to see from him?[129] Part of the reason for this was that race differences in intelligence was always a side issue for him. He had no research profile on it to speak of and defend, and no great plans to change this.[130] Defending his popular survey of the literature hardly rated as a scientific priority. He almost completely avoided engaging his peers on the technicalities of race and IQ in print. Nor did he show much zest for debating the issue in popular forums after the debacle of the *Controversy* programme. For instance, Eysenck turned down an invitation to participate in a panel discussion on John Baker's 1974 *Race* book, which he had glowingly reviewed.[131] 'I don't feel that public meetings are a good place for discussing scientific findings … [they] are more likely to give rise to a "punch-up" than to enlightenment', he replied.[132]

---

[129] J. Philippe Rushton, interview, 3–4 July 2002.

[130] Eysenck did propose limited investigations of this nature late in his career, plans that were subsequently blocked. However, in the overall context of his career, the question of racial differences in intelligence (or in anything else) hardly appeared a consuming one. While Eysenck was drawn into commenting on the downstream controversy he helped create with 'that little book', he published only a handful of pieces dedicated to addressing the question in upstream scientific terms—and only well after things had simmered down. In fact, his only other lengthy treatment of the issue was Eysenck, H.J. (1984*b*), The effect of race on human abilities and mental test scores, in *Perspectives on bias in mental testing* (ed. C.E. Reynolds and R.T. Brown), pp. 249–91, Plenum Press, New York. In this chapter, Eysenck again echoed Arthur Jensen's work, particularly Jensen, A.R. (1980), *Bias in mental testing*, Free Press, New York. Eysenck's other shorter treatments of the issue were: Eysenck, H.J. (1985*b*), The nature of cognitive differences between Blacks and Whites, *Journal of Behaviour and Brain Science* **8**, 229; Eysenck, H.J. (1988*d*), New light on racial differences in intelligence, *MENSA Research Journal* **24**, 13–14; Eysenck, H.J. (1989*a*), Ground rules for scientific debates: application to the study of racial differences [paper presented at the 10th International Conference in Human Ethology], *Ethology and Sociobiology* **10**, 387; and, Eysenck, H.J. (1991*d*), Race and intelligence—an alternative hypothesis, *Mankind Quarterly* **32**, 123–5.

[131] Eysenck, H.J. (1974*a*), Races apart, Review of *Race* by John Baker, *New Society*, 14 February 1974, 298. See also Eysenck, H.J. (1974*b*), Race and equality, Review of *Race* by John Baker, *Books and Bookmen*, March 1974, 17–20.

[132] Hans J. Eysenck to Robert Moore, 12 March 1974, Section E Anthropology and Race, CSAC 114.5.86/E.98, Papers and Correspondence of John Randal Baker, Special Collections and Western Manuscripts, Bodleian Library, Oxford University, p. 1.

Several other factors also appeared to ensure that Eysenck dialled down on race and IQ. To reprise a point, Eysenck was no ordinary scientist. Most would find his attraction to controversy far too risky and would self-censor to avoid making too much trouble for themselves.[133] However, Eysenck appeared to have an unusually high threshold for opprobrium, a quality Sybil Eysenck described as 'broad-shouldered'.[134] Being hated by those he felt were misguided and wrong—be they his scientific peers or the general public—was only to be expected. While he could not have enjoyed getting his glasses punched off, unpopularity *per se* didn't appear to faze him nearly as much as it might many others.

Nevertheless, the public hounding had flow-on effects that Eysenck didn't quite allow for, taking a toll on his family that must have been hard to shrug off. While Hans and Sybil retained the Eysenck name in professional contexts, they otherwise began to travel under pseudonyms. The harassment and threats finally made life so difficult that Eysenck changed the family name by deed poll.[135] Of Hans and Sybil's children, only Connie changed it back.[136] Moreover, it appeared that the research students tended to stop coming for a time in the early to mid-1970s. Being seen as close to Eysenck apparently carried a degree of stigma that put prospective students off. It can't have made collaborations with some of his colleagues any easier either. Since these personnel resources were one of the keys to his immense and broad-ranging output, his research programme appeared to suffer as a result. This too would have been an unforeseen effect that would have been difficult to discount.

---

[133] Jensen contrasted Eysenck's intellectual courage in this sense with that of his peers. According to Jensen, many colleagues wrote to him supportively but would hastily add a 'don't quote me' caveat. Arthur Jensen, interview, 1 July 2002.

[134] Sybil Eysenck, interview, 21 October 2001; William Yule, interview, 22 July 2003.

[135] Sybil Eysenck, interview, 21 October 2001; Eysenck (1991*f*), Introduction: science and racism, pp. 37–9. Eysenck admitted that what upset him more than anything was that his children were 'made to suffer by their teachers … Whenever my name got into the papers, they would make pointed remarks in class, putting my children in an impossible position.' (p. 38.) To illustrate the 'ignorant tittle-tattle' they had to endure, Eysenck recalled how one of his daughter's lecturers made her the infant victim and her father the 'evil behaviourist' in the infamous baby-in-a-box story—a story Eysenck branded as 'pretty representative of the Left-wing nonsense spouted by all-too-many militant lecturers and professors of what was laughingly called "social science" at the time.' (p. 39.) However, this scurrilous and misleading tale arose from the career of American B.F. Skinner, with its bizarre exaggerations apparently aimed at de-legitimizing radical behaviourism.

[136] Eysenck's first son Michael, from his first marriage to Margaret Davies, never changed his name.

While Eysenck freely admitted to backing off in deference to those closest to him, he never made mention of difficulties it might have created at work. Eysenck took on the mantle of noble victim but refused to admit that this mother of all controversies was in any way professionally damaging. The race and IQ controversy ensured he became the household name he always wanted to be, never mind the costs. Eysenck recalled:

> ...the attempt to silence me was of course quite counter-productive. It made my name, and my theories, much more widely known than they would otherwise have been. Taxi drivers, passport control officers, and even spectators at football matches recognized me, and talked to me; the media came to consider me *the* representative of modern psychology; and in general I became the hero rather than the villain of the piece.[137]

At the time, however, the need to demonstrate unbowed defiance clashed with the demands of noble martyrdom, and that created dilemmas of spin. Far from wanting to give comfort to irrational opponents, Eysenck affected an unflustered pose with the media, just as he did when faced with chanting demonstrators. He would acknowledge the unpleasantness of the whole experience only up to a point. In the *Mail* article immediately following the LSE incident, he admitted that when his race and IQ book was published 'the roof fell in. I entered a nightmare world in which I could never be sure of what would happen next.'[138] Yet he was always careful never to express regret or any sense that he should or even could have done things differently.

Indirect payback

Years after the LSE incident, IoP clinician William Yule asked Eysenck how he had felt when the protesters set upon him. Eysenck responded in characteristically dry fashion: 'Feelings don't come into it.' He was not terribly intimidated, he added, because he could tell they were 'not well organised'.[139] Instead of fear, a devilishly rational thought sprang into his head:

> Here at last is conclusive evidence for the existence of the Left-wing fascists I had predicted in my *Psychology of Politics*, and whose existence had been vociferously denied by the sociologists at this very school of the University.[140]

---

[137] Eysenck (1991*f*), Introduction: science and racism, p. 37.

[138] Mather, I. (1973), They call ME a racialist ... But these 'students' are the new Hitlers, Interview with Hans J. Eysenck.

[139] William Yule, interview, 22 July 2003. Yule admitted he had made a mistake by asking Eysenck how he 'felt', for he was a man who didn't 'do' introspection or feelings.

[140] Eysenck (1991*f*), Introduction: science and racism, p. 37. Eysenck appeared to be conceding he had not demonstrated the 'reality of left-wing fascism' up to that point—in contrast to his stance in the 1950s.

In the subsequent media coverage, Eysenck would make this interpretative leap look easy. Time and again he reiterated how these attempts to silence him were just an unpleasant fulfilment of his model of left–right extremism. To back up his foresight, Eysenck made sure some significant evidence was put before his peers again.

Eysenck had finally chosen to write up the late Thelma Coulter's dissertation. Nearly twenty years after she completed it, he submitted it to *The Journal of Social Psychology* in 1971, just as *Race, intelligence and education* came out. In effect, Eysenck took out insurance on any extreme reactions that might follow. Nevertheless, Eysenck felt obliged to excuse the long delay. 'Dr. Coulter's tragic death' made joint publication of this 'unique piece of research' impossible, and the senior author's 'growing interest in other matters did not leave him time to write up the research for publication on his own.'[141] Given that time constraints had never seemed to hinder Eysenck in the past, it is hard to take this explanation at face value.

The paper itself was a strategic mixture of careful elucidation and sharp argumentation. Eysenck went through the tests used, including his revised R (radical/conservative) and T (tough-/tender-mindedness) scales, and the California F scale, along with the samples of fascists and communists and soldier controls Coulter managed to obtain. Intriguingly, 61 (rather than 43) working class communists were actually tested. However, 11 were omitted because they were actually middle-class, and seven Jews were excluded because one of the tests used, the ethnocentrism scale, had anti-Semitic items that made it 'invalid for Jewish subjects'.[142] The main part of the analysis was given to testing the differences between the groups on the various tests. For each measure, Eysenck presented the results of analysis of variance across all three groups and paired *t*-tests, along with the relevant *p*-values. And that was all. Group means, the kind Christie was forced to laboriously calculate from the scatter plots Eysenck presented in *The psychology of politics*, were still not given. No one would be able to recalculate a thing from *these* data.

The results, baldly presented, did little to disabuse any reader familiar with Christie's interpretation of the data 16 years before. Fascists did indeed score much higher than control subjects on the F scale. They were also more tough-minded on the R scale, with communists in between, suggesting that there

---

[141] See p. 59 of Eysenck, H.J. and Coulter, T. (1972), The personality and attitudes of working-class British communists and fascists, *Journal of Social Psychology* **87**, 59–73.

[142] Ibid., p. 60. Implicitly defending this decision, Eysenck wrote that the same criteria were applied to the fascist sample: they were all active members of the party, working class males, and non-Jewish.

were important differences at play. To rescue his notion of left-wing authoritarianism, Eysenck made several debatable moves. The fact that fascists were more tough-minded than communists was merely:

> an artifact resulting from the presence of many Jews in the Communist group; when the revised T scale scoring was used, omitting the antisemitism items, no significant differences were found to exist between Communists and Fascists. Both were more tough-minded than the controls ...[143]

In the past, Eysenck had linked tough-mindedness with authoritarianism, and authoritarianism with F scale definitions of fascism. But why take anti-Semitism out of this equation? Apparently he now thought this feature of many varieties of fascism had little to do with authoritarianism. Even though Jews were seemingly excluded from the analyses (though it was not entirely clear whether they actually were), Eysenck suggested the presence of Jews within communist ranks made their comrades' tolerance for them less than authentic. No other justification was given for abridging the T scale Coulter had originally used. And just what 'revised T scale scoring' really entailed was anyone's guess.[144]

As for the F scale, Eysenck did allow that fascists scored higher than the other two groups. Nonetheless, he reminded readers that communists also scored higher than his admittedly non-random control group. This, Eysenck suggested, meant that communists 'had a potential for prejudice' too.[145] Eysenck's reliance on significance testing tended to obscure the pattern and magnitude of differences—for example, of how much higher the fascists were on the F scale than both communists and control subjects. The data also indicated some quite interesting complementarities. For instance, fascists showed a penchant for open aggression, communists for what he termed 'indirect aggression.'[146] Eysenck put this down to the greater control he claimed the Communist Party of Great Britain exercised over its member's lives, inhibiting actions that would reflect badly on the party image.

Eysenck still found room to emphasize political attitudes as part and parcel of more global personality structures. But, as an attempt to underline a point made decades ago, it was a decidedly partisan, if not patronizing exercise.

---

[143] Ibid., p. 64.

[144] While it was hard to tell, this may have Eysenck's way of reconciling Coulter's data with the diagrammatic representation published in Eysenck, H.J. (1957a), *Sense and nonsense in psychology*, Penguin Books, London, p. 303.

[145] Eysenck and Coulter (1972), The personality and attitudes of working-class British communists and fascists, p. 64.

[146] Ibid., p. 68.

Left-wing anti-racism should not be taken at face value, it seemed. The tolerance expressed by Coulter's communists was not genuine because it was born of personal engagement; they were, or were close to, the very 'outgroups' to be tolerated. What of Eysenck himself, one wonders: was his *Weltanschauung* similarly compromised by his own somewhat mixed heritage, or the denial thereof?

At one level, Eysenck was shadow boxing with his past. While his write-up of Coulter's work was cited by over 30 other research papers, neither Christie nor Rokeach replied directly.[147] The intellectual debate had begun to move on. Attention shifted to the psychological dimensions of authoritarianism, and the hunt for a specifically left-wing version was beginning to look both fruitless and inconsequential. However, at another level, the Coulter paper worked beautifully, helping to fulfil an empirical prophecy at the same time as pathologizing protests against him. This was Eysenck at his passive-aggressive best. With the trap set, all that was needed was a few chanting, flailing activists to stumble into it. When they did, it wasn't just Eysenck making 'I-told-you-so' noises. Tony Gibson, for one, risked defending Eysenck to his fellow anarchists along these very lines. In a tirade that drew on his experiences at the LSE, Gibson suggested that the left in England had taken considerable umbrage in being called authoritarian. 'But what could they do at that time? If they had resorted to their usual hooligan tactics of breaking up meetings … they would merely have drawn attention to the fact that Eysenck was right.'[148] Eysenck's writings on race were just a 'thin pretext' for revenge, Gibson argued.

However, 20 years is a long time in radical politics, especially when students are involved. The generational turnover in personnel, in the causes they hold dear, and in the vendettas they wage is typically rather rapid. Twenty years was certainly a long time for CPE-ML founder Hardial Bains, who died in 1997. In the mid-1950s, Bains had not yet landed in England and the CPE-ML was but a twinkle in his eye. The leaders of the Birmingham student movement were still in short pants.[149] While many on the left surely hated being called authoritarian, it was far from clear that *any* of the old guard Gibson fingered as

---

[147] For a retrospective viewpoint, see Stone, W.F., et al. (eds.) (1993), *Strength and weakness: the authoritarian personality today*, Springer Verlag, New York.

[148] Gibson (1973a), Professor Eysenck's nose. Gibson also wrote to various newspapers repeating this interpretation. See Gibson, T. (1975), Eysenck is not a racist—or an authoritarian, *Wildcat* 7, 1975.

[149] News reports named Derrick Pounder, a 24-year-old medical graduate, as a central figure in the Birmingham group, along with a number of other young activists. See Chaytor (1973), Revolutionaries on the campus.

itching for retribution were involved at Birmingham, Leicester, or the LSE.[150] It seemed Eysenck was the one with the longest memory. Having baited the more extreme, hot-headed youth of the day, Eysenck could stand back for the most part as sympathizers like Gibson obliged with uncompromising versions of the payback interpretation.

Speaking of settling of scores, Eysenck would soon suggest that the dispositions of his two-factor theory of political attitudes were more born than made. Eysenck had begun collaborating with Lindon Eaves at Birmingham in the early 1970s, picking up on the heritable basis of personality where he left off in the 1950s. Interestingly, Eysenck also included R and T measures in the multi-faceted study design. Eysenck and Eaves were still only able to test relatively simple genetic versus environment models.[151] Even so, the results they were able to place in *Nature* in May 1974 were, Eysenck recalled, 'rather sensational'.[152] The most impressive h^2 estimates were for R and T (at 0.65 and 0.54, respectively), exceeding those for neuroticism (0.49), extraversion (0.48), and psychoticism (0.35). With the circumspect Eaves as co-author, the paper emphasized the sample-dependent nature of these estimates and their relatively low intergenerational predictive power. Nonetheless, over half the variation in political attitudes could be accounted for by genetic factors, they concluded, with environmental influences making a potentially important but unspecified contribution. This was obviously a sweet set of numbers for Eysenck. They suggested that variation in political convictions was genetically determined to a degree that Eysenck's 'environmentalist' critics would find hard to swallow. The tacit point was unmistakable: their objections to his work could be accounted for in the very terms they objected to. And it reinforced his contention that communists and fascists were alike, both 'born spiteful, hateful and full of aggression'.[153] While the method, personnel, and publication venue all suggested this was top-notch science, this was also science as a weapon—a restrained kind of payback. Eysenck could argue that the nature–nurture of social attitudes was an important scientific question, but there was

---

[150] Gibson mentioned Harold Laski and pointed to the influence of the Communist Party of Great Britain at the LSE in the 1950s, but was otherwise vague about their specific connections to attacks on Eysenck in the 1970s. See Gibson (1973*a*), Professor Eysenck's nose.

[151] Eaves, L. and Eysenck, H.J. (1974), Genetics and the development of social attitudes, *Nature* **249**, 288–9.

[152] Eysenck (1997*b*), *Rebel with a cause*, p. 87.

[153] See p. 29 of Deary, I. (1985–6), Eysenck at nightfall, *Bethlem and Maudsley Hospital Gazette* **31** (winter 1985–86), 28–31.

no significant precedent for an investigation of the heritability of political attitudes in the research literature. Eysenck had chosen to blaze this new path on his own initiative.[154]

The paradox here was that Eysenck was beginning to subtly modify his symmetric model of political attitudes in ways less confronting to left-wing sensibilities.[155] In a short series of papers revisiting the topic in the 1970s, Eysenck reaffirmed the central aspects of his two-factor model with better samples but began to acknowledge the complex, context-dependent effect on his 'same but different' characterization of left and right extremes.[156] In his last important statement on political attitudes in 1978, tough-mindedness was positioned as a function of psychoticism, with an admixture of extraversion thrown in.[157] It thus correlated with the dogmatism, aggressiveness, and Machiavellianism other researchers had identified, Eysenck added, finally recognizing Milton Rokeach's contribution to the field. He conceded that authoritarianism may well be correlated with right-wing conservatism, as Rokeach had long argued. Nevertheless, Eysenck still did not rule out a left-wing version being found in Britain in the late 1970s.

History has had the last laugh. Researchers might not have quite closed the book on left-wing authoritarianism but no consistent conception of it emerged.[158] In any case, the ideological map was redrawn after the collapse of

---

[154] Others would follow in his wake, most notably the Pioneer-funded Minnesota twins study led by Thomas Bouchard. See for example, McCourt, K. *et al.* (1999), Authoritarianism revisited: genetic and environmental influences examined in twins reared apart and together, *Personality and Individual Differences* **27**, 985–1014.

[155] Nevertheless, see the critical appreciations of Eysenck model in Stone, W.F. (1974), *The psychology of politics*, Free House, London, and Altemeyer, B. (1981), *Right-wing authoritarianism*, University of Manitoba Press, Manitoba.

[156] Eysenck, H.J. (1971*d*), Social attitudes and social class, *British Journal of Social and Clinical Psychology* **10**, 201–12; Eysenck, H.J. (1975*b*), The structure of social attitudes, *British Journal of Social and Clinical Psychology* **14**, 323–31; Hewitt, J.K., Eysenck, H.J., and Eaves, L. (1977), The structure of social attitudes after 26 years: a replication, *Psychological Reports* **40**, 138–88.

[157] Eysenck, H.J. and Wilson, G. (1978*b*), Conclusion: ideology and the study of social attitudes, in *The psychological basis of ideology* (ed. H.J. Eysenck and G. Wilson), pp. 303–12, University Park Press, Baltimore.

[158] Recent work has suggested that those classed as left-wing tend to score lower on authoritarianism than those classed as right-wing, even on measures specifically designed to assess it. See Roiser, M. and Willig, C. (2002), The strange death of the authoritarian personality: 50 years of psychological and political debate, *History of the Human Sciences* **15**, 71–96, and Wolfe, A. (2005), 'The authoritarian personality' revisited, *The Chronicle Review* **52**, B12.

Soviet communism. New research indicated Russian authoritarians to be similar to those in the West, except that they pined for the order of the old communist regime. Eysenck claimed this backed up his original thesis, but it also illustrated the problematic character of political measurement that made definitions of left and right so context-dependent.[159] Political attitudes research occupies a relatively peripheral place in contemporary social psychology. However, ongoing research into the potential for fanaticism in both individuals and governments has been made all the more apposite by some of the more diabolical events of this new century.

The Burt affair

Eysenck left the topic of intellectual differences alone for a spell in the mid-1970s. While Eysenck had set out a road map for future research in 1967, he was yet to take to it personally. What helped get him back to the nature and nurture of intelligence, at least in public forums, was a growing storm that again piqued his sense of disciplinary responsibility.

Concerted attacks on the reputation of Sir Cyril Burt had begun soon after his death in October 1971. Leon Kamin, a researcher in animal conditioning and learning at Princeton University, had first raised concerns about curious aspects of Burt's data in 1972. Two years later, Kamin published an influential book that damned Burt as well as genetic explanations of IQ differences.[160] That year, Alan and Ann Clarke also went public with their criticisms of their former teacher's work.[161] Other colleagues pointed to Burt's missing research assistants and his shifty ways.[162] Attempting to control the damage, Jensen quickly admitted there were problems with Burt's numbers.[163] These concerns were gathered together to form the basis for a fraud charge launched in October 1976 by Oliver Gillie, medical correspondent for Eysenck's *bête noire* broadsheet *The Sunday Times*.[164] What many insiders had suspected or half-known for years now became a full-blown public scandal.

---

[159] See Winter, D.G. (2001), A tough look at tough-mindedness, *Contemporary Psychology* **46**, 486–9.

[160] Kamin, L.J. (1974), *The science and politics of IQ*, Erlbaum, Potomac, Maryland.

[161] Clarke, A. and Clarke, A. (1974), *Mental deficiency*, Methuen, London, Chapter 7.

[162] See Tizard, J. (1977), The Burt affair, *University of London Bulletin*, May 1977, 4–7.

[163] Jensen, A.R. (1974), Kinship correlations reported by Sir Cyril Burt, *Behavior Genetics* **4**, 24–5. See Jerry Hirsh's commentary on Jensen's rush to print in Hirsh, J. (1981), To 'unfrock the charlatans', *Sage Race Relations Abstracts* **6**, 20.

[164] Gillie, O. (1976), Pioneer of IQ faked his research findings, *The Sunday Times*, 24 October 1976.

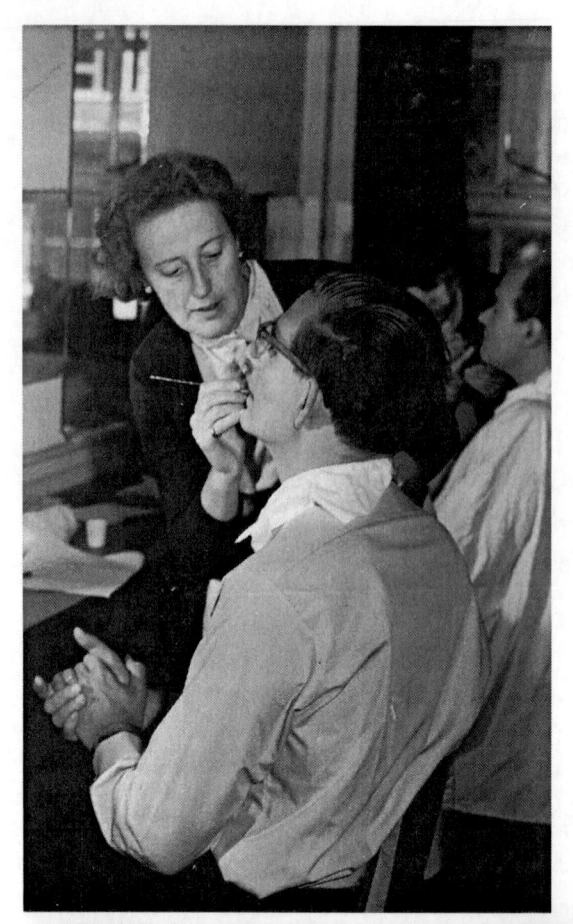

Fig. 8.1 An early UK TV appearance.
Source: Centre for the History of Psychology, Staffordshire University.

Fig. 8.2 Cover of the UK edition of 'that little book'.

Fig. 8.3 Under attack at the LSE – the long-haired figure to the right struggling to keep the protesters away from Eysenck is J. Philippe Rushton, then a student at the LSE.
Source: Express Newspaper Group.

Fig. 8.4 On tour in Australia with Arthur Jensen in 1977.
Source: Centre for the History of Psychology, Staffordshire University.

Fig. 8.5 On tour in Australia – Eysenck appears on *Monday Conference*, ABC TV, broadcast 26 September 1977. Host and moderator Robert Moore is on the far left. The handpicked Eysenck antagonists are John Gibson (left), Professorial Fellow in population biology at the ANU, and Robin Winkler (right), clinical psychologist at the University of Western Australia.
Source: Centre for the History of Psychology, Staffordshire University.

Eysenck had at least two reasons for taking these attacks personally. First, Burt was his doctoral supervisor and an intellectual role model. Although they had not been particularly close since Eysenck ascended to head the IoP department, Eysenck's work retained the imprint of his former mentor. Second, Burt's data were a central element in the case for the high heritability of IQ differences. They had a number of significant features, as unusual as they were valuable. They included the largest collection of twins reared apart of any comparable study, with the correlation between the socio-economic status of the adoptive parents of these twins being close to zero.[165] Eysenck and Jensen had cited these data reverently when assessing the arguments for the genetic basis for racial differences.

Both Eysenck and Jensen got straight on to the front foot. In a letter to *The Sunday Times* on 7 November 1976, Eysenck rejected the fraud charge as 'sensationalized'. He wrote again to *The Times* a few days later to add that these 'absurd accusations ... hardly need rebuttal.'[166] Jensen followed suit, arguing that Burt could only really be faulted for carelessness.[167] Eysenck also wrote to Marion Burt, Cyril's sister, reassuring her that the 'whole affair is just a determined effort on the part of some very Left Wing environmentalists determined to play a political game with scientific facts.'[168] In a lengthy article in *Encounter* early the next year, Eysenck conceded that Burt treated his data cavalierly, but drew the line against outright fabrication.[169] Omitting Burt's data made little difference anyway, Eysenck said; he could point to many other similar findings offering ample support for the 'genetic determination of IQ'. What seemed to rile him most was the unwelcome involvement of outsiders, those he saw as unsympathetic to hereditarian psychometrics. Omitting the whistle-blowing role of Kamin completely, Eysenck implied that intelligence researchers could keep their own house in order thank you very much.[170] Gillie's 'unspeakably

---

[165] See Tucker, W.H. (2007), Burt's separated twins: the larger picture, *Journal of the History of the Behavioral Sciences* **43**, 81–6 for a discussion of these features of Burt's data and their 'improbability'.

[166] Eysenck, H.J. (1976*d*), Letter to the editor, *The Sunday Times*, 7 November 1976.

[167] Jensen, A.R. (1976), Letter to the editor, *The Times*, 9 December 1976.

[168] Hans J. Eysenck to Marion Burt, 16 November 1976, cited in Fletcher, R. (1991), *Science, ideology and the media: the Cyril Burt scandal*, Transaction, New Brunswick, New Jersey, p. 177.

[169] See p. 24 of Eysenck, H.J. (1977*g*), The case of Sir Cyril Burt, *Encounter* **48**, 19–24.

[170] Continuing the damage control campaign, Eysenck also refused to give Kamin any credit in Eysenck, H.J. (1977*a*), Letter to the editor, *American Psychologist* **32**, 674–5, and Eysenck, H.J. (1977*f*), Letter to the editor, *Bulletin of the British Psychological Society* **30**, 22.

mean' attack was inappropriate and unjustified, akin to a McCarthyist witch hunt.[171]

By poking his head above the parapet, Eysenck revived the controversy over race and IQ and his part in it. Public protests flared again when Eysenck visited Leeds early in 1977 and followed Eysenck and Jensen through a joint speaking tour of Australia in September that year. Eysenck spoke on clinical and personality topics when he was down under, but he did take part in a high-profile ABC TV programme that touched on the race issue.[172] The tour had to be organized well in advance, so student political groups had plenty of time to mobilize. They were certainly better prepared than the academics hosting the pair. Scenes reminiscent of the LSE were repeated with even more vigour in Sydney and Melbourne. At Sydney University, Eysenck was prevented from speaking by chanting demonstrators throwing water bombs, fireworks, and fists.[173] At Melbourne University, riot police had to cordon off Wilson Hall to ensure each of their talks went ahead. Eysenck was greeted by Aboriginal and student activists giving Nazi salutes and shouting *Sieg Heil*. Choosing to take it literally, Eysenck concluded this was the best demonstration of left-wing fascism he had ever witnessed.[174] Jensen said the Melbourne University disturbances were the worst he had experienced outside the US, and equal to anything within.[175]

Eysenck argued that Jensen had already picked up the problems in Burt's data and had dealt with them in a dignified manner.

[171] Eysenck (1977g), The case of Sir Cyril Burt, p. 19. See also Oliver Gillie's reply in Gillie, O. (1982), The Burt scandal [letter to the editor], *The Listener*, 6 May 1982 and Gillie, O. (1980), Burt: the scandal and the cover-up, in 'A balance sheet on Burt' (special supplement), *Bulletin of the British Psychological Society* 33, 9–16, in which Gillie accused Eysenck *et al.* of a deliberate cover-up.

[172] The topic of discussion was: What makes us what we are?' and questions addressed the nature versus nurture of sex, and race differences in intelligence. *Monday Conference*, ABC TV, broadcast 26 September 1977.

[173] University of Sydney (1978), Statement by committee of enquiry, The disruption of meetings with Professor Eysenck, *The University of Sydney News*, 6 March 1978, 224–77. I have to thank the late Alison Turtle for making her personal archival material and newspaper clippings available, which helped bring this episode back to life.

[174] Eysenck, H.J. (1977c), Storm troopers of the left: the Australian case, *Quadrant*, November 1977.

[175] Pearce, M. (1977), Professors, protestors and police, *Farrago*, 16 September 1977. For more of the flavour of the press coverage of Eysenck and Jensen's Australian tour, see Monks, J. (1977), Campus drowns out free speech, *The Australian*, 16 September 1977;

Just before Eysenck left for Australia in September 1977, he elaborated on the disastrous effects politics was having on his kind of science in his news vehicle of choice, the *Daily Mail*.[176] Political pressures had found their way into the heart of the scientific enterprise, Eysenck claimed, frustrating researchers interested in heredity with an insidious form of censorship. Books he and Jensen authored were given to reviewers known to be hostile to their views, or not reviewed at all. Papers supporting genetic determinism were given an unreasonably critical ride. In a world where the left 'had science by the throat', pro-heredity researchers were being made to feel like pariahs. From Eysenck's point of view, his science was not the expression of a repressive brand of politics; rather, the negative reaction of his peers was.

The Burt scandal took on a life of its own. While Eysenck, Jensen, and company strove to divorce it from the substantive nature–nurture issues, their critics did the very opposite. In the meantime, Leslie Hearnshaw had begun an authorized biography of Burt. In contact with Hearnshaw late in 1977 and forewarned of his negative judgement, Eysenck hastily began to shift ground. But the experience confirmed for Eysenck the kind of ideologically driven hostility hereditarian psychometrics now attracted, and provided a new set of adversaries to battle with.

Fronting the Front?

It would become harder and harder for Eysenck to steer a seemingly neutral course between the extremes he had engaged. Eysenck's ideological journey from a Marxist youth had apparently left him with only one good eye. His warning about the ideological distortion of science reflected his take on the broader political scene, and came across as just as myopic as the 'Studies in prejudice' research he took exception to in the 1950s. He saw the threat of extremism as almost solely emanating from the left. 'The militant right',

---

Sydney Morning Herald (1977), Shameful, dangerous [editorial], *The Sydney Morning Herald*, 17 September 1977; and, Knopfelmacher, F. (1977), The Eysenck–Jensen scandal, *Nation Review*, 22–28 September 1977. Knopfelmacher had much in common with Eysenck. A wartime Jewish émigré, Knopfelmacher was a researcher in Eysenck's department for a short time in the early 1950s before joining the psychology department at Melbourne University. He went on to become a noted conservative commentator on the Australian political scene.

[176] Eysenck, H.J. (1977*b*), How the left has science by the throat, *Daily Mail*, 2 September 1977.

Eysenck and Wilson wrote in 1978, 'has almost disappeared' from national affairs.[177]

In mid-1970s Britain, however, the National Front was a force to be reckoned with. While not particularly successful in gaining parliamentary representation, the party had a noisy street presence in the industrial heartlands of London and the West Midlands. Clashes with hard left groups like the Socialist Workers Party were commonplace. However, the campaigning efforts of media commentators and more mainstream anti-fascist groups—notably those centred on the magazine *Searchlight*—helped destabilize the National Front by exposing its neo-Nazi links. A populist, Strasserite faction of the Front's membership led by John Read was forced out, forming the short-lived National Party in 1975.[178] By the end of the 1970s, the National Front was a party in decline, wracked by internal squabbles and eclipsed by the hard-line conservatism of Maggie Thatcher's reinvigorated Tories. Even so, it seemed a little inaccurate to describe the Front and other right-wing groups as invisible, and this put Eysenck in a tricky spot.

In 1978, social psychologist Michael Billig published a study of British fascism that would help focus attention on Eysenck's alleged links with right-wing groups.[179] Billig was not alone in warning that the National Front's preoccupation with race and immigration masked a genuinely fascist ideology rooted in the archaic paranoia of Zionist conspiracy. More to the point, Billig helped make it widely known that Eysenck's work (e.g. *Race, intelligence and education, The inequality of man*) was on the National Front's list of recommended reading. In addition, interviews with Eysenck had been published in the National Party's *Beacon*, and were subsequently republished in other forums like the US-based neo-fascist journal *Steppingstones*.[180]

---

[177] Eysenck and Wilson (1978*b*), Conclusion: ideology and the study of social attitudes, p. 307.

[178] The National Party of the mid-1970s should not be confused with the more successful British National Party currently still operating. The lineage of today's British National Party can be traced directly back to the old National Front.

[179] Billig, M. (1978), *Fascists: a social psychological view of the National Front*, Harcourt Brace, Jovanovich, New York.

[180] *Beacon* (1977), Interview with Hans Eysenck. A similar interview with Eysenck had been published the year before: *Neue Anthropologie* (1976), Interview mit Hans-Jürgen Eysenck, *Neue Anthropologie*, January/March, 1976, 16–17. *Neue Anthropologie* was a kind of sister publication to *Mankind Quarterly*, having similar contributors and sometimes sharing the same articles.

Readers were asked to join the political dots. Those inclined to believe the worst about Eysenck readily did so, with one reviewer of Billig's book dismissing Eysenck with the comment: 'so much for scientific detachment'.[181] Nonetheless, being approvingly cited by the National Front demonstrated very little about Eysenck's political motives. Moreover, Billig had not noticed that the *Beacon* interview had actually taken place sometime earlier, judging by the context. By my estimate, it dated back to late in 1973 or early 1974, well before the National Party and its journal were established.[182] Thus when neurobiologist Steven Rose wrote an open letter to *Nature* in August 1978 that mentioned the interview Eysenck had allegedly granted to *Beacon*, Eysenck said this was untrue, 'although I am sure he [Rose] made the allegation in good faith'.[183] Given this chronology, Eysenck's oddly worded denial made some sense. But just whom he had given an interview to and how it had ended up in hands of the *Beacon* editors would remain a mystery. Later Eysenck sounded an even more defiant note: 'What is wrong in being interviewed by, and hopefully spreading some light among, even advocates of darkness?' he asked, pointing out that it should not be assumed that interviewers and interviewees share political convictions.[184] In turn, Rose suggested Eysenck should complain to the Press Council if the interview had taken place under false pretences.[185] Eysenck never took this any further. Nor did he take any action against Billig for that matter. Instead, Eysenck defended the value of free speech and again insisted that there was no necessary connection between science and politics. His vagueness about the circumstances in which this interview took place may

---

[181] See Marsh, P. (1979), Fascist mythology, Review of *Fascists: a social psychological view of the National Front* by Michael Billig, *New Society*, 18 January 1979.

[182] Tucker has already, independently, noted this point. According to Tucker, the interviewer was probably someone associated with *Mankind Quarterly*. See Tucker (1994), *The science and politics of racial research*, p. 342 (footnote 330).

[183] See Rose, S. (1978), Letter to the editor, *Nature* **274**, 738, and Eysenck's reply, Eysenck, H.J. (1978a), Letter to the editor, *Nature* **274**, 738.

[184] Eysenck, H.J. (1979b), Letter to the editor, *New Society*, 15 February, 1979. This letter was a response to Peter Marsh's aside 'so much for scientific detachment' in his review of Billig's book. See Marsh (1979), Fascist mythology, Review of *Fascists: a social psychological view of the National Front* by Michael Billig. True to his word, Eysenck spoke to a great range of groups and published in a wide variety of outlets. Plenty of low- and middlebrow publications like *Reader's Digest*, *New Horizons*, *The Listener*, and *Family Doctor* litter his bibliography. Eysenck even published several articles in *Penthouse* and allowed himself to be interviewed by *Mayfair*—without seeing himself as endorsing what these magazines represented.

[185] Rose, S. (1979), Letter to the editor, *New Society*, 15 March 1979.

have had something to do with protecting the identity of the interviewer. However, the charge that he knowingly gave an interview (and tacit endorsement) to a far right group seemed unfair.[186]

Before going to print, Billig had made sure Eysenck knew the National Front thought his work harmonized with their political viewpoint, sending him a 1976 copy of the Front's publication *Spearhead*. Aware of his vulnerability to accusations of guilt-by-association, Eysenck wrote to *The Times* in March 1978 to deplore such appropriations.[187] While some thought the tone of his *Times* letter uncharacteristically circumspect, Eysenck was still forthright and unambiguous.[188] Speaking on behalf of Jensen and himself, Eysenck maintained they were both 'strongly opposed to any form of racialism including that advocated by the National Front.' Eysenck said he hoped they would 'cease to quote Jensen and myself as giving support to their views.'[189] Emphasizing the overlap in the IQ distributions of various racial groups, Eysenck argued their work could not be used to justify differential treatment on the basis of race. Despite their preoccupation with explaining group differences, Eysenck argued that the individual was still paramount when it came to deciding matters of social policy. Their findings did not support racial doctrines, he said, and such doctrines should be dismissed on moral grounds in any case.

Billig continued his campaign to highlight the intrusion of reactionary politics in the academic world, with more bad press for Eysenck. His 1979 *Searchlight* pamphlet also called into question Eysenck and Jensen's association with a number of right-wing periodicals with a history of publishing racialist science.[190] Eysenck had joined the advisory board of one such journal, *Mankind Quarterly*, in 1974 on the invitation of Richard Lynn. Eysenck had used the journal for a critical review of the Ashley Montagu edited volume, *Race and IQ*. Perhaps he was exploiting the readiness of the editor to publish such a piece, for Eysenck had already reviewed the same book in at least two

---

[186] See various materials published online by the Institute for the Study of Academic Racism http://www.ferris.edu/ISAR/ some of which takes its cues from Billig's work and is quite critical of Eysenck.

[187] Eysenck, H.J. (1978*b*), Letter to the editor, *The Times*, 16 March 1978.

[188] See Ian Low's commentary on Eysenck's *Times* letter, Low, I. (1978), Pursuing truth, *New Scientist*, 6 April 1978.

[189] Eysenck (1978*b*), Letter to the editor.

[190] Billig, M. (1979), *Psychology, racism and fascism*, Searchlight, London. Having limited circulation, this pamphlet has been republished online at http://www.ferris.edu/ISAR/archives/billig/homepage.htm.

other periodicals.[191] *Mankind Quarterly* had been launched in 1960 in Edinburgh, with the eccentric Robert Gayre at the helm. Its mission was to reunite biology and anthropology, with a special focus on race and inheritance. Reviled by sections of the scientific community for supporting racial science and social policies, those associated with *Mankind Quarterly* attempted to reinvent it as a more mainstream academic vehicle in the late 1970s.[192] However, the appointment of Roger Pearson as the new editor struck trouble when a number of newspapers highlighted Pearson's involvement with far right groups such as the Northern League and World Anti-Communist League.[193] When questioned by reporters in 1979 about his association with the journal, Eysenck said he would have to reconsider his position. His name was promptly removed from the masthead, as was Roger Pearson's. Pearson still remained heavily involved behind the scenes. The same could not be said of Eysenck, however, who appeared to keep his distance.[194] Eysenck had other fish to fry, and would soon start his own journals and associations that more directly reflected his intellectual agenda in these areas.

---

[191] Eysenck, H.J. (1976a), Review of *Race and IQ*, ed. Ashley Montagu, *Mankind Quarterly*, **17**, 149–50. Not surprisingly, Eysenck's review was quite critical. Focusing on Montagu's introduction almost exclusively at the expense of the 14 other pieces, Eysenck suggested the book was a one-sided, politically motivated misrepresentation of the issues Jensen had raised. Eysenck had already published two similar reviews of *Race and IQ* in *Books and Bookmen* and the *New Humanist*. Montagu complained about this doubling-up in print. See Montagu, A. (1978), The ethics of book reviewing, *Current Anthropology* **19**, 385.

[192] *Mankind Quarterly* had an especially poor reputation amongst anthropologists. See various contributions in *Current Anthropology* (1962) under the heading '*Mankind Quarterly* under heavy criticism', *Current Anthropology* **3** (1962): 154–8, especially Ehrenfels, U.R. (1962), Critical paragraphs deleted, *Current Anthropology* **3**, 154–5. Robert Gayre, of Gayre and Nigg, was a physical anthropologist who described himself as Knight of St. Lazarus and an honorary Colonel in the militias of Alabama and Georgia.

[193] Beresford, D. (1979), Right-wing editor for journal, *The Guardian*, 9 April 1979. See also Tucker (1994), *The science and politics of racial research*, pp. 255–60.

[194] Eysenck published only three more short commentary pieces in this journal, and only after his retirement. These were Eysenck, H.J. (1987c), Thomson's 'bonds' or Spearman's 'energy': sixty years on, *Mankind Quarterly* **27**, 257–74; Eysenck, H.J. (1990c), Freud, science and Professor Joseph S. Jacob: a reply, *Mankind Quarterly* **31**, 183–4; and, Eysenck (1991d), Race and intelligence—an alternative hypothesis. The latter piece, a review of Richard Lynn's work on racial differences in intelligence, was his last on race and IQ. Eysenck's name had been added to the *Mankind Quarterly* masthead in January 1974, but it had been removed by the end of 1979.

Science and/or politics

While Billig's efforts may have contributed to the demise of the National Front, the effects on Eysenck and his public image were more ambiguous. For those looking to thoroughly demonize Eysenck, his links with far right groups revealed his true political sympathies. Not only were Eysenck's writings *overtly* racist, they were being promulgated by a scientist wilfully misrepresenting a dark political agenda.[195] Yet the tip-of-the-iceberg metaphor implicit in this accusation appears to be seriously misleading in Eysenck's case. More than most, what you saw was what you got. He spread himself too thin to be harbouring much beneath the surface. For example, at the height of the race and IQ controversy in the mid-1970s, Eysenck was also selling behaviour therapy, consolidating his bio-behavioural model, investigating crime, gender, and personality, and beginning to dabble in parapsychology and astrology, smoking and personality, and more—all while directing a large graduate department programme. Moreover, in juggling these multiple research strands, he was never particularly close to anyone, save a few key collaborators. The game he played was very much an overt one, spread high and low across topics, sites, and forums. There appeared to be no hidden agenda to Hans Eysenck. He was too self-absorbed, too preoccupied with his own aspirations as a great scientist to harbour specific political aims.[196]

Harder to brush off was the impression that Eysenck was insensitive, even wilfully blind to the way his work played out in a wider political context. He did not want to believe, almost to the point of utter refusal, that his work gave succour to right-wing racialist groups. But there is little doubt that Jensen and Eysenck helped revive the confidence of these groups. Billig's interviews with National Front leaders made it clear that the emergence of Jensen in the US in 1969 gave them considerable heart, and Eysenck only reinforced this feeling closer to home.[197] It was unexpected vindication from a respectable scientific quarter. The cautionary language of Eysenck's interpretation of the evidence

---

[195] Again see materials at http://www.ferris.edu/ISAR/. Likewise, William Tucker described Eysenck as 'Jensen's dark doppelganger'. See Tucker (1994), *The science and politics of racial research*, p. 234.

[196] A contrast could be drawn with Cattell, for example, who tended to shield his social and political views from his colleagues in psychology. See the up-coming biography of Cattell by William Tucker.

[197] Martin Webster, National Front organizer, wrote in the Front's *Spearhead* in April 1973: 'The most important factor in the build-up of self-confidence amongst racists ... was the publication in 1969 of Professor Arthur Jensen in the *Harvard Educational Review*.' Quoted by Billig (1979), *Psychology, racism and fascism*, Chapter 2.

made little difference. To the racialist right, a genetic basis for group differences in intelligence bore out racialist claims of inherent, immutable hierarchy. Never mind Eysenck's caveat that intelligence was not everything. No one denied it was a socially important variable, least of all those psychologists who spent much of their working lives attempting to isolate and measure it. Never mind his injunctions to look at the individual and refrain from group-based policies. Eysenck's claim that the high overlap in group distributions prohibited racially based decision-making was summarily dismissed by prominent members of the National Front, who argued that it did not 'invalidate the proven fact of genetic differences between the races.'[198]

Since Eysenck maintained there was no necessary connection between science and politics, a scientist's first and only duty was to tell the truth.[199] It contrasted sharply with the views of critics such as Leon Kamin, who saw the science and politics of hereditarian psychometrics as 'not clearly separable'.[200] However, as Billig argued, Eysenck would say one thing, but his science was consistently taken to point in another direction. While it is difficult to tell whether this was his deliberate intention, it was certainly not unwitting. When Billig sent Eysenck material making crystal clear that National Front leaders saw his work as very supportive, Eysenck sounded genuinely exasperated: 'the devil can quote scripture, and malevolent people can always misquote factual evidence ... These things are sent to try us and there is very little that can be done about it.'[201]

Eysenck could agree to disagree with interpretations made of his work, but the fact that some extreme right-wing groups *thought* his work supported their cause remained a festering sore that politicized his wider public reputation. To critics like Billig, it did not seem to bother Eysenck as much as *they* thought it should, his blithe resignation a betrayal of his responsibility as a disciplinary leader. Yet Eysenck was still willing to disassociate himself from and condemn overtly racist and fascist groups. For instance, he took considerable steps to ensure—in his mind at least—that American segregationists could not cite his race and IQ book as a scientific prop. While his disclaimers might not have been as vigorous or as frequent as some might have liked, they can more

---

[198] Verrall, R. and Reed-Herbert, A. (1978), Letter to the editor, *The Times*, 20 March 1978.

[199] Eysenck, H.J. (1975a), The ethics of science and the duties of scientists, *British Association for the Advancement of Science* 1, 23–5. This was Eysenck's position statement on a scientist's social responsibility when dealing with controversial topics.

[200] Kamin (1974), *The science and politics of IQ*, p. 18.

[201] The quotation comes from Billig (1979), *Psychology, racism and fascism*, Chapter 6.

consistently be interpreted as a genuine attempt to distance himself from the extreme right rather than as some sort of covert, conspiratorial strategy. Those asking Eysenck to conform to their notion of politically engaged science had a better idea of what was done with some of his material than he did. He did so many interviews and wrote so many articles that he did not even manage to keep his publication record up to date and was not always up to speed on all that was attributed to him, said about him, or said on his behalf.[202]

Eysenck would complain that virtually *all* attacks against him in the nature–nurture wars were politically motivated, which struck some on-lookers as hypocritical.[203] But like his accusers, Eysenck had become a *political* actor, even if he refused to acknowledge it. If the appropriation of his work by right-wing groups brought him baggage that would be hard to shake off, how then did he construe his relationship with them? Curiously enough, he saw himself as a kind of enlightened scientific shepherd, guiding the blinkered and ignorant toward more sensible positions.[204] Good research would eventually help temper all social wrongs and excesses. The trouble for Eysenck was that empirical science was clearly taken to be part of the problem as well as the solution. Its very impartiality was itself held up to question. The lack of consensus on the technical issues fed open-ended arguments about truth, social justice, and how we should live. Thus the controversy ran on and on.[205]

Keeping the faith

As the 1970s drew to a close, life began to resemble business as usual for Eysenck. Nonetheless, the more limited research initiatives of his last years as IoP head reflected the fact that his reputation as some kind of 'IQ warrior' had narrowed the spectrum of potential collaborators and students. Those attracted to the old professor now tended to be interested in individual differences in intelligence—a venerable and traditionally nativistic field. Reinvigorating it

---

[202] Eysenck did subscribe to a newspaper cutting service to track commentary about himself. See material housed at Centre for the History of Psychology, Staffordshire University, Stoke-on-Trent.

[203] See, for example, p. 246 of Fancher, R. (1985), Review of *Hans Eysenck: the man and his work* by H.B. Gibson, *Journal of the History of the Behavioral Sciences* **21**, 245–7. See also Fancher, R. (1990), Look at me, Review of *Rebel with a cause* by Hans Eysenck, *London Review of Books*, 28 June 1990, 22.

[204] See, for example, Eysenck, H.J. (1979*b*), Letter to the editor.

[205] I am drawing on Roger Smith's summary of the controversy in Smith, R. (1997), *The Fontana history of the human sciences*, Fontana, London, pp. 634–5.

would also help provide Eysenck with an empirical arsenal to go with his public status as Britain's leading advocate of hereditarian psychometrics.

Eysenck's central concern was to make good on the promises and prescriptions he had outlined in his 1967 manifesto.[206] Eysenck had produced an edited volume on intelligence in 1973, but it contained very little research from his labs.[207] It was not until 1979 that Eysenck's first full-length book on the topic appeared, *The structure and measurement of intelligence*.[208] Attempting to resolve the age-old dispute between the general ability factor of Spearman's *g* and specific primary abilities, Eysenck's solution was to place *g* at the apex of a hierarchical structure that incorporated specific abilities at lower levels. It amounted to a restatement of the position he first put forward as a graduate student in 1939, and it could accommodate all major models of intellectual differences except Guilford's, which denied the existence of *g* almost entirely. *The structure and measurement of intelligence* was a laudable attempt at integration in a field in danger of being fossilized. Thus his most promising move was to focus on the problem of speed versus power in IQ testing. Inspired by Furneaux's research once again, Eysenck argued that test performance could best be split into three components: speed, persistence, and error-checking. While research on speed already had a considerable history, it remained to be seen whether the latter two factors could more properly be regarded as personality factors.

Three chapters of *The structure and measurement of intelligence* co-written with colleague David Fulker focused on nature–nurture. Applying the simple model to a range of other researchers' twin and kinship studies (while carefully excluding Burt for the most part), they came up with an estimate of heritability for IQ of 0.69—or 0.79 when a dubious correction for unreliability was made.[209] However, the major assumptions of this simple model could and already had been severely criticized. Kamin, amongst others, questioned

---

[206] Eysenck (1967*b*), Intelligence assessment: a theoretical and experimental approach.

[207] Eysenck, H.J. (ed.) (1973*f*), *The measurement of intelligence*, Medical and Technical Publishers, Lancaster.

[208] Eysenck, H.J. (1979*c*), *The structure and measurement of intelligence*, Springer Verlag, New York.

[209] See ibid., pp. 117–18. The corrected figure of 0.79 was based on assumptions that would now be questioned. The simple model Eysenck was using made a distinction between common family environment and specific non-family environment. Eysenck arrived at his corrected figure by assuming that the smallest parcel of variance in his model—that labelled specific environmental influences—was *all* due to measurement unreliability. More recent research has pointed to specific non-family environmental influences as crucial—as more important than common family influences, if anything. However,

whether all family members could be safely assumed to share the same common environment to the same degree, and whether adoptive parents were a representative cross-section of the community. On the genetic side, the simple model made no allowance for non-additive effects, for assortative mating (i.e. the tendency of like to marry like) or gene–environment interactions and covariation. While Eysenck and Fulker conceded that more complex *genetic* models were possible, the good fit of the simple model was taken to strongly imply all these effects were minimal. The authors repeatedly broke off to conclude that the 'evidence relating to a strong heritable component to IQ is overwhelming' and remark on the 'strikingly consistent picture', reinforcing the message readers seemed to take from this book of a high, relatively fixed genetic estimate.[210] For Eysenck, it all but confirmed his long-held formula of four-parts-nature, one-part-nurture, a formula he and Fulker claimed was consistent with the phenomenon of regression to the mean.[211]

Having thus fortified his intellectual battle-mounts, Eysenck was ready to deal with the fallout from the Burt scandal. Eysenck fell into line with Hearnshaw's acceptance of the fraud charge and his diagnosis of Burt's pathology.[212] Eysenck duly recalled many examples of Burt's eccentric and unethical behaviour.[213] Much later, Eysenck would cautiously return to his original position when several defences of Burt appeared, and offer an apologia for the kind of sins his

---

it was the corrected figure, one Eysenck habitually rounded up to 0.8, that he emphasized in his later writings.

[210] See the Fulker and Eysenck joint chapter, 'Nature and nurture: heredity,' in Eysenck (1979c), *The structure and measurement of intelligence*, p. 127. For a critical appreciation of Eysenck and Fulker's position see Mackintosh, N.J. (1998), *IQ and human intelligence*, Oxford University Press, Oxford, Chapter 3.

[211] Fulker and Eysenck's joint chapter, 'Nature and nurture: heredity', in Eysenck (1979c), *The structure and measurement of intelligence*, p. 118. Eysenck hammered home this 80% heritability formula in the sole-authored epilogue in *The structure and measurement of intelligence*, and would do so in the years to come.

[212] See Appendix A of *The structure and measurement of intelligence* for Eysenck's more damning assessment of Burt, and his review of Hearnshaw's biography, Eysenck, H.J. (1979a), Psychology's Jekyll and Hyde, Review of *Cyril Burt: psychologist* by Leslie Hearnshaw. *New Scientist*, 26 July 1979, 302.

[213] See Eysenck, H.J. (1980f), Psychology of the scientist: XLIV. Sir Cyril Burt: prominence versus personality, *Psychological Reports* **46**, 893–4, and Eysenck, H.J. (1980e), Professor Sir Cyril Burt and the inheritance of intelligence: evaluation of a controversy, *Zeitschrift für Differentielle und Diagnostische Psychologie* **3**, 183–99.

mentor appeared to indulge in.[214] What remained to be done in 1980, however, was to repair any damage done to the hereditarian position and deal directly with Burt's accusers.

Eysenck had briefly debated Steven Rose in 1979 in the pages of *New Scientist*, each agreeing to disagree on the validity of IQ tests, the heritability of IQ, and race differences.[215] Leon Kamin required more attention, however, having played a principal role in Burt's fall from grace. Eysenck readily agreed to book-length debate. For those doubting Eysenck enjoyed such confrontations, he told friends that he was looking forward to facing the Princeton professor.

The result was *Intelligence: the battle for the mind*, consisting of two lengthy antagonistic essays, followed by short replies.[216] While Eysenck and Kamin each put their take on intelligence and IQ testing, the heritability of intellectual differences was the main bone of contention. Eysenck spent a large part of his opening statement giving a primer on the concept of intelligence and its measurability, rehashing old material in a popular vein, before making his case for the high heritability of IQ. In turn, Kamin spent most of his allotted space trying to demolish Eysenck's case, heaping Burt's fraudulence on the specific and inherent limitations of twin and kinship studies. Kamin saw the Galton–Spearman approach to intelligence as all but sterile, yet full of political dangers.

The way Eysenck and Kamin talked past each other illustrated the gulf between empirical data and personal convictions. As Nicholas Mackintosh commented, both examined the same data, but came to remarkably different conclusions.[217] By way of example, Mackintosh cited the evidence generated by the golden egg of separated identical twins. Any resemblance in the IQ was taken as a genetic rather than environment effect. The discredited Burt had the most impressive collection of such twins, though other smaller studies had been carried out. But, as Kamin pointed out and Eysenck conceded, the

---

[214] See Eysenck, H.J. (1989b), Sensitive intelligence issues, Review of *The Burt affair* by Robert Joynson, *The Spectator*, 2 September 1989, and Eysenck, H.J. (1995a), Burt and hero and anti-hero: a Greek tragedy, in *Burt: fraud or framed?* (ed. Nicholas J. Mackintosh), pp. 111–29, Oxford University Press, Oxford. For another defence of Burt, see Fletcher (1991), *Science, ideology and the media: the Cyril Burt scandal*.

[215] Eysenck, H.J. and Rose, S. (1979), Race, intelligence and education, *New Scientist*, 15 March 1979, 849–52.

[216] Eysenck, H.J. and Kamin, L. (1981a), *The battle for the mind*, Macmillan/Pan, London. The American title for this book was *The intelligence controversy*; see Eysenck, H.J. and Kamin, L. (1981b), *The intelligence controversy*, John Wiley, New York.

[217] Mackintosh, N.J. (1981), Psychological sound and fury, Review of *The battle for the mind* by Hans Eysenck and Leon Kamin, *The Times Educational Supplement*, 13 March 1981.

upbringing of these twins was likely to be systematically similar. This allowed Kamin to argue that the high correlation of the twins' IQs might be entirely environmentally induced, while Eysenck merely adjusted his estimate of heritability downward slightly to take this into account.

Both Eysenck and Kamin finished with unedifying attacks on the other's credibility. Eysenck highlighted what he claimed were 'incredible statistical errors' in Kamin's critical analyses, while Kamin made light of Eysenck's close paraphrasing of others and his selective citation practices.[218] Kamin had simply set out to undermine Eysenck's position rather than advance a position of his own, and this seemed to irritate Eysenck no end. Still, it was a shock to see Eysenck castigating Kamin for not playing by the rules of cooperative science, for being too adversarial and negative—defensive gestures reminiscent of Eysenck's unfortunate opponents of years past.[219]

The result was a stand-off, a 'score draw' that enabled each side able to claim that their man had won.[220] This book-length debate also left the key intellectual issues unresolved. To old London school advocates, including Eysenck, there was no question that genetics made the substantial contribution to variation in intellectual ability. All that was needed was a more definitive calculation—nothing that bigger and better twin and kinship studies couldn't fix. While biometric psychometricians on both sides of the Atlantic spent the next two decades doing just that, the political intensity of the nature–nurture wars began to perceptibly wane. Allied researchers in human genetics turned to models that undermined the binary characteristics of nature–nurture questions, not to mention the singular social implications of the answers. Eysenck had produced a few counterexamples to the nativism-to-the-right, environmentalism-to-the-left rubric that Kamin exploited, and the broad sweep of history now seems to be on Eysenck's side.

---

[218] Fulker, D. (1975), Review of *The science and politics of IQ* by Leon Kamin, *American Journal of Psychology* **88**, 505–19.

[219] For Eysenck's laments, see Eysenck and Kamin (1981a), *The battle for the mind*, pp. 157–8. Elsewhere, Eysenck had taken the unusual step of complaining of Kamin's review of his book *The structure and measurement of intelligence*. See Eysenck, H.J. (1981b), Reply to critical notice. *Journal of Child Psychology and Psychiatry* **22**, 299–300.

[220] See Nicholson, J. (1981), A score draw, *New Society*, 12 March 1981. Those at the IoP—Jeffrey Gray, for example—thought Eysenck had got the better of it. Other onlookers did not see it that way. See, for example, Stott, D.H. (1983), *Issues in the intelligence debate*, NFER-Nelson, Windsor, Berkshire.

Channelling Galton

In the early 1980s, Eysenck finally knuckled down to work on intellectual differences. His contribution came in two main forms. He repeatedly stepped in to arbitrate theoretical debates, offering synthetic accounts of current knowledge and scouting the road ahead. He also attempted to lead by example, weighing in with some key research in the areas he regarded as most promising.

In 1980, Eysenck founded the journal *Personality and Individual Differences* (*PAID*), and in 1983 became founding president of the International Society for the Study of Individual Differences. Along with the American journal *Intelligence*, *PAID* gave 'homes to a homeless topic'—the quasi-experimental study of intellectual differences.[221] These two journals were key destinations for Eysenck's latter-day intelligence research. The first issue of *PAID* set out the journal's philosophy, including Eysenck's opposition to peer-review. With Eysenck in control as editor, each *PAID* contributor would be free to 'tell his own story, without referees acting as nannies'.[222]

Eysenck's attempts to direct intelligence research pushed a dual line. While praising the achievements of the psychometric approach, he stressed the need to push beyond its circular, descriptive limitations. From 1981, Eysenck began to harp on the notion of biological intelligence as the primary learning capacity of the individual.[223] Dubbing it 'intelligence A', Eysenck argued this biological potential was largely dictated by genetic endowment. 'Intelligence B', in contrast, was the expression of this potential in everyday life, 'greatly contaminated by educational, cultural and socioeconomic factors'.[224] 'Intelligence C', finally, was defined by performance on psychometric tests.[225]

---

[221] This phrase comes from Deary, I. (1997a), Intelligence and information processing, in *The scientific study of human nature: tribute to Hans J. Eysenck at eighty* (ed. H. Nyborg), pp. 282–310, Pergamon, Oxford, p. 292.

[222] See Eysenck, H.J. (1980d), Editorial, *Personality and Individual Differences* 1, 1–2. For more on Eysenck's opposition to peer review see Eysenck, H.J. (1989c), Refereeing in psychology journals: a reply from Hans Eysenck, *The Psychologist* 3, 98–9.

[223] Eysenck, H.J. (1981a), The nature of intelligence, in *Intelligence and learning* (ed. M.P. Friedman, J.P. Das, and N. O'Connor), pp. 67–85, Plenum Press, New York.

[224] Eysenck, H.J. and Barrett, P. (1984), Psychophysiology and the measurement of intelligence, in *Methodological and statistical advances in the study of individual differences* (ed. C.R. Reynolds and V.L. Willson), pp. 1–49, Plenum Press, New York, p. 1.

[225] Eysenck was following the terminology of Vernon, P.E. (1962), *The structure of human abilities*, Methuen, London.

Eysenck noted that psychologists had concentrated on psychometric intelligence C and its relation to everyday intelligence B. While of some practical short-term value, he concluded, it had been a long-term mistake, one he laid squarely at Binet's doorstep. Having had 40 years to familiarize himself with Oxbridge put-downs of testing as a crude and superficial technology, Eysenck was more than ready to promote alternatives.[226] And, while he was still an ardent proponent of *g*, he and his fellow IQ researchers had been obliged to downgrade its status to that of 'useful concept'—careful to avoid the accusation they were treating it as a reified entity.[227] Eysenck still wanted to make it 'real', and one way of doing so would be to show that it reflected a general material property of the brain. 'Culture-fair' tests were still culturally loaded, and were far too close to a naturalistic approximation of everyday intelligence to be of much analytic value. Eysenck wanted to exorcise social 'contamination' altogether, to strip intellectual differences down to their neurophysiological 'under-ware'. Much of his subsequent work on intelligence, stretching well into his retirement, could be summarized as an attempt to both impress this view on his peers and demonstrate its empirical power.

For Eysenck, there were two main ways of getting to biological intelligence, one old and one new. The idea of correlating measures of mental speed with intellectual differences had been around since the inception of the modern era in the late nineteenth century.[228] Disappointing results and the rise of Binet-style testing saw reaction-time research languish until it was revived in the post-war era. Digging up the 1960s work of Roth and the Erlangen group in Germany, Eysenck pointed to burgeoning research in the US, Australia, and Scotland in the late 1970s that focused on the relationship between simple

---

[226] See Eysenck's concluding chapter in Eysenck (ed.) (1973*f*), *The measurement of intelligence*.

[227] See for example, Eysenck's final, somewhat defensive point of the Epilogue in Eysenck (1979*c*), *The structure and measurement of intelligence*. Eysenck was pre-emptively constructivist in tone, arguing on p. 228: 'The concepts used in this book are human inventions, like all scientific concepts; they are abstractions which cannot be said to exist or not to exist. Concepts are useful or useless; they cannot be true or false. It is only by this criterion that the theories here discussed should be judged.'

[228] Eysenck argued that his forebear Galton retained intellectual priority over this idea, ahead of the practical research of James McKeen Cattell and others across the Atlantic. See Eysenck's exchange with historian Ray Fancher (Eysenck, H.J. (1983*f*), Letter to the editor, *Encounter*, June 1983, 91–2, and Fancher, R. (1983), Letter to the editor, *Encounter*, June 1983, 91) wherein he gave Galton priority in developing the modern distinctions between intelligences A and B, and between general intelligence and special abilities. In addition, Eysenck took issue with historical accounts of early reaction-time research and Galton's legacy.

cognitive tasks and intelligence. These tasks included the simple and choice reaction time paradigms (e.g. one response to one stimulus versus several possible responses to several stimuli) and inspection time paradigms (e.g. choosing between stimuli, or short-term memory tests), each showing promising but variable negative correlations with IQ.[229] Some of the most important research was being done by Eysenck's former students and colleagues, such as Jensen and Ian Deary, but much of it was not. Here Eysenck was a smaller fish in a bigger pond, a follower as much as a research leader.

A second way of getting at biological intelligence was distinctly new-fangled. It used the technology of electroencephalography (EEG)—the electrical activity of the brain measured as a series of voltage changes and recorded via electrodes attached to the head. Eysenck had introduced EEG measures in *Race, intelligence and education*, pointing to Canadian psychologist John Ertl's late 1960s research on evoked potentials (EP).[230] Evoked potentials differ from 'at rest' EEG traces insofar as they are recorded just after the subject is presented with a sudden stimulus, such as a tone. The stimulus is typically presented every one to two seconds, perhaps 50 to 100 times, with a particular time-segment of these responses (e.g. 0–260 ms) averaged to get an interference-free picture. The basic idea was that characteristic differences in EP wave patterns would reflect differences in intelligence. Nevertheless, demonstrating reliable correlations between these differences and other measures of intelligence would prove an enormous challenge.

Ertl's research suggested there might be a strong relationship between high IQ scores and EP latencies (the interval between waves), especially initial response latencies. Quickly picking up on the possibilities, Eysenck encouraged one of his students, Elaine Hendrickson, to attempt to replicate this work. By 1973, Eysenck was citing her unpublished 1972 Ph.D. thesis as a 'serious step ... in the direction of identifying the physiological basis of intelligence'.[231] Elaine Hendrickson was joined by husband Alan, who brought in his expertise in neurochemistry and physiology. Together they worked at better ways to analyse the EP waveform data to boost correlations with psychometric tests.

---

[229] Eysenck gave his version of the history of this research tradition in Eysenck, H.J. (1986a), Inspection time and intelligence: a historical introduction, *Personality and Individual Differences* **7**, 603–7. See also Deary, I. (1986), Inspection time: discovery or rediscovery, *Personality and Individual Differences* **7**, 625–31.

[230] Ertl, J. and Schafer, E. (1969), Brain response correlates of psychometric intelligence, *Nature* **223**, 421–2. Evoked potentials have otherwise been termed event-related potentials.

[231] Eysenck (ed.) (1973f), *The measurement of intelligence*, p. 429.

However, other experimenters using different stimulus parameters and analyses—even those working at the IoP—were having trouble replicating these correlations. Jensen's 1980 review described the field as a 'thicket of seemingly inconsistent and confusing findings ...'[232] That year, though, the Hendricksons finally published their extension of Ertl's work as the lead article in the first edition of *PAID*.[233]

By 1981, Eysenck was promoting EP indices as the royal road to biological intelligence. In his joust with Leon Kamin, Eysenck claimed they rivalled IQ tests as measures of intellectual differences. Dull subjects tended to produce shallower waves, he wrote, exhibiting greater latency and amplitude.[234] Not everyone agreed, not Kamin certainly. Like many a critic, he was able to point to the divergence in results to dismiss EP indices as techno-faddish myth.[235]

Retiring but hardly shy

In 1982, Eysenck brought together some of the leading figures in the search for biological intelligence in an edited volume *A model for intelligence*. Judging by citations counts, the book made a significant impact on the field, with Jensen's chapter on reaction times and Brand and Deary's chapter on inspection times both heavily referenced. The book also carried twin contributions from the Hendricksons. They had hit upon a new measure of the EP waveform complexity dubbed 'string length', likening EP traces to pieces of string. The more intelligent the individual, the *longer* his/her EP string lengths. According to Elaine Hendrickson, the correlation between string length and WAIS scores was an impressive 0.72. She also reported a measure of variability (defined as the level of deviance of individual trace recordings from the composite, average trace) to correlate at a similarly high, *negative* level. According to Alan Hendrickson, these consistently longer and more complex EP traces represented a greater propensity for error-free transmission.

---

[232] Jensen (1980), *Bias in mental testing*, p. 709.

[233] Hendrickson, D.E., and Hendrickson, A.E. (1980), The biological basis of individual differences, *Personality and Individual Differences* **1**, 3–33.

[234] Eysenck and Kamin (1981a), *The battle for the mind*, pp. 70–3.

[235] Kamin pointed to the negative EP research of John Rust, an IoP student who had tried to extend on Elaine Hendrickson's work. See Rust, J. (1975), Cortical evoked potential, personality and intelligence, *Journal of Comparative and Physiological Psychology* **89**, 1220–6. In a footnote to his concluding chapter in Eysenck (ed.) (1973*f*), *The measurement of intelligence*, Eysenck argued that Rust's failure to replicate Hendrickson's promising results should not be used to argue against the value of the EP approach, since Rust used a different experimental set-up.

Eysenck's introduction and conclusion to *A model for intelligence* held it all together. Reprising his role in the promotion of behaviour therapy, he spelt out the collective significance of these findings with a true believer's zeal. Now the 'best tests of individual differences in cognitive ability are non-cognitive' Eysenck wrote.[236] Reaction and inspection time tasks and EP measurements were simple and direct, yet they showed correlations with IQ at '0.8 and above, without correction for attenuation'.[237] For Eysenck, the data suggested the existence of a fundamental biological property of the central nervous system underlying measures of IQ, reaction and inspection time, and errorless information processing—the latter 'presumably being the most fundamental biological correlate (and probably [the] cause) of intelligence B'.[238]

Eysenck would ratchet up the hype in 1983, announcing that a 'revolution' had occurred in the theory and measurement of intelligence.[239] His favourite measure of biological intelligence, evoked potentials, had more than just potential. The high correlations the Hendricksons had achieved between EP string length and the g-loading component of psychometric tests meant it came 'close to being a perfect measure of genotypic intelligence'.[240] Other claims were just as bold. For instance, he suggested that, for mildly retarded subjects, the inspection time–IQ correlation was in the –0.8 to –0.9 range.[241]

Eysenck also retired in 1983. He had no plans to vacate the research scene, though. With the support of his successor Jeffrey Gray, Eysenck was still able to assemble an effective research team. He continued to supervise graduate students, such as Warwick Frearson, and he still enjoyed data-processing and administrative support. Post-doctoral researcher Paul Barrett became Eysenck's right-hand man in this period, taking care of the nuts-and-bolts of the increasingly elaborate IT requirements. Having vacated his professorial office and

---

[236] Eysenck (ed.) (1973*f*), *The measurement of intelligence*, p. 9.

[237] Ibid.

[238] Ibid., p. 259.

[239] Eysenck, H.J. (1983*b*), Revolution dans la theorie et la measure de l'intelligence, *La Revue Canadienne de Psycho-Education* 12, 3–17. He also wrote up a lengthier screed that Jensen called 'a veritable masterpiece of theoretical integration' a year later that apparently did not see the light of published day. See Jensen, A.R. (1986), The theory of intelligence, in *Hans Eysenck: consensus and controversy* (ed. S. Modgil and C. Modgil), pp. 89–102, Falmer Press, Philadelphia, p. 100.

[240] Eysenck (1983*b*), Revolution dans la theorie et la measure de l'intelligence, p. 15.

[241] For instance, Eysenck also claimed that simple reaction time correlated with g at 0.45, and yielded progressively higher correlations when more complex choice elements were added to the task. Eysenck (1983*b*), Revolution dans la theorie et la measure de l'intelligence.

needing more space for his experimental facilities, Eysenck set himself up in separate offices 300 metres from the Institute. The financial backing for this phase of his research career would come from a variety of unlikely sources—including the Pioneer Fund and the tobacco industry.

The long morning after

Eysenck's retirement marked the zenith of his promotion of the biological basis of intellectual differences. However in the years that followed, the big promises he made confronted the cold light of empirical day. Divergent results were clarified and a clearer picture emerged. But, while the search for the physiological basis for intellectual differences continued, some of the optimism Eysenck had helped instil began to dissipate. His disappointing results with evoked potentials in the early 1990s saw that particular paradigm almost grind to a halt—an indecisive note to finish on as the 70-something Eysenck's energies began to wane.

After stepping down in 1983, Eysenck continued to pump out theoretical integrations of intelligence research every few years, constantly extolling the virtues of the direct physiological approach to a wide audience.[242] The reaction time and EP research Eysenck contributed to in this period was more intriguing. It was some of the most thorough follow-up work of Eysenck's career; not coincidently, it took place after he had flown the 'beehive'. Working closely with a small group of young researchers, the retired professor tended to adopt a secondary, team-oriented role. Few articles named Eysenck as first author, unusual for him. Moreover, these publications had a more moderate, enjoined tone—in keeping with the conspicuously mixed results they reported.

As Eysenck noted in 1986, reaction and inspection time data had been positive but they lacked convergence.[243] When studies using small samples were

---

[242] Many of these articles were in foreign journals or popular publications, with a high level of doubling up. In 1986, Eysenck revised the intelligence A, B, and C concepts as a series of concentric circles. First and most narrowly there was biological intelligence. Next was psychometric intelligence, which was highly dependent on genetic endowment but also influenced by environmental factors. Most generally, there was social intelligence, reflecting success in those aspects of life that required intelligent behavior. Social intelligence was dictated by psychometric intelligence but was also said to be enhanced or hindered by personal events, practical habits, and emotional propensities. Eysenck, H.J. (1986e), Toward a new model of intelligence, *Personality and Individual Differences* 7, 731–6.

[243] With researchers working on a variety of experimental set-ups, correlations with IQ scores had ranged from 'insignificant' (e.g. –0.1 to –0.2) to 'overpowering' (e.g. –0.9). Eysenck (1986a), Inspection time and intelligence: a historical introduction, p. 604.

eliminated and range of ability effects were controlled, a more coherent set of figures began to emerge. Simple reaction time showed reliable but small correlations with IQ of around –0.2 and were of little interest, Eysenck allowed. Choice reaction time generally exhibited slightly higher correlations, which Eysenck put in the –0.3 to –0.5 range.[244] Other workers, notably Jensen in 1987, suggested somewhat weaker figures. Correlations strengthened slightly but progressively as more choices were added, reaching a ceiling level of around –0.35.[245]

The reaction and inspection time research Eysenck had a hand in furnished some interesting data, but failed to reproduce several foundational results. For example, a 1986 choice reaction time study undermined two results that had strongly implicated speed in intellectual differences. Eysenck and his colleagues found that not everyone followed Hick's law: about 20% of their subjects' reaction times did not progressively slow as more choices were added. More choices did not slow duller subjects more than the brighter ones either, another blow to the speed hypothesis.[246] Conversely, Eysenck's research group did manage to strengthen reaction time correlations with IQ considerably (i.e. up to –0.48) using the more complex 'odd-man-out' choice procedure.[247] Later research suggested this was because the odd-man-out procedure shared features with inspection time tasks, which would continue to show similarly high correlations.[248]

These more complex tasks could hardly be considered pure measures of speed, as Eysenck was well aware. Even though in 1986 he still clung to the idea that the corrected 'true' reaction time–IQ figure might be as strong as –0.7 or –0.8, Eysenck had begun to switch tack.[249] He had to concede that speed of

---

[244] Ibid., p. 605.

[245] Jensen, A.R. (1987), Individual differences in the Hick paradigm, in *Speed of information processing and intelligence* (ed. P.A. Vernon), pp. 101–75, Ablex, Norwood, New Jersey. Jensen described how it was possible to break through this 0.35 barrier using an aggregate of reaction time and elementary cognitive tasks. See Jensen, A.R. (1997b), The psychometrics of intelligence, in *The scientific study of human nature: tribute to Hans J. Eysenck at eighty* (ed. H. Nyborg), pp. 221–39, Elsevier, Oxford.

[246] Eysenck, H.J. et al. (1986), Reaction time and intelligence: a replication study, *Intelligence* **10**, 9–40.

[247] Frearson, W. and Eysenck, H.J. (1986), Intelligence, reaction time (RT) and a new 'odd-man-out' RT paradigm, *Personality and Individual Differences* **7**, 807–17.

[248] Bates, T.C. and Eysenck, H.J. (1993b), Intelligence, inspection time and decision time, *Intelligence* **17**, 521–31.

[249] Eysenck (1986a), Inspection time and intelligence: a historical introduction, p. 605.

processing was not the central factor in intellectual differences.[250] Reaction time and inspection time tasks did not show particularly strong correlations with each other, suggesting different processes were involved, and considerable debate centred on just how much of their correlation with IQ was *g* loaded or related to specific, peripheral perceptual capacities.[251] In its place, Eysenck began to emphasize the notion of error-less transmission—prompted by the Hendricksons' work, his own nerve conductivity results, and the moderately high correlations found between intra-individual variability in reaction time and IQ. Greater speed was a byproduct of clearer and more efficient neural pathways, and Eysenck turned to the EEG evoked potentials paradigm to confirm this.

This was expensive, highly technical research, but the resourceful Eysenck was able to use his connections to procure the necessary funds. The tobacco company R.J. Reynolds had sponsored Eysenck's reaction time work in the mid-1980s, interested in any positive effects nicotine might have on performance.[252] Philip Morris subsequently came on board, helping to underwrite his evoked potential programme. From 1986, the Pioneer Fund was also tapped for money, reportedly handing over £200, 000 in the 1986 to 1990 period.[253] Most of this money went toward purchasing the necessary testing, recording, and computing facilities. Space requirements were so great they outgrew anything on offer on Denmark Hill and in 1989 took over the old animal labs down at Bethlem Royal Hospital. Mass computation real-time Unix computers

---

[250] Eysenck's joint research on speed of nerve conductivity further relegated the importance of speed. No correlation between nerve conduction velocity and IQ was found in two studies Barrett, Eysenck, and others performed. The 0.37 correlation between velocity and EPQR psychoticism found in the first 1990 study failed to replicate in the second 1993 study. However, Barrett and Eysenck did manage to partially replicate a negative correlation between nerve conduction variability and IQ. Barrett, P., Daum, I., and Eysenck, H.J. (1990), Sensory-nerve conduction and intelligence: a methodological study, *Journal of Psychophysiology* 4, 1–11; Barrett, P. and Eysenck, H.J. (1993), Sensory nerve conduction and intelligence: a replication, *Personality and Individual Differences* 15, 249–60.

[251] See Jensen (1997*b*), The psychometrics of intelligence, as well as the discussion of these points in Mackintosh (1998), *IQ and human intelligence*, Chapter 7.

[252] The tobacco industry sponsored research on the effects of nicotine more for public relations than legal reasons. I discuss these funding issues in more detail in the next chapter.

[253] Littlewood, R. (1995), *Mankind Quarterly* again, *Anthropology Today* 11, 17–18; Lane, C. (1944), The tainted sources of *The bell curve*, *The New York Review of Books*, 1 December 1994, 14–19. For an inside account of their activities, see materials at the Pioneer Fund website: http://www.pioneerfund.org/.

replaced the old rat cages, as one era in psychological research ended and a new one began. They assembled 'one of the finest electrophysiological labs in the UK', Barrett recalled, able to test 20 subjects at a time in 'room space beyond the dreams of avarice'.[254]

Whether this was money well spent depends on one's view on the meaning and importance of non-replications. Other researchers had only limited success duplicating the Hendricksons' remarkably strong string length–IQ correlations. It was clear from their articles that considerable judgement had gone into whittling their data down to a 'good' set of analysable records. Ian Deary called their presentations 'suspiciously complete', implying a *post facto* gloss had been put on an open-ended search for effects and explanations.[255] Leaping to the Hendricksons' defence in 1983, Eysenck specified very narrow conditions, suggesting, for example, that high string length–IQ correlations might be possible only with an 85 db sound stimulus.[256] Such restrictions would not make replication easy.[257] Much discussion centred on the nature of stimulus used and recording electrode placement, as well as analytic variables such as latency, variability, amplitude, and string length. However, the large and well-controlled 1987 study of Vogel *et al.* failed to produce consistent or substantial EP–IQ correlations.[258]

The jury was still out on evoked potentials and intelligence as the 'decade of the brain' began in 1990. Eysenck remained convinced this paradigm held great promise. Through the late 1980s, he and Paul Barrett worked their way through practical problems of EEG research, focusing on the procedural rules recommended by the Hendricksons. Barrett recalled that Eysenck 'did not get

---

[254] See p. 14 of Barrett, P. (2001), Hans Eysenck at the Institute of Psychiatry—the later years, *Personality and Individual Differences* **31**, 11–15.

[255] Deary (1997*a*), Intelligence and information processing, p. 292.

[256] See Eysenck (1983*b*), Revolution dans la theorie et la measure de l'intelligence; Eysenck, H.J. (1983*c*), Revolution dans la theorie et la measure de l'intelligence: reponse a quelques critiques, *La Revue Canadienne de Psycho-Education* **12**, 144–7. Despite this, Eysenck maintained that *uncorrected* string length–IQ correlations as high as 0.84 could still be obtained. However, Carlson and Widaman labelled such correlations 'so high as to challenge credibility'. See Carlson, J.S. and Widaman, K.F. (1986), Eysenck on intelligence: a critical perspective, in *Hans Eysenck: consensus and controversy* (ed. S. Modgil and C. Modgil), pp. 103–32 and 135–6, Falmer Press, Philadelphia, p. 135.

[257] See Stankov, L. (1998), H. Eysenck on intelligence: biological correlates and polemics, *Psihologija* **31**, 257–70.

[258] See Vogel, F., *et al.* (1987), No consistent relationships between oscillations and latencies of visual evoked EEG potentials and measures of mental performance, *Human Neurobiology* **6**, 173–82. Barrett and Eysenck would cite this study as a key non-replication.

involved with the technical features of what we were doing [but] he was there every step of the way examining how we were achieving solutions ... digesting results, thinking about the consequences to theory and his on-going writings ...'[259] However, the more research they did, the murkier the picture became. Barrett and Eysenck confirmed the negative relationship between EP variability and IQ. But string length showed low to moderate correlations with IQ, and these tended to be in the 'wrong' (i.e. *negative*) direction. The Hendrickson paradigm was 'not a paradigm at all', they concluded. The Hendrickson rules were wholly unhelpful or irrelevant.[260] Any well-controlled EP study might yield correlations with IQ. After this disappointment, Barrett was just about ready to give up on the topic. No evoked potential parameter had been shown to be 'consistently replicable'.[261] Divergent methodologies were no longer the problem, since high-specification rules had not helped.[262] Perhaps differences in test instructions might account for flip-flopping EP–IQ correlations. However if this was so, it would imply EP differences were more effect than cause.[263]

---

[259] Barrett (2001), Hans Eysenck at the Institute of Psychiatry—the later years, p. 14.

[260] See p. 377 of Barrett, P. and Eysenck, H.J. (1992b), Brain evoked potentials and intelligence: the Hendrickson paradigm, *Intelligence* **16**, 361–81.

[261] Barrett, P. and Eysenck, H.J. (1992a), Brain electrical potentials and intelligence, in *Handbook of individual differences: biological perspectives* (ed. A. Gale and M.W. Eysenck), pp. 255–85, Wiley, London, p. 279. Reviewing spontaneous (as opposed to evoked potential) EEG research, Barrett concluded there was 'very little use in continuing with this approach as an attempt to elucidate the causal mechanisms with regard to IQ correlates.' Even though he wrote this review with Eysenck, Barrett made it clear this was strictly 'the opinion of the first author' (p. 276).

[262] Barrett and Eysenck (1992b), Brain evoked potentials and intelligence: the Hendrickson paradigm, p. 377. Eysenck and Barrett had used just such an explanation of the scattered EP results in Eysenck and Barrett (1984), Psychophysiology and the measurement of intelligence.

[263] Instructions to attend to the task might produce negative string length–IQ correlations because brighter subjects needed fewer resources to do it; conversely their greater surplus neural activity might explain positive correlations when they were instructed just to 'sit quietly'. Bates and Eysenck did indeed demonstrate a negative string length–IQ correlation when subjects were told to pay attention, and they were backed up by other researchers employing different EP variables. But this implied more intelligent people simply did it with less effort. See Bates, T.C. and Eysenck, H.J. (1993a), String length, attention, and intelligence: focussed attention reverses the string length IQ relationship, *Personality and Individual Differences* **15**, 363–71, and Bates, T.C. et al. (1995), Intelligence and the complexity of the average evoked potential: an attentional theory, *Intelligence* **20**, 27–39.

Scientific retrenchment might be too strong a word to describe the effect on the EP field, but it came close.[264] With various elaborate models undermined by inconsistent data, Deary and Caryl suggested that the next phase of research should be atheoretical.[265] Barrett and Eysenck's 1992 review recommended researchers compute every possible EP parameter. The last investigation they undertook on the topic did just that, but they again came up with largely negative results.[266]

True to his London school roots, Eysenck had tried to account for performance differences rather than elucidate the common features of that performance—in contrast to the work of the equally prolific Robert Sternberg, for example. Even his strongest supporters would fault him for not taking advantage of the information processing models. In pursuing elusive correlation between EP and IQ, Eysenck and his co-workers diverted attention from what actually happened to EEG waveforms when particular tasks were encountered.[267] Other neuro-imagining techniques loomed on the horizon and Eysenck was not the only one to suggest a marriage with EP work.[268] Reaction and inspection time research had also reaching an impasse of sorts by the

---

[264] More optimistically, Barrett and Eysenck described the area as 'still in a state of flux'. Barrett and Eysenck (1992b), Brain evoked potentials and intelligence: the Hendrickson paradigm, p. 377.

[265] Deary, I. and Caryl, P.G. (1987), Intelligence, EEG, and evoked potentials, in *Biological approaches to the study of human intelligence* (ed. P.A. Vernon), pp. 259–315, Ablex, Norwood, New Jersey.

[266] Barrett, P. and Eysenck, H.J. (1994), The relationship between evoked potential component amplitude, latency, contour length, variability, zero-crossings, and psychometric intelligence, *Personality and Individual Differences* **16**, 3–32.

[267] Barrett conceded that using EPs in the way they had—as a general index of mental activity/acuity with quantitative scalar properties—was bound to end in failure. Instead, he suggested that EPs needed to be analysed more closely, perhaps down to each trace. Barrett's last work on the topic was an attempt to do a kind of waveform analysis using 'electronic fingers'—a huge computational exercise that was never finished because by then Eysenck had lost interest. Paul Barrett, personal communication, 11 September 2008. Barrett cited the early 1990s work of David Robinson as particularly important because it cast doubt on just what an EP was thought to be. In a recent study, Robinson divided EP traces into components said to be characteristic of different parts of the brain and brainstem. He compared the patterns of different personality types, predicting differences in line with a Pavlovian excitation–inhibition framework. Robinson's results would surely have gladdened Eysenck's heart. Robinson, D.L. (2001), How brain arousal systems determine different temperament types and the major dimensions of personality, *Personality and Individual Differences* **31**, 1233–59.

[268] See, for example, Barrett and Eysenck (1992a), Brain electrical potentials and intelligence.

mid-1990s. Speed was obviously only part of the story. Higher correlations with psychometric measures had come at the cost of greater complexity, so much so that such tasks resembled those of the psychometric tests they were supposed to de-compose. The biological basis of intellectual differences remained a work in progress. It had certainly not lived up to promises of its most enthusiastic advocates, Mackintosh wrote in 1998, implicitly taking Eysenck to task for overselling the 'revolution'.[269]

Eysenck had envisaged biological intelligence as the ultimate comeback to all those who doubted hereditarian psychometrics and the Galton-Spearman legacy. Locating *g* differences in brain substrates would allow hereditarian psychometricians to re-assert its corporeality. But the big promises Eysenck had made had the unfortunate effect of making some of his reaction time and EP research look like embarrassing failures. Looking back, Paul Barrett saw it all as a necessary process. It took a great deal of carefully controlled work, Barrett recalled, not to mention 20 years of advances in neuroscience, to see 'just how feeble this approach really was ...We learned a lot about how simple physiological indices do not relate replicably to complex behavioural performance—and just how fragile are those links between "chronometric" indices and constructs like "psychometric intelligence".'[270] Eysenck could still take a great deal of satisfaction in helping generate some promising new leads in intelligence research and seeing experimental and correlational perspectives connected in a way he was never able to achieve with his biosocial approach to personality. But then again, the quasi-experimental study of intellectual differences had a much longer history, making for less formidable integrative barriers.

Eysenck's last book, published posthumously, was titled *Intelligence: a new look*.[271] Having first planned it in the late 1980s, it took him almost a decade to actually write. Eysenck surveyed a wide selection of intelligence-related topics, invoking the ideas of past heroes backed by new results to show they were on the right track. He still held on to his version of the error-less transmission model even though the Hendricksons' neurochemical model had

---

[269] Mackintosh (1998), *IQ and human intelligence*, pp. 234 and 250. The more sympathetic Ian Deary still bemoaned the dead-ends and lack of theoretical integration. Deary (1997a), Intelligence and information processing, pp. 303–4.

[270] Paul Barrett, personal communication, 9 September 2008.

[271] Eysenck, H.J. (1998), *Intelligence: a new look*, Transaction Press, New Brunswick, New Jersey.

fallen out of empirical favour.[272] It might still prove productive in the future. DNA-dependent biological intelligence was still the first cause in his one-way, feed-forward model for the development of intellectual capacities. While neurocognitive modelling was only given lip-service, this was still an impressively broad effort for a man of 80.

Whither nature versus nurture?

With his attention focused on biological intelligence in his twilight years, Eysenck had little to contribute on the nature–nurture issue. In any case, the frantic pace of developments in behavioural genetics began to undermine the oppositional character of these concepts he had thrived on. While a significant group of psychologists never lost their enthusiasm for computing the relative contributions of nature and nurture, they freely admitted that this was no longer enough.[273] Specialists in behavioural genetics began to focus on ongoing, developmental interactions, on specific, complex gene action in response to particular environments, which all made it more a case of nature *and* nurture.

After Eysenck retired, Fulker left the IoP and Eaves went to North America, and his collaborations in behavioural genetics petered out. A major summary statement of this work in the genetics of personality appeared in 1989.[274] Containing a lengthy critique of the simple twin study model, *Genes, culture and personality: an empirical approach* was a quantum leap in conceptual sophistication and design methodology. It made Eysenck's initial research with Prell look primitive, not to mention a little misleading. With the respected Lindon Eaves as the first of three principal authors, it also included contributions from another five specialists. Its findings were hard to summarize,

---

[272] See Barrett and Eysenck (1992*b*), Brain evoked potentials and intelligence: the Hendrickson paradigm.

[273] See Vernon, P.A. (1997), Behavioural genetic and biological approaches to intelligence, in *The scientific study of human nature: tribute to Hans J. Eysenck at eighty* (ed. H. Nyborg), 240–58, Pergamon, Oxford. Since there was no question about the substantial heritability of intelligence differences *per se*, Vernon wrote, many other more complex and interesting questions should be addressed. These included the precise kinds of gene action involved, the influence of assortative mating, the specific nature of environmental influences, and the joint effects of genes and environment (p. 241). See also Plomin, R. (1994), *Genetics and experience: the interplay between nature and nurture*, Sage, Thousand Oaks, California, and Pigliucci, M. (2001), *Phenotypic plasticity: beyond nature and nurture*, Johns Hopkins University Press, Baltimore.

[274] Eaves, L., Eysenck, H.J., and Martin, N.G. (1989), *Genes, culture and personality: an empirical approach*, Academic Press, New York.

provoking as many questions as they did answers. Twin and kinship data collected in London, Europe, the US, and Australia, confirmed that genetic factors contributed significantly to 'family resemblances' in Eysenck's personality dimensions. However, the mid-range estimates obtained still left plenty of room for environment influences—'upwards of 50%' Eaves *et al.* allowed.[275] Interestingly, they felt the source of such influences could be largely traced to an individual's unique experiences rather than shared family environment, a product of the 'slings and arrows of outrageous fortune' rather than the actions of parents and teachers. No one model could explain the transmission of every variable, either.[276]

The authors of *Genes, culture and personality* admitted their analysis of gene–environment interactions was still weak. Many current environmental indices behaved just like inherited personality variables in twin studies, making them hard to sort out. Likewise, the process of mate selection and cultural inheritance was too complex to be captured by conventional studies of families or twins. One finding they also labelled as 'surprising' was the relatively high genetic contribution for social attitudes, reprising Eysenck's *Nature* article of a decade previously. While the notion of genes for voting left or right was patently 'absurd', they wrote, this result may reflect sensitivity to certain situations or rewards.[277] The social attitudes data appeared to demonstrate the strong effect of cultural inheritance, but significant assortative mating effects were also readily apparent. All in all, they concluded, 'the causes of variation in personality and attitude are more complex than we had thought fifteen years ago.'[278] The picture hinted at was that of a branching and compounding partnership of genes and environment, each modifying the expression or realization of the other.[279]

The nuanced subtlety of this collective work highlighted the widening disparity between Eysenck's unreconstructed nativism and contemporary behavioural genetics. Left to his own devices, Eysenck's sole publications dealing with the nature and nurture of intelligence looked less refined. Brushing aside the limitations of the simple twin study model, he would cite his four-to-one

---

[275] Ibid., p. 121.

[276] For instance, Eaves *et al.* were able to account for the heritability of neuroticism with a relatively simple additive model. However, a far more interactive model was needed to explain the genetic transmission of introversion–extraversion.

[277] Ibid., p. 405.

[278] Ibid., p. 415.

[279] For a recent, very readable treatment of these issues, see Rutter, M. (2006), *Genes and behavior: nature–nurture interplay explained*, Blackwell, Malden, Massachusetts.

genetic versus environment ratio to the very end.[280] In contrast, many others who calculated this ratio put it somewhat lower, at something closer to a 50:50 split.[281] Much depended on the data set analysed nonetheless. For example, the direct estimates derived from twin and nuclear family studies tended to be higher than indirect estimates derived from more distant kin relationships.[282] Age of testing also made a huge difference, given that heritability appeared to (counter-intuitively) increase with age. A task force set up by the APA at the end of 1994 in response to *The bell curve* brouhaha estimated the heritability of IQ differences at 'around 75%' for *adult* samples. [283] Conversely, Nicholas Mackintosh suggested the heritability of IQ fell within the range 0.30 to 0.75 for modern industrialized societies.[284] There was no point in being more precise, he argued. Different studies used different models and assumptions, making them hard to compare, and all estimates were subject to sampling and measurement error, as well as the effects of time and place of testing. Besides these methodological factors, estimate variability probably reflected genuine fluctuations in heritability as a population-specific measure. Different groups, with a greater or lesser range in their social circumstances, have produced demonstrably different heritability estimates. However, Eysenck's latter-day statements on the nature and nurture of intelligence conveyed a somewhat fixed idea of what a fixed estimate was and meant, with little discussion of the qualifications and uncertainties involved.[285] Eysenck would continue to stress

---

[280] See Eysenck (1998), *Intelligence: a new look*, p. 47.

[281] See for example, the inclusive modelling approach of Chipuer, H.M. *et al.* (1990), LISREL modelling: genetic and environmental influences on IQ revisited, *Intelligence* **14**, 11–29. See also Carlson and Widaman (1986), Eysenck on intelligence: a critical perspective, and Vernon (1997), Behavioural genetic and biological approaches to intelligence.

[282] Plomin, R. and Loehlin, J. (1989), Direct and indirect IQ heritability estimates, *Behavior Genetics* **19**, 331–42.

[283] Neisser, U. *et al.* (1996), Intelligence: knowns and unknowns, *American Psychologist* **51**, 77–101. This conclusion was echoed by Plomin, R., *et al.* (2001), *Behavioral genetics*, Freeman, New York. Eysenck followed suit in emphasizing adult estimates, but tended to cite a higher figure.

[284] Mackintosh (1998), *IQ and human intelligence*, p. 93.

[285] Eysenck did not make a great deal of effort to spell out the qualifications surrounding heritability estimates and did not tend to acknowledge criticism of the assumptions they required. For example, in his last book, Eysenck (1998), *Intelligence: a new look*, he pointed to the disparities generated by age-related effects to account for the (misleadingly in his opinion) lower figures others cited. But this discussion was buried in his 'Endnotes, references and comments' section (p. 200). See also his cavalier discussion of the topic in Eysenck and Barrett (1984), Psychophysiology and the measurement of intelligence, p. 4.

the singular contribution of genetic factors to intellectual and personality differences, amidst ritual incantations of the sins of 'environmentalism'. IQ had a heritability of 80%, Eysenck would write, a figure he represented as *the* summary estimate across the field. Many other hereditarians regarded this as an optimistic upper limit.[286]

Even so, the idea of nature *and* nurture had a way of reframing the implications the issue generated, and Eysenck was a little more reserved on the social policy front in the latter part of his career. His scepticism towards compensatory education never diminished, but it did soften ever so slightly. In 1981, Eysenck had argued that profound improvements had to be clearly demonstrated before reforms were widely implemented. 'Simply to press for greater equality in education, in salaries, and in similar matters would not greatly alter the observed differences in IQ.'[287] The mediocre results of government-driven reforms left him cold. He bitterly regretted sending two of his children to the 'anti-learning environment' of comprehensive schools.[288] However in his last book, Eysenck argued that a focus on achieving IQ change in programmes like Head Start was misplaced. It obscured the 'real advantages enjoyed by children affected and the general educational achievements often reported'.[289] Eysenck went on to suggest that more empirical research on novel quasi-environmental measures—such as better nutrition—was needed.

It also appeared that Eysenck's classic eugenic warnings of a degenerative decline in intelligence would not come to pass. In 1979, Eysenck and Fulker had conceded that their analyses showed home background and general social advantage had pronounced effects 'capable of producing mean differences of around 15 IQ points'.[290] They were quick to add that neither of these factors accounted for IQ variation better than the predominantly genetic plus environment model. Conversely, Eysenck and Fulker's assessment of class, differential fertility, and IQ implied 'a rapid decline in mean IQ is not to be expected'.[291]

---

[286] See Loehlin (1986), H.J. Eysenck and behaviour genetics: a critical view and Lindon Eaves's exasperated reply, Eaves, L. (1986), Eaves replies to Loehlin, in *Hans Eysenck: consensus and controversy* (ed. S. Modgil and C. Modgil), pp. 58–9, Falmer Press, Philadelphia.

[287] Eysenck (1981a), The nature of intelligence, p. 74.

[288] The quotation comes from a PR piece on Eysenck's broad-audience book *Mindwatching*, co-authored by son Michael. See Hastings, M. (1981), Putting their eggheads together, *New Standard*, 28 April 1981.

[289] Eysenck (1998), *Intelligence: a new look*, p. 101.

[290] Eysenck (1979c), *The structure and measurement of intelligence*, p. 153.

[291] Ibid., p. 174.

Again they hastened to add this was no reason to be complacent about downward 'dysgenic trends where they exist'.[292] But their results implied that measured IQ may even be rising, a widespread phenomenon that is now well accepted as the 'Flynn effect'.[293] When canvassing this phenomenon near his death, Eysenck was reluctant to admit that this meant that average intelligence was actually increasing, as were many other researchers in the area. Perhaps it was only a 'not perfectly accurate measure of it' that was doing so.[294] This latter-day concession of the limitations of IQ testing represented a strategic retreat from the psychometric orthodoxy Eysenck had espoused decades earlier, and it provided an additional background element to his search for alternative biological measures.

Late into his retirement, Eysenck wrote extensively on genius and creativity and had a predictable say in *The bell curve* debate.[295] In the early 1990s, Eysenck even got involved in attempts to raise intelligence with vitamins, ever willing to 'play with fire' as wife Sybil put it.[296] The suggestive results of his studies were, however, contested by competing researchers with data countering the IQ boost Eysenck's team claimed. No one covered themselves in glory in this episode, with the profiteering of commercial pharmaceutical companies and

---

[292] Ibid.

[293] The name stems from James Flynn, the researcher generally credited with identifying this phenomenon in the mid-1980s. Measured IQs have apparently risen 15–20 points across the developed world over the past 50 to 75 years. It is an effect not totally understood but it is widely attributed to environmental factors, being far too rapid to have a genetic basis. This increase has occurred in spite of the consistently negative correlation between measured IQ and fertility. Rising class and education levels, better health and nutrition, smaller families, greater environmental complexity, and increasing degrees of test sophistication have all been put forward as possible explanations. More controversial are suggestions that the Flynn effect may have peaked in some countries and/or is still masking dysgenic trends in others. For more, see Neisser, U. (ed.) (1998), *The rising curve: long term gains in IQ and related measures*, American Psychological Association, Washington, DC, and Flynn, J.R. (2007), *What is intelligence? Beyond the Flynn effect*, Cambridge University Press, Cambridge.

[294] Eysenck (1998), *Intelligence: a new look*, p. 25. Not wishing to talk up the implications, Eysenck regarded the Flynn effect as a purely performance effect. Increases in test sophistication, greater cultural complexity, and better nutrition were the likely causes, he opined.

[295] Eysenck gave *The bell curve* an enthusiastic reception. Eysenck, H.J. (1994b), Much ado about IQ, Review of *The bell curve* by Richard Herrnstein and Charles Murray, *The Times Higher Education Supplement*, 11 November 1994.

[296] Sybil Eysenck, interview, 25 July 2003.

backroom media wheeling and dealing very much in evidence.[297] While Eysenck might have been pleased to still have an audience and an influence, the way some of those in the media and the medical fraternity quickly arranged themselves against him spoke volumes for the negative intellectual karma he had built up over the years. Overall, the controversy seemed to generate more heat than light, with most interpreting the various findings to suggest substantial IQ gains were only possible amongst those suffering significant nutritional deficits. Eysenck continued to take this agreeably biological environmental intervention seriously, more seriously than more conventional compensatory measures. In a *volte-face* of sorts, he even suggested nutritional factors might account for race differences.[298]

Conspiracy theories and the 'pariah effect'

While the nature–nurture issue had a largely conventional disciplinary history, Jensen's re-introduction of race differences was a game-changer, adding a volatile political aspect that could not be ignored. Race was a scientific category like no other, the basis for antipathies that played out in war and genocide, lynching and ethnic cleansing. With a high-stakes moral flavour exacerbated by popular media attention, each side of the race and IQ debate demonized the other. Moderation went by the board. Only those most comfortable with the increasingly caricatured extremes and/or the thickest of hides were willing to risk publicly identifying themselves with either viewpoint.

---

[297] In 1988, David Benton and Gwilym Roberts reported suggestive nutrition-driven IQ gains, and this research was accompanied by substantial media coverage. See Benton, D. and Roberts, G. (1988), Effect of vitamin and mineral supplementation on intelligence of a sample of schoolchildren, *The Lancet* **331**, 140–3. Eysenck subsequently helped organize a bigger follow-up study, with a team led by Stephen Schoenthaler. Their findings suggested modest but significant gains, especially in non-verbal IQ measures. These findings were presented on the TV science show *QED* on the 27 February 1991 as well as in *PAID*. See Schoenthaler, S.J., Amos, P., Eysenck, H.J., Peritz, E., and Yudkin, J. (1991), Controlled trial of vitamin–mineral supplementation: effects on intelligence and performance, *Personality and Individual Differences* **12**, 351–62. A commercial pharmaceutical company chimed in with an ad campaign for a 'smart pill' they had prepared. However, to Eysenck and his team's surprise, immediately following the *QED* broadcast the British Medical Research Council was allowed to issue a statement endorsing the results of other studies it had supported that found no significant effects. For a later addition to this research, see Schoenthaler, S.J. and Eysenck, H.J. (1997), Raising IQ level by vitamins and mineral supplementation, in *Intelligence, heredity, and environment* (ed. R.J. Sternberg and E. Grigorenko), pp. 363–92, Cambridge University Press, Cambridge.

[298] Eysenck (1991*d*), Race and intelligence—an alternative hypothesis.

In coming to Jensen's aid, Eysenck made himself the centre of attention for political extremes. When the militant left singled Eysenck out, their protests helped bring him to the attention of right-wing groups attracted by the alleged racism of his views as well as the image of courageous oracle speaking unpalatable truths. One group's hate figure became the other's saviour.[299] It drew some odd and extreme people to him, making for the very appropriations and associations the watchdogs of the left made so much of. They were the byproduct of a polarizing debate as much as dark conspiracy.

Much has been written about the history and racialist origins of Pioneer Fund of late.[300] In the post-Jensen era, the Fund attempted to reinvent itself as a benevolent provider for mainstream behavioural science. Nevertheless, it appeared to place a priority on funding research into bio-genetically grounded psychological differences, including racial differences. Affiliate journal *Mankind Quarterly* was a favoured destination for many Fund recipients. According to William Tucker, Eysenck was one of seven social scientists the Fund added to its roster in the 1980s and 1990s. Only two of these scientists, Phil Rushton and Richard Lynn, focused directly on racial differences. The work of the rest (Eysenck included) tended to be 'shaded' by an old-school nativism that emphasized the hard-nosed social implications of nature over nurture.[301] Pioneer tended to pick up on scientists with agreeable views and findings and support them to do more of the same. What made Eysenck so attractive, it seemed, was the high profile he had already generated on the race and IQ issue, as well as his public authority and academic prestige.

Whatever the agenda of the Pioneer Fund, one should not assume it was shared by all researchers it supported.[302] Pioneer's largest single research grant

---

[299] This 'pariah effect' also helped draw the attention of related groups like animals rights activists, who daubed his house in anti-vivisectionist graffiti at one point—despite the fact that the rat research he oversaw was relatively small in scale and not the kind of mass caging/dissection work that such groups usually target.

[300] A good place to start would be Kenny, M.G. (2002), Toward a racial abyss: eugenics, Wickliffe Draper, and the origins of the Pioneer Fund, *Journal of the History of the Behavioural Sciences* **38**, 259–83. Pioneer stalwart Harry F. Weyher put up a vigorous defence of the Fund, Weyher, H.F. (1999), The Pioneer Fund, the behavioural sciences, and the media's false stories, *Intelligence* **26**, 319–36.

[301] See Tucker, W.H. (2002), *The funding of scientific racism*, University of Illinois Press, Urbana, pp. 179–82.

[302] For a taste of the furious debate over Pioneer's objectives, see the exchange between Paul Lombardo and current director of the Fund J. Philippe Rushton in Lombardo, P.A.(2002), 'The American breed': Nazi eugenics and the origins of the Pioneer Fund, *Albany Law Review* **65**, 743–830, Rushton, J.P. (2002), The Pioneer Fund and the scientific study of

went to Thomas Bouchard—hardly an extremist—heading the huge Minnesota twin study. The origins of Eysenck's relationship with the Fund followed a similar pattern to that of another outsider lobby group looking for a way in—Big Tobacco. (See the next chapter for more details on Eysenck's dealings with the tobacco industry.) Pioneer representatives particularly welcomed Eysenck's rehash of Jensen's *Harvard Educational Review* article and came to regard him as a potentially useful ally.[303] However, it wasn't until the mid-1980s that Eysenck applied for support from this source. Eysenck used (or intended to use) the money mainly for his biological correlates of intelligence work, allowing him to maintain and extend a programme already running.[304]

Nevertheless, the association between the Pioneer Fund and Eysenck was actually rather short-lived. After concerns about Pioneer were raised at the IoP, an internal review saw Eysenck's funding halted. In 1990, the University of London stepped in to also ban members of affiliated institutions from receiving support from Pioneer. So what the Fund didn't advertise, and many of Eysenck's critics didn't know, was that he was forcibly weaned off Pioneer and made to return some of the money the Fund had allotted him.[305] He had to look elsewhere to get his intelligence research done. Thus Eysenck's research on evoked potentials in the early 1990s cited both Philip Morris and Pioneer as backers.[306]

---

human differences, *Albany Law Review* **66**, 207–62, and the mixed appreciation of Neisser, U. (2004), Serious scientists or disgusting racists? *Contemporary Psychology* **49**, 5–7.

[303] According to Tucker, those associated with the Pioneer Fund may have helped distribute Eysenck's race and IQ book in the US. See Tucker (2002), *The funding of scientific racism*, p. 265 (footnote 128).

[304] Pioneer sources suggested the Fund also helped Eysenck maintain the Maudsley Twin Registry, and supported his collaborative investigations into the heritable basis of intelligence, personality, social attitudes, and crime in the late 1980s and early 1990s. See materials at the Pioneer Fund website: http://www.pioneerfund.org/ and Weyher (1999), The Pioneer Fund, the behavioural sciences, and the media's false stories.

[305] Michael Rutter alludes to this episode in Rutter (2006), *Genes and behavior: nature–nurture interplay explained*, p. 7. Prompted by public relations headaches such as this, large institutions like the IoP now give far greater consideration to the ethics of research funding. Rather than respond on a *post hoc* basis, they have moved to develop guidelines for their researchers. See, for example, http://admin.iop.kcl.ac.uk/randd/downloads/policy_on_external_research_funding.doc.

[306] See the concluding acknowledgements in Barrett and Eysenck (1994), The relationship between evoked potential component amplitude, latency, contour length, variability, zero-crossings, and psychometric intelligence.

As for the 'cavorting with devils' charge, Eysenck welcomed sympathetic contacts amongst his peers after his retirement, just as he welcomed any financial support he could garner. He became closer to a range of scientists from related fields than he ever might have, united as they were by their nativistic viewpoint and shared sense of oppression. Elevated to hero status on the right, Jensen was courted by conservative forces both at home and on the continent, and in turn brought the likes of A.J. Gregor, Roger Pearson, and Billy Shockley to Eysenck.[307] Psychologists Richard Lynn and Eysenck's former student Phil Rushton—both of whom were also controversial figures in their own right—were also key allies. Eysenck did several tours and conferences with Jensen, and frequently conferred with other hereditarian scientists like John Baker, Cyril Darlington, and Oxford physicist Kurt Mendelssohn.[308] In private they would repeatedly use the term 'we', evoking a sense of solidarity and mutual support normally reserved for close collaborators.[309] But one cannot lump these researchers all in the same boat; their science and their political opinions

---

[307] For more on the intersecting careers of Gregor, Pearson, Shockley, and Jensen see the critical works of Billig (1979), *Psychology, racism and fascism*; Tucker (1994), *The science and politics of racial research*, pp. 233–61, and Tucker (2002), *The funding of scientific racism*, pp. 140–79. According to Tucker, leading representatives of the American racialist right saw science as the new road to legitimacy, and their European counterparts took note. Tucker suggested the process was largely one of secondary cultivation from without. Scientists seen as sympathetic were brought together and supported, with the Pioneer Fund especially important in this respect. These moves provided a bridge between the old academic segregationists and New Right think tanks with Washington influence.

[308] Mendelssohn had written to Eysenck saying he was planning, much to his publisher's horror, a book titled 'The basis for White domination'. Eysenck wrote back: 'Your new book sounds very exciting ... No doubt you are boarding up your house ... I hope your book does not meet the same fate as mine; in America SDS (Students for a Democratic Society) went around to bookshops and wholesalers, warning them that if they stocked the book their premises would be set on fire. As a consequence, it is almost impossible to get hold of my book on "Race", or the one on "The Equality of Man" [sic] in the United States.' Hans J. Eysenck to Kurt A.G. Mendelssohn, Section H. General Correspondence, CSAC 93.4.83/H.13, Papers and Correspondence of Kurt Alfred Georg Mendelssohn, Special Collections and Western Manuscripts, Bodleian Library, Oxford University, p. 1.

[309] I am partly basing this point on the correspondence I have cited between Eysenck, John Baker, and Kurt Mendelssohn, as well as interviews with key figures. The way hereditarian scientists used the collective term 'we' conveyed a sense of shared mission and/or a siege mentality in the face of hostile outside forces. While latter-day commentators saw this as an organized intellectual–political conspiracy, hereditarian scientists saw their informal associations simply as mutual support. J. Philippe Rushton, interview, 3–4 July 2002.

differed significantly. In any case, it seemed Eysenck's engagement with right-wing political forces was at this level, never stretching far beyond the traditional borders of science.

In a broad political context, psychologists like Eysenck and Jensen were arguably more important and more useful to the racialist right than it seems these groups were to them. The further to the right one went, the stranger the bedfellows. For example, in order to adopt the likes of Jensen, Eysenck, and even Richard Herrnstein as scientific mascots, neo-fascist groups had to gloss over some of their political views and ethnic heritage. Jensen made no bones of his partial Jewishness and expressed notably liberal views on many social issues, while Herrnstein was, of course, unambiguously Jewish.

However, it appeared the bargain Eysenck struck involved simply not wanting to know. Anyone or anything that could help further his research programme couldn't be all bad. He never wanted to look a gift horse in the mouth—be it Pioneer, Big Tobacco, or some other controversial patron. And he obviously did not ask many questions about the background of some of his associates. While some found Eysenck's blinkered innocence unconvincing, it was in keeping with his detached, hands-off approach to nearly all facets of his research. Thus, if Eysenck can be faulted for the counterproductive effects of this approach when it came to building trust, he could be given the benefit of the doubt when he appeared surprised to learn of the uses being made of his work and the company he was keeping.[310] With nothing more in the way of archival or personal material to go on, I feel obliged to leave it at that.

In any case, Eysenck's emphasis on nature over nurture looked very much like old-fashioned obstinacy, a polemicist still locked in an oppositional framework. Given the ordeal he endured in the public media, Eysenck was at pains to assert that he was right all along, *and right in his original terms*.[311] His intellectual

---

[310] See Winston, A. (1998), Science in the service of the far right: Henry E. Garrett, the IAAEE, and the Liberty Lobby, *Journal of Social Issues* 54, 179–210. Winston told of an encounter with an elderly Eysenck, whereupon Winston furnished Eysenck with a dossier on the activities of some of Eysenck's associates. Winston wrote: '[Eysenck] appeared quite unaware and shocked when I presented him with the details of [Roger] Pearson's career, and he assured me than he would not contribute to the *Mankind Quarterly* again or meet with Pearson in the future (H.J. Eysenck, personal communication, August 2, 1996). He had no difficulties with the material on race in *Mankind Quarterly*, but he considered himself a lifelong opponent of fascism' (p. 198).

[311] Jensen's obsessive focus on race differences—as if determined to salvage his original scientific point one way or another—has perplexed many of his colleagues. Other examples of this phenomenon can be found in Segerstråle (2001), *Defenders of the truth: the sociobiology debate*.

quirks were preserved in the process. At the height of the race and IQ controversy and for some time after, Eysenck would inappropriately relate the heritability of IQ differences to the heritability IQ for a particular individual.[312] He still treated gene–environment interactions in additive terms until the end, as the sum of discrete nature and nurture parcels. He still thought of regression to the mean as some sort of biological law indicating heritability, despite a consensus that it was a statistical phenomenon with no genetic implications.[313] Either Eysenck was exploiting lay misconceptions for other purposes, or he did not fully understand what he was talking about. He had not been involved in any extensive work on the nature and nurture of intelligence since 1979. His last effort, *Intelligence: a new look*, contained more than a faint echo of his first extended treatment of the topic in *Race, intelligence and education*. In it, we still find him repeating the 80% nature 20% nurture dictum from beyond the grave, still implying that those who did not accept this were afraid of the truth.[314]

Racial science

Currently, psychologists and geneticists, anthropologist and sociologists are only slightly closer to arriving at a common template for human subpopulation groupings. The problematic concept of 'race' lies at the very centre of these disagreements. Interestingly, biomedical and population perspectives have merged lately, sparking renewed interest in exploring genetic variation at the level of geographically defined subpopulations. Some structural patterning in genetic variation has indeed been found to be associated with geographical origin.[315] However, any equivalence between genetic patterning and self-ascribed race

---

[312] See Eysenck (1971c), *The IQ argument*, p. 67, or Eysenck and Kamin (1981a), *The battle for the mind*, p. 59.

[313] For more examples of Eysenck' s scientific howlers, see Velden, M. (1999), Vexed variations. Review of *Intelligence: a new look* by Hans Eysenck, *The Times Literary Supplement*, 16 April 1999, 34. For a contrary opinion on this book, see Arthur Jensen's glowing review, Jensen, A.R. (2000), Hans Eysenck's final thoughts on intelligence. Review of *Intelligence: a new look* by Hans Eysenck, *Personality and Individual Differences* **28**, 191–4.

[314] Eysenck (1998), *Intelligence: a new look*, p. 32.

[315] For an overall view on human biogenetic diversity see Cavalli-Sforza, L.L. (2000), *Genes, peoples, and languages*, University of California Press, Berkeley.The watershed paper, however, is Rosenberg, N. *et al*. (2002), Genetic structure of human populations, *Science* **298**, 2381–5, which won *The Lancet* paper of the year prize for 2003. Neither Rosenberg nor Cavalli-Sforza—the father of modern population genetics—tended to use the politically loaded 'r-word'.

has not been shown to be particularly exact and/or reversible.[316] Aware of the immense political sensitivities involved, contemporary researchers have otherwise tended to avoid using the 'r-word'. The spectre of invidious racial profiling even prompted the American Medical Association to issue guidelines in 2003 that cautioned against the use of self-ascribed race as a proxy for genetic variation.[317]

Those of a more philosophical bent have suggested that everyday racial self-ascription will always be misleading, that such labelling perpetuates self-fulfilling social effects that go far beyond any biological reality.[318] For example, in 2003 Massimo Pigliucci and Jonathan Kaplan suggested that truly meaningful biological groupings will likely be smaller and more numerous than their everyday racial counterparts. Given these two category sets will not closely map on to each other, the adduced phenotypic characteristics of biological races will not feed the social effects of everyday race labelling one way or another.[319] Nevertheless, a small group of psychologists—Jensen and Rushton prominent amongst them—have taken this fast-breaking genetic research as a vindication

---

[316] It has long been recognized that, whatever biological definition of subpopulation grouping is employed, the pattern of variation is gradual rather than discrete, with within-group variation tending to overwhelm between-group variation. More recent genetic investigations have attempted to fit structural models to these overlapping patterns. Such work has been far from definitive about the correspondence between self-ascribed race and genetically defined categories. One study that suggested genetic markers could be used to predict self-designation was Tang, H. *et al.* (2005), Genetic structure, self-identified race/ethnicity, and confounding in case-control association studies, *American Journal of Human Genetics* **76**, 268–75. Other research implied that the correspondence is far from robust and reversible. See Barnholtz-Sloan, *et al.* (2005), Examining population stratification via individual ancestry estimates versus self-reported race, *Cancer Epidemiology, Biomarkers and Prevention* **14**, 1545–51. They found 'that significant [genetic] population substructure differences exist that self-reported race alone does not capture' (p. 1545).

[317] Kaplan, J. and Bennett, T. (2003), Use of race and ethnicity in biomedical publications, *Journal of the American Medical Association* **289**, 2709–12. See also Kenny, M.G. (2006), A question of blood, race, and politics, *Journal of the History of Medicine and Allied Sciences* **61**, 456–91.

[318] For a sceptical view on the reality of the biological concept of race, see Tate, C. and Audette, D. (2001), Theory and research on 'race' as a natural kind variable in psychology, *Theory and Psychology* **11**, 495–520.

[319] See Pigliucci, M. and Kaplan, J. (2003), On the concept of biological race and its applicability to humans, *Philosophy of Science* **70**, 1161–72. See also various contributions to a special issue of *American Psychologist*, 'Genes, race, and psychology in the genome era'. Anderson, N.B. and Nickerson, K.J. (eds.) (2005), *American Psychologist* **60**, January 2005.

as they strive to keep the race differences in intelligence controversy alive with a programme still based on self-ascribed designations.[320]

The battle lines in the nature–nurture debate have been redrawn for the new century. Looking back, the strongly moral readings given to Eysenck and Jensen's work have a sepia-toned quality. Particular scientific positions were equated with specific social and political consequences—a view that now looks overly deterministic as well as woefully ahistorical. While genome mapping and the resurrection of evolutionary psychology sparked new fears of biological determinism amongst some veteran critics, their protests tended to be greeted with a sense of weary *déjà vu*. The march of biotechnology has otherwise split the left and the right. Old-guard radicals worried by such things as genetic engineering and cybernetic implants now find they have religious conservatives for company, while left-leaning 'post-humanists' celebrate these developments as an opportunity for enhancement and emancipation, as do many right-of-centre commentators and scientists.[321]

Science and politics may well be linked, but these links are obviously formed and reformed anew across time and place. But even this knowledge has been the subject of political point-scoring, with some conservative hereditarians arguing that this was a lesson only their radical critics needed to learn.[322] Yet in hindsight, it is clear that each side chose to connect different aspects of the science and politics of their opponents' work, rewarded for doing so in a debate that went so far and so fast downstream. No one was above the ideological fray it seemed, least of all the scientists themselves.

And despite all the protests and critical reflexivity, the discipline of psychology remains stubbornly lily-white. In the US in particular, the anxious inward gaze of its individualistic assumptions are still seen as the expression of a

---

[320] Rushton, J.P. and Jensen, A.R. (2005), Thirty years of research on race differences in cognitive ability, *Psychology, Public Policy and Law* 11, 235–94. Acknowledging but dismissing some well-worn critical points, Rushton and Jensen's response to the lack of direct evidence was to broaden the net of plausible pointers to the 'possibility' that some genetic component was involved.

[321] For a recent meditation on these biopolitical issues, including the new, more emancipatory concepts of race and tailored medicine, see Rose, N. (2007), *The politics of life itself: biomedicine, power, and subjectivity in the twenty-first century*, Princeton University Press, Princeton, NJ.

[322] See Nyborg, H. (2003), The sociology of psychometric and bio-behavioral sciences: a case study of destructive social reductionism and collective fraud in 20th century academia, in *The scientific study of general intelligence: tribute to Arthur R. Jensen* (ed. H. Nyborg), pp. 441–502, Pergamon, Amsterdam.

European culture alien if not hostile to African American concerns.[323] Eysenck's race and IQ book addressed the reader as 'you', but wrote about 'negroes' as 'they'. It was a book directed at 'Whites' about 'Blacks'; those of African descent played only marginal roles in the subsequent scientific debate on both sides of the Atlantic. Not much has changed.[324]

Public image politics

Eysenck's race and IQ book was a fateful intervention. While sensitive to public perceptions and uncharacteristically restrained in his responses, he still treated the issue as an essentially intellectual conundrum, a key demonstration of the principle of scientific freedom. But the controversy that little book triggered went well beyond the cerebral confines of journals and data. It brought Eysenck much wider fame, but he and his family paid a high price.

Eysenck remained defiantly upbeat about the effects the controversy had on his reputation. He saw it as legitimating his status as Britain's 'Mr Psychology', despite the harassment he had to endure. He became, he thought, the everyman-rebel the public could identify with, an outsider they could rely on to represent the real message of his discipline. Yet Eysenck was always quick to point out that this was mainstream science all along, claiming he was simply a spokesman for a misrepresented and publicly suppressed majority.[325]

---

[323] See Guthrie, R.V. (1976), *Even the rat was white: a historical view of psychology*, Allyn and Bacon, Needham Heights, Massachusetts. Recent attempts to address the deficit model of Eurocentric psychology include Belgrave, F.Z. and Allison, K.W. (2006), *African American psychology*, Sage, Thousand Oaks, California.

[324] This was point discussed at the time by Mallory Wober. See Wober (1971), Race and intelligence.

[325] Eysenck (1997b), *Rebel with a cause*, pp. 266–8. In many respects, Eysenck was correct. The assumptions and methodology of hereditarian psychometrics *was* mainstream science within psychology. A majority agreed with the major contentions it put forward, including the capacity of IQ tests to measure intelligence and the substantial genetic basis for observed differences. Eysenck's views only became a minority position when it came to racial differences, with most of his peers not convinced that genetics played a role. See Snyderman, M. and Rothman, S. (1988), *The IQ controversy: the media and public policy*, Transaction Books, New Brunswick, New Jersey. Prompted by *The bell curve* debate, Pioneer-funded psychologist Linda Gottfredson, put together a relatively cautious statement to this effect: Gottfredson, L. (1994), Mainstream science on intelligence, *Wall Street Journal*, 13 December 1994. As for race differences, Gottfredson wrote: 'Most experts believe that environment is important in pushing the bell curves apart, but that genetics could be involved too.' (p. A18). The statement was endorsed by 52 experts in intelligence and allied fields—many of whom were also Pioneer-funded—and was later published in *Intelligence*, **24** (1997). See also Neisser *et al.* (1996), Intelligence: knowns

He remained a feted insider within the subspecialities he contributed to or helped set up. To actually judge him an outsider would be to buy into the kind of romantic rebel image he constructed for himself in the wake of the controversy. By the same token, it certainly made him a pariah amongst liberal activist groups within psychology and the elite echelons of the scientific community. Of all the barriers to Fellow of the Royal Society (FRS) status—and there were several—this was the biggest. Two attempts to get Eysenck elected an honorary fellow of the BPS also failed for the same reasons.[326] Being black-balled by the elected representatives of science had its compensations, however dubious. It left him 'free' to explore the wilder fringes of psychological nature and he, in turn, attracted a range of scientific outsiders and eccentrics in his twilight years and retirement. His interests in astrology and parapsychology, personality and cancer took off in the mid- to late 1970s, after race and IQ had left its mark.[327] While he came to these fields late, he made a considerable impact. As ever, Eysenck made much of what was available; given lemons, he made good lemonade.

Above all, race and IQ made sure Eysenck became one of the most divisive figures in the history of modern psychology. Capping-off a host of contentious interventions, it made it hard for anyone—lay person or eminent researcher—to be neutral about him. Eysenck became a political litmus test: being pro or con Eysenck inevitably implied something about one's ideological convictions. Conversely, *his* political views remain enigmatic and contradictory. While he vehemently opposed the Labour Party version of socialism of the 1970s, Eysenck was no fan of Thatcher's policies on science, medicine, and transport.[328] Nonetheless, he objected when critics labelled her government 'fascistic', for it made light of the suffering inflicted upon victims of real

---

and unknowns; Weidman, N. (1997), Heredity, intelligence and neuropsychology; or, why the bell curve is good science, *Journal of the History of Behavioral Sciences* 33, 141–4.

[326] Jeffrey Gray, interview, 25 July 2003; Michael Rutter, interview, 26 April 2002.

[327] Well into his retirement, Eysenck took a fresh look at the psychology of prejudice in tandem with Ronald Grossarth-Maticek, with maverick Australian researcher John Ray also involved in related work. See, for example, Grossarth-Maticek, R., Eysenck, H.J., and Vetter, H. (1989), The causes and cures of prejudice: an empirical study of the frustration–aggression hypothesis, *Personality and Individual Differences* 10, 547–58.

[328] Eysenck (1997b), *Rebel with a cause*, pp. 270–1. Eysenck maintained he had never been a member of a political party. However, he did support Thatcher's government during the crippling miner's strike, drawing a familiar parallel with his experience in Hitler's Germany. See Eysenck, H.J. (1984a), Scargill and the fascists of the left, *Daily Express*, 19 April 1984.

fascism, he said. Eysenck could not abide overt, emotive racism either. Anti-Semitism especially cut him personally—even if he kept a little quiet about it—and he also made a point of distancing himself from those he saw as having appeased Hitler.[329] Soon after he retired in the early 1980s, when the National Front had been pushed back to the margins, Eysenck claimed they had never really read him anyway.[330] Perhaps this was wishful thinking. Implicitly responding to the 'cavorting with devils' charge, he would point to the many communists he had worked with.[331] Backed by his formidable intellect, Eysenck positioned himself above it all, insulated from the political winds that blew others. He was his own counterexample, proof that race differences research did not necessarily equate with fascistic politics. He had lived through and rejected the most famous attempt to link the two. However, it was not the kind of inner certainty he could easily share, and simply demonstrating it made him look arrogant.

Becoming a public intellectual meant constructing some kind of relationship with a general lay audience. Maintaining public trust, that ineffable and transitory quality *moral* capital, required more than just good scientific credentials. Eysenck understood this well enough to shift tack in popular forums, always ready to call on biographical material to underwrite his own neutrality. As the stakes rose, however, this tactic wore thin, depending as it did on repeated exhortations to take him at his word. Eysenck's capacity to sustain the faith of a broad audience, as well as that of his publishers and his peers, was undermined by the combative streak that had served him so well in upstream scientific forums. When it came to staying popular (and the pun is intended), a lighter touch was needed, not to mention a greater generosity of spirit.[332]

---

[329] For example, Nobel prize-winner Konrad Lorenz visited the IoP in 1957. Alerted to Lorenz's pro-Nazi record by Else Frenkel-Brunswik, Eysenck snubbed Lorenz's lecture and refused to meet him. See Gibson (1981), *Hans Eysenck: the man and his work*, pp. 115–16.

[330] Ian Deary quoted him as saying: 'I didn't give any comfort to them [the National Front]. They never read me. They couldn't care less [about scientific facts].' Deary (1985–6), Eysenck at nightfall, p. 29.

[331] See Eysenck (1997b), *Rebel with a cause*, p. 101.

[332] Admittedly this would be a hard ask amidst the trench warfare of the race and IQ controversy. Others managed it a little better, however. The folksy humour of Stephen J. Gould, arch-enemy of creationist scientists and the hereditarian right, allowed him to get away with some thinly researched polemics and fairly personal attacks. Towards the end of his life and after his death, Gould's popular work has been slammed as biased and inaccurate by former opponents and their sympathizers. In part, this illustrated the constant need for upkeep a popular reputation demanded. See again the critique in Nyborg

He made too many enemies and became too divisive a figure to remain a figurehead. After 'that little book', Eysenck's career as a spokesman for mainstream psychological science stalled. New versions of his test-yourself 'amusements' still sold well, and a popular text he penned with son Michael was moderately well-received.[333] However, he and Penguin parted company after 1972 and no follow-up series of generalist paperbacks would emerge.[334]

Towards the end of his life Eysenck attended his last International Society for the Study of Individual Differences meeting, 21 July 1997. His health failing, he was quietly wheeled in at the back of the room. As word of his presence filtered forward, those in attendance rose to give him a spontaneous standing ovation knowing this would probably be the last time they would see him in any professional capacity.[335] According to Jeffrey Gray, there was a real sense of adulation and gratitude in the room the likes of which he could only compare to the reception accorded another battle-scarred champion at an APA conference, one Leon J. Kamin.[336] So there they were: venerated by their colleagues, united as opponents. Both had fought the good fight as leading advocates of their respective faiths.

Perhaps Kamin was right: psychological science *was* insuperably political. But this was politics in a narrower, rarefied stratum. What I have laboured long and hard to highlight was a particularly strategic level to the us-versus-them confrontations Eysenck engaged in. From political attitudes to race and IQ,

---

(2003), The sociology of psychometric and bio-behavioral sciences: a case study of destructive social reductionism and collective fraud in 20th century academia.

[333] Eysenck, H.J. and Eysenck, M.W. (1981), *Mindwatching*, Michael Joseph, London.

[334] Eysenck's 1972 book *Psychology is about people* was his last Penguin title. Aside from the test-yourself books, Eysenck wrote several other books for general audiences. They tended to be on off-beat or relatively specialized topics, however. There were the paranormal titles: Eysenck, H.J. and Sargent, C. (1982), *Explaining the unexplained: mysteries of the paranormal*, Weidenfeld and Nicholson, London; Eysenck, H.J. and Sargent, C. (1986), *Are you psychic?* Prion, London; and, Eysenck, H.J. and Nias, D. (1982), *Astrology—science or superstition?* Maurice Temple Smith, London. Then there were his middle-brow books on sex and personality: Eysenck, H.J. (1976e), *Sex and personality*, Open Books, London, and Eysenck, H.J. and Wilson, G.D. (1979), *The psychology of sex*, Dent, London. Finally, there were his popular works that combined elements of test-yourself and self-help: Eysenck, H.J. (1983d), *... I do. Your guide to a happy marriage*, Multimedia Publications, London, and Eysenck, H.J. (1977e), *You and neurosis*, Maurice Temple Smith, London.

[335] See Ian Deary's emotional account of this occasion at the University of Aarhus, Denmark, in Deary, I. (1997b), Tribute to Hans Eysenck, *Personality and Individual Differences* 23, 713–14.

[336] Jeffrey Gray, interview, 25 July 2003.

Eysenck took on what he saw as the suffocating left-wing orthodoxy of post-war social science, typecasting its representatives as the misguided do-gooders and schoolyard bullies of their time. In doing so, one can infer a sly bait-and-switch arc on Eysenck's part that was more competitive, more *personally* political than he—or even Kamin—would care to allow.

Eysenck fell into this consuming oppositional role as much by accident as design. Clearly, though, the emphasis on the political pathology of the right in the reconstruction years did not sit well with his boyhood memories of the streets of Berlin. The subsequent hammering he took from Christie and Rokeach both confirmed his view of 'liberal' social science and left him looking for ways to even up. Race and IQ provided just such an opportunity—an issue too controversial to resist, tangential to his main focus but a logical extension of it nonetheless. By this reckoning, Eysenck was not so much a scientific reactionary as a man who kept score, keeping his friends close and his enemies even closer in mind.

Chapter 9

Smoking, cancer, and the final frontier

Time has a curious way of passing judgement on the actions of men and women, a capricious capacity for turning out heroes and villains after the fact. Time may well be relatively kind on Hans Eysenck and his race and IQ intervention. Future science may well bear out the biogenetic basis of cognitive differences, much more so than his pronouncements on smoking, cancer, and personality. The increasingly dogmatic medical outlook on the dangers of smoking has been woven with modern day tales of the evils of Big Tobacco and the bought advocate, typecasting those involved to the right or wrong side of history. However, a careful reconstruction of this final chapter in the Eysenck story—*sans* hindsight where possible—should provide an antidote to any rush to judgement.

Much of this story takes place during the latter part of Eysenck's life and involves one of his last collaborations. For over 30 years, Eysenck claimed that medical scientists and the public health lobby had got it wrong. They had greatly exaggerated the ill effects of tobacco. In their haste to condemn smoking, they had ignored the possibility that genetic and personality factors were largely to blame. Increasingly pilloried for this heterodoxy, Eysenck held out for evidence that would back his position.

In the early 1980s, Yugoslav sociologist Ronald Grossarth-Maticek offered Eysenck a lifeline. Grossarth-Maticek's data strongly suggested that personality coping styles greatly affected the course of various physical diseases. In a series of papers published in the late 1980s and early 1990s, Eysenck and Grossarth-Maticek reported a variety of results that demonstrated a strong association between particular personality types and cancer, coronary heart disease, and other ailments. Although these were mostly write-ups of studies set in train by Grossarth-Maticek more than a decade earlier, Eysenck's input helped fine-tune the presentation, theoretical explanations, and analyses. A number of intervention studies were also carried out, each suggesting that particular forms of psychotherapy that targeted cancer sufferers or unhealthy personalities could have remarkable therapeutic or preventive effects.

As in the race and IQ brouhaha, a raft of questions arise from Eysenck's actions. Why did he go out of his way to defend a suspect product like cigarettes in the face of mounting evidence? In the past, critics have pointed vaguely but darkly to his links with the tobacco industry, implying his views were influenced by the money he received. Why too did he take up with the notorious Grossarth-Maticek, a man short on support and allies within the scientific community? This scepticism only intensified with Eysenck's unabashed promotion of Grossarth-Maticek's research in the English-speaking world. A diverse chorus of opinion suggested Grossarth-Maticek's astonishing results had little credibility. So, where did this leave Eysenck? He was Grossarth-Maticek's most ardent and prominent supporter, the co-author of many papers reporting these results.

Eysenck's sceptical stance on the connection between smoking and illness appeared relatively unremarkable in the beginning. However, his continued defence of this stance, the strategies he invoked, and the criticism he took, all had a way of drawing him toward someone like Grossarth-Maticek and his wondrous results. Their fateful embrace saw the Eysenck story go from the controversial, touch on the scandalous, and, finally, descend into the tragic.

Smoking, personality, and disease

Eysenck wrote that his interest in the topic of smoking and health, and the disease-prone personality, began when 'I had shown that cigarette smoking was significantly correlated with personality, specifically with extraversion, and followed a lawful, predictable course.'[1] In what could be seen as simple extension of his biosocial perspective, Eysenck took seriously the ancient folk wisdom linking temperament with physical health. Eysenck had become a smoker as a student in London, a habit that was cemented during the war, and one that escalated during his 1949 sojourn in the US. When he returned to England, Eysenck began to notice the effect his pack-a-day smoking was having on his sense of well-being and fitness, not to mention his hip pocket. Moreover, as a good behaviourist, he felt ashamed to be a slave to this habit. So Eysenck promptly—and effortlessly he said—gave up cigarettes. Thus he claimed he rid himself of the weed before the link between smoking and disease had begun to be widely advertised by medical authorities in the mid-1950s.[2] Pity those who

---

[1] See p. 297 of Eysenck, H.J. (1991c), Reply to criticisms of the Grossarth–Maticek studies, *Psychological Inquiry* **2**, 297–323.

[2] See Eysenck, H.J. (1997b), *Rebel with a cause* [revised and expanded], Transaction Press, New Brunswick, New Jersey, Chapter 5, and his introduction to Eysenck, H.J. (1965b), *Smoking, health and personality*, Basic Books, New York.

did not, he added, for alarmist health warnings would make it an increasingly guilty pleasure.

Concern about the adverse health effects of smoking can be traced all the way back to the importation of tobacco to Europe. These warnings became commonplace in Western countries but were not taken seriously until the late 1940s, when it became clear that lung cancer was increasing dramatically—and not just as a function of changing patterns of diagnosis.[3] With acute infectious diseases on the wane, chronic conditions like cancer and heart disease loomed as a new mountain to climb for modern medicine. Cancer posed a special challenge, however, with its long latencies and potentially multiple aetiologies. Research into causes had mainly focused on individual susceptibility, possibly as some kind of imperfection at the cellular level mediated by host (i.e. genetic) propensity. Nevertheless the steep rise in the incidence of lung cancer saw a number of environmental triggers also fall under suspicion—smoking and atmospheric pollution amongst them.

Teasing out causality from a multitude of candidate factors would not be easy. While experimentation with human subjects was not practical, the use of animals was also fraught. It was not clear whether humans and animals reacted to carcinogens in the same way or whether cancers developed in the same manner. Appropriate exposure methods were also a problem. At the time, epidemiology was a young and fragile ancillary field, typically utilizing cross-sectional studies of morbidity to shed light on the spread of infectious conditions but not on their causes. All this would change by the early 1950s when the results of several ambitious longitudinal studies were published—most notably those of Ernest Wynder and Evarts Graham in the US and A. Bradford Hill and Richard Doll in the UK. Collectively, this data raised the possibility of a strong link between smoking and lung cancer.[4]

Hill and Doll's work was particularly influential, with their ground-breaking 1950 report setting a new standard in case control methods. Supported by the Medical Research Council (MRC), Hill and Doll's 1954 follow-up helped shift

---

[3] See Burnham, J.C. (1989), American physicians and tobacco use: two Surgeons General, 1929 and 1964, *Bulletin of the History of Medicine* **63**, 1. For an update on the fate of lung cancer treatments see Timmermann, C. (2007), As depressing as it was predictable? Lung cancer, clinical trials, and the Medical Research Council in postwar Britain, *Bulletin of the History of Medicine* **81**, 312–34.

[4] For an excellent historical account of this debate, see Talley, C. *et al.* (2004), Lung cancer, chronic disease epidemiology, and medicine, 1948–1964, *Journal of the History of Medicine and Allied Sciences* **59**, 329–74.

scientific opinion and made the Ministry of Health take notice.[5] It also spurred those in the British tobacco industry into defensive action. Initially the industry funded research into the causes of cancer with donations to the MRC. Dissatisfied with the results, British tobacco formed the Tobacco Research Council (TRC) in 1956, originally designated the Tobacco Manufacturer's Standing Committee. This industry-representative group was entrusted to perform 'research on smoking and health ... and make information available to scientific workers and the public.'[6] However, the overriding and obvious function of the TRC was to fight science with science and, at a public relations and legal level, to combat warnings about the dangers of smoking.

By the late 1950s, biochemical analyses and animal experimentation began to reinforce the smoking–cancer link repeatedly suggested by epidemiological research. Most scientists were convinced cigarette consumption could at least partly explain the upsurge in lung and bronchial cancers and the statistical profile of heart disease. Even so, the epidemiological field remained a contested space. The distinguished, pipe-smoking statistician R.A. Fisher was not alone in doubting whether such studies could demonstrate causality. In his 1955 trip to Berkeley, Eysenck said he mused on Fisher's suggestion that genetic factors might underlie the relationship between smoking and disease. Since Eysenck was sure personality had a basis in genetics, he reasoned that personality might be an intervening variable in this link. Those with a certain personality might be predisposed to smoke and might also be predisposed to get lung cancer and other diseases. Eysenck was subsequently approached by the social research group Mass Observation to do media work on smoking and they, in turn, had been sounded out by the TRC to study smoking patterns. From 1958, Mass Observation did a series of studies directed by Eysenck that suggested there were 'genotypic differences' between smokers and non-smokers. Smokers scored higher on extraversion though not necessarily on neuroticism.[7] Eysenck claimed that later studies revealed that the smokers, especially women, did tend to be more neurotic.[8]

---

[5] Doll, R, and Hill, A.B. (1950), Smoking and carcinoma of the lung: preliminary report, *British Medical Journal* 2, 739.

[6] Review of past and current activities, Tobacco Research Council, January 1963, http:// tobaccodocuments.org/ctr/11297751-7780.html, p. 5.

[7] Eysenck, H.J., Tarrant, M., and Woolf, M. (1960), Smoking and personality, *British Medical Journal* 11, 1456–60; Review of past and current activities, Tobacco Research Council, p. 16.

[8] Eysenck (1997b), *Rebel with a cause*, Chapter 5.

From this work, Eysenck went on to look at smoking-related diseases. He was 'lucky enough to meet' Dr David Kissen, an oncologist who ran a chest clinic in Edinburgh and a psychosomatic research programme at the University of Glasgow.[9] Kissen also did research financed by the TRC and appeared at conferences and conventions that the British industry helped organize. Collaborative work with Kissen demonstrated an association between cancer and personality, with low scores on neuroticism associated with cancer, and high scores with the absence of cancer.[10] Kissen suggested this bore out the old idea that cancer was associated with the suppression of emotion or a difficulty in its expression. Kissen died in the mid-1960s of heart disease and Eysenck cast about amongst medical researchers for new collaborators. No one was willing to work with him, Eysenck said, or even let him test patients. It seemed Kissen's approach did not have a great deal of support at Glasgow, and his psychosomatic research group went with him.

In Britain, the Royal College of Physicians had intervened in the debate in 1959. In 1962, they issued their first report pointing out the dangers of smoking. Smoking was linked specifically to lung cancer, and it was also implicated in other forms of cancers and coronary heart disease. The US Surgeon General followed suit in 1964. Smoking was now officially regarded as a health risk—even if the carcinogenic action of cigarette smoke was still to be thrashed out. Given that the statistical data could not meet the highest standard of medical proof (i.e. Koch's postulates), a new standard of causality had been introduced in tandem with the burgeoning field of chronic disease epidemiology.[11] The scientific debate was all but over, but the legal and PR war between the tobacco companies and the public health lobby had only just begun. Hans Eysenck came in at the tail end of this scientific debate and made little impact. However, Eysenck would revisit the topic time and time again over the years, recapitulating the old doubts and adding a few of his own to become a significant player in this new phase of the tobacco wars.

To this end, he ramped-up his public profile on this issue in 1965 with the book *Smoking, health and personality*. Written in an accessible manner, presumably for the educated lay person, *Smoking, health and personality* was a triumph of discursive even-handedness. The book was tailored to a public

---

[9] Eysenck (1991c), Reply to criticisms of the Grossarth–Maticek studies, p. 297.

[10] Kissen, D.M. and Eysenck, H.J. (1962), Personality in male lung cancer patients, *Journal of Psychosomatic Research* **6**, 123–7.

[11] Harkness, J.M. (2006), The US Public Health Service and smoking in the 1950s: the tale of two more statements, *Journal of the History of Medicine and Allied Sciences* **62**, 171–212.

debate that had already begun to polarize. With his distinctive blend of provocation and equivocation, Eysenck presented his version of the history of the dispute, his interest in it, and the cases for and against the link between smoking and disease. He argued that the causal role of cigarettes in various diseases, specifically lung cancer, had not been convincingly proven. Eysenck did not present a great deal of his own data; he had done little relevant research. The evidence he did present was tied to the suggestion that host factors (i.e. genetics and personality) might play a mediating role in the smoking–disease link.[12] Eysenck remained cautious, however, for available data did not appear to bear this hypothesis out. While extraverts did tend to smoke more, they seemed only slightly more likely to get cancer or heart disease. 'It cannot by any stretch of the imagination, be held responsible for the total effect', Eysenck concluded.[13] Neither could neuroticism explain the link, since cancer victims appeared to be less neurotic while smokers, if anything, were more so. Even so, Eysenck could point to many holes, inconsistencies, and unexpected patterns in the epidemiological data, just as Fisher had.[14] Most of the research was retrospective, vulnerable to recall and reporting problems. Mortality statistics were unreliable and demonstrated historical fluctuations. Finally, smokers and non-smokers could not be considered to be matched groups; they might systematically differ in ways other than just tobacco use. Alternative or multifactorial causation of cancer was thus a possibility, deflecting the blame away from smoking.

Overall, *Smoking, health and personality* did still give considerable credence to the ill-effects of smoking, though in a most ambiguous way. For example, Eysenck assigned his review of the experimental animal research to the section outlining the case against the link between smoking and disease, with the section detailing the case for the link dominated by a review of the statistical/data. In doing so, he gave the impression that the experimental evidence was overwhelmingly negative. This was especially evident in his lengthy summary of Mayo Clinic epidemiologist Joseph Berkson's critical review of the evidence. But in an intriguing endnote to this section, Eysenck indicated that he thought Berkson's account was selective and incomplete. Eysenck went on to say that, while other research did not 'give completely convincing direct evidence of a

---

[12] It was a hypothesis Eysenck had rehearsed two years before. See Eysenck, H.J. (1963), Personality and cigarette smoking, *Life Science* **3**, 777–92.

[13] Eysenck (1965*b*), *Smoking, health and personality*, p. 118.

[14] See Fisher, R.A. (1959), *Smoking: the cancer controversy*, Oliver and Boyd, Edinburgh, and Stolley, P.D. (1991*a*), When genius errs: R.A. Fisher and the lung cancer controversy, *American Journal of Epidemiology* **133**, 416–25.

relationship, it cannot be denied, I think, that they do strongly support the statistical evidence'.[15] No doubt Eysenck might have defended his editorial decisions based on the format of his book, but this important point was buried in a footnote.

In his final summing up, given in mock legal style, Eysenck opined that the case against smoking was 'strong'. Nevertheless, he warned that the evidence was only circumstantial rather than causally direct. 'Clearly smoking is not likely to promote anyone's health... However, in science, as in law, the accused is presumed innocent unless proven guilty...'[16] With a hint of self-congratulation, Eysenck gave smoking the benefit of the doubt.

Eysenck's studied agnosticism in *Smoking, health and personality* gave him plenty of rhetorical wiggle-room and laid logical booby-traps for opponents to stumble into. Moreover, the message was layered to the point where, if necessary, it could be sharpened to further provoke or toned down to be eminently defensible. In self-consciously balancing the evidence pro and con, Eysenck created the impression of reasonable doubt while apparently avoiding a fixed position himself. Of course, Eysenck did take a position, at least implicitly, in his selection and interpretation of a diverse range of evidence, but it was buried in ostentatious displays of even-handed discussion. What was left to dispute were his grounds for rejecting any judgement as premature or unwarranted, and his authority and judgment in doing so.

While *Smoking, health and personality* did cause a stir, its rebel halo took on a brighter hue in hindsight. When it first appeared it got many favourable reviews and sold quite well. Moreover, Eysenck was not alone amongst scientists in questioning the evidence. His views were echoed by a number of more tendentious popular writers, notably C. Harcourt Kitchin in his 1966 book *You may smoke*.[17] Together these works helped promote the public perception that warnings from medical authorities were alarmist and misleading. While

---

[15] Eysenck, H.J. (1965*b*), *Smoking, health and personality*, p. 157, note 11.

[16] Ibid., p. 151.

[17] Apart from the dissenting scientists already mentioned, such as Fisher and Berkson, see Brownlee, K.A. (1965), A review of 'Smoking and health', *Journal of the American Statistical Association* **60**, 722–39. Later publications questioning the link included Seltzer, C.C. (1967), Constitution and heredity in relation to tobacco smoking, *Annals of the New York Academy of Sciences* **142**, 322–30; Seltzer, C.C. (1972), Critical appraisal of the Royal College of Physician's report on smoking and health, *The Lancet* **300**, 243–8; and, Burch, P. (1974), Does smoking cause lung cancer? *New Scientist*, 21 February 1974, 458–63. For popular writers, see Kitchin, C.H. (1966), *You may smoke*, Award Books, New York, and the notorious industry-sponsored pieces by Stanley Frank: Frank, S. (1968), To smoke or not to smoke, *True Magazine*, 15 January 1968. and (under the

Eysenck disagreed with the certainty expressed by the British and American medical communities, his cautiously dissenting message provoked little response because the scientific debate was largely seen as settled while the anti-smoking public health campaign had barely got off the ground.

Eysenck gave the impression that the controversy surrounding *Smoking, health and personality* was enough to ensure that the TRC stopped funding his work.[18] This is an odd point, since it might be assumed they would welcome any such stir. In any case, tobacco industry documents reveal that the TRC continued to fund Eysenck's research, such as that on the positive, stimulating effects of nicotine, until the summer of 1970.[19] By this stage other, more lucrative funding avenues had opened up.

Eysenck published very little research or commentary in this area again until the 1980s.[20] Even so, he continued various programmes aimed at elucidating the hidden variables in the smoking–disease link, especially genetic and personality factors. While Eysenck happily disclosed his funding source as the tobacco industry, he did not care to detail his deep involvement with the powerful American companies.

Big Tobacco, big money

> When the law is not on your side, try the facts; when the facts are not on your side, argue a point of law; when neither facts nor the law is on your side, try the plaintiff. (Legal defence maxim)

> The key defence strategy [for the tobacco industry] is to try the plaintiff (Crist, P.G. et al. (1986), Re: Jones/Day Liability Summary ('Corporate Activity Project'), http://tobaccodocuments.org/landman/37575.html)

It is not easy to trace how Eysenck became involved with the American tobacco industry. It is clear, however, that those within the American industry were well aware of him and had been monitoring his writings on smoking and

---

pseudonym Charles Golden) Golden, C. (1968), Cigarette cancer link is bunk, *National Enquirer*, 3 March 1968.

[18] Eysenck (1997*b*), *Rebel with a cause*, Chapter 5.

[19] Tobacco Research Council, Review of activities 670000—690000, http://www.tobaccodocuments.org/lor/00500025-0085.html.

[20] The most notable piece Eysenck published in this period proposed a model of smoking behaviour. See Eysenck, H.J. (1973*c*), Personality and the maintenance of the smoking habit, in *Smoking behaviour: motives and incentives* (ed. W.L. Dunn), pp. 113–46, Winston/Wiley, Washington.

health since the early 1960s.[21] Just as clear is the trail of money Eysenck had begun to receive from the Americans, beginning in the late 1960s.

Like their British counterparts, the American-based tobacco giants battled to turn the negative tide against smoking. However, the American companies were engaged in an even tougher fight in their domestic market; they had to endure intense scrutiny in Congress and the never-ending threat of litigation from those who had used their products. It made the actions and attitudes of US industry leaders more ruthlessly instrumental and, crucially, these attitudes and actions were harnessed to immense resources. The Americans were quick to seize on anyone or anything that might prove useful in shoring up their legal-cum-commercial position.

In an attempt to influence public opinion and gain legal purchase, those within or associated with the American tobacco industry began to monitor scientific debates and develop a role in them. Beginning in the mid-1950s, they sponsored outside research projects and set up some in-house research programmes. Yet when it came to science, they quickly realized that any rapacious acquisition of evidence and experts had to be tempered with a consideration of what made science trustworthy and persuasive. To be legally and politically bullet-proof, scientific research had to be seen to be independent.

In January 1954, the US tobacco industry responded to adverse publicity with a widely publicized 'Frank statement to cigarette smokers'. At the time, this statement was more or less in tune with mainstream scientific opinion, namely, that it had not been scientifically established that smoking caused diseases like cancer. After 1964, however, only the tobacco industry insisted that the question was still unresolved. In order to keep the appearance of controversy alive the American industry went out of its way to sponsor and promote those few scientists still prepared to dispute the smoking–disease link. One of the main ways of doing so was through the Council for Tobacco Research (CTR), originally the Tobacco Industry Research Council. Set up in 1954 by five of the six American companies, the CTR and its Scientific Advisory Board (SAB) was made up of non-industry scientists and was supposed to be hands-off. Within certain parameters it usually was. Most researchers were given a free rein once they received the go-ahead, but they had to pass through the eye of a needle to do so. Industry representatives routinely passed over any researcher with a negative attitude on the health issue and screened out

---

[21] See Henry H. Ramm to James I. Erickson, 27 July 1965, http://www.tobaccodocuments.org/rjr/500887047-7048.html.

'dangerous project proposals'.[22] Moreover, they monitored the work subsequently done, halting some projects (e.g. central nervous system research), and censoring and harassing the few who did uncover and attempt to publish damaging results (e.g. Dr Frederic Homberger). Over the years they learnt to hide research that had the potential for trouble, by funding them 'off the books'.[23]

For many years (at least since the early 1960s) American tobacco industry leaders privately acknowledged that they could not disprove the case against smoking; an aggressive counter-attack was never on. Instead, industry representatives threw up their hands over conflicting evidence. While reluctantly fulfilling their duty to warn and to investigate, their main tactic was to find ways to muddy the scientific waters that made the see-no-evil public stance more credible. It was the scientific arm of a much bigger PR effort that critics have since dubbed a 'disinformation machine'.[24] The CTR was ostensibly intended to resolve the 'open question' on tobacco and health but was, in the words of founding SAB chairman Dr Clarence Little, actually created to build a research foundation to 'arrest continuing or future attacks'.[25] Thus some in-house and CTR research was conducted on a need-to-know basis; however, most of it was carried out because it might serve to cloud or complicate the causal link between smoking and disease. It was a tactic they executed brilliantly, but it was a time bomb. It bet against the results of future research and it assumed that industry confidentiality could be maintained indefinitely.

The American companies' relationship with Eysenck started in the late 1960s, soon after *Smoking, health and personality* was published. Up to the mid-1960s, the American industry had been content to fund domestic researchers. Early victories in smoking health litigation had led to a sense of complacency. Industry leaders were given a rude shock in the mid-1960s, however, when the negative assessment of the Surgeon General was quickly followed by congressional hearings on possible health warnings. Legal counsel for the industry were in dire need of expert witnesses to testify that association was not equal to causation and that you can't extrapolate from animals to humans.

---

[22] One of the most definitive reports documenting the strategies the industry used and the precarious legal position that the industry eventually found itself in is Crist, P.G. *et al.* (1986), Re: Jones/Day Liability Summary ('Corporate Activity Project'), http://tobacco-documents.org/landman/37575.html.

[23] See Hilts, P.J. (1997), *Smokescreen: the truth behind the tobacco industry cover-up*, Addison-Wesley, Reading, Massachusetts, Chapter 2.

[24] See Ibid.

[25] Crist *et al.* (1986), Re: Jones/Day Liability Summary ('Corporate Activity Project'), p. 18.

However, the industry had no relevant research to present to congress. Tobacco research had become unpopular; regular journals no longer published it, and government funding and CTR applications had dried up. Thus, CTR Special Projects was born in 1966, narrower in scope than the regular CTR grant programmes, and usually directed toward specific legal, and political applications.[26] Industry insiders actively solicited researchers and projects, with the underlying aim being the development of expert witnesses. Since it was difficult to find scientists who did not think tobacco was harmful, CTR Special Projects served as a valuable recruiting tool. New witnesses were always needed to supplement the same old roll call that judges tired of and prosecuting advocates could learn to dissect.

Eysenck became part of the CTR team in 1969. Noted American cancer researcher and CTR member Dr Arthur Furst helped secure Eysenck's initial $10, 000 Special Projects grant for an investigation of tumour induction in 'emotional' versus 'non-emotional' rats. Eysenck had wanted to explore personality theories in cancer, but was advised at the time that the SAB was not very sympathetic to psychological theories of cancer. Nevertheless, Furst's opinion was that there was 'still a long shot that this work may prove useful'.[27]

Eysenck thus entered a truly nether world of science, full of highly partisan interests, heavy secrecy, and unspoken obligations. But the Americans surely held out one great attraction to him: they had boatloads of money, especially for the right kind of research. While Eysenck might have recognized the potentially corrupting effects of tobacco money, he must have felt immune in some sense. For Eysenck, it was simply business as usual, merely a shift in funding source. As he commented to Furst in 1970, 'we have built up an excellent research unit with unusual facilities for experimental work on nicotine, smoking etc. [and] it would seem a pity if this had to be disbanded.'[28]

The relationship was maintained by both sides, mostly at a respectful distance, given the *quid pro quo* involved. Eysenck took many proposals to the American tobacco industry during the 1970s and early 1980s. It appeared that many, if not most, were accepted and supported. Much of this work was funded as part of CTR Special Projects, as well as under the regular CTR banner. The projects were typically related and often overlapping. One large

---

[26] Ibid., pp. 142–3.

[27] Arthur Furst to David R. Hardy, 24 December 1969, http://tobaccodocuments.org/tplp/MNATPRIV00013508.html, p. 1.

[28] Hans Eysenck to Arthur Furst, 20 May 1970, http://www.tobaccodocuments.org/ctr/CTRSP-FILES008794-8794.html, p. 1.

project that commenced in the early 1970s was based on data derived from the twin registry Eysenck had set up. With the help of Oxford biometrician Lindon Eaves, the idea was to investigate the 'inheritance of the smoking habit'.

As well as funding for specific research, Eysenck received additional funds from the American tobacco industry that indicated he had been earmarked for an exceptional role. From the late 1970s on, Eysenck received generous payments from special account #4. These special accounts were a secret source of funds operated by industry lawyers. They were used to fund projects that otherwise did not gain approval or fit CTR guidelines, projects judged to be primarily of legal (as opposed to scientific) value. And because these accounts were controlled by the lawyers, the work they financed could be retained as a privileged legal product. This avoided unwanted disclosure to the Federal Trade Commission and made results 'non-discoverable'. In other words, these special accounts allowed those in the industry to hide research from public view, where it was expedient, *especially* from plaintiff lawyers. Using this scientific evidence in court, incidentally, meant placing favourable results in the hands of an expert witness, for legal counsel could not present such evidence themselves.

Eysenck was paid at an annual rate of $US 19, 000 in 1977 from special account #4. Payments were progressively boosted up to $40, 000 in the years following to at least 1983, for what was described as 'consultative research'. It is difficult to ascertain just what this amounted to but it appeared that Eysenck gave assistance of a strategic nature, advising the industry as to what research might be worth doing and building contacts with willing researchers. Eysenck was enlisted to help set up research projects in Europe and the US, some of which were 'strictly litigation-oriented'. He helped devise studies done by students trained by him, such as Kieron O'Connor.[29] Noted American psychologist Charles Spielberger also received special account #4 industry funding in his attempt to replicate the twin-study and personality correlates reported by Eysenck. The tobacco industry specifically commissioned Spielberger's study to head off any legal challenge that Eysenck's findings might not apply in

---

[29] 'Hans J. Eysenck—Special Account #4 Project,' 20 October 1993, http://tobaccodocuments.org/pm/2025502365-2369.html; 'CTR Special Projects,' 2 February 1979, http://tobaccodocuments.org/bliley-pm/20975.html. For some of the published results, see Eysenck, H.J. and O'Connor, K. (1979), Smoking, arousal and personality, in *Electrophysiological effects of nicotine* (ed. A. Redmond and C. Izard), pp. 147--57, Elsevier, Amsterdam.

an American context.[30] Special account #4 also helped finance Eysenck's incidental expenses. He was treated to numerous expenses-paid trips to speak at conferences that were organized by the industry or on industry-relevant topics. A considerable largesse was involved; subsidized Concorde air travel, for example, was not out of the question.

Special account #4 also bankrolled Eysenck's late 1970s survey on alternative satisfactions sought by smokers who gave up, and later twin-study investigations of the relationship between heart disease, smoking, and stress. By the early 1980s, Eysenck had also begun advising the industry on personality factors and psychological motivations behind smoking, the better to construct and market their product. For instance, Eysenck suggested ways to position various brands to appeal to introverts or extraverts, and how the characteristics of these cigarettes could be matched to their target market.[31]

In return, Eysenck was to appear as a key witness in several court cases and political hearings—including the 1983 'Waxman' US congressional hearings.[32] He consistently testified that the link between smoking and disease was still an open question, a link possibly moderated by biogenetic factors. A number of books and papers by Eysenck that utilized research supported by American tobacco all helped promote this message.

Through most of this period Eysenck's industry 'handler' was Ed Jacob, a lawyer with a background in science. Jacob was a key legal strategist for R.J. Reynolds, working at the New York law firm of Jacob, Medinger, and Finnegan. Although Eysenck was supported by a range of industry sources, he reported directly to Jacob. While Jacob helped steer CTR Special Projects, he also played a key role in vetting proposals funded by the secret special accounts.

---

[30] For an account of this work and Spielberger's positive opinion of Eysenck's research, see Spielberger, C. (1986), Smoking, personality and health, in *Hans Eysenck: consensus and controversy* (ed. S. Modgil and C. Modgil), pp. 305–15, Falmer Press, Philadelphia.

[31] See for example, J.D. Weber, 'Meeting with Dr. Hans Eysenck,' 9 July 1982, http://tobaccodocuments.org/rjr/505004727-4730.html. Eysenck had discussions with industry representatives about brand image promotion for different personality types and how it might relate to cigarette composition. For instance, he suggested that introverts needed a stronger initial nicotine 'hit' that should decline with continued puffing. Conversely, extraverts smoked to avoid boredom and thus needed to get an increasing 'hit' from each successive puff.

[32] See Hearings Before the Subcommittee on Health and the Environment of the Committee on Energy and Commerce House of Representatives Part 2 of 3, 9 March 1983, http://tobaccodocuments.org/lor/03637300-7630.html, and Eysenck's, 'Lung cancer, coronary heart disease and smoking', Statement Regarding S. 772 Submitted to the Committee Labor and Human Resources, 6 May 1983.

In this capacity Jacob seemed to play a significant role overseeing the research Eysenck did. From notes of meetings it appeared Eysenck would present ideas to Jacob and Jacob, in turn, would comment on their desirability from an industry point of view. In these meetings, Eysenck demonstrates his keenness to explore alternative factors (i.e. genetics and personality) that might explain the smoking–cancer link. Here and in public, Eysenck maintained that nicotine was not a drug of dependence but one that met pre-existing needs exacerbated by environmental stresses.[33] Although Eysenck generally disclosed his financial sources for specific research, he did not detail this 'special' avenue of tobacco industry funding and the level of consultative intimacy that went with it.

An intellectual cart and a legal horse

The first of Eysenck's books that detailed CTR-sponsored research was *The causes and effects of smoking*, published in 1980.[34] The first half briefly rehashed *Smoking, health and personality*. However, the detached tone of Eysenck's earlier book had given way to more tendentious argument. Eysenck again pointed to the flaws in the epidemiological evidence. He noted the provisos the Royal College of Physicians should have acknowledged in their 1977 report, despite the fact that this report gave serious attention to the constitutional hypothesis. Eysenck focused almost exclusively on the evidence and arguments that disputed the smoking–disease link, notably those of University of Leeds's Philip R.J. Burch, who also advanced a constitutional hypothesis, and Harvard's Carl C. Seltzer, who was also supported by CTR funds. Eysenck also reviewed the evidence linking personality, smoking, and disease to develop a set of hypotheses for the new research recounted in the latter part of the book.

The second part of *The causes and effects of smoking* presented twin data analyses done with Lindon Eaves. This was serious scientific research, densely packed with quantitative, biometric data. Although some attempt was made to explain and summarize the methods and measures used, it was clearly aimed at presenting sufficient data to convince fellow scientists, rather than a simplified review aimed at the educated lay-person. In so doing, Eysenck put more on the line: not just his capacity to explain and critique the research of others but also his reputation as a scientist capable of redirecting the cutting edge of research. Eysenck and Eaves's data seemed to suggest there was a significant

---

[33] Edwin Jacob, 'Notes of discussion with Dr. Eysenck—10/22/76,' 11 November 1976, http://tobaccodocuments.org/bliley_bw/521029803-9808.html.

[34] Eysenck, H.J. (with Lindon Eaves) (1980c), *The causes and effects of smoking*, Maurice Temple Smith, London.

genetic factor in smoking uptake and maintenance, but left considerable room for environmental factors as well. Moreover, Eysenck and Eaves had to admit that previous assumptions about smoking and personality were not strongly supported. There was little 'evidence that neuroticism makes any overall contribution to differences between smokers and non smokers', while the role of the extravert personality 'does not emerge in our data.'[35] Only the Eysenck Personality Questionnaire's P (psychoticism) and Lie scale scores seemed to relate to smoking behaviour. Indeed, the addition of psychoticism as a dimension of personality appeared to 'have led to the incorporation in P of some of the elements of E [extraversion] which previously mediated the correlation with smoking.'[36] As George Hill noted, *The causes and effects of smoking* demonstrated that personality and genetics might be implicated in smoking, but 'not strongly'. There was 'nothing in this to skittle them down at the Royal College…'[37], Paul Kline agreed. The evidence in *The causes and effects of smoking* left Eysenck's hypothesis of a factor predisposing both smoking and illness as viable, and that was all.[38] While Eysenck had concluded that some form of interaction of genes and environment was the most promising model for future investigations, details that went beyond this sketch were still to be filled in.

Other reviews of *The causes and effects of smoking* were generally more negative than those for his earlier work, the scientific and public health landscape having changed considerably. The biochemical mechanisms underlying the smoking–cancer link were still not entirely clear. Researchers had produced skin carcinomas in mice and rats with isolated components of tobacco smoke— even though they still had trouble agreeing on exposure methods and animal models. Fearing the worst, the tobacco industry embarked on a concerted attempt to discredit animal research, for it might be harder to argue against than the statistical data. It was not until the 1990s that scientific opinion closed around experimental evidence *directly* demonstrating human-type lung cancer caused by whole tobacco smoke. However by 1980, most scientists with an interest in the area believed the epidemiological evidence had clinched the causal connection between smoking and disease long ago. In any case, the public health risk was of such a magnitude that absolute proof was deemed

---

[35] Ibid., p. 314.

[36] Ibid.

[37] Hill, G. (1980), Review of *The causes and effects of smoking* by Hans Eysenck (with Lindon Eaves), *The Times*, 12 December 1980.

[38] Kline, P. (1981), No smoking. Review of *The causes and effects of smoking* by Hans Eysenck (with Lindon Eaves), *London Review of Books*, 19 February–4 March 1981.

a luxury. Certainly medical authorities on both sides of the Atlantic saw the evidence as sufficient to warrant action, and the public education campaigns were gathering pace. For example, the British Medical Association and Ministry of Health had condemned smoking since the mid-1960s, and the government had progressively restricted tobacco advertising.[39]

Eysenck downplayed the animal studies in *The causes and effects of smoking*. He also ignored many more recent epidemiological studies in the relatively brief first section of the book. More revealingly, his point of view had shifted from interested bystander to frank advocate, one more implicated in the debate. With an aggressively anti-smoking message now an integral part of the public health agenda, Eysenck now had the look of a quixotic figure tilting at the windmills of orthodoxy. The net effect of *The causes and effects of smoking* was to redouble the negative chorus from health spokespeople. For example, the passionate Richard Peto gate-crashed the press conference for Eysenck's book, chiding him for ignoring contradictory evidence, including the important follow-up review he and Sir Richard Doll had published in 1976.[40] Eysenck explained uncertainly that it was too late at the time of publication to include this evidence and that he had included very recent studies favourable to his point of view only because he had received pre-prints. According to tobacco industry observers, those present found this explanation unconvincing; in fact, it was greeted with howls of laughter. Yet Peto went on to plead with the media to not report on Eysenck's 'heresy', an invocation for suppression that the *New Scientist* labelled 'injudicious' and 'unworthy'.[41] On a subsequent BBC radio talk show, Peto continued to attack Eysenck, with Eysenck on the defensive: 'I have no doubt smoking is not a healthy habit. All I am saying is ... that I am not completely convinced by the evidence ... [of] causal factors to the extent that is suggested.'[42] Eysenck might be seen to be rhetorically retreating but his

---

[39] No form of public health policy action directed at the tobacco companies came into play in Britain until the mid-1980s. See British Medical Association, Public Affairs Division (1986), *Smoking out the barons: the campaign against the tobacco industry*, Wiley, Chichester.

[40] See Doll, R. and Peto, R. (1976), Mortality in relation to smoking: 20 years' observations on doctors, *British Medical Journal* **2** (6051), 1525–36.

[41] *New Scientist* (1980), Smoking out censorship [editorial], *New Scientist* **88** (18/25 December 1980), 756.

[42] Quoted in Susan von Hoffman to Joan Mebane, 3 March 1981, http://tobaccodocuments.org/ness/1855.html, p. 2.

high profile on the airwaves still helped to promote the idea of disagreement between experts—just the sort of thing Peto feared.[43]

Eysenck can't have enjoyed being the butt of jokes. He was now seen as quite out of step with the public health lobby—those he likened to a 'mafia' spreading 'propaganda'—and was now very visibly locked into a position that questioned the smoking–disease link.[44] However, he lacked good counterevidence. Support for his alternative genetics-and-personality hypothesis was hardly overwhelming, undermined by merely suggestive or disappointing results.

By the early 1980s, Eysenck was emphasizing the notion of environmental stress as well as personality in the maintenance of the smoking habit, also pointing to other research linking stress with the aetiology of cancer.[45] For Eysenck, smoking was a function of boredom and emotional strain. With their lower level of cortical arousal, extraverts smoked to spice up what for them were non-stimulating environments. In contrast, neurotics were characterized by higher levels of emotional lability, and they tended to smoke to cope with what for them were stressful circumstances. Finally, those high in P had their smoking reinforced by a need to express non-conforming social tendencies. The model was complicated by evidence of the biphasic properties of nicotine: while small amounts of nicotine raised cortical arousal, large amounts tended to suppress autonomic nervous system activity. Thus (somewhat neurotic) introverts might smoke heavily to calm themselves and reduce social anxiety.

Nevertheless, there was still much work to be done and Eysenck was nearing the end of his professional career. Retirement loomed; the connections and privileges of a professorship were about to disappear. Already marginalized by the race and IQ controversy, access to research funds would only become more difficult. Moreover, lengthy and large-scale longitudinal research must have seemed impractical at his age.

Eysenck had apparently heard of the Yugoslav researcher with longitudinal evidence linking personality and disease soon after the publication of *The causes and effects of smoking* in 1980. At this stage, Grossarth-Maticek was

---

[43] On another occasion, Eysenck came in for criticism from David Simpson, director of anti-smoking pressure group ASH, that almost amounted to ridicule: 'Frankly he [Eysenck] ought to turn to geography next. By missing out similar research there he could prove the earth is flat. It is honestly on that sort of level when we deal with this book.' BBC Radio I, 'Newsbeat', 11 December 1980.

[44] See, for example, Fletcher, C. (1981), Plea for the guilty, *Times Higher Education Supplement*, 16 January 1981, and Jill Turner, 'Mind if I Smoke?' *Books*, 1 January 1981.

[45] Eysenck, H.J. (1984c), Stress, personality and smoking behaviour, in *Stress and anxiety*, Vol. 9 (ed. C.D. Spielberger *et al.*), pp. 37–49, Hemisphere, Washington.

largely unknown in the English-speaking world. Within European scientific circles, especially the medical and psychological fields in Germany and the Netherlands, Grossarth-Maticek had an ambiguous reputation. A few thought him a visionary, but many distrusted him. According to Eysenck, he found a man at 'the end of his tether'—a dedicated and charismatic researcher with radical ideas and the courage to explore them.[46] Empathizing with Grossarth-Maticek's status as a scientific heretic, Eysenck portrays himself as duty bound to assist a man who through no fault of his own had run out of friends and money.

The Heidelberg connection

> Dr. Grossarth-Maticek is a gifted and creative researcher... Rarely have I been so impressed by a young research investigator in this difficult field. (Dr George F. Solomon, Professor of Psychiatry, UCSF and UCLA Schools of Medicine, 15 November 1976—Letter of reference for Ronald Grossarth-Maticek)

Today Ronald Grossarth-Maticek performs consultative work and research for pharmaceutical companies and is affiliated with the European Centre for Peace and Development, part of the University of Peace established by the UN in Costa Rica. Living in a fine house overlooking Heidelberg Castle in Germany on the steep embankments of the Neckar river, he has come a long way from the town of Sobor in war-torn former Yugoslavia where he grew up. Precise details of Grossarth-Maticek's qualifications and work history are hard to pin down, however, and still subject to dispute. He was born in 1940 and educated at the University of Belgrade. Grossarth-Maticek received a D.Phil. in sociology from the University of Heidelberg in 1973 with a thesis dealing with student behaviour.[47] Psychology was the second discipline in his Rigorosum, the oral defence made to faculty. It was not clear whether he had extensive formal training in psychotherapy or much experience in the construction of personality questionnaires. However, he soon began working in the field of psychosomatic epidemiology, a notoriously difficult new field.

The relationship between affective states and physical disorders had been debated since antiquity. Integrating mind into medicine had particular appeal in the US, however, where psychosomatic medicine developed as a distinct entity mid-century that bordered on internal medicine, psychiatry, and psychology. Inspired by psychoanalysis, key figures such as Helen Flanders

---

[46] Eysenck (1991c), Reply to criticisms of the Grossarth–Maticek studies, p. 298.

[47] These details come courtesy of the CV Grossarth-Maticek supplied, translated for me by Hermann Vetter.

Dunbar and Franz Alexander investigated the bodily effects of emotional reactions and the chain linking psychological stimulus and organic end-result. Greater medical specialization in the 1960s and 1970s tended to marginalize attention to mind–body interactions. Internal medicine opted for a strictly material, reductionist approach, while psychiatrists pursued non-analytic psychotherapies, drug regimes, and neurobiological research. Psychosomatic research was increasingly performed by psychologists, who shifted the focus from the exploration of unconscious conflict to psychometric studies of maladaptive behaviour patterns. This approach was extended to the psychological concomitants of ailments like heart disease, hypertension, and stroke, as well as cancer—still seen as essentially somatic conditions, notwithstanding their complex aetiologies.[48] And, following the example of chronic disease epidemiology, one way of getting a handle on these psychological concomitants was through longitudinal surveys.

After Grossarth-Maticek moved to Germany in the 1970s, he was employed as a research assistant in the Institute of Social Medicine for a short time and then worked as a private scholar.[49] Most significantly, Grossarth-Maticek initiated a very ambitious prospective longitudinal research programme in 1973. In the spring of 1977, he turned up at the University of Heidelberg psychology department to present a 100-page manuscript for the purpose of his Habilitation—a German degree equivalent to a second doctoral title and a prerequisite for a full professorship. When he was informally told by the review panel of Professors Amelang, Weinert, and Wottawa that his Habilitation application would be refused, he requested the manuscripts be returned, citing a need to distribute them to other interested parties. One member of the panel, Manfred Amelang, refused to give his up because he had made comments on his copy.[50]

According to Amelang, the document was rejected largely because the claims made were so extraordinary. They seemed 'too good to be true'—a judgement that has dogged Grossarth-Maticek ever since. Written in German and prepared in his own institute, the manuscript detailed a prospective study

---

[48] See Theodore M. Brown's general account 'The rise and fall of American psychosomatic medicine,' at http://www.human-nature.com/free-associations/riseandfall.html.

[49] Even these simple details are difficult to definitively ascertain. According to Manfred Amelang, the University of Heidelberg wrote to him on 3 December 2004 to say they had no record of Grossarth-Maticek's employment as a research assistant in this period or any other. Grossarth-Maticek may have been employed privately by Professor Blohmke. Manfred Amelang, personal communication, 15 January 2008.

[50] Manfred Amelang, interview, 1 February 2003.

undertaken in Crvenka, Yugoslavia from 1965. Using a sample of 1353 people, predictions of future mortality were made with the aid of the 88-item Ronald Grossarth-Maticek (RGM) questionnaire. The questionnaire measured factors like reactions to stress and the inhibited expression of psychological needs. Based on a series of analyses, Grossarth-Maticek was able to predict 38 cases of fatal cancer. When follow-up data were obtained ten years later, 37 of 38 had indeed died of cancer, a strike rate of 97.3%, 'even 100% if suspected cancer diagnosis is included.'[51] The strike rate for coronary heart failure was also a remarkably high 92.1%. Moreover, the RGM questionnaire used an unusual system of differential weighting to achieve this predictive success, with individual item weights varying from 1 to 100. No other psychological measures were mentioned in the Habilitation document.

The review panel at Heidelberg was left in a quandary as to what to make of Grossarth-Maticek. Apart from the stunning results, they were concerned about an insufficient separation of the predictor and follow-up data. Knowledge of predictions might have in one way or another influenced the gathering of mortality data (or even vice-versa). The following year, the panel advised Grossarth-Maticek to deposit the predictor scores of his ongoing Heidelberg studies in order to head off the criticism of criterion contamination. Grossarth-Maticek did so, registering names and a subject coding list, but not before the first follow-up had been done. Some but not all predictor data were also deposited. However, there was no complete, official registration of the personality predictor scores.

In the late 1970s and early 1980s, Grossarth-Maticek struggled to get English-language versions of his ideas and research in peer-reviewed journals. Nevertheless, he appeared at various conferences outside Germany and started to become better known in the Netherlands, the UK, and the US. Various aspects of the Yugoslav study were published in the early 1980s; none detailed the typological approach that would characterize his later publications with Hans Eysenck. One of the first English-language publications on the Yugoslav study was in the relatively obscure journal *Psychotherapy and Psychosomatics*. The focus was on two major psychological factors, rationality and anti-emotionality,

---

[51] Grossarth-Maticek, R. (1977), Social scientific aspects in the aetiology of organic diseases: perspective, method, and results of a prospective study, Habilitation thesis, University of Heidelberg. [Translated by Manfred Amelang, University of Heidelberg, 1977. Original German title: Veröffentlichung im Rahmen des interdisziplinären Forschungsprojektes sozialwissenschaftlicher Onkologie.], p. 23.

and life events causing chronic hopelessness and depression.[52] Using the RGM questionnaire (that had grown to 109 items), it was reported that high rationality and anti-emotionality scores predicted coronary heart disease, stroke, and cancer, while life events leading to hopelessness and depression increased the risk of cancer.

Some time in the early 1980s Eysenck became interested in Grossarth-Maticek's work, having read the 1980 *Psychotherapy and Psychosomatics* piece. According to Grossarth-Maticek, Eysenck was intrigued and excited by his work. He had immediately phoned Grossarth-Maticek, and promptly turned up in Heidelberg with the express view of joining forces. However, there is no concrete evidence that Eysenck became involved with Grossarth-Maticek before 1984.

The reasons behind Eysenck's interest in this work were obvious enough, given that Grossarth-Maticek's work appeared to reinforce his own ideas in the area of personality and disease. Grossarth-Maticek had demonstrated that easily measured individual differences in personality were perhaps *the* primary variable, combining and interacting with other risk factors (e.g. smoking, drinking, and diet) to determine long-term disease propensities. These prospective studies made more detailed and specific the kind of speculation and arguments he had long engaged in, and then some. According to Grossarth-Maticek, it also pleased Eysenck greatly to be able to work with and admire someone in Germany, contrasting with the distant relationship he had experienced with his former Nazi father.[53]

Reconciling the differences and doubts

When interviewed about their work together, Grossarth-Maticek was at pains to stress that this was not a blind embrace; Eysenck came in with his eyes open to the scepticism Grossarth-Maticek's work had aroused. Their collaboration therefore had a critical condition, Grossarth-Maticek recalled. Eysenck demanded: 'You must let me check your data, for if you deceive me I will never forgive you.'[54] To this end, Eysenck proposed 'three steps to trust'. The first was to check mortality data for some extreme cases, such as when a whole family was recording as having died of cancer, by visiting homes of next of kin. The second was to match Grossarth-Maticek's mortality data with that of the

---

[52] Grossarth-Maticek, R. (1980a), Psychosocial predictors of cancer and internal diseases: an overview, *Psychotherapy and Psychosomatics* **33**, 122–8.

[53] Ronald Grossarth-Maticek, interview, 31 January-1 February 2003.

[54] Ibid.

City of Heidelberg records. The final check was to compare the subject name and code assignment lists that had been deposited with the mayor of Heidelberg in 1978 with the data files used in subsequent analyses.[55]

There were also some quite significant differences in the backgrounds and intellectual world-views of Eysenck and Grossarth-Maticek that had to be reconciled, minimized, or glossed over. Grossarth-Maticek's ideas grew out of a European psychodynamic tradition. While his views on the role of personality in the genesis or promotion of disease were not dissimilar to Eysenck's, they were not wholly in accord as to what kind of personality characteristics mediated which particular diseases. For example, Eysenck and Kissen's work had suggested a personality profile of cancer sufferers as low in neuroticism and the expression of emotion. Grossarth-Maticek's cancer type presented a very different picture, with those expressing feeling of hopelessness and depression being more prone to cancer, and more neurotic, if anything. However, the uncertain and contradictory results Eysenck had obtained on this score probably made him receptive to a new theoretical understanding of relationship he was still sure existed. As he said to others when questioned about his uncritical support for Grossarth-Maticek: 'I like the data, and I like the theory.'[56]

According to Grossarth-Maticek, Eysenck had very little say over the formulation of personality type concepts. Indeed, Eysenck had struggled to banish typological concepts in favour of continuous dimensions for most of his career. However, the kind of synergistic thinking that dominates the Crvenka and Heidelberg projects, wherein risk factors combine to greater effect than in isolation, clearly predated Eysenck's involvement. Nevertheless, Eysenck emphasized quasi-experimental intervention as a way of teasing out causality; he pushed the empirical aspects of the project at the cost of minimizing its theoretical underpinnings; and, he fine-tuned and polished the presentation of the work so that some of the more glaring methodological mistakes (and the more outrageous claims said the critics) no longer appeared. Eysenck was even able to adjust to and adjust Grossarth-Maticek's approach to psychotherapy. Grossarth-Maticek initially termed it 'social psychotherapy' and later changed this to 'creative novation (behaviour) therapy'.[57] For Eysenck, Grossarth-Maticek's

---

[55] Ibid.

[56] Henk van der Ploeg, interview, 11 February 2003.

[57] Grossarth-Maticek, R. (1980*b*), Social psychotherapy and course of the disease: first experiences with cancer patients, *Psychotherapy and Psychosomatics* **33**, 129–38.

approach was acceptable insofar as it was a reformulation of psychoanalytic ideas in 'meaningful learning terms'.[58]

It was a curious collaboration, by turns appearing one-sided, team-based, or that of mentor–student. It must be remembered that Eysenck played no role in the initiation of these studies, nor had much influence over the process of most of the data gathering. During the peak of their collaboration in the late 1980s, Eysenck would visit Heidelberg for a few days, several times a year. In Grossarth-Maticek's telling of the story, Grossarth-Maticek would do most of the talking, expounding on his ideas and the data supporting them. Eysenck would listen closely, asking lots of critical questions until he could finally satisfy himself that things were sufficiently clear. It was not just a matter of taking and translating ready-made results, however. Eysenck would suggest analyses given the data that existed and even suggested that certain variables be more systematically explored. There was ample opportunity to select, tease out, or redirect attention—given a data set that was apparently sprawling and chaotic, but also rich and ambitious. It included many personality and physiological measures taken over a long period, accompanied by much personal information. Some of the analyses were done by Eysenck's statisticians in London and some by statistical consultant Hermann Vetter in Heidelberg. However, from the mid-1980s, Eysenck did virtually all the writing for publication in English and presumably exerted a strong editorial control.

Eysenck first co-authored an article with Grossarth-Maticek on the Crvenka data in 1985, accompanied by a piece rehashing his doubts about orthodox theories linking smoking and disease. In the years to follow, a bewildering set of studies and measures was progressively brought forth and published. Yet at no stage were these reports greeted with widespread acceptance and acclaim.

Big Tobacco hedges its bets

Tobacco industry scientists had been well aware of Grossarth-Maticek's work. It would be no exaggeration to say that many were extremely sceptical. For those surprised that the cigarette companies did not embrace him, one could emphasize that there were risks for them, risks magnified by their pariah status. Not only were they wary of the charge of buying results and advocates, they also appeared to value only research that was not 'attackable'. So even the tobacco industry had stayed clear of Grossarth-Maticek for the most part, despite the seemingly favourable message his work conveyed. R.J. Reynolds' chief scientist, Frank Colby, had such a low opinion of this work that he had

---

[58] Ronald Grossarth-Maticek, interview, 31 January-1 February 2003.

made no attempt to meet Grossarth-Maticek. When he finally did in 1980, he came away more sceptical than ever, 'unsure whether to feel sorry for him or completely distrust him… I definitely recommend against any involvement between Dr. Grossarth-Maticek and us, or the industry in general.'[59] Scientists associated with Phillip Morris, Tom Osdene and William L. Dunn, apparently shared similar opinions.[60]

At first, only the *Verband der Deutschen Ziagarettenindustrie*—a German body roughly equivalent to the CTR—offered Grossarth-Maticek some assistance. However, by the early 1980s, they were reviewing their funding arrangements and only agreed to support a further study with the assurance that Professor Jan Bastiaans was involved. Bastiaans was a somewhat controversial Dutch psychiatrist, famously treating holocaust survivors with unorthodox forms of psychotherapy and LSD. At the time he was probably Grossarth-Maticek's strongest supporter outside Germany. Internal *Verband* documents from 1984 reveal that industry representatives were especially concerned about the integrity of his data.[61] They allude to unorthodox methodological practices, such as minimizing missing data by filling in the gaps in a subject's record by matching them with an otherwise similar subject.[62] The results of this practice were spotted in 1993 by Bastiaans' former research assistant Henk van der Ploeg, who later added that this was 'bad news for trustworthiness of the data'.[63]

It appears that Eysenck's prestige and authority, and his keenness to collaborate with Grossarth-Maticek was enough to paper over the doubts and bring R.J. Reynolds on board in 1985. Still, they had to be dragged kicking and screaming to fund their collaborative efforts. Those within the tobacco industry

---

[59] Frank Colby, Report on a trip to Germany, 1 May 1980, http://tobaccodocuments.org/rjr/503794792-4795.html, p. 14.

[60] See W.L. Dunn to T.S. Osdene, The Grossarth-Maticek paper on the psychosomatics of cancer, 2 February 1982, http://tobaccodocuments.org/pm/1000039191-9193.html.

[61] See Minutes of VDC Scientific Committee Meeting, 17 October 1984, http://tobaccodocuments.org/pm/2023539765-9768.html, p.4.

[62] See for example, J.H. Robinson to A. Wallace Hayes, Articles authored by R. Grossarth-Maticek,' 12 December 1985, http://tobaccodocuments.org/rjr/504231829-1830.html.

[63] For van der Ploeg's critiques (authored in conjunction with Hermann Vetter and Wim Chr. Kleijn) and Eysenck's reply (with contributions from Grossarth-Maticek), see van der Ploeg, H. and Vetter, H. (1993), Two for the price of one: the empirical basis of the Grossarth-Maticek interviews, *Psychological Inquiry* **4**, 65–6 [quote on p. 66]; van der Ploeg, H. and Kleijn, W.C. (1993), Some further doubts about Grossarth-Maticek's data base, *Psychological Inquiry* **4**, 68–9; and, Eysenck, H.J. (1993*b*), Reply to van der Ploeg, Vetter and Kleijn, *Psychological Inquiry* **4**, 70–3.

could not help notice the inconsistency of Eysenck's position. When asked to comment on the two 1985 articles on smoking, personality, and health, J.H. Robinson and D.G. Gilbert at R.J. Reynolds commended Eysenck's clarity in pointing to the flaws in studies that linked smoking and cancer, and admitted the attraction of a theory relating personality to cancer. Yet they were 'disturbed by Dr. Eysenck's total acceptance' of Grossarth-Maticek's data. 'While the author [Eysenck] spends about 80% of his time stressing the pitfalls of assuming causation based on correlational data, he seems to make this leap in logic himself in several places.'[64]

Running almost parallel to Eysenck's involvement with Grossarth-Maticek was that of a re-analysis project. The tobacco companies agreed to support Eysenck only if a 'rival' team was set up to check Grossarth-Maticek's data set and the conclusions drawn from it.[65] Clearly, industry representatives were covering themselves. Grossarth-Maticek was obliged to share data and furnish details of his methods and the analyses. Led by Charles Spielberger, who had been introduced to Grossarth-Maticek by Eysenck, the team also included the more critical van der Ploeg and Boston University psychologist Bernard Fox. While Grossarth-Maticek saw this move as an act of bad faith, he cooperated with the re-analysis team (albeit rather reluctantly according to van der Ploeg). However, the critical reports of various members of the re-analysis team made sure that this tobacco industry patronage was relatively short-lived.

The Eysenck influence

Once Hans Eysenck came on board, the direction of Grossarth-Maticek's research appeared to shift appreciably. A 1987 article in the *European Journal of Psychiatry* introduced English-speaking audiences to their notion of psychological types.[66] Even though it centred on the Crvenka data, Eysenck was, oddly, the sole author. Four types were outlined: (1) understimulation; (2) overarousral; (3) ambivalence; and, (4) personal autonomy. Gone was the prediction of disease from (weighted) item scale scores that characterized

---

[64] J.H. Robinson and D.G. Gilbert to A. Wallace Hayes, Interoffice memo, 10 July 1985, http://tobaccodocuments.org/rjr/504226599-6600.html, p.2.

[65] See Charles D. Spielberger and Henk van der Ploeg, Proposal for further review and evaluation of the Grossarth-Maticek studies, 6 January 1986, http://tobaccodocuments.org/rjr/515805043-5047.html.

[66] Eysenck, H.J. (1987*b*), Personality as a predictor of cancer and cardiovascular disease, and the application of behaviour therapy in prophylaxis, *European Journal of Psychiatry* 1, 29–41. Grossarth-Maticek had published his typological theories in German the previous year.

Grossarth-Maticek's first write-ups of the Crvenka data, despite the unusually successful results achieved. Gone also was any discussion of the concept of rationality and anti-emotionality, and the importance of depression and hopelessness tended to be de-emphasized in favour of personal autonomy. However, the type-approach managed to bring Grossarth-Maticek's ideas in line with contemporary psychosomatic work, especially the cognitive–behavioural typologies emanating from the US. Grossarth-Maticek's type 1 roughly matched up with the type C personality identified by Lydia Temoshok as cancer-prone. His type 2 corresponded with the type A personality that Meyer Friedman, Ray Rosenman, and David Jenkins had earmarked as prone to heart disease. Sure enough, follow-up mortality analyses revealed that those with type 1 and 2 personalities had a far greater likelihood of dying of cancer or coronary heart disease, respectively. The inventory used to assess type membership was not published till 1988, and it appeared that it functioned as the basis for structured interviews. Items were very wordy and cumbersome, which often necessitated explanations from interviewers. Groups of 11 or more items formed subscales for each type. They were yes/no in format, with the highest subscale score (augmented by interviewer judgement) typically used to assign subjects to type.[67]

An expanded six-type approach was used to analyse the more extensive Heidelberg data. In the six-type approach, types 1 and 2 were again conceptualized as cancer- and heart disease-prone, type 3 was linked with psychopathic behaviour, and type 4 was regarded as healthy and autonomous. Eysenck and Grossarth-Maticek added two new types: type 5 was characterized by (the return of) rationality and anti-emotionality, and type 6 by anti-social behaviour and drug addiction. Again strong findings in relation to predicted mortality were put forward, backing up the pivotal role personality had in the course of physical disease and destructive behaviour. The inventory that assessed these six types was not published until 1990, despite first being used in 1974 (or 1973).[68] It contained 182 yes/no items that were far more direct and succinct than the four-type inventory, with more sophisticated rules for assigning type membership. Eysenck and Grossarth-Maticek also claimed the need for a more satisfactory psychometric device prompted the construction of the six

---

[67] Grossarth-Maticek, R., Eysenck, H.J., and Vetter, H. (1988), Personality type, smoking habit and their interactions as predictors of cancer and coronary heart disease, *Personality and Individual Differences* **9**, 479–95.

[68] Grossarth-Maticek, R. and Eysenck, H.J. (1990), Personality, stress and disease: description and validation of a new inventory, *Psychological Reports* **66**, 355–73.

type inventory in the early 1970s—even though it appeared Grossarth-Maticek had only limited feedback on his earlier measures at that time.

However, not all were convinced by the data and the conclusions drawn. Re-analysis group member Bernard Fox was one of the first to go public with his doubts in 1988, stating that he could not believe in the validity of the Crvenka data he had been shown.[69] Others in the Spielberger-led group, such as Henk van der Ploeg, also began making disbelieving noises, as did medical epidemiologist and statistician P.N. Lee in his industry-commissioned analyses of the data. This was enough for the tobacco companies; in 1988, they took the decision to cease sponsoring Eysenck's work with Grossarth-Maticek, citing a need to shift research directions.[70] While this was not the end of Eysenck's relationship with the American tobacco industry, it appeared to be the end of the industry's dealings with Grossarth-Maticek.

1991: a sceptical peak

Eysenck was unbowed by the flak his collaboration with Grossarth-Maticek had begun to attract and completed another short book on smoking and health in 1991.[71] Again he dismissed the epidemiological evidence, suggesting that by itself smoking was a relatively weak risk factor. Only when combined with the more potent effects of unhealthy personalities and life stress did it become dangerous. For Eysenck it followed that only those with high-risk personalities— perhaps a quarter and no more than a third of the population—should be pinpointed for interventions that changed their lifestyle and habits. Those with healthy personalities could puff away with relative impunity. Eysenck's emphasis on the combined effects of personality and stress allowed him to add a new line of attack on the insidious anti-smoking 'mafia', one that he had first suggested in 1989.[72] He argued the public health campaigns were unlikely to have a positive net effect. Even if they helped reduce smoking, they might

---

[69] Fox, B.H. (1988),Psychogenic factors in cancer, especially its incidence, in *Topics in health psychology* (ed. S. Maes *et al.*), pp. 37–55, Wiley, Chichester.

[70] A. Wallace Hayes to Hans J. Eysenck, 28 January 1988, http://tobaccodocuments.org/rjr/509882911-2911.html.

[71] Eysenck, H.J. (1991g), *Smoking, personality and stress: psychosocial factors in the prevention of cancer and coronary heart disease,* Springer-Verlag, New York.

[72] Grossarth-Maticek, R. and Eysenck, H.J. (1989), Is media information that smoking causes illness a self-fulfilling prophecy? *Psychological Reports* **65**, 177–8. See also Eysenck, H.J. (1986c), Consensus and controversy: two types of science, in *Hans Eysenck: consensus and controversy* (ed. S. Modgil and C. Modgil), pp. 375–98, Falmer Press, Philadelphia.

unnecessarily stress those who didn't manage to quit with potentially fatal consequences.

In contrast, Eysenck claimed the best way to lower cancer and heart disease rates was by giving those at risk a brief course of psychotherapy that targeted personality change and stress reduction. Here and in other articles he produced data that apparently demonstrated this point. Grossarth-Maticek's 'creative novation behaviour therapy' was able to markedly improve the life expectancy of those with cancer- or heart disease-prone personalities, and even greatly improve the prospects of those already suffering from these diseases.[73] Even an abbreviated form of 'bibliotherapy', which amounted to written material accompanied by a short session of explanation and coaching, could produce excellent results. Other data from Eysenck and Grossarth-Maticek suggested that those identified with unhealthy personalities could, even by their own efforts, move their personality styles in the healthy direction. When they did so, they significantly improved their long-term health outlook.

These were remarkable results—given that the targets for change were relatively enduring behavioural characteristics, outcomes of genetic predispositions. Eysenck and Grossarth-Maticek's use of psychotherapy had taken up a promising new research line in cancer care, largely initiated by Stanford's David Spiegel. Using a controlled intervention design, Spiegel's 1989 *Lancet* article suggested psychotherapy could have strong positive effects for those suffering from metastatic breast cancer.[74] Given the attention and hope this work generated, it was hardly surprising that many other researchers jumped aboard and set similar studies in train. What made Eysenck and Grossarth-Maticek stand out, though, was how quickly they were able to do likewise.

Reviewers of Eysenck's 1991 book opined that, if such work could be independently replicated, 'it would cause us to revise substantially all our current notions of personality, disease and the relation of mind to body. The results

---

[73] Therapy intervention studies had been outlined in a number of previous articles by Grossarth-Maticek. However, in 1988 this topic was put front and centre with a number of pieces by Eysenck *et al.* in various forums. See for example, Eysenck, H.J. (1988*a*), Personality as a predictor of cancer and cardiovascular disease, and the application of behaviour therapy in prophylaxis, *British Journal of Clinical and Social Psychiatry* **6**, 4–12. A series of articles in 1991 in Eysenck's high-impact journal *BRAT* gave them extensive, prominent coverage. For a start, see Grossarth-Maticek, R. and Eysenck, H.J. (1991), Creative novation behaviour therapy as a prophylactic treatment for cancer and coronary heart disease. I. Description of treatment, *Behaviour Research and Therapy* **29**, 1–16.

[74] Spiegel, D., Bloom, J.R., Kraemer, H.C., and Gottheil, E. (1989), Effect of psychosocial treatment on survival of patients with metastatic breast cancer, *The Lancet* **334**, 888–901.

he reports are that extraordinary.'[75] It is impossible not to discern an undercurrent of extreme scepticism here; however, only those closest to the work tended to articulate the reasons for their doubts explicitly.

This scepticism reached a crescendo that year in a special issue of *Psychological Inquiry*. Eysenck presented a target article outlining results of prospective studies in Heidelberg (and Crvenka), the predictive power of the four- and six-type approach, and the effectiveness of psychotherapeutic intervention.[76] Over a dozen senior researchers were invited to comment, including leading figures Temoshok and Spiegel. Almost all were critical or extremely critical. Yet it is extremely difficult to summarize their critiques succinctly. Like the proverbial blind men feeling the elephant, they each focused on different parts of the sprawling project that had been put in front of them.

Nevertheless, all these researchers seemed to concur on a number of points. All agreed these were stunning results. As P.N. Lee commented, the effect sizes were 'absolutely mammoth... so outside my experience as an epidemiologist that I find it very difficult indeed to accept them as real.'[77] Estimates of relative risk conferred by personality type membership for particular fatal disease were all high to very high, depending on what data was being analysed and how. For example, Bernard Fox's analysis of the Crvenka data put the relative risk for cancer-prone personalities at 23.7, one that surely dwarfed all known other factors like smoking or poor diet.[78]

Various respondents pointed to insufficient data controls, including the possibility that prior knowledge of predictions might have compromised the independence and reliability of the mortality data. Lydia Temoshok went further in suggesting a *post hoc* matching of personality prediction and mortality criteria had occurred, adding that Eysenck's influence must have been behind the latter shift toward the kind a typological thinking that partly borrowed from her work. Fox pointed to unlikely characteristics of the data, including analyses that showed validity exceeding reliability and 113% success rates.

---

[75] Baker, T.B. and Fiore, M.C. (1992), Elvis is alive, the mafia killed JFK, and smoking is good for you, *Contemporary Psychology* 37, 1014–16. Even Eysenck struggled to incorporate these findings within his, by now, well-articulated biosocial theory of personality. See Eysenck (1991c), Reply to criticisms of the Grossarth–Maticek studies and Eysenck (1986c), Consensus and controversy: two types of science.

[76] Eysenck, H.J. (1991b), Personality, stress and disease: an interactionist perspective, *Psychological Inquiry* 2, 221–32.

[77] Lee, P.N. (1991), Personality and disease: a call for replication, *Psychological Inquiry* 2, 251–3.

[78] Fox, B.H. (1991), Quandaries created by the unlikely numbers in some of Grossarth-Maticek's studies, *Psychological Inquiry* 2, 242–7.

P.N. Lee did likewise with other aspects of the data, while Amelang related his personal dealings with Grossarth-Maticek dating back to his initial negative appraisal of Grossarth-Maticek's Habilitation document. More worrying still, Grossarth-Maticek's statistician, Hermann Vetter, outlined his suspicions that some sort of fudging was involved in the figures he was being fed.[79]

Perhaps the most eye-popping critique came from van der Ploeg.[80] He recounted a bizarre episode of subject record substitution, fake names, and research assistant blaming. When sent a set of subject records for analysis, van der Ploeg found they lacked any predictive validity. When Grossarth-Maticek was informed, he claimed a mistake had occurred. Mortality and personality data were matched with a set of (deliberately) fictitious subject names that were nonetheless crossed out to respect Germany's privacy laws. Attempts to match the correct name lists with the correct data went back and forth. Some subjects still appeared to match up with lists they shouldn't, while others seemed to have 'died' several times from different illnesses. The matter has never been fully explained, and the conclusions drawn differ depending on whom you ask. For Grossarth-Maticek, it was an excusable mix-up by an (unnamed) assistant that was finally resolved with a trip to the former city mayor's house to check the subject lists deposited there. Eysenck apparently concurred. For van der Ploeg and for Amelang, the trip settled nothing.[81]

Dr Rainer Frentzel-Beyme, an epidemiologist at the German Cancer Research Centre, was virtually alone amongst those in *Psychological Inquiry* in defending Grossarth-Maticek. A co-author on several Grossarth-Maticek papers, Frentzel-Beyme argued that in epidemiological research 'every insider knows that no satisfactory, flawless and impeccable large studies' exist. The critics carp over errors and 'faulty' data because 'the whole direction of creative thinking is terrifying to them.'[82] He singled out Amelang's criticisms as a prime example of this envious and fearful attitude in German academia. The controls

---

[79] See Amelang, M. (1991), Tales from Crvenka and Heidelberg: what about the empirical basis? *Psychological Inquiry* 2, 233–6; Temoshok, L. (1991), Assessing the assessment of psychosocial factors, *Psychological Inquiry* 2, 276–80; and, Vetter, H. (1991), Some observations on Grossarth-Maticek's data base, *Psychological Inquiry* 2, 286–7.

[80] van der Ploeg, H. (1991), What a wonderful world it would be: a reanalysis of some of the work of Grossarth-Maticek, *Psychological Inquiry* 2, 280–5.

[81] Manfred Amelang, interview, 1 February 2003; Henk van der Ploeg, interview, 11 February 2003. As Amelang recently told me: 'I have never understood what the aim of that enterprise was and at what time the lists had been given to the Mayor.' Manfred Amelang, personal communication, 16 January 2008.

[82] See p. 292 of Frentzel-Beyme, R. (1991), Levels of interest in an epidemiological approach of identifying psychomental risk factors for cancer, *Psychological Inquiry* 2, 290–3.

for ensuring the separation of predictor and criterion data that Grossarth-Maticek was forced to submit by those at Heidelberg were, for Frentzel-Beyme, unprecedented and based on 'impossible' demands.

Eysenck attempted to respond to each critique with some input from Grossarth-Maticek. Yet much was left unanswered. How might the interested reader react, Eysenck asked, with 'Perplexity? Incomprehension? Bewilderment? Mystification?...'[83] In the face of this fearsome assault Eysenck (uncharacteristically) distanced himself from Grossarth-Maticek, though only strategically: 'There were many aspects of the work about which I could only say that I would not have done it that way, and regretted that I had not been there when decisions about methodology and statistical analysis had been made.'[84] While acknowledging that mistakes had occurred, Eysenck still argued that this research should be taken very seriously. But Eysenck was left with a delicate balancing act. In defending the overall value of the studies he pointed out that, despite the quibbles, they demonstrated the overriding importance of personality in determining who succumbed to particular physical diseases. Conversely, in heading off the too-good-to-be-true charge, he was forced to argue that these results were merely in line with other previous research. For once he appeared to be talking down his data rather than talking it up.

The following year, British psychiatrists Tony Pelosi and Louis Appleby launched a scathing broadside at the Heidelberg intervention studies in the *British Medical Journal*. More saliently, they accused Eysenck and Grossarth-Maticek of not understanding the significance of their own data. According to the British pair, Eysenck and Grossarth-Maticek had actually outlined risk factors that were 'the highest ever identified in non-infectious disease epidemiology'. They calculated that type I personalities were 121 times more likely to die of cancer than healthy personalities. Furthermore, they had described intervention effects which, 'if correct, would make creative novation therapy a vital part of the public health policy throughout the world.'[85] However, Pelosi and Appleby questioned the reliability of the four-type approach, the spotty description of methods and measures used, and the immense amount of time Grossarth-Maticek must have necessarily put in to fulfil the therapeutic

---

[83] Eysenck (1991c), Reply to criticisms of the Grossarth–Maticek studies, p. 316.

[84] Ibid., p. 298.

[85] See p. 1296 of Pelosi, A.J. and Appleby, L. (1992), Psychological influences on cancer and ischaemic heart disease, *British Medical Journal* **304**, 1295–8. Taking the mortality rates of the healthy autonomous types as a baseline, they also calculated that Eysenck and Grossarth-Maticek's type 2 subjects were 27 times more likely to die of heart disease.

interventions described, a point raised previously by David Spiegel.[86] Eysenck's reply defended some of the charges but also raised further problems.[87] Concluding the exchange, Pelosi and Appleby poured scorn on Eysenck's claim that the studies were in line with other research in the area.

A denouement of sorts

Other questions from the critics lingered. What relationship did the published inventories and scales have to those that were actually used in Crvenka and Heidelberg in the 1960s and 1970s? Most of these inventories and scales have never been made available in the language they were originally presented in decades before. Why did the concept of rationality and anti-emotionality disappear and reappear as it did? It was first linked with cancer but later, inexplicably, with rheumatoid arthritis. More intriguingly, how was Grossarth-Maticek able to achieve such a breakthrough—given that he was not 'burdened' by a formal education in psychometrics, epidemiological statistics, and psychotherapeutics, nor an intimate knowledge of most of the relevant psychological literature since he barely spoke English?

Attempts to investigate and replicate ran into obstacles of various kinds. Amelang has made the most effort on this score. His 1992 finding that Grossarth-Maticek's type 1 and type 2 were almost psychometrically indistinguishable was seen as very damaging.[88] How could measures of what was essentially the same thing predict different outcomes with such power and accuracy? Grossarth-Maticek countered by arguing that one needed to use additional items and measures. Pre-warned of this replication work, Eysenck and Grossarth-Maticek also published two papers at this time that highlighted the importance of the administration method when using the type inventories.[89] The best (i.e. most reliable and predictively valid) results were achieved when the inventories were presented in an atmosphere of rapport and trust,

---

[86] Spiegel, D. (1991), Second thoughts on personality, stress, and disease, *Psychological Inquiry* **2**, 266–8.

[87] Eysenck, H.J. (1992a), Psychosocial factors, cancer and ischaemic heart disease, *British Medical Journal* **305**, 457–9; Pelosi, A.J. and Appleby, L. (1993), Personality and fatal diseases, *British Medical Journal* **306**, 1666–7.

[88] Amelang, M. and Schmidt-Rathjens, C. (1992), Psychometric properties of modified Grossarth-Maticek and Eysenck inventories, *Psychological Reports* **71**, 1251–63.

[89] Grossarth-Maticek, R., Eysenck, H.J., and Barrett, P. (1993), Prediction of cancer and coronary heart disease as a function of questionnaire administration, *Psychological Reports* **73**, 943–59; Grossarth-Maticek, R., Eysenck, H.J., and Boyle, G.J. (1995), Method of test administration as a factor in test validity: the use of a personality questionnaire in the

accompanied by explanations that ensured understanding. It was a point that had taken a curiously long time to come to light, and it made these studies far more difficult, if not impossible, to duplicate. For Amelang, it was the final step on the part of Grossarth-Maticek and Eysenck to insulate their work from criticism.

Nonetheless, Eysenck became increasingly enraged, exasperated, and even dejected by what he saw as the one-sidedness of the critique. His reaction to the volley of criticism in *Psychological Inquiry* was possibly the most defensive of his career. While he was back on the front foot in responding to the Pelosi and Appleby attacks, an air of apostasy becomes apparent in his last dealings with the subject. Nearing the end of his life, Eysenck told Grossarth-Maticek that the controls put in place to ensure the integrity of the data were not working.[90] Clearly not; some saw them as insufficient, while others saw them as evidence of a less than benign form of secretiveness, or even as a combination both. So by the end, Eysenck had had enough. According to Grossarth-Maticek, Eysenck felt it was time for a new phase, to move on to replicating or extending these studies. Uncharacteristically perhaps, Eysenck began to withdraw from the debate and ceased to engage with critics. For example, he refused to listen to a 1996 conference paper on this work given by Amelang in Ghent, even though they had once been firm friends and collaborators.[91] His last contribution on the subject was added to an edited volume with the understanding that it would not be subject to peer review.[92]

The implications of Eysenck's involvement

Much has been left unresolved following Eysenck's death. Some commentators—even Eysenck himself—argued Grossarth-Maticek's research was still worthy, even if he had cut corners and made silly errors. Grossarth-Maticek's personal investment might have also exerted an unconscious, biasing effect

---

prediction of cancer and coronary heart disease, *Behaviour Research and Therapy* 33, 705–10.

[90] Ronald Grossarth-Maticek, interview, 31 January–1 February 2003.

[91] Manfred Amelang, interview, 1 February 2003.

[92] See Eysenck, H.J. (2000), Personality as a risk factor in cancer and coronary heart disease, in *Stress and health: research and clinical applications* (ed. D.T. Kenny and J.G. Carlson), pp. 291–318, Harwood, Amsterdam. This information came from Dianna Kenny via Michael Eysenck, personal communication, 20 November 2001. See also the posthumous article Marusic, A., Gudjonsson, G., Eysenck, H.J., and Starc, R. (1999), Biological and psychosocial risk factors in ischaemic heart disease: empirical findings and a biopsychosocial model, *Personality and Individual Differences* 26, 285–304.

over his own judgement and that of his assistants. Questions regarding the reliability of the mortality data might be seen to fall into this category, along with suggestions that student interviewers falsified or adjusted records through laziness or an effort to please their director.

Those who were more critical suggested the matter went beyond sloppiness, that there were grounds to suspect something more serious was involved. More specifically, they questioned the independence of the mortality and personality data. Certainly the 1977 Habilitation document, with the idiosyncratic item weights that guaranteed almost perfect prediction, struck those at the University of Heidelberg as a case of retrospective fitting masquerading as genuinely blind prediction. Other problems, such as the dodgy descriptions of subjects, numbers that didn't add up or were not consistent from report to report, also fell into this category. Some of those closest to the research—such as Hermann Vetter, Manfred Amelang, and Henk van der Ploeg—even harboured doubts about Grossarth-Maticek's raw numbers.[93]

The Heidelberg study was extraordinary in terms of its breadth and ambition. It must have been a huge undertaking. Grossarth-Maticek had to track large numbers of people for a long time, requiring immense resources and the cooperation of all sorts of agencies and individuals. Over 20,000 people were said to have been tested in a prospective design—a bigger, richer data base than any other comparable study in the field. Such studies are typically the province of major institutions and big research teams. Grossarth-Maticek did it all as an independent scholar, and on a shoe-string. It was never clear where the money came from, especially for financing the expensive labour-intensive phase of initial data gathering. Grossarth-Maticek said he had private family sources (i.e. overseas relatives), but otherwise preferred to keep his funding arrangements private.

It seemed Grossarth-Maticek had supporters in the medical fields and was able to obtain, one way or another, genuine mortality data. However, the names (and the reality) of the thousands of subjects tested by Grossarth-Maticek have remained a mystery, a situation confounded by strict German privacy laws. Likewise, the student assistants employed on the project were hard to track down. To my knowledge, none have been questioned on record, although private accounts of their working experience are said to exist. In addition, the co-therapists who apparently helped Grossarth-Maticek perform the thousands of hours of psychotherapy were never clearly credited. Eysenck and Grossarth-Maticek were ambiguous on this point. On some occasions they

---

[93] Manfred Amelang, interview, 1 February 2003.

used the terms 'we' when describing the performance of therapy, while on others they appeared to indicate that Grossarth-Maticek alone performed the treatment. Both the Crvenka and Heidelberg studies were also remarkable for their foresight, apparently made possible by Grossarth-Maticek's prescient incorporation of ideas and measures that anticipated the future direction of psychological research (e.g. high-risk personality types) and advances in knowledge of drug, disease, and lifestyle interactions.

If one takes a dim view of this work, it poses many questions about the involvement of Hans Eysenck. Clearly, the ageing professor came in with his eyes open knowing there were problems with this research that went beyond the methodological. The 'three steps to trust' he reputably imposed were testament to this. Eysenck thought he could straighten out the data as well as polish up its presentation. Many of those who took a close interest in this case were very curious about Eysenck's role in the reconstruction of the inventories. Although never fully explained, it appears he had a major role in translating and cleaning them up for publication. Sybil Eysenck confirmed this, recalling that the original German language versions of the questionnaire(s) were 'a mess'.[94] She also told me that her late husband thought he could make sure mortality information was only gathered at the appropriate juncture—all of which pointed to Eysenck's involvement in an enterprise he knew was questionable, even if his intentions were good.

In any event, good intentions may not have mattered, such was the Yugoslav's control over the whole project. According to Amelang, Grossarth-Maticek saw himself as the project's *spiritus rector*, publishing whatever details he cared to at a time of his choosing.[95] No one besides Grossarth-Maticek could vouch for the integrity of the central methodological feature of this research—the matching of personality predictions to mortality. Eysenck was never in a position to fulfil the role of guarantor he set himself up in. Eysenck's 'three steps to trust' were ultimately irrelevant, if indeed they were taken seriously at all. For example, a check of subject lists deposited with the former mayor was only provoked by the alarm bells van der Ploeg set off. Otherwise, why would Eysenck participate in a ritual he had already performed? No one else involved in this work, not even the re-analysis team led by Spielberger, could confirm whether Eysenck actually performed the other checks. It is only on Grossarth-Maticek's assertion that these checks existed. And even if Eysenck did perform them, it would have still left open the integrity of the predictor score data. Tobacco industry

---

[94] Sybil Eysenck, interview, 25 July 2003.

[95] Manfred Amelang, interview, 1 February 2003.

documents indicate that Eysenck also wanted to use his own personality measures (the EPQ presumably) to compare with and corroborate Grossarth-Maticek's. This work was apparently not done either. There was no one else on hand and on record to corroborate many of the details of his work. Eysenck could only ever deal with the most easily disposed of accusations and invariably called upon Grossarth-Maticek as the final witness. Since it was Grossarth-Maticek's credibility at issue in the first place, this never convinced the sceptics.

Despite the damning judgements, Eysenck continued to hold Grossarth-Maticek in high esteem and talked of him with sincere reverence as he did of his other heroes like Galton and Pavlov.[96] Eysenck was so entranced by the potential of the data he took Grossarth-Maticek at his word. Having met Ronald Grossarth-Maticek, I can only say that he can be highly persuasive. In person, he presents as a man under siege from jealous rivals who persecute him because of his radical but successful ideas. In his own defence, he says he has neither the cynical outlook nor the mathematical sophistication to produce the effects he is accused of, and takes great delight showing off his neat folders of raw data to anyone who is interested. When questioned further, Grossarth-Maticek is able to furnish all sorts of explanations and anecdotes to cover the quirky aspects of his work. Clearly committed to his science, he is a charming and expansive theorizer who apparently has the Midas touch—almost every research area he has gone into has turned up empirical gold.[97]

Nevertheless, the whole affair suggested Eysenck's judgement left something to be desired. Excusing him as merely naïve in taking Grossarth-Maticek's data at face value would seem inappropriate as well as inaccurate. At the very least, Eysenck appeared guilty of not questioning when he should have, of not exercising the kind of scientific rectitude one might expect from a man of his stature.

Stop press!!

Given that the case had some potentially sensational ingredients, it is curious why it never made news headlines. After all, important scientific claims were involved, and several thousand subjects and massive amounts of data were

---

[96] Jeffrey Gray, interview, 25 July 2003.

[97] See, for example, an account of Grossarth-Maticek's on the effects of vitamin supplementation and autonomy in Nias, D. (1997), Psychology and medicine, in *The scientific study of human nature: tribute to Hans J. Eysenck at eighty* (ed. H. Nyborg), pp. 92–108, Elsevier, Oxford.

at issue. Plus there was the mystery surrounding the therapy collaborators and the scores of assistants, not to mention the uncertain patronage. The superficial explanation for this inattention would be that the case has never been conclusively adjudicated. And unlike some other well-publicized controversies, no crusading journalist has ever taken up the cause. In the mid-1990s, Tony Pelosi and *British Medical Journal* editor Richard Smith had hoped the news media would take this matter further. Pelosi talked to one reporter, but no story subsequently appeared. Perhaps the technicalities of this research were too difficult to turn into a clear and concise story, Pelosi suggested.[98] Comment within the more popular outlets was otherwise relatively light, confined to a few cautious news reports and the likes of a *Psychology Today* article and a recent *Der Spiegel* article.[99]

Why then has so little been resolved? One of the lessons drawn in recent times was that the necessary checks and balances can be undermined by an intimidating, powerful figure. Certainly, Grossarth-Maticek had almost sole control over the generation and analysis of the data. The English language versions of this research were mainly funnelled through journals in which Eysenck had a strong editorial role.[100] Eysenck's collaboration with Grossarth-Maticek could be seen as a somewhat counterproductive attempt to bring him in from the cold. His astounding claims cut across, undermined, or seriously diminished the more modest efforts of other more mainstream research programmes. It forced attention on the work and it forced a judgement from those who might simply have preferred to ignore it. However, Grossarth-Maticek's outsider status put these fellow researchers in a difficult position. In making their negative assessments, they were all too aware of being seen to be rounding on an inspired but poorly resourced underdog. Thus sympathetic allies like Rainer Frentzel-Beyme could argue that Grossarth-Maticek was a man unfairly maligned by a narrow-minded scientific establishment. Consequently, most commentators kept their distance, citing only rational-empirical, public domain reasons for not trusting this work. Tellingly, it was those closest to

---

[98] Anthony Pelosi, personal communication, 29 September 2003. Nevertheless, Simon Wessely, then at the IoP, wrote a sceptical evaluation: Wessely, S. (1993), Is cancer all in the mind? *The Times*, 22 June 1993.

[99] See Fishman, J. (1988), The character of controversy, *Psychology Today*, December 1988, and Kurz, F. (2002), Akademisches Schattenreich, *Der Spiegel*, **37**, 9 October 2002.

[100] These Eysenck-controlled journals were *Behaviour Research and Therapy* and *Personality and Individual Differences*.

Grossarth-Maticek—van der Ploeg, Vetter, and Amelang—who produced the most damning and frankly accusatory appraisals.[101]

The whole saga highlighted the ineffable but essential ingredients of science—of cooperation, openness, and good faith that underwrite trust. According to Amelang, 'no scientifically satisfactory evaluation of the claims made by Grossarth-Maticek and by Eysenck can be achieved through critical readings of their published works.' Only an 'independent and methodologically incontestable' corroboration could resolve the matter, he wrote.[102] But a close and therefore convincing replication would require far greater levels of cooperation than Grossarth-Maticek was apparently willing to display. And while a successful replication might support the validity of this kind of approach, it would not dispel *all* the doubts about Grossarth-Maticek's conduct and research. (See the concluding chapter for an extended discussion of this point.)

Sanctions and brand name protection

It seems that all the ink given to the scientific review process in the last couple of decades hasn't helped much. Various checks and safeguards have been put in place, including guidelines for data retention and sharing. While increased responsibility has been sheeted home to employing institutions, independent watchdog bodies have also been set up in several countries, including the US. Despite these developments, the uncertain status of Grossarth-Maticek's work demonstrated some glaring weaknesses in the scrutiny process. This work was only subjected to the standard peer review process, and only in some publications at that. And anyway, peer review was not capable of dispelling the sceptics' doubts because, as a recent study concluded, peer review provides no absolute barrier to misrepresentation and malpractice.[103]

Moreover, the internationalism of science did not mesh easily with the national and institutional structure of its policing mechanisms, leading to an unfortunate form of buck passing. Grossarth-Maticek's outsider position and émigré status were barriers to his credibility but they did tend to head off potentially damaging independent investigations. The recent war-torn history

---

[101] Apart from van der Ploeg and Vetter's critiques already mentioned, see also Vetter, H. (1993), Further dubious configurations in Grossarth-Maticek's psychosomatic data, *Psychological Inquiry* **4**, 66–7.

[102] Amelang (1991), Tales from Crvenka and Heidelberg: what about the empirical basis?, p. 235.

[103] See Hagan, P. (2003), Review queries usefulness of peer review, *The Scientist*, 28 January 2003.

in Grossarth-Maticek's native Yugoslavia made any inquiries there very difficult. Grossarth-Maticek did not belong to any mainstream academies or professional associations in Germany and England. While he had left the Institute of Social Medicine at Heidelberg in the mid-1970s after failing to share data with Professor Maria Blohmke, once he set himself up as private scholar, he was Teflon-coated.[104] For example, the German psychological hierarchy was unable to pursue Amelang's complaint about potential scientific misconduct because it did not have the jurisdiction to investigate. Following the damaging published critiques of the early 1990s, those at the IoP sought to distance the Institute from Grossarth-Maticek. The IoP's Dean, Stuart Checkley, wrote to the *British Medical Journal*, 31 July 1993, stating that, to the best of his knowledge, Maudsley ethics committees had never had the opportunity to vet this research, nor had the Institute ever awarded a title to Grossarth-Maticek, nor was this research part of the joint research strategy of the Maudsley and the Institute.[105] But that was all they could do. There were also some moves to censure Eysenck, with talk of stripping him of his emeritus professorship, but this came to nothing.[106] In 1995, Tony Pelosi even presented an ethics complaint to the BPS relating to Grossarth-Maticek's use of subjects with high blood pressure in a case control therapy trial. No action was taken, however, for reasons the Society did not care to reveal.[107]

In lieu of any special investigation of Grossarth-Maticek's work, the published record was left to speak for itself. It seemed that sanctions functioned to protect the brand name of employing institutions and the reputation of professional bodies; whether they ensured the integrity of the science was an altogether different question. Grossarth-Maticek continues to do research at the margins of science investigating unusual drug remedies and promoting his lifestyle and health strategies. He is still cited approvingly by cancer patient groups and the self-help movement that goes with this. Amazingly, Vetter

---

[104] A. Wallace Hayes, Meeting with Professor Blohmke, Dr. Kulessa, and Mr. Stelzer, 4 November *c.* 1983, http://tobaccodocuments.org/bliley_rjr/505743303-3315.html.

[105] Checkley, S. (1993), Letter to the editor, *British Medical Journal* **307**, 329.

[106] Louis Appleby, personal communication, 22 September 2003; Anthony J. Pelosi, personal communication, 29 September 2003.

[107] The BPS Investigatory Committee deemed it 'inappropriate' to set up an investigatory panel to look into the material Pelosi had sent them, and henceforth considered the matter closed. Pelosi disagreed, of course, but was left with little recourse. Anthony J. Pelosi, personal communication, 29 September 2003. See also Anthony J. Pelosi, The responsibilities of academic institutions and professional organisations after accusations of scientific misconduct, The COPE Report 1998 at BMJ.com, and Richard Smith, Editorial misconduct: time to act', at http://www.pitt.edu/~super1/lecture/lec14631/index.htm.

continues to work with the personable Yugoslav, even though the profound reservations he expressed in print have been only 'partially resolved'.[108]

The negative judgements have had a discernible effect; very few textbooks in health psychology and epidemiology mention Grossarth-Maticek's work favourably, if they mention it at all. Exceptions to this are the work of Spielberger in the US and Rocio Fernandez-Ballesteros in Spain, who have sought to extend or adapt aspects of this work.[109] In the meantime, the psychologists, oncologists, epidemiologists, and psychiatrists involved have simply moved on. Few were prepared to do the detective work required, or persist with liasing with Grossarth-Maticek. They would not be paid or rewarded for this. It would not, ultimately, be in their interests as scientists. Only Amelang maintains an active scepticism, and only because he has been able to incorporate personal investigations and critical replications as part of an on-going research programme. Amelang's most recent work has offered only limited support for the personality–disease link, hardly vindicating the particulars and the strength of the causal connection Grossarth-Maticek claimed.[110] Moreover, his specific appraisals of the Grossarth-Maticek questionnaires hardly supported their incremental or predictive validity.[111]

More generally, the interaction between personality and psychological stress has been shown to have a small but significant role in the development of heart disease. Intervention aimed at changing coping and lifestyle patterns is now a relatively well-established part of heart patient care. The picture is murkier

---

[108] Hermann Vetter, interview, 31 January–1 February 2003. For an example of these profound reservations, see Vetter (1993), Further dubious configurations in Grossarth-Maticek's psychosomatic data, p. 67.

[109] See, for example, Fernandez-Ballesteros, R. *et al.* (1997), Assessing emotional expression: Spanish adaptation of the rationality/emotional defensiveness scale, *Personality and Individual Differences* 22, 719–29.

[110] See Stürmer, T. *et al.* (2006), Personality, lifestyle, and risk of cardiovascular disease and cancer: follow-up of a population based cohort, *British Medical Journal* 332, 1359. Only certain personality traits (e.g. internal locus of control and time urgency, rather than anger control, psychoticism, and depression) were weakly associated with a reduced risk for cancer and heart disease.

[111] The most recent study is Amelang, M. *et al.* (2004), Personality, cardiovascular disease, and cancer: first results from the Heidelberg Cohort Study of the Elderly, *Zeitschrift für Gesundheitspsychologie* 12, 102–15. See also Amelang, M. (1997), Using personality variables to predict cancer and heart disease, *European Journal of Personality* 11, 319–42, and Amelang, M. *et al.* (1996), Personality, cancer and coronary heart disease: further evidence on a controversial issue, *British Journal of Health Psychology* 1, 191–205.

when it comes to cancer, however.[112] A cancer-prone personality has not been consistently established, nor is it clear whether psychological intervention prolongs survival. Positive outcomes, like Spiegel's 1989 study, have not been consistently corroborated.[113]

The field of psychosomatic medicine has all but split up. Perhaps the most successful offshoot is the new subfield of psychoneuroimmunology, which explores the interrelationships among the mind, the central nervous system, the immune system, and the outcome of immunologically mediated diseases. It is research in these areas, and in the genetics of disease propensities, that is now regarded as the most promising.[114]

Smoking guns and public disclosure

> I was unwilling to accept the so often heard argument that smoking is addictive, because neither then [i.e. the late 1950s] nor now does the term have any scientific content. I like playing tennis and writing books on psychology; does that mean I am addicted to tennis and book writing? (Eysenck, H.J. (1997b), *Rebel with a cause*, p. 168)

Hans Eysenck claimed that nicotine was not addictive but tobacco money may well have been, and controversy likewise. He received upwards of £ 800, 000 (in today's terms) from Big Tobacco in the US and UK. It was probably Eysenck's biggest single funding source and *the* means for supporting for his post-retirement research.[115] Tobacco money enabled him to maintain the intellectual lifestyle and the research profile he became accustomed to in his twilight years. It fuelled the kind of media juggernaut his career had become, but it also had its drawbacks (pun again intended).

Can we therefore lumber Eysenck with the dreaded title 'bought advocate'? Well, that depends on what is meant by the term. Despite trawling through a massive number of confidential industry documents, I found nothing to suggest tobacco industry representatives directly influenced Eysenck's views. Moreover, there are good reasons to believe they would not have even tried to do so. Those in the industry who understood the science also understood the

---

[112] Garssen, B. (2004), Psychological factors and cancer development: evidence after 30 years of research, *Clinical Psychology Review* 24, 315–38.

[113] Stephen, J. et al. (2007), What is the state of the evidence on the mind–cancer survival question, and where do we go from here? *Supportive Care in Cancer* 15, 923–30.

[114] Genes too have been shown to play a key role in the development of many cancers, including cancer of the lung, partially bearing out what was a standard hypothesis for one branch of cancer research, not just Hans Eysenck.

[115] Michael Eysenck, personal communication, 12 December 2001.

importance of its demonstrable independence. Unlike the more interventionist style of other players in the funding business—the pharmaceutical industry springs to mind—the American tobacco industry kept the outside researchers they supported at a distance and were ambivalent about the usefulness of in-house research. For example, legal advisor Ed Jacob nearly resigned over the R.J. Reynolds in-house programme. Tobacco insiders were also very sensitive about accusations of censorship, doing their best to (paradoxically) cover up the few instances where it was deemed necessary. Tobacco industry leaders were able to maintain their canny hands-off stance because they were not interested in clarifying the tobacco–disease link but in obscuring it. They made sure they did not have to intervene by mounting an open-ended quest to look elsewhere, and by handpicking the personnel and the projects to do just that.

Without a doubt, Eysenck allowed himself to be used to defend the legal and commercial interests of a ruthless industry. He and many other prominent and respected scientists were an essential part of an industry strategy of public denial, despite private acknowledgement, of the hazards of smoking.[116] Up until around the time of Eysenck's death, the tobacco industry had never lost a single one of the over 800 liability cases brought in the US. One must give industry representatives credit for this, for they played a bad hand very well. To do so, they needed people like Eysenck. They parlayed his credibility as a scientist to reassure their customers and foil liability claims.

It was always unclear how much knowledge Eysenck had of industry thinking. It is hard to believe that Eysenck was not aware of the reasons those involved in the tobacco industry valued him, given his deep involvement with Jacob at R.J. Reynolds. He may not have shared their cynically instrumental perspective, however. Eysenck and many others were used to create the impression that scientists still disagreed about the smoking–disease link. For Eysenck, it apparently *was* an open question. He gave sworn testimony to the US Congress to this effect, emphasizing his impartiality into the bargain.[117] He claimed he came to these conclusions despite the money he received from

---

[116] See Glantz, S.A. et al. (1996), *The cigarette papers*, University of California Press, Berkeley.

[117] See Hearings Before the Subcommittee on Health and the Environment of the Committee on Energy and Commerce House of Representatives Part 2 of 3, 9 March 1983, http://tobaccodocuments.org/lor/03637300-7630.html, and Eysenck's, Lung cancer, coronary heart disease and smoking, statement regarding S. 772 submitted to the Committee on Labor and Human Resources, 6 May 1983. See also Eysenck's testimony in a Finnish liability case. Hans J. Eysenck, 'Transcript of Testimony As Checked by the Court Exhibit 11 Subsection 1,400,' January 1990 (est.), http://tobaccodocuments.org/pm/2501072480-2506.html.

the industry and, ironically, he might have genuinely disagreed with the private opinions of industry insiders who supported him. Yet here was a distinguished scientist richly sponsored by a powerful industry because he was *seen* as authoritative, simply because his work had legal, and therefore commercial, value.

While any attempt to directly influence Eysenck's views may be discounted, a *prima facie* case can be made for the contention that tobacco money appeared to affect *what* he researched, if not the *outcome* of this research. Jacob and Eysenck appeared to work together on what questions Eysenck should research, of what would be useful in a 'litigation-oriented' sense.[118] In these discussions, Eysenck directly and indirectly suggested ways that the case against smoking could be attacked. A tacit form of complicity can't be ruled out in the other, more 'above board' projects Eysenck proposed. The CTR only tended to support projects with likely favourable outcomes, and advance notice of this helped secure grants.[119] In other scientists' projects, tobacco industry lawyers not only had a say in *what* was researched, they also had some say in *how* it was researched—tinkering with experimental designs to yield the most useful results.[120] While no such shenanigans were apparent in Eysenck's case, the suggestion remains that he tailored his research proposals to appeal to the tobacco industry. He also appeared quite willing to lend a hand to help them sell their product—although it appeared industry types took him less seriously on this front.

Eysenck always gave smoking the benefit of the doubt, even though it become less and less clear why he thought it deserved it. Having sincere reservations about the case against smoking was one thing—Eysenck was certainly not

---

[118] Here is a sample quote: 'Jacob reported that Eysenck's work indicates that people begin to smoke due to environmental factors (peer pressure) but continue to smoke due to genetic factors. He thinks that a psychological study showing that behavioral patterns are genetically founded would be of great use. He wants to explore with Eysenck the possibility of getting a group of people in the US who have trained under Eysenck to work up material that might be useful in litigation. This would not be a CTR project but would be strictly litigation-oriented. Jacob has no real idea of the money involved but thinks that $25, 000 to $50, 000 would be involved for a pilot project and that the whole project could involve $100, 000 to $200, 000 over a couple of years. Jacob also noted that Eysenck will be looking for support for a twin registry project in about 1 1/2 years but that this would be appropriate for consideration by CTR.' A.H. 'Request From FTC For Input Regarding Carbon Monoxide Testing,' no date, c.1980, http://tobaccodocuments.org/ness/43461.html, p.3.

[119] Crist *et al.* (1986), Re: Jones/Day Liability Summary ('Corporate Activity Project').

[120] See Glantz *et al.* (1996), *The cigarette papers*, Chapter 8.

alone in this—but feeling compelled to broadcast them from the rooftops for decades was another. Eysenck's career path on this issue was one of acquired dependency and diminishing options. Following in Fisher's footsteps, Eysenck's initial stance of disengaged scepticism seemed rational and relatively unexceptional. Courted by the American industry in the 1970s, minimal intellectual compromise led to a lot more research money at a time when his notoriety was beginning to work against him. By the early 1980s, however, as the data linking cigarettes and disease accumulated, Eysenck must have been aware that his position was becoming increasingly untenable.

In downplaying the animal studies and preferring to argue the point in the epidemiological realm, he only succeeded in painting himself into a corner. As one of a minority of scientists highlighting the shortcomings of retrospective investigations, Eysenck prompted epidemiological researchers to adopt more sophisticated and more convincing designs. It seemed that he (of all people) would never accept the new standard of causality this research generated. But it also meant that Eysenck needed better evidence for his alternative genetic-and-personality explanation. As P.N. Lee pointed out, personality could be the key intervening variable explaining most of the association of cigarettes and cancer only if the relationship between personality and cancer was at least as strong—a point Eysenck later acknowledged explicitly.[121] Grossarth-Maticek's data, with its startling relative risk estimates, were a mathematical necessity, Eysenck's get-out-of-jail card. Even some of his closest colleagues saw Eysenck's championing of Grossarth-Maticek as a notable example of his critical double-standard—applying the most rigorous scepticism to rival accounts, while protecting his own work and that of others supporting it.[122] The contradictions were more nuanced than this, however. On the one hand, Eysenck continued to question whether statistical epidemiology could demonstrate a causal connection between smoking and disease, in the process aligning himself with an old-style medical framework dominated by a focus on individual susceptibility. On the other hand, he argued that this

---

[121] See the exchange between Eysenck, Stolley, and Vandenbroucke regarding Fisher's contribution to the tobacco and cancer debate. Eysenck, H.J. (1991a), Were we really wrong? *American Journal of Epidemiology* **133**, 416–25; Stolley (1991a), When genius errs: R.A. Fisher and the lung cancer controversy; Stolley, P.D. (1991b), Author's response to 'How much retropsychology?', *American Journal of Epidemiology* **133**, 428; and Vandenbroucke, J.P. (1991), How much retropsychology? and, reply to Eysenck, *American Journal of Epidemiology* **133**, 426–7.

[122] Jack Rachman, interview, 24 June 2002.

kind of epidemiological research could reveal just such a connection between personality and disease.

Grossarth-Maticek's results gave Eysenck renewed licence to continue his distinctive line in pro-smoking advocacy through the late 1980s and early 1990s. Eysenck's increasingly certain pronouncements on the *qualified and specific* risks of smoking and the non-addictiveness of nicotine were widely interpreted as providing a licence to smoke. His input had little bearing on an intellectual debate that had long since closed. As Baker and Fiore put it in their review of *Smoking, personality and stress*, the (non-)addictiveness of nicotine had 'virtually no bearing on whether smoking causes coronary heart disease or lung cancer. ... Rather, he [Eysenck] seems more interested in justifying and defending tobacco use. It is hard to discern a theoretical or scholarly intent here.'[123] Instead Eysenck made himself a champion of consumer choice who went out of his way to undermine the public health message. For example, in a 1995 piece Eysenck made much of the semantic ambiguity of the 'smoking kills' warning. Somewhat pedantically he argued that, as a relatively weak, synergistic factor, smoking *per se* did not kill anyone. He remained a thorn in the side of public health advocates, who declared him unqualified and irresponsible but could not afford to ignore him.[124]

Eysenck carried on with the fight, despite and perhaps because of denunciations he endured. As with the race and IQ debate, pig-headedness should be added to the explanatory mix. He was certainly not a man who liked to admit he was wrong. In his 1991 exchange with Vandenbroucke and Stolley, his refusal to concede any ground was truly breath-taking. He wrote: '*at the time of writing* the evidence was clearly insufficient ... we never engaged in prophecy as to what future evidence might disclose... our statements were correct at the time.' [orig. italics][125] The public image drove the private scholar.[126] It made it hard for Eysenck to change tack or leave the debate alone once he had staked out a position. It made for a paranoid style, finding vindication in frustrating a litany of opponents instead of embracing new approaches.[127]

---

[123] Baker, T.B. and Fiore, M.C. (1992), Elvis is alive, the mafia killed JFK, and smoking is good for you, p. 1016. For another example of this kind of promotion, see Eysenck, H.J. (1986*d*), Nicotine's not all nasty, *The Best of Health*, February 1986.

[124] Fletcher, C. (1965), Eysenck v. anti-smokers, *The Observer*, 27 June 1965.

[125] Eysenck (1991*a*), Were we really wrong?, p. 429.

[126] Fletcher (1981), Plea for the guilty.

[127] As with his flirtations with parapsychology and astrology, race and IQ, Eysenck denied that he was simply playing the *enfant terrible*. Nonetheless, he did allow himself a

Eysenck always defended himself by saying that it was all valid research, no matter who paid for it. And he was by no means alone in taking such money. The tobacco industry supported a great range of scientists; many were reputable and influential, some were recipients of Nobel prizes. It may also be unfair to condemn Eysenck in this manner, given that many scientists in many fields face perennial funding crises. With only finite means available in a competitive tendering environment, they must have an eye on the main chance, the most likely opportunities to get money. Moreover, the tobacco industry may have used him, but Eysenck also used the industry to fund projects that went beyond smoking and health. Along with support from the Pioneer Fund, he used tobacco money to piggyback his extensive post-retirement research on the biology of intelligence by adding in a study of the effects of nicotine on performance. Without such 'tainted' funds this important research would probably not have been possible. As Eysenck's successor as head of psychology Jeffrey Gray noted, it was always extremely difficult to get money for basic personality and individual differences research from more mainstream sources such as the MRC or the Wellcome Trust.[128] Those at the IoP felt these funding bodies tended to favour more traditional experimentalism and practical, medically oriented research. Thus Eysenck was forced to fall back on more unorthodox sources. Eysenck cheerfully disclosed much of his tobacco sponsorship, though he was often vague about specific amounts. When it was revealed toward the end of his life that he had received funds from accounts the industry preferred to keep secret, Eysenck's evasive response was that 'we get a lot of research money', implying that he could never keep track of where it all came from.[129] Like many scientists, the ideas and the ambition did not stop when he reached 65. He did not want to put his feet up and felt no obligation to do so. The tobacco industry enabled Eysenck to maintain a vigorous post-retirement career and lifestyle—a reversal of the usual effect the industry's product has on its customers.

As he did in the race and IQ controversy, Eysenck always maintained that the duty of a scientist is to advance human knowledge no matter whose

---

contrary streak of critical rationalism, coupled with a sneaking sympathy for the underdog. See Eysenck, H.J. (1986c), Consensus and controversy: two types of science.

[128] Jeffrey Gray, interview, 25 July 2003. Gray was himself a recipient of tobacco money, having been introduced to Ed Jacob by Eysenck. When he took over in 1983, Gray said he was shocked to learn just how little money the IoP psychology department had been receiving from established sources.

[129] Freeman, J. (1997), The pugnacious psychologist [obituary for Hans Eysenck], *The Guardian*, 8 September 1997.

interests it may serve. 'Scientists look at any factual question as a legitimate topic for science', he testified.[130] Given the evidence he saw before him, this was the truth, the most reasonable position—and the consequences were not for him to dictate or imagine. He turned Claude Bernard's dictum, 'when in doubt, abstain' in his favour, arguing that anti-smoking measures were unjustified. Smokers should relax and continue to smoke if they wanted to; it was the health care do-gooders who needed to lay off. Eysenck's position was a use/abuse model carried to an extreme, with science conceptualized as occurring in some sort of neutral, politically valueless space. For some, this is science at its most idealized, as it should be. For others, it is science at its most cynical, as it too often is. Certainly, it was a stance that offered Eysenck plenty of freedom of choice—of topics and collaborators, patrons and supporters. But it conflicted sharply with the world-view of those who saw science and its consequences as more intimately intertwined, as an enterprise necessarily harnessed to social goals. Eysenck may have convinced himself that no one told him what to think, that in his heart of hearts he was no one's lackey. Yet once again, his insensitivity to conflicts of interest and guilt-by-association made him a target. As Jeffrey Gray agreed, it tended to confirm his critics' suspicions that there was 'something sleazy about Hans.'[131]

Today the large US tobacco companies are still vehemently contesting the compensation claims of some of their best customers. Insider revelations in the mid-1990s saw their see-no-evil posture exposed as conscious deception. Having been forced to admit what they had known all along—that nicotine is addictive and cigarettes kill—they are fighting a rearguard action against individual and class-based liability claims.

How Hans Eysenck might have responded to the recently forced admissions from the major tobacco companies one can only speculate. He certainly would have found it more difficult to use them as a funding source. Most scientists and scientific institutions in Britain were forced to wean themselves off tobacco money by the late 1990s. In the case of the IoP, the pressure came first from below—from conscientious junior staff members, and psychiatric trainees, rather than from senior management who tended to take a more benign view.[132] Sniffing the political wind, many of the established funding bodies (including the Wellcome and the MRC) responded to this kind of activism by

---

[130] Hans J. Eysenck, 'Transcript of testimony as checked by the court exhibit 11 Subsection 1,400,' c. January 1990, http://tobaccodocuments.org/pm/2501072480-2506.html, p.24.

[131] Jeffrey Gray, interview, 25 July 2003.

[132] Ibid.

refusing to sponsor anyone supported by the tobacco industry—relegating tobacco money to the renegades and the outsiders.[133]

A sad postscript

Although Eysenck was an extraordinarily focused, rational man, the last few years of his life were distracting and demanding. The results were fatal, if you take Grossarth-Maticek's word for it, a final confirmation that personality and stress were intimately related to the development of serious diseases.[134] When Eysenck contacted Grossarth-Maticek with the dark news, Grossarth-Maticek urged him to undergo the kind of autonomy training the Yugoslav was sure would prolong his life. Ever the empiricist, Eysenck thought the number of cases linking unresolved stress to cancer was too low to trust. He was dead within months.

For Grossarth-Maticek, this was the loss of a key ally. His publications in English tapered off, and so did the controversy. Still it was not plain sailing. It transpired that Eysenck had apparently left a great deal of money (i.e. several million Deutschmarks!) to Grossarth-Maticek as a parting gift to allow him to continue his research. Grossarth-Maticek said he elected to forgo this money, given that pressing for it might damage relations with the Eysenck family. However, his account of the situation was sharply contradicted by several other people I discussed it with, Sybil Eysenck amongst them.[135]

---

[133] For one of the latest instalments in the history of tobacco industry cover-ups, see Rego, B. (2009), The polonium brief: a hidden history of cancer, radiation, and the tobacco industry, *Isis* **100**, 453–84.

[134] Ronald Grossarth-Maticek, interview, 31 January–1 February 2003.

[135] Sybil Eysenck, interview, 25 July 2003.

Chapter 10

Conclusions

Hans Eysenck did many extraordinary things in his lifetime. Failing health hardly seemed befitting. Frail and dependent, he especially hated the fact that cancer had attacked his brain. It was, David Nias sadly recalled, the only time he had ever seen Hans Eysenck afraid.[1] At a meeting of the International Society for the Study of Individual Differences Eysenck attended in July 1997, the Society presented him with the last *Festschrift* he received in his lifetime. It was subtitled *A tribute to Hans J. Eysenck at eighty*.[2]

London, 4 September 1997

When Eysenck passed away in September 1997 at the age of eighty-one his death was overshadowed by the unprecedented mourning for Princess Diana, who met her untimely end a few days earlier on August 31. With Mother Theresa also dying the day after Eysenck, the global media allocated him a comparatively brief run. His friends and colleagues, those he mentored and inspired, gave Eysenck much more. While he was not an easy man to get close to, the tributes were heart-felt. It was the kind of affection reserved for heroes. Gordon Claridge noted that Eysenck was not a talkative man in private, but 'we will all, in our own ways, miss him'.[3] Jeffrey Gray felt confident that his 'position as the most highly cited psychologist of his generation will, I believe, be matched in the history books.'[4] Fed by Frank Farley, Daniel Schacter, and Arthur Jensen, *The New York Times* emphasized Eysenck's contentious views

---

[1] David Nias, interview, 24 October 2001.

[2] Nyborg, H. (ed.) (1997), *The scientific study of human nature: tribute to Hans J. Eysenck at eighty*, Pergamon, Oxford. The others were Lynn, R. (ed.) (1981), *Dimensions of personality: papers in honour of H.J. Eysenck*, Pergamon Press, Oxford, and the *Festschrift* (of sorts) as part of Modgil, S. and C. Modgil (eds.) (1986), *Hans Eysenck: consensus and controversy*, Falmer Press, Philadelphia.

[3] See p. 394 of Claridge, G. (1998), Contributions to the history of psychology: CXIII. Hans Jürgen Eysenck (4 March 1916–4 September 1997), an appreciation, *Psychological Reports* **83**, 392–4.

[4] Gray, J. (1997b), Obituary: Hans Jürgen Eysenck (1916–97), *Nature* **389**, 794.

on psychotherapy, race, and smoking, rating him the British equivalent of B.F. Skinner. Even those not inclined to think well of him acknowledged Eysenck's importance. *Guardian* obituarist Joan Freeman, dubbed him the 'people's psychologist', noting that Eysenck was the 'name everyone knew. It was his books they bought.'[5]

There are many ways to think about a scientist's legacy, many ways to take a metaphorical slide rule to their achievements. The way we single out great scientists inevitably implies something about what we think the ultimate goals of science should be and how we should get there. In light of this, I'd like to reiterate the way Eysenck went about his work—the better to evaluate his accomplishments in their own terms, as well as to highlight what might have been.

Big picture man

Late in his career Eysenck characterized himself as a revolutionary romantic, as someone willing to take up ideas that ran counter to accepted wisdom.[6] However, this self-assessment can hardly be accepted without qualification, for it was part of an attempt to fashion himself as the rebel outsider. In general, we tend to value works of greatness for their novelty as well as their anticipatory status. Especially creative works are seen to be at odds with the context in which they first appear; but the strangeness that repels or baffles critics comes to appear prescient and profound. Yet that was not often the case with Eysenck. Most of the elements and ideas he used were already familiar; what was unusual was the way they were combined, pressed into service in an increasingly inclusive conceptual scheme.

It would have to be said that Eysenck also went about his work in a manner quite different to that of the 'romantic'. His was not the kind of consuming thirst for knowledge that meant pouring oneself into nature. This did not suit him, nor strike him as profitable. Instead, Eysenck positioned himself as radically separate from what he was investigating, unmoved and unchanged by the focus of his curiosity. His abiding detachment ruled out any kind of deep immersion based on firsthand, touch-and-feel experience. These two ways of seeing and doing science have coexisted since the Enlightenment, though not

---

[5] Freeman, J. (1997), The pugnacious psychologist [obituary for Hans Eysenck], *The Guardian*, 8 September 1997.

[6] Eysenck, H.J. (1986c), Consensus and controversy: two types of science, in *Hans Eysenck: consensus and controversy* (ed. S. Modgil and C. Modgil), pp. 375–98, Falmer Press, Philadelphia, p. 377.

always happily. Advocates for each approach have often sought to discredit or repress the other. Eysenck was no exception. He repeatedly sought to show that the human condition could and should be thought of as an object puzzle. Close contact was unnecessary; personal involvement potentially misleading. One should read the literature, do the experiments, analyse the results, and draw one's conclusions.

More than once the elderly Eysenck was asked whether he was a genius—an apposite question given that he spent much of his twilight years investigating this topic. In all instances it seems he replied that he thought not. In his estimation, genius required more than just intelligence and more than just a specific aptitude. What was needed was a touch of madness—moderately elevated P scores in Eysenckian terms.[7] According to Eysenck, he possessed little of such a trait. He only did things, he said, that anyone else could have done if they had only opened their eyes to the obvious.[8] But perhaps the point is that they didn't. Eysenck's perspective helped him to do this, and do it easily. Paul Barrett probably put it best:

> One thing that always stood out with Hans ... was that he could always see the big picture. That is, his grasp of theory was such that many diverse results in different domains were construed as part of some integrated picture in his own representational system. Thus, he was able to suggest propositions and provide insights that might sometimes have eluded others ... [9]

Being cool and aloof allowed Eysenck to make connections others had missed, to slice and dice his way through the verbiage and ambiguity. His dimensions of personality were just such a breakthrough. Eysenck did not come up with a new view on temperamental differences, but a new basis for them, a far greater confidence for believing suggestions that dated back to antiquity. Long-time friend and colleague Desmond Furneaux argued that Eysenck was not strong on creative leaps, but he 'could sprint down a logic tree faster than anyone else'.[10] He could quickly spot the implications and possibilities of diverse viewpoints and data. Eysenck helped clean up the proliferation of traits and attributes characterizing the emerging field of personality psychology.

---

[7] See, for example, Eysenck's target article, Eysenck, H.J. (1993a), Creativity and personality: suggestions for a theory, *Psychological Inquiry* **4**, 147–78 and the critical comment that follows, and one of his last books Eysenck, H.J. (1995b), *Genius: the natural history of creativity*, Cambridge University Press, Cambridge.

[8] Arthur Jensen, interview, 1 July 2002.

[9] See p. 14 of Barrett, P. (2001), Hans Eysenck at the Institute of Psychiatry—the later years, *Personality and Individual Differences* **31**, 11–15.

[10] Desmond Furneaux, interview, 26 October 2001.

As Jeffrey Gray put it, 'the argument was not about maths, but about which version of the maths best applied to the true "structure of personality".'[11] The dimensional trinity he quickly arrived at proved a powerful organizing scheme, a beachhead for British psychology. It enabled Eysenck and those following in his wake to link up vast tracts of psychological work. For example, almost all task performance or learning models could be subdivided in dimensional terms, and any psychological variable where personality was thought to be a factor could be analysed in these individual difference terms. The standardized scales and measures he developed allowed for convergent, comparable, and cumulative data gathering.

Eysenck never compromised on the simplicity of three dimensions as sufficient to describe the underlying, culturally universal structure of personality. However, Eysenck's three 'super-factors' have always been countered by more complex descriptive systems in the US, particularly the 16 personality factors of Cattell. Five factors are currently seen as the most defensible; the fact that two are similar to Eysenck's I–E and N bears out his push for simplification along the lines he advocated.

Eysenck's eye on the big picture also made obvious the need to push beyond factorial psychometrics, to anchor his dimensions in something more concrete, more corporeal. He was almost alone in asking the question: if there is a descriptive structure to personality, what gives rise to it? Eysenck pioneered the link to neurophysiological processes and introduced testable theoretical accounts into an area that had appeared to avoid them. It has now become customary to acknowledge the importance of biological and genetic factors on personality, even if cognitive neuroscientists don't necessarily draw from Eysenck directly.

To coin a cinematic metaphor, Eysenck's fondness for the wide-angle shot also encouraged him to vault intellectual barriers that must have looked less formidable to him than they did to others. His detachment from his material also ensured that he did not get too caught up in the miniature, nor overcommitted to small details and trapped in blind alleys. As he remarked to friends, he rarely, if ever, allowed his feeling to intrude on his work.[12] He was able to keep moving on, hopping from topic to topic, investigating whatever seemed promising.

Eysenck's attempt to bridge Cronbach's two disciplines of psychology was one of his least controversial hobby-horses. Yet, as Gordon Claridge noted,

---

[11] Gray (1997b), Obituary: Hans Jürgen Eysenck (1916–97), p. 794.

[12] Claridge (1998), Contributions to the history of psychology: CXIII. Hans Jürgen Eysenck (4 March 1916–4 September 1997), an appreciation, p. 393.

it was met with indifference bordering on hostility. The physiological aspects of his work alienated many personality and social psychologists. Only a handful of Eysenck's protégés and colleagues, such as Gray, Claridge, Robert Stelmack, Jan Strelau, Marvin Zuckerman, and Martin Seligman, and more distant contemporaries like J.A Brebner, C.R. Cloninger, and Robert Plomin, wanted to follow his lead. Conversely, experimentalists did not appreciate Eysenck's insistence on accounting for individual differences in their search for general basic mechanisms, mostly ignoring his suggestion that such variation could help explain why there was so much 'error' variance in their data.[13]

In this sense, Eysenck could not transcend disciplinary specialization. But in another sense, he depended on it and exploited it. While specialization generates focus and depth, it also creates a need for the kind of player who can pull these distinct strands together—the interpretative generalist. As a federation of pure and applied subfields bordering on several adjacent disciplines, modern psychology has always ached for a sense of identity and cohesion. At the same time, its practitioners have had to generate knowledge, skills, and technology useful to business, government, and the military, and they have had to sell themselves to the general public. The interpretative generalist helps meet these needs, helps psychologists do all these things, fulfilling a role that often segues into that of public intellectual.

For the ambitious Eysenck, a career spent buried in any one subspeciality would be far too limiting. Even so, Eysenck's decision to become a big picture generalist was partly forced upon him. The 'shallow-broad' approach he adopted early on was a way of making the most of the student army he had at his disposal. But he soon found this role fitted him like a glove. Eysenck was a master at drawing together other people's research and adapting them to his own ends. He worried less about the provenance of data than most, and was unabashed about intervening in any area or debate. He came to think of it as his responsibility to his science as much as his due. He fronted all manner of collaborative efforts, and became the public face of IoP psychology, the spokesman for a number of international movements and intellectual trends reaching out beyond south London. He produced papers for conferences on order, and could present them compellingly. He spoke to schoolchildren and lay groups, and wrote for periodicals and magazines of all sorts. He was the one the media turned to because he was good at giving them what they wanted. Like his mentor Cyril Burt before him, Eysenck was British psychology's

---

[13] See ibid. and Corr, P.J. (2000), Reflections on the scientific life of Hans Eysenck, *History and Philosophy of Psychology* **2**, 18–35.

leading advocate and educator, and he had no equivalent as an interpretive synthesiser.

Eysenck provided a programmatic context for the work of many specialist researchers, especially those subsumed within the individual differences tradition. He provided a deeper significance to the narrow problems they were working on, and he took risks on their behalf. Attempting to ground personality in neurophysiology—his first typological postulate especially—was a big risk. Ditto his promotion of chronometric and 'brain-wave' measures of intellectual differences. These specialists welcomed Eysenck's grandness. They found his integrative books and articles helpful and stimulating. They will miss him the most. When one hears of complaints about the muddled state of the biological bases of personality or intelligence research, one can detect nostalgia for the super-ordinate clarity he provided.

At the same time, some of Eysenck's peers read an implicit arrogance into his behaviour, an undeserved claim to eminence. Burt voiced disquiet about the young Eysenck's publication proclivities and the risks this presented in preempting the work of other students and researchers. As his first biographer Gibson suggested, but didn't explore, that there was something very English about these qualms, an effect no doubt heightened by Eysenck's 'otherness' as a German-born émigré. Eysenck seemed to embody something of a lack of proper respect for disciplinary pecking orders. One must wait one's turn, taking care not to seem overly ambitious or boastful. Ideally, lofty aspirations should be cloaked in humorous self-depreciation. These were niceties the young Eysenck worried over early in his career as he struggled to find his place as a research player. He soon stopped worrying. While shy in person, he was hardly self-effacing in print, playing up to the unspoken rules of English civility with brazen self-promotion. He didn't tolerate fools gladly, and made it clear he saw plenty amongst his fellow psychologists, and even more amongst philosophers, psychiatrists, and sociologists.

Many experimentalists could be accused of being too preoccupied by their own narrow concerns to pay attention anyway. For those who did, Eysenck had to overcome the perception that he was an interloper, that he was spreading himself too thin to keep up, that he didn't understand their field well enough to forge connections with it.[14] Likewise, many clinical specialists doubted his credibility as a spokesman for behaviour therapy, given that Eysenck rarely went anywhere near a patient. Those who had taken on advisory roles for government or business—often the elite of the discipline—also

---

[14] Claridge (1998), Contributions to the history of psychology: CXIII. Hans Jürgen Eysenck (4 March 1916–4 September 1997), an appreciation, p. 393.

doubted his credentials as public policy advocate, since he had very little formal experience in such roles. Outside his coterie, his practical recommendations came across as dangerously speculative.

Eysenck tried to lead by example, to put his money where his mouth was. He talked up particular approaches and imagined possibilities in a way that mobilized his peers. But it also made it easy for those assessing his contributions (including the present writer, I have to say) to point to the occasions on which he ended up with egg on his face. Never mind, he wore it well. He even seemed to welcome the attention, for it enabled him to display some truly remarkable critical skills.

Fig. 10.1 Eysenck flanked by his intellectual heroes.
Source: Centre for the History of Psychology, Staffordshire University

Competitive advocate, loyal nurturer

For many of his fellow psychologists, Eysenck was a prickly beast. However, all scientific disciplines need at least a few contrary types as a counter to complacency. Even when arguing for positions most of his colleagues thought wrong, Eysenck performed a service to psychology. He forced them to re-examine their assumptions, their reasons for subscribing to the orthodox, mainstream view. Post-war US clinical psychology was particularly ripe for the shake-up Eysenck gave it. His 1952 psychotherapy paper ushered in a much needed sense of empirical accountability, despite its shortcomings. He gave many other areas a similar jolt, generally to beneficial effect. To update John Stuart Mill, Eysenck was a 'necessary intellectual heretic'.

Eysenck's willingness to partake in combative debate served to divide support and generate motivated opposition. Those on the wrong end of his pen or tongue felt Eysenck's style was less than conciliatory, resembling a prosecuting lawyer selectively marshalling facts for a preferred point of view. Even close colleagues suggested that he would have done well in the legal profession.[15] Often thought of as self-interested bias, Eysenck's partisan advocacy reflected a genuinely held *real politic* understanding of his mission. For Eysenck, science was a zero-sum game; it had to be played ruthlessly, for there could only be one winner. Critic and friend Cyril Franks went as far as to suggest that Eysenck was not afraid of 'denting the truth' in order to win, even if his 'heart was in the right place and his intentions were good'.[16] Not only was he selective in the way he presented evidence, Franks recalled, he was consciously so—even if he only admitted this in private. 'Oh you noticed that move did you?' he would say, as if engaging in an enjoyable joust of chess or tennis. However, in public confrontations, Eysenck always played it straight. To acknowledge the strategic level of his thinking would give too much ammunition to opponents. Jeffrey Gray likewise said he quickly learned that you had to read any review Hans wrote 'very carefully', despite being generally sympathetic to his aims.[17] It was the case for this, the argument against that. In many such interpretive pieces Eysenck seemed loath to embrace the heterogeneity of the field, to represent all sides sympathetically for the reader to judge.

Eysenck competitiveness hardly made him a lone wolf, however. He had enough company to suggest that such an approach, even in its most negative

---

[15] Furnham, A. (1998), Contributions to the history of psychology: CXIV. Hans Jürgen Eysenck, 1916–1997, *Perceptual and Motor Skills* **87**, 505–6.

[16] Cyril Franks, telephone interview, 13 July 2002.

[17] Jeffrey Gray, interview, 25 July 2003.

forms, is an integral part of the scientific enterprise. Yet, as sociologist Robert Merton once pointed out, scientists are also guided by a communal spirit. Indeed, they greatly depend on sharing results and information, suggesting that open cooperation is just as important as competitive individualism. It seems that not only is science able to tolerate these two approaches, it utterly depends on both to survive and thrive. While Eysenck was capable of great intellectual generosity, particularly amongst his inner circle, others embodied an ideal of collective knowledge-building far more readily. Amongst his British peers, Donald Broadbent would spring to mind; outside psychology, Nobel-prize-winning American physicist Richard Feynman could be cited. Both were more inclined to share their knowledge and their doubts, as well as the credit and the honours, as part of a collective quest with their fellow scientists.

Many assumed Eysenck waded into sensitive issues because he wanted the attention, that he was addicted to the limelight. Jeffery Gray recalled that he had once tried to get Eysenck interested in gender differences he had found in his animal research in the early 1960s. According to Gray, Eysenck could not have been less interested at the time. Yet he became much more interested in this very issue just a few years later. Why the change of heart? Gray was sure it was not because the data had become more compelling and the ideas more intriguing. Rather, the social context for them had changed. Feminism had turned this into a hot button issue, a 'limelight attracting topic'.[18] Yet Eysenck claimed he never deliberately courted attention by provoking controversy, at least not for its own sake, and maintained he would much rather have had his views accepted without dispute.[19] It is a view Sybil Eysenck strenuously defends. Certainly, Eysenck could not have enjoyed the more personal attacks he and his family endured. But he clearly wished to have his ideas taken seriously and actively discussed, if only negatively. A favourite index of his standing was his massive citations counts, and this blunt measure of influence doesn't necessarily discriminate the positive from the negative, the substantive from the perfunctory. It was a measurable means by which he could demonstrate success. As Eysenck noted in his memoirs, his citations far exceeded those of his British contemporaries.[20] Only the likes of Freud and Skinner could hold a candle to him.

---

[18] Ibid. See also Gray, J. (1997a), Foreword to *The scientific study of human nature: tribute to Hans J. Eysenck at eighty* (ed. H. Nyborg), pp. xi–xiii, Elsevier, Oxford.

[19] Eysenck, H.J. (1986c), Consensus and controversy: two types of science, p. 375.

[20] See the concluding sections of Eysenck, H.J. (1997b), *Rebel with a cause* [revised and expanded], Transaction Press, New Brunswick, New Jersey, and Rushton, J.P. (2001), A scientometric appreciation of H.J. Eysenck's contribution to psychology, *Personality and Individual Differences* 3, 17–39.

More than this, Eysenck's attraction to controversy appeared to stem from a need to test himself according to his own assumptions. If one is working according to an adversarial model, go for a contentious, adversarial issue. Eysenck and controversy were a natural marriage. It gave him the opportunity to display his rhetorical skills, affirming his way of doing science in the process. It was the game rather than the limelight he was addicted to. He loved to challenge himself and those around him. Colleagues like Paul Barrett and Adrian Furnham reported that, in idle moments, Eysenck would goad them into an argument, deliberately take an extreme position, and then win in a canter.[21] William Yule recalled how Eysenck once said he loved picking up his mail, for it gave him the chance to pick and choose which debates to engage in. Eysenck's more gratuitous interventions can be partly read as playful intellectual cheek, a need to challenge himself tethered to a discursive dialogical ideal.

Eysenck enjoyed robust debate, and the way he baited certain groups helped create it. His passive–aggressive pose cleverly exploited the politely vicious side to British intellectual life. He presented a big target that was hard to resist. The complaints of those foolhardy or brave enough to take him on could be taken as sour grapes—Eysenck was simply better at the same kind of adversarial sport they had willingly engaged in. By this reckoning, he was a massive success. Despite some of the more damaging episodes I have highlighted, Eysenck's win–loss record was exceptional. He could defend his record brilliantly. Nonetheless, playing it this way inevitably turned him into a man inflexibly wedded to his past, making it harder for him to be part of the scientific future. As Michael Rutter opined: 'What was striking with Eysenck was that he had a formidable intelligence, but instead of harnessing this to move things forward, he came increasingly to defend the indefensible.'[22]

Eysenck wanted to win, and to win over everyone, if possible. He made as much effort with schoolchildren, special interest groups, and the general public as he did with the upper echelons of his discipline, maybe more. His popular works often poked fun at established orthodoxy, playing to an educated laity at the expense of some of his eminent fellow scientists. In turn, Eysenck's very visible commitment to a variety of unpopular positions would give them plenty of reasons to reject him in his lifetime, and a lofty shingle upon which to hang decidedly measured appraisals when he was gone. Those he offended or disappointed, those who did not care for Eysenck's competitive ethic, were offended

---

[21] Furnham (1998), Contributions to the history of psychology: CXIV. Hans Jürgen Eysenck, 1916–1997, p. 505.

[22] Michael Rutter, personal communication, 15 January 2008.

or disappointed *on behalf of their science*. 'If only... a man of his gifts could have done so much more...'

Eysenck's adversarial style was the flip-side to his detached praxis, and both reinforced contradictory perceptions of him. Eysenck was competitively destructive in his outreach, but positively constructive within his coterie. His colleagues, former students, collaborators, and friends saw a very different man from the public image. They found him quiet and friendly, stimulating and supportive, inclusive and trusting. While he was devastating in his criticism of rivals, he was consistently supportive of the work of those he saw as allies. And despite portraying himself as a data-driven empiricist trusting numbers rather that people, Eysenck was personally faithful in practice. He gave almost unconditional loyalty, and he inspired it in return. He came to Jensen's aid when no one else would, took up with Ronald Grossarth-Maticek, defended 'race Professor' Philippe Ruston's right to speak when he was under pressure. But to adapt Richard Herrnstein's comment about intelligence, loyalty might best be regarded as a trait not a virtue. While commonly seen as a touching or noble, loyalty still prompts questions about its basis. Adrian Furnham observed that Eysenck appeared to make his judgments only on rational, cognitive grounds. If you had good ideas, if you could argue your case well, then Eysenck decided you 'were a good egg'.[23] It was a rule of thumb that sometimes landed him in trouble, for it seemed he was not always a good judge of character.

Many former students and research assistants, such as Frank Farley, Philippe Rushton and Paul Barrett, felt they owed their careers to Eysenck. He gave them their big break, taught them how to think, how to do research, how to succeed. During Eysenck's reign at the IoP, at least 200 research graduates were turned out, mostly doctorates. Many reached very senior positions in the UK and across the world. While he was not every young student's cup of tea, most felt genuinely empowered by their experience with him. To be part of an immense, ongoing programme that had the potential to re-shape personality psychology and cognate areas was hugely exciting. Eysenck nurtured by example and, within certain constraints, gave them the freedom to grow as researchers. Quietly prodding them from time to time, he showed great confidence in their capacity to get the job done. They, in turn, knew hard work would be rewarded. Eysenck knew how to access journals, knew how to organize sessions at conferences, knew how to make it all count, and he could help them do the same. Eysenck's name and work commanded the attention of his peers, and their work in his programme would too.

---

[23] Adrian Furnham, interview, 23 July 2003.

Eysenck's students are one very important component of his legacy and continuing influence, one he was quite consciously proud of. He used to keep a tote board mounted outside his office, with a continually updated listing of the progress and destination of his former students and protégés. Many had come from far-flung places and later went off to spread the word. He was an internationalist in this sense, appearing to prefer non-British outsiders like himself. They returned the favour, spreading his word and defending his good name.

Quantification, objectivity, and trust

Lurking like a poltergeist in and around contradictory perceptions of Eysenck was the notion of trust. If recent scholarship tells us anything, it is that trust is a subtle but crucial ingredient in the scientific enterprise. The rise of quantitative methods in the social sciences, for example, has been identified with a kind of mechanical, depersonalized objectivity, a way of overcoming the drawbacks associated with a dependence on learned judgement. Quantitative data is not subject to personal guarantees or capricious whim, and it can be extended far beyond the particular circumstances and personnel producing it.[24] But wordless numbers still need to be spoken for—an act of expert ventriloquism best performed when hardly noticed. And they still require trust, however hidden by de-contextualizing methodologies and the erasure of personal authorship.

Post-war British psychology had no greater spokesman, no bigger salesman, than Hans Eysenck. And therein lay the catch. By appointing himself sage for the hard-nosed empiricism of individual differences, Eysenck called attention to himself in a way that heightened the uncertainties his approach was supposed to dispel. Those most inclined to put their faith in Hans were those who knew him, and knew him in an up close and personal way. And, apart from a few well-documented exceptions, those least inclined to take him at his word tended to be those who did not know him, *but who came to think that they did*. For a significant chunk of his fellow psychologists, neither the man nor his numbers could be trusted. After a while, it was impossible to tell which came first. As one anonymous critic told Eysenck's first biographer Gibson, 'I do not think evaluations of Eysenck's work can be separated from Eysenck as

---

[24] As Lorraine Daston argued, the adoption of such methods was an acknowledgement of political weakness rather than strength, a surrendering of practitioners' authority to judge to the visible rigour of empirical protocols and suppress any dependence on craft or bodily skill. See Daston, L. (2001), Scientific objectivity with and without words, in *Little tools of knowledge: historical essays on academic and bureaucratic practices* (ed. P. Becker and W. Clark), pp. 259–84, University of Michigan Press, Ann Arbor, Michigan.

a person.'²⁵ Time and again I encountered this kind of arch scepticism. Few had worked closely with Eysenck, but many fancied they knew enough about him to write him off as some kind of self-serving show-pony or cheat. Bad faith built up. Eysenck's more contentious latter-day interventions, his pragmatic approach to funding, his associations with some dubious outsiders, all helped to feed an increasingly negative interpretation of his character. His back catalogue was re-interpreted accordingly. Every small Eysenck error, every recorded mistake or fudge in his earlier work took on an ever darker significance, evidence of original sin.

In an otherwise sympathetic appraisal, *The Guardian's* Joan Freeman felt obliged to note that 'in spite of constant accusations of manipulating figures to produce his desired results, no one ever proved that he did'.[26] Perhaps Eysenck was not a suspiciously 'lucky' researcher—as some critics privately suggested to me. Certainly what came out of his labs and went into his books and articles were the edited highlights. However, one could just as easily argue that this form of selectivity reflected an implicit understanding of what was demanded and rewarded in post-war psychology. With his finger in so many pies, with all those students under his wing, he simply had more to choose from than most. The uncharitable would be better off marking him down as a one-eyed champion of the one-off result, for he did not always do follow-up well. And if they thought of him as a monster, then he was a monster of the discipline's own making—a cheerfully efficient adaptation to the institutionalized pathologies said to bedevil post-war psychology.[27] He dashed off his publications at a furious rate, packing off his intellectual offspring before they had a chance to grow up. While Eysenck's dimensional scheme stood the test of time, the wafer-thin basis for many of his more speculative conjectures could be one reason why they lacked robustness. Such a line would hardly characterize all his empirical resources—certainly not the painstaking data generated in his

---

[25] Gibson, H.B. (Tony) (1981), *Hans Eysenck: the man and his work*, Peter Owen, London, p. 42.

[26] Freeman (1997), The pugnacious psychologist [obituary for Hans Eysenck].

[27] Selective reporting has long been a part of modern psychology, especially in relation to the practice of null-hypothesis significance testing. Critics have argued that these statistical protocols greatly distorted publication practices and inhibited theory development because they enshrined the non-reporting of non-significant results. There is a wealth of literature on this point stretching back to the 1960s. A good place to go for a more general outline of the pathologies of modern psychology would be Lykken, D.T. (1991), What's wrong with psychology anyway? in *Thinking clearly about psychology: matters of public interest*, Vol. 1 (ed. D. Cicchetti and W.M. Grove), pp. 3–39, University of Minnesota Press, Minneapolis.

partnerships with Irene Martin, Lindon Eaves, Peter Broadhurst, Paul Barrett, and wife Sybil, for example. But like a linen suit, *some* of the specifics didn't travel well. If follow-up was a problem, then so too was openly acknowledging and adjusting to these failures. The frustration this produced fuelled his critics' doubts.

The scurrilous whisperings of Eysenck's detractors accompanied and even partially substituted for the more usual forms of critical public discourse. As William Yule explained, Eysenck's intimidating reputation went 10 feet in front of him. Those he troubled learned not to take him on openly. As a result, Eysenck's career was conducted against a distant background rumble of scepticism and scorn. It went beyond fear and beyond envy, however, pointing to the notion of trust as a performative, constructive process.

Gaining and maintaining trust is hard work, but it does work. It helps consolidate claims, close gaps in understanding, and build consensus. The 'virtual witnessing' of scientific papers and experimental reports and journal papers depends on it. It is the ultimate guarantee that you did what you said you did, the endpoint to the documentation protocols designed to make research more accountable and replicable. In this way, trust and fraud go hand-in-hand.

What should now be readily apparent was that many of Eysenck's work practices demanded and even *increased* the need for trust. Eysenck's attempts to shore up support for his grander theoretical integrations demanded great belief, requiring his peers to redirect a lifetime's work. But his cursory methodological descriptions and casual referencing, his habit of writing popular even on most technical of subjects, all required his audience take it from him. These habits made replications and extensions of his many promising results much more difficult. Unless one was working closely with him, one would get little sense of the contingency of his conclusions, little idea of ways it might be possible to duplicate or improve on his surveys and experiments. The fact that he avoided the nitty-gritty of experimentation made it hard for him to communicate a shared framework and bring outsiders into his orbit. Instead, a kind of siege mentality characterized the latter part of his career, and that tended to cut him off from new people, ideas, and alliances.

Eysenck's insensitivity to the myriad demands for trust was rooted in a remarkable self-belief. Here was a man, Jeffrey Gray wistfully recalled, who was absolutely certain of his own greatness.[28] Arthur Jensen concurred, allowing that Eysenck lacked the capacity for self-criticism.[29] Such a trait may well be a

---

[28] Jeffrey Gray, interview, 25 July 2003.

[29] Arthur Jensen, interview, 1 July 2002.

prerequisite for doing big things. Eysenck was certainly not alone amongst correlational psychologists to have grand, systematizing impulses. However, he seemed to want to press the claim that his system was more compelling than any other. *His* dimensions were not just summarizing variables or organizing principles; they represented the underlying structure of psychological reality. The latent Platonic realism in Eysenck's thinking would, on occasions, rear its head for all to see. In Plato's cave parable, ordinary folk may only be able to see the shadows in the cave, the visible imperfect flickerings of perfect forms. Data was the imperfect and sometime deceptive or misleading representation of this reality. It was as if Eysenck—and perhaps few other gifted thinkers—had chanced a glimpse of the magisterial order that lay beyond. The nominalism of his later career—wherein the status of his personality factors and his version of general intelligence *g* were downgraded to that of useful theoretical 'constructs'—hid deeper convictions dating back to when these ideas were formulated. Eysenck's search for their biological underpinnings reflected an abiding faith that such certainty could and would be found.

Much of this is speculation, of course, for Eysenck was not keen on self-examination. There is not much in the way of introspection in his memoirs, for example, especially when it didn't suit him. However in one particular latter-day piece, an apologia for Burt, Eysenck gave an intriguing insight to his thought processes. In discussing 'Why do scientists cheat?' Eysenck wrote:

> their finest and most original discoveries are rejected by the vulgar mediocrities filling the ranks of orthodoxy ... The figures do not quite fit, so why not fudge them a little bit to confound the infidels and unbelievers? Usually, the genius is right, of course (if he were not, we should not regard him as a genius), and we may in retrospect excuse the childish games, but clearly this cannot be regarded as a licence for non-geniuses to foist their absurd beliefs on us. Such, then, are the conditions in which leading scientists may be led to falsify their data, or invent them. Convinced (rightly as it happens) that they have made a stupendous discovery, ... they see that discovery threatened by enemies or hostile people with the power to destroy them. The first findings in trying to substantiate a theoretical discovery are never clear cut ... But enemies would seize upon these anomalies to destroy his theory. Obviously the way to overcome this problem is simply to make sure the data fit the theory![30]

Eysenck was at pains to compare towering scientific figures, such as Isaac Newton, with those he regarded as charlatans, such as Sigmund Freud.

---

[30] Eysenck, H.J. (1995*a*), Burt and hero and anti-hero: a Greek tragedy, in *Burt: fraud or framed?* (ed. Nicholas J. Mackintosh), pp. 111–29, Oxford University Press, Oxford, p. 126. I have to thank Michael Eysenck for directing my attention to this intriguing piece.

Both felt the need to massage data to fit their theories, Eysenck claimed, and were caught doing so. For geniuses like Newton, cheating was an understandable foible, a way of avoiding giving 'priceless ammunition to the hostile continental physicists jeering at him'.[31] For the deluded and unscrupulous Freud, it was an unforgivable sin that derailed scientific progress. Eysenck's analysis was alarmingly pragmatic. Cheating only really mattered if the cheater was wrong. Newton was on the right track, just as Burt was; Freud was not. Thus, only Freud deserved infamy. The moral seemed to be that one must *know* one is a genius—and therefore correct—before one cheats. It seemed odd that, for a man not given to introspection, the greatest scientific sin would be to not follow the injunction: 'know thyself'.

Perhaps this passage tells us as much about Eysenck's view of his place in the scheme of things as it did about his attitude to Burt's alleged fraud. It can certainly be read as a recipe for scientific individualism wherein singular flair should trump mediocre collectivism and be excused its occasional waywardness. The enjoined, communal nature of science—with all its checks and balances—was a cumbersome burden for the impatiently gifted. Those touched by special powers should not be constrained by the usual rules; in fact, they could probably write their own.

Judging from the impression Eysenck left with both friends and foes, he assumed he possessed a level of self-awareness that matched his scientific brilliance. As Eysenck often argued, a scientist's primary responsibility was to tell the truth.[32] As he just as often suggested, the evidence is always incomplete, a set of clues. One must pick a side, he would tell his colleagues, even though a partial picture is only ever available. Eysenck thought of himself as someone who could see what others couldn't. And what mattered was that he knew he had that power, and that furnished him with an overriding responsibility to his science. Thus he felt at liberty to take up other people's work, the better to alert them to its true significance. Thus he was unabashed about appropriating whatever suited his purposes. He had glimpsed the truth and this gave him a licence to prosecute the facts aggressively. Partisan advocacy was not just an option, it was a *duty*. He may have played it as a game, but he truly believed in the positions he defended. Eysenck's selectivity and interpretative leaps might be seen as cynical stratagems, but they hid a profound faith in his capacity to carve human nature at its joints.

---

[31] Ibid., p. 125.

[32] Eysenck, H.J. (1975a), The ethics of science and the duties of scientists, *British Association for the Advancement of Science* **1**, 23–5.

It was a sophisticated version of an ends-justifies-the-means argument based on the power of one. Many of Eysenck's supporters appeared to follow his lead—especially when confronted with the spectre of doubtful research and Eysenck's role in it. Ronald Grossarth-Maticek's work was cited as a case in point. If this research has 'legs', they argued, it will replicate. One need not care if corners were cut, if some procedures were a bit dodgy, if reporting practices were a bit economical with the truth. Yet if one believes in the ideal of Science at all, the notion that anything goes as long as it replicates *won't work* and *can't work*. Surprising results would always leave researchers in a dilemma: is it worth following it up? There has to be trust to head off any suspicion they were being played for fools, to justify the investment follow-up would demand. An overriding faith that it will all be sorted out in the end absolutely depends on the honest toil of all parties. No scientist, no matter how clever, should be above such strictures. No scientist can be sure of his or her own righteous prescience. And no scientist can really know just how history will treat them, not even Hans Eysenck.

A prophet without honours

So far, history has been relatively kind to him. Without qualification, Eysenck was the most prominent psychologist in post-war Britain, and probably the most influential. Yet during his lifetime he received only belated acknowledgement in the US and was never truly honoured in his adopted homeland.[33] As Gordon Claridge remarked, 'the distance was too great between him and

---

[33] Aided by the support of former students and colleagues who had risen to positions of power in the States—Martin Seligman and Frank Farley in particular—various American and international bodies gave Eysenck his due recognition. The American Psychological Association presented Eysenck with the Distinguished Scientist Award in 1988, a Presidential Citation in 1993, and the Clinical Division Centennial Award for Outstanding Contributions to Clinical Psychology in 1996. The American Psychological Society awarded him its highest honour, the William James Fellow Award, in 1994. The International Society for the Study of Individual Differences gave Eysenck the Distinguished Contribution Award in 1991. The only real recognition the British Psychological Society gave him came after he was gone, with the establishment of an annual memorial lecture. Much of the above information came from the various obituaries already cited, plus Frank Farley's somewhat belated obituary, Farley, F. (2000), Hans J. Eysenck (1916–1997), *American Psychologist* **55**, 674–5. A few variations in detail arise that are hard to resolve. For example, the exact title of the first APA award mentioned is unclear, even after a perusal of the records on their website, while the year of his Presidential Citation has been variously put at 1993 or 1994.

the Oxbridge clubbishness that determines such matters.'[34] Always a reluctant social networker, he remained a foreigner in many senses—too ambitious, too controversial, too much the non-conformist. Moreover, his habit of playing to the masses sacrificed support at the big end of town. While Eysenck was made a fellow of British Psychological Society, he never attained honorary status. He was never elected Fellow of the Royal Society either, a snub Michael Rutter said particularly irked him.[35] Official respect was something Eysenck valued, whether he could afford to admit it or not. Early in his career, a small piece on him as a 'man to watch' had appeared in *Time*. Arthur Jensen recalled how proud he was of this.[36] But, as the controversies mounted and the enemies list grew, Eysenck had to forgo the kind of recognition a prominent figure might expect as his due. Instead he managed to turn this on its head and remodelled himself as a rebel, outwardly proud of his non-acceptance in established circles.

At an organizational level, Eysenck gave up on dominating or re-modelling existing disciplinary bodies and instead created his own. People joined him rather than the other way round. Despite a reluctance to push the idea of a dogmatic 'Eysenckian school', his ideas and research approach continue to evolve in the hands of an international network of admirers. This network centres around journals like *Personality and Individual Differences* and *Behaviour Research and Therapy*, and the International Society for the Study of Individual Differences—all of which Eysenck was pivotal in founding. They are full of his intellectual descendants, largely 'correlational' researchers. While no one can emulate his breadth, different groups of researchers still give more than a passing nod to this or that aspect of his immense *oeuvre*. Each edition of *PAID*, for example, is still a bumper issue with reference lists dominated by the Eysenck name—testimony to the vitality of his version of the London school tradition it upholds.

Coda

I am well aware that I have not covered everything in Eysenck's prodigious career. There will be more stories to tell about his work on tests, crime and personality, astrology and parapsychology, genius and creativity, and so on. The stories historians tell are often dictated—sometimes embarrassingly

---

[34] Claridge (1998), Contributions to the history of psychology: CXIII. Hans Jürgen Eysenck (4 March 1916–4 September 1997), an appreciation, p. 394.

[35] Michael Rutter, interview, 26 April 2002.

[36] Arthur Jensen, interview, 1 July 2002.

so—by access to historical information. Although the emphases I have given to some topics blessed with archival or obscure material may look awkward, I have tried to use them to serve more general purposes, to illustrate a larger truth about Eysenck's career and the practice of science. Even so, I know there will be arguments about the factual record, and my interpretations, accents, and emphases. I suspect this will only be the beginning of a new debate about Hans Eysenck's legacy and influence.

Primary sources

Archival sources

Hans J. Eysenck Collection, Centre for the History of Psychology, Staffordshire University, Stoke-on-Trent.
Milton Rokeach Papers, Archives of the History of American Psychology.
Personenakten der Reichkulturkammer, Bundesarchiv, Berlin.
NSDAP-Gaukartei, Bundesarchiv, Berlin.
Bibliothek für Bildungsgeschichtliche Forschung, Landesarchiv, Berlin.
Cyril Burt Papers, Special Collections and Archives, University of Liverpool.
Aubrey Lewis Papers, Bethlem Royal Hospital Archives and Museum.
Records of the British Psychological Society, History of Psychology Centre, BPS, London.
Papers and Correspondence of John Randal Baker, Special Collections and Western Manuscripts, Bodleian Library, Oxford University.
Papers and Correspondence of Kurt Alfred Georg Mendelssohn, Special Collections and Western Manuscripts, Bodleian Library, Oxford University.

Interviews

By the author:

Manfred Amelang, interview, 1 February 2003.
Michael Berger, interview, 28 October 2001.
Gordon Claridge, interview, 24 July 2003.
Alan and Ann Clarke, interview, 25 April 2002.
Sidney Crown, interview, 22 April 2002.
Alan R. Dabbs, interview, 13 October 2003.
Alan R. Dabbs, telephone interview, 23 June 2005.
Michael Eysenck, interview, 28 October 2001.
Sybil Eysenck, interview, 21 October 2001; 25 July 2003; 13 September 2004.
Frank Farley, interview, 12 July 2002.
Cyril Franks, telephone interview, 13 July 2002.
Chris Frith, interview, 26 October 2001.
Desmond Furneaux, interview, 26 October 2001.
Adrian Furnham, interview, 23 July 2003.
Jeffrey Gray, interview, 25 July 2003.
Ronald Grossarth-Maticek, Interview, 31 January–1 February 2003.

Gisli Gudjonsson, interview, 25 October 2001.
David Hemsley, interview, 22 July 2003.
Arthur Jensen, interview, 1 July 2002.
A.R. Jonckheere, interview, 22 April 2002.
Donald Kendrick, interview, 13 October 2003.
Isaac Marks, telephone interview, 5 October 2005.
Irene Martin, interview, 27 October, 2001.
Ralph McGuire, telephone interview, 11 August 2005.
David Nias, interview, 24 October 2001.
Richard Passingham, interview, 24 July 2003.
Jack Rachman, interview, 24 June 2002.
J. Philippe Rushton, interview, 3–4 July 2002.
Michael Rutter, interview, 26 April 2002.
Martin Seligman, interview, 12 July 2002.
David Shapiro, interview, 14 October 2003.
Henk van der Ploeg, interview, 11 February 2003.
Hermann Vetter, interview, 31 January–1 February 2003.
Glenn Wilson, interview, 23 October 2001.
William Yule, interview, 22 July 2003.

By others:

Peter L. Broadhurst, interviewed by H.B. Gibson, 30 January 1979.
Hans J. Eysenck, interviewed by H.B. Gibson, 1 March 1979.
Victor Meyer, interviewed by Maarten Derksen, 26 April 1999.
Linford Rees, interviewed by H.B. Gibson, 28 February 1979.

Personal communication and secondary correspondence

Manfred Amelang, personal communication, 15 January 2008; 16 January 2008.
Louis Appleby, personal communication, 22 September 2003.
Paul Barrett, personal communication, 9 September 2008; 11 September 2008.
Brigitte Berg, personal communication, 3 December 2003.
Michael Eysenck, personal communication, 20 November 2001; 12 December 2001; 3 December 2003.
Robert Green, personal communication, 9 February 2002.
Eric Hobsbawm, personal communication, 21 September 2004.
Arthur Jensen, personal communication, 23 January 2008.
Penelope Leach, personal communication, 8 October 2003.
John Mollon, personal communication, 24 June 2003.
Anthony J. Pelosi, personal communication, 29 September 2003.

Pat Rabbitt, personal communication, 5 September 2003; 8 September 2003; 10 September 2003.

Michael Rutter, personal communication, 9 May 2002; 15 January 2008.

Hans J. Eysenck to Ann Clarke, 19 November 1979; copy supplied by Ann Clarke.

Bob Payne to Alan R. Dabbs, October 1994; copy supplied by Alan R. Dabbs.

Bibliography

Atticus (1970). The row which is splitting psychology in two. *The Sunday Times*, 25 January 1970.

Abrahams-Sprod, M.E. (2007). Life under siege: the Jews of Magdeburg under Nazi rule. Ph.D. thesis, University of Sydney http://hdl.handle.net/2123/1627.

Ackerman, N.W. and Jahoda, M. (1950). *Anti-semitism and emotional disorder: a psychoanalytic interpretation*. Harper and Row, New York.

Adorno, T.W., Frenkel-Brunswik, E., Levinson, D.J., and Sanford, R.N. (1950). *The authoritarian personality*. Harper and Row, New York.

Akass, J. (1971). Come outside, stupid, and say that again. *The Sun*, 18 June 1971.

Albino, R.C. (1953). Some criticisms of the application of factor analysis to the study of personality. *British Journal of Psychology* **44**, 164–8.

Allderidge, P. (1991). The foundation of the Maudsley Hospital. In *150 years of British psychiatry, 1841–1991*, Vol. 1 (ed. G.E. Berrios and H. Freeman), pp. 79–88. Royal College of Psychiatrists, London.

Allport, F.H. and Hartman, D.A. (1925). The measurement and motivation of atypical opinion in a certain group. *American Political Science Review* **14**, 735–60.

Allport, G. (1937). *Personality: a psychological interpretation*. Holt, New York.

Allport, G (1954). *The nature of prejudice*. Addison-Wesley, Reading, Massachusetts.

Allport, G. and Vernon, P.E. (1930). The field of personality. *Psychological Bulletin* **27**, 677–730.

Allport, G. and Vernon, P.E. (1931). *Study of values: a scale for measuring the dominant interests in personality*. Houghton Mifflin, Boston.

Altemeyer, B. (1981). *Right-wing authoritarianism*. University of Manitoba Press, Manitoba.

Amelang, M. (1991). Tales from Crvenka and Heidelberg: what about the empirical basis? *Psychological Inquiry* **2**, 233–6.

Amelang, M. (1997). Using personality variables to predict cancer and heart disease. *European Journal of Personality* **11**, 319–42.

Amelang, M. and Schmidt-Rathjens, C. (1992). Psychometric properties of modified Grossarth-Maticek and Eysenck inventories. *Psychological Reports* **71**, 1251–63.

Amelang, M., Schmidt-Rathjens, C., and Matthews, G. (1996). Personality, cancer and coronary heart disease: further evidence on a controversial issue. *British Journal of Health Psychology* **1**, 191–205.

Amelang, M., Hasselbach, P., and Stürmer, T. (2004). Personality, cardiovascular disease, and cancer: first results from the Heidelberg Cohort Study of the Elderly. *Zeitschrift für Gesundheitspsychologie* **12**, 102–15.

American Psychiatric Association Task Force on Behavior Therapy (1973). *Behavior therapy in psychiatry*, report 5. American Psychiatric Association, Washington, DC.

Anderson, N.B. and Nickerson, K.J. (2005). Genes, race, and psychology in the genome era: an introduction. *American Psychologist* **60**, 5–8.

Auerbach, C. and Beale, G.H. (1969). Letter to the editor. *New Scientist*, 29 May 1969, 491–2.

Babington Smith, B. (1941). Discussion: the validity and reliability of group judgements. *Journal of Experimental Psychology* **29**, 420–6.

Bagnall, N. and Taylor, P. (1973). Student rowdyism: is there a link? *Sunday Telegraph*, 24 June 1973.

Baistow, K. (2001). Behavioral approaches and the cultivation of competence. In *Psychology in Britain: historical essays and personal reflections* (ed. G.C. Bunn, A.D. Lovie, and G.D. Richards) pp. 309–29. BPS Books, Leicester.

Baker, T.B. and Fiore, M.C. (1992). Elvis is alive, the mafia killed JFK, and smoking is good for you. *Contemporary Psychology* **37**, 1014–16.

Ball, D., Corbett, A., Greatbach, S., Gretton, J., Morton, J., Overy, P., Watson, P., and White, D. (1971). Protest by *New Society* staff [letter to the editor]. *The Sunday Times* 27 June 1971.

Bannister, D. (1970). Review of *The biological basis of personality* by Hans Eysenck. *British Journal of Psychiatry* **116**, 103.

Barbrack, C. and Franks, C. (1986). Contemporary behaviour therapy and the unique contribution of H.J. Eysenck: anachronistic or visionary? In *Hans Eysenck: consensus and controversy* (ed. Sohan Modgil and Celia Modgil), pp. 233–46. Falmer Press, Philadelphia.

Barnholtz-Sloan, J.S., Chakraborty, R., Sellers, T.A., and Schwartz A.G. (2005). Examining population stratification via individual ancestry estimates versus self-reported race. *Cancer Epidemiology, Biomarkers and Prevention* **14**, 1545–51.

Barrett, P. (2001). Hans Eysenck at the Institute of Psychiatry—the later years. *Personality and Individual Differences* **31**, 11–15.

Barrett, P. and Eysenck, H.J. (1992*a*). Brain electrical potentials and intelligence. In *Handbook of individual differences: biological perspectives* (ed. A. Gale and M.W. Eysenck), pp. 255–85. Wiley, London.

Barrett, P. and Eysenck, H.J. (1992*b*). Brain evoked potentials and intelligence: the Hendrickson paradigm. *Intelligence* **16**, 361–81.

Barrett, P. and Eysenck, H.J. (1993). Sensory nerve conduction and intelligence: a replication. *Personality and Individual Differences* **15**, 249–60.

Barrett, P. and Eysenck, H.J. (1994). The relationship between evoked potential component amplitude, latency, contour length, variability, zero-crossings, and psychometric intelligence. *Personality and Individual Differences* **16**, 3–32.

Barrett, P., Daum, I., and Eysenck, H.J. (1990). Sensory-nerve conduction and intelligence: a methodological study. *Journal of Psychophysiology* **4**, 1–11.

Bartlett, F.C. (1930). Experimental method in psychology. *Journal of General Psychology* **30**, 49–66.

Bates, T.C. and Eysenck, H.J. (1993*a*). String length, attention, and intelligence: focussed attention reverses the string length IQ relationship. *Personality and Individual Differences* **15**, 363–71.

Bates, T.C. and Eysenck, H.J. (1993*b*). Intelligence, inspection time and decision time. *Intelligence* **17**, 521–31.

Bates, T.C., Stough, C., Mangan, G., and Pellet, O. (1995). Intelligence and the complexity of the average evoked potential. An attentional theory. *Intelligence* **20**, 27–39.

Beacon (1977). Interview with Hans Eysenck. *Beacon*, February 1977, 8.

Beezhold, F.W. (1953). On criterion analysis. *Journal of the National Institute for Personnel Research* **5**, 176–82.

Belgrave, F.Z. and Allison, K.W. (2006). *African American psychology.* Sage, Thousand Oaks, California.

Benton, D. and Roberts, G. (1988). Effect of vitamin and mineral supplementation on intelligence of a sample of schoolchildren. *The Lancet* **331**, 140–3.

Beresford, D. (1979). Right-wing editor for journal. *The Guardian*, 9 April 1979.

Berg, B. Les indépendants du 1er siècle—biographie de Max Glass. http://www.lips.org./Bio_GlassM_GB.asp.

Berger, M. (2000). Monte Shapiro [obituary]. *The Independent*, 1 May 2000.

Berger, M. and Yule, W. (1979). Retirement of an enthusiast—Dr. M.B. Shapiro. *Bethlem and Maudsley Gazette* winter 1979, 16–17.

Bergin, A. (1963). The effects of psychotherapy: negative results revisited. *Journal of Counseling Psychology* **10**, 244–50.

Bergin, A. and Garfield, S. (eds.) (1971). *Handbook of psychotherapy and behaviour change.* Wiley, New York.

Berrios, G.E. and Freeman, H. (eds.) (1991). *150 years of British psychiatry, 1841–1991,* Vol. 1. Royal College of Psychiatrists, London.

Berrios, G.E. and Porter, R. (eds.) (1995). *A history of clinical psychiatry: the origin and history of psychiatric disorders.* Athlone, London.

Berry, D. (1985). In a mind field. *New Statesman* 6 September 1985.

Bersh, P.J. (1980). Eysenck's theory of incubation: a critical analysis. *Behaviour Research and Therapy* **18**, 13–17.

Bersh, P.J. (1983). The theory of incubation: comments on Eysenck's reply. *Behaviour Research and Therapy* **21**, 307–8.

Billig, M. (1978). *Fascists: a social psychological view of the National Front.* Harcourt Brace, Jovanovich, New York.

Billig, M. (1979). *Psychology, racism and fascism.* Searchlight, London.

Binet, A. and Henri, V. (1895). La psychologie individuelle. *L'Psychologique* **2**, 411–65.

Bird, R. (1969). Letter to the editor. *New Scientist* 15 May 1969, 376.

Black, S.L. (2003). Cannonical [sic] confusions, an illusory allusion, and more: a critique of Haggbloom *et al.'s* List of eminent psychologists (2002). *Psychological Reports* **92**, 853–7.

Blackburn, R. (1970). Review of *The biological basis of personality* by Hans Eysenck. *British Journal of Social and Clinical Psychology* **9**, 398–9.

Blackman, D. (1976). Be on your 'best' behaviour. *Times Higher Education Supplement,* 22 October 1976.

Blair, J. (1973). The assault on Eysenck at the LSE by Jon Blair, an eye-witness. *The Listener,* 17 May 1973.

Blewett, D.B. (1953). An experimental study of the inheritance of neuroticism. Ph.D. thesis, University of London.

Bos, J., Park, D.W., and Pietikainen, P. (2005). Strategic self-marginalization: the history of psychoanalysis. *Journal of the History of the Behavioural Sciences* **41**, 207–24.

Bourdieu, P. (1975). The specificity of the scientific field and the social conditions of the progress of reason. *Social Science Information* **14**, 19–47.

Bowen, R. (1969). Letter to the editor. *New Scientist*, 15 May 1969, 376.

Bowlby, J. (1940). *Personality and mental illness*. Kegan Paul, London.

Breger, L. and McGaugh, J. (1965). Critique and reformation of 'learning theory' approaches to psychotherapy and neurosis. *Psychological Bulletin* **63**, 338–58.

Breger, L. and McGaugh, J. (1966). Learning theory and behaviour therapy: reply to Rachman and Eysenck. *Psychological Bulletin* **65**, 170–3.

British Medical Association, Public Affairs Division (1986). *Smoking out the barons: the campaign against the tobacco industry*. Wiley, Chichester.

Broadhurst, P.L. (1959). Application of biometric genetics to behaviour in rats. *Nature* **184**, 1517–18.

Brody, N. (1987). Controversy and truth: can scientists' contributions be evaluated by studying the controversies they have generated? Review of *Hans Eysenck: consensus and controversy*, ed. Sohan Modgil and Celia Modgil. *Personality and Individual Differences* **8**, 983–4.

Brown, R. (1965). *Social psychology*. Collier-Macmillan, London.

Brown, T.M. The rise and fall of American psychosomatic medicine. At http://www.human-nature.com/free-associations/riseandfall.html.

Brownlee, K.A. (1965). A review of 'Smoking and health'. *Journal of the American Statistical Association* **60**, 722–39.

Buchanan, R. (1997). Ink blots or profile plots: the Rorschach versus the MMPI as the right tool for a science-based profession. *Science, Technology and Human Values* **21**, 168–206.

Buchanan, R. (2002). On *not* 'giving psychology away': the MMPI and public controversy over testing in the 1960s. *History of Psychology* **5**, 284–309.

Buchanan, R. (2003). Legislative warriors: american psychiatrists, psychologists and competing claims over psychotherapy in the 1950s. *Journal of the History of the Behavioral Sciences* **39**, 225–49.

Bunn, G.C. (2001). Introduction. In *Psychology in Britain: historical essays and personal reflections* (ed. G.C. Bunn, A.D. Lovie, and G.D. Richards), pp. 1–29. BPS Books, Leicester.

Bunn, G.C., Lovie, A.D., and Richards, G.D. (eds.) (2001). *Psychology in Britain: historical essays and personal reflections*. BPS Books, Leicester.

Burch, P. (1974). Does smoking cause lung cancer. *New Scientist*, 21 February 1974, 458–63.

Burnham, J.C. (1989). American physicians and tobacco use: two Surgeons General, 1929 and 1964. *Bulletin of the History of Medicine* **63**, 1.

Burt, C. (1915). The general and specific factors underlying the primary emotions. *Report to the British Association for the Advancement of Science* **69**, 45.

Burt, C. (1937). Correlations between persons. *British Journal of Psychology* **28**, 59–96.

Burt, C. (1940). *Factors of the mind: an introduction to factor-analysis in psychology*. University of London Press, London.

Burt, C. (1969). Intelligence and heredity. *New Scientist* 1 May 1969, 226–8.

Burt, C. and Myers, C.S. (1946). Charles Edward Spearman. *Psychological Review* **53**, 67–71.

Carlson, J.S. and Widaman, K.F. (1986). Eysenck on intelligence: a critical perspective. In *Hans Eysenck: consensus and controversy* (ed. S. Modgil and C. Modgil), pp. 103–32 and 135–6. Falmer Press, Philadelphia.

Cavalli-Sforza, L.L. (2000). *Genes, peoples, and languages.* University of California Press, Berkeley.

Cavalli-Sforza, L.L. and Bodmer, W.F. (1971). *The genetics of human populations.* Freeman, San Francisco.

Chambers, E.G. (1932). Statistical psychology and the limitations of the test method. *British Journal of Psychology* **33**, 189–99.

Chaytor, R. (1973). Revolutionaries on the campus. *Birmingham Evening Star*, 16 July 1973.

Checkley, S. (1993). Letter to the editor, *British Medical Journal* **307**, 329.

Chipuer, H.M., Rovine, M., and Plomin, R. (1990). LISREL Modelling: genetic and environmental influences on IQ revisited. *Intelligence* **14**, 11–29.

Chomsky, N. (1959). Review of *Verbal behavior* by B.F. Skinner. *Language* **35**, 26–58.

Christie, R. (1955). Review of *The psychology of politics* by Hans Eysenck. *American Journal of Psychology* **68**, 702–4.

Christie, R. (1956*a*). Eysenck's treatment of the personality of communists. *Psychological Bulletin* **53**, 411–30.

Christie, R. (1956*b*). Some abuses of psychology. *Psychological Bulletin* **53**, 439–51.

Clare, A.W. (1973). Eysenck the controversialist. *The Spectator*, 21 April 1973.

Claridge, G. (1961). Arousal and inhibition as determinants of the performance of neurotics. *British Journal of Psychology* **52**, 53–63.

Claridge, G. (1986). Eysenck's contribution to the psychology of personality. In *Hans Eysenck: consensus and controversy* (ed. S. Modgil and C. Modgil), pp. 73–85. Falmer Press, Philadelphia.

Claridge, G. (1997). Eysenck's contribution to understanding psychopathology. In *The scientific study of human nature: tribute to Hans J. Eysenck at eighty* (ed. H. Nyborg), pp. 364–88. Pergamon, Oxford.

Claridge, G. (1998). Contributions to the history of psychology: CXIII. Hans Jürgen Eysenck (4 March 1916–4 September 1997), an appreciation. *Psychological Reports* **83**, 392–4.

Clarke, A.D.B. (1950). The measurement of emotional stability by means of objective tests: an experimental inquiry. Ph.D. thesis, University of London.

Clarke, A. and Clarke, A. (1974). *Mental deficiency.* Methuen, London.

Coard, B. (1971). Review of *Race, intelligence and education* by Hans Eysenck. *The Times Educational Supplement*, 18 June 1971.

Cohen, D. (1977). Interview with Hans Eysenck. In *Psychologists on psychology* (ed. D. Cohen), pp. 101–25. Routledge and Kegan Paul, London.

Cohn, T.S. (1953). The relation of the F-scale response to a tendency to answer positively. *American Psychologist* **8**, 335.

Collini, S. (2006). *Absent minds: intellectuals in Britain.* Oxford University Press, Oxford.

Collins, A. (2001). The psychology of memory. In *Psychology in Britain: historical essays and personal reflections* (ed. G.C. Bunn, A.D. Lovie, and G.D. Richards), pp. 150–68. BPS Books, Leicester.

Corcoran, D.W.J. (1961). Individual differences in performance after loss of sleep. Ph.D. thesis, University of Cambridge.

Corcoran, D.W.J. (1964). The relation between introversion and salivation. *The American Journal of Psychology* 77, 298–300.

Corcoran, D.W.J. (1965). Personality and the inverted-U relation. *British Journal of Psychology* 56, 267–73.

Corr, P.J. (2000). Reflections on the scientific life of Hans Eysenck. *History and Philosophy of Psychology* 2, 18–35.

Costa, P.T. and McCrae, R.R. (1986). Major contributions to the psychology of personality. In *Hans Eysenck: consensus and controversy* (ed. S. Modgil and C. Modgil), pp. 63–72 and 86–7. Falmer Press, Philadelphia.

Costa, P.T. and McCrae, R.R. (1992). The five factor model of personality and its relevance to personality disorders. *Journal of Personality Disorders* 6, 343–59.

Costall, A. (1992). Why British psychology is not social: Frederic Bartlett's promotion of the new academic discipline. *Canadian Psychology* 33, 633–93.

Costall, A. (2001). Pear and his peers. In *Psychology in Britain: historical essays and personal reflections* (ed. G.C. Bunn, A.D. Lovie, and G.D. Richards), pp. 188–204. BPS Books, Leicester.

Coulter, T.T. (1953). An experimental and statistical study of the relationship of prejudice and certain personality variables. Ph.D. thesis, University of London.

Craik, K.H., Hogan, R., and Wolfle, R.N. (eds.) (1993). *Fifty years of personality psychology*. Plenum Press, New York.

Cramer, J.L. (1996). Training and education in British psychiatry, 1770–1970. In *150 years of British psychiatry, 1841–1991*, Vol. 2 (ed. H. Freeman and G.E. Berrios), pp. 209–36. Athlone Press, London.

Crist, P.G., Marple, W.E., Kaczynski, S.J., and Abrams, T.L. (1986). Re: Jones/Day Liability Summary ('Corporate Activity Project'). http://tobaccodocuments.org/landman/37575.html.

Crown, S. (1965). Review of *Causes and cures of neurosis* by Hans J. Eysenck and Stanley Rachman. *British Journal of Psychiatry* 111, 1234–5.

Crown, S. (1968). Criteria for the measurement of outcome in psychotherapy. *British Journal of Medical Psychology* 41, 31–7.

Current biography (1972). Biographical entry. In *Current biography* (ed. M.D. Candee and C. Moritz), pp. 18–21. H.W. Wilson, London.

Current Contents (1980). This week's citation classic. *Current Contents* 46 (11 August 1980), 275.

Current Contents (1983). This week's citation classic. *Current Contents* 46 (14 November 1983), 22.

D'Adamo, A. (1971). Review of *The IQ argument* by Hans Eysenck. *Commonwealth*, 8 October 1971.

Daily Mirror (1973). Throw them out [editorial]. *Daily Mirror*, 10 May 1973.

Danziger, K. (1997). *Naming the mind: how psychology found its language*. Sage, London.

Danziger, K. (2006). Universalism and indigenization in the history of modern psychology. In *Internationalizing the history of psychology* (ed. Adrian C. Brock), pp. 208–25. New York University Press, New York.

Darlington, C.D. (1971). Review of *Race, intelligence and education* by Hans Eysenck. *The Sunday Times*, 27 June 1971, 29.

Daston, L. (1992). Objectivity and the escape from perspective. *Social Studies of Science* **22**, 597–619.

Daston, L. (2001). Scientific objectivity with and without words. In *Little tools of knowledge: historical essays on academic and bureaucratic practices* (ed. P. Becker and W. Clark), pp. 259–84. University of Michigan Press, Ann Arbor, Michigan.

Davies, C. (1968). What is behaviour therapy? *The Listener*, 15 February 1968.

Davies, M. (1938). A statistical study of individual preferences for olfactory stimulus. MA thesis, University of London.

Davies, M. (1939). The general factor in correlations between persons. *British Journal of Psychology* **29**, 404–21.

Davies, M. (1944). An experimental and statistical study of olfactory preferences. *Journal of Experimental Psychology* **39**, 246–52.

Deary, I. (1985–6). Eysenck at nightfall. *Bethlem and Maudsley Hospital Gazette* **31** (winter 1985–86), 28–31.

Deary, I. (1986). Inspection time: discovery or rediscovery. *Personality and Individual Differences* **7**, 625–31.

Deary, I. (1997a). Intelligence and information processing. In *The scientific study of human nature: tribute to Hans J. Eysenck at eighty* (ed. H. Nyborg), pp. 282–310. Pergamon, Oxford.

Deary, I. (1997b). Tribute to Hans Eysenck. *Personality and Individual Differences* **23**, 713–14.

Deary, I. (1999). The origins of Eysenck's dimensions. *Contemporary Psychology* **44**, 318–19.

Deary, I. and Caryl, P.G. (1987). Intelligence, EEG, and evoked potentials. In *Biological approaches to the study of human intelligence* (ed. P.A. Vernon), pp. 259–315. Ablex, Norwood, New Jersey.

Dehue, T. (1995). *Changing the rules: psychology in the Netherlands, 1900–1985*. Cambridge University Press, Cambridge.

Dehue, T. (1997). Managing distrust. Review of *Trust in numbers: the pursuit of objectivity in science and public life* by Theodore Porter. *Theory and Psychology* **7**, 417–20.

Derksen, M. (2000). Clinical psychology and the psychological clinic: the early years of clinical psychology at the Maudsley. *History and Philosophy of Psychology* **2**, 1–17.

Derksen, M. (2001). Science in the clinic: clinical psychology at the Maudsley. In *Psychology in Britain: historical essays and personal reflections* (ed. G.C. Bunn, A.D. Lovie, and G.D. Richards), pp. 267–89. BPS Books, Leicester.

Desai, M. (1969). The function of clinical psychologists in relation to treatment. *Bulletin of the British Psychological Society* **22**, 197–9.

Desmond, A. and Moore, J. (1991). *Darwin*. Michael Joseph, London.

Doll, R, and Hill, A.B. (1950). Smoking and carcinoma of the lung: preliminary report. *British Medical Journal* **2**, 739.

Doll, R. and Peto, R. (1976). Mortality in relation to smoking: 20 years' observations on doctors. *British Medical Journal* **2** (6051), 1525–36.

Duhrssen, S. and Jorswieck, E. (1963). Zur Korrektur von Eysenck's Berichterstattung über psychoanalytische Behandlungsergebnisse. *Acta Psychotherapeutica* **10**, 329–42.

Eaves, L. (1986). Eaves replies to Loehlin. In *Hans Eysenck: consensus and controversy* (ed. S. Modgil and C. Modgil), pp. 58–9. Falmer Press, Philadelphia.

Eaves, L. and Eysenck, H.J. (1974). Genetics and the development of social attitudes. *Nature* **249**, 288–9.

Eaves, L., Eysenck, H.J., and Martin, N.G. (1989). *Genes, culture and personality: an empirical approach.* Academic Press, New York.

Ehrenfels, U.R. (1962). Critical paragraphs deleted. *Current Anthropology* **3**, 154–5.

Epstein, S. and O'Brien, E.J. (1985). The person–situation debate in historical and current perspective. *Psychological Bulletin* **98**, 513–37.

Ertl, J. and Schafer, E. (1969). Brain response correlates of psychometric intelligence. *Nature* **223**, 421–2.

Erwin, E. (1980). Psychoanalytic therapy: the Eysenck argument, *American Psychologist* **35**, 435–43.

Evans, P. (1971). Review of *Race, intelligence and education* by Hans Eysenck. *The Times*, 16 June 1971.

Eysenck, H.J. (1939*a*). The validity of judgements as a function of the number of judges. *Journal of Experimental Psychology* **25**, 650–4.

Eysenck, H.J. (1939*b*). Critical notice of 'Primary mental abilities' by L.L. Thurstone. *British Journal of Educational Psychology* **9**, 270–5.

Eysenck, H.J. (1940*a*). Some factors in the appreciation of poetry, and their relation to temperamental qualities. *Characteristics of Personality* **9**, 160–7.

Eysenck, H.J. (1940*b*). The general factor in aesthetic judgements. *British Journal of Psychology* **31**, 94–102.

Eysenck, H.J. (1940*c*). Processes of perception and aesthetic appreciation. Ph.D. thesis, University of London.

Eysenck, H.J. (1941*a*). Reply: the validity and reliability of group judgements. *Journal of Experimental Psychology* **29**, 427–34.

Eysenck, H.J. (1941*b*). Personality factors and preference judgements. *Nature* **148**, 346.

Eysenck, H.J. (1941*c*). The empirical determination of an aesthetic formula. *Psychological Review* **48**, 83–92.

Eysenck, H.J. (1941*d*). An experimental study of the improvement of mental and physical functions in the hypnotic state. *British Journal of Medical Psychology* **18**, 304–16.

Eysenck, H.J. (1941*e*). A critical and experimental study of colour preferences. *American Journal of Psychology* **54**, 385–94.

Eysenck, H.J. (1942*a*). The experimental study of the 'good gestalt'—a new approach. *Psychological Review* **49**, 344–64.

Eysenck, H.J. (1942*b*). The appreciation of humour: an experimental and theoretical study. *British Journal of Psychology* **32**, 295–309.

Eysenck, H.J. (1944*a*). Types of personality: a factorial study of seven hundred neurotics. *Journal of Mental Science* **90**, 851–61.

Eysenck, H.J. (1944*b*). General social attitudes. *Journal of Social Psychology* **19**, 207–27.

Eysenck, H.J. (1947a). *Dimensions of personality*. Routledge and Kegan Paul, London.

Eysenck, H.J. (1947b). The measurement of socially valuable qualities. *Eugenics Review* **39**, 103–7.

Eysenck, H.J. (1947c). Primary social attitudes. 1. The organization and measurement of social attitudes. *International Journal of Opinion and Attitude Research* **1**, 49–84.

Eysenck, H.J. (1948). Some recent studies of intelligence. *Eugenics Review* **40**, 21–2.

Eysenck, H.J. (1949). Training in clinical psychology: an English point of view. *American Psychologist* **4**, 173–4.

Eysenck, H.J. (1950a). Function and training of the clinical psychologist. *Journal of Mental Science* **96**, 710–25.

Eysenck, H.J. (1950b). Criterion analysis: an application of the hypothetico-deductive method to factor analysis. *Psychological Review* **57**, 38–53.

Eysenck, H.J. (1951a). Primary social attitudes as related to social class and political party. *British Journal of Sociology* **2**, 198–209.

Eysenck, H.J. (1951b). Psychology Department, Institute of Psychiatry, Maudsley Hospital, University of London. *Acta Psychologica* **8**, 63–8.

Eysenck, H.J. (1952a). *The scientific study of personality*. Routledge and Kegan Paul, London.

Eysenck, H.J. (1952b). Letter to the editor. *Quarterly Bulletin of the British Psychological Society* **3**, 97–8.

Eysenck, H.J. (1952c). The effects of psychotherapy. *Proceedings of the Royal Society of Medicine* **45**, 447.

Eysenck, H.J. (1952d). The effects of psychotherapy. *Quarterly Bulletin of the British Psychological Society* **3**, 41.

Eysenck, H.J. (1952e). The effects of psychotherapy: an evaluation. *Journal of Consulting Psychology* **16**, 319–24.

Eysenck, H.J. (1952f). Uses and abuses of factor analysis. *Applied Statistics* **20**, 345–84.

Eysenck, H.J. (1953a). The application of factor analysis to the study of personality: a reply. *British Journal of Psychology* **44**, 169–72.

Eysenck, H.J. (1953b). *Uses and abuses of psychology*. Penguin Books, London.

Eysenck, H.J. (1953c). The logical basis of factor analysis. *American Psychologist* **8**, 105–14.

Eysenck, H.J. (1953d). On criterion analysis: a reply to F.W. Beezhold. *Journal of the National Institute for Personnel Research* **5**, 183–7.

Eysenck, H.J. (1953e). Primary social attitudes: a comparison of attitude patterns in England, Germany and Sweden. *Journal of Abnormal and Social Psychology* **48**, 563–8.

Eysenck, H.J. (1953f). Letter of resignation. Reproduced in the *Quarterly Journal of Experimental Psychology* **5**, 39, http://www.eps.ac.uk/society/eysenck.html.

Eysenck, H.J. (1954a). *The psychology of politics*. Routledge and Kegan Paul, London.

Eysenck, H.J. (1954b). A note on the review. *British Journal of Psychology (Statistical Section)* **6**, 44–6.

Eysenck, H.J. (1954c). A reply to Luborsky's note. *British Journal of Psychology* **45**, 132–3.

Eysenck, H.J. (1954d). The science of personality: nomothetic! *Psychological Review* **61**, 339–42.

Eysenck, H.J. (1955a). A dynamic theory of anxiety and hysteria. *Journal of Mental Science* **101**, 28-51.

Eysenck, H.J. (1955b). Cortical inhibition, figural after effect, and a theory of personality. *Journal of Abnormal and Social Psychology* **51**, 94–106.

Eysenck, H.J. (1955c). Psychiatric diagnosis as a psychological and statistical problem. *Psychological Reports* **1**, 3–17.

Eysenck, H.J. (1955d). La validité des techniques projectives: une introduction. *Revue de Psychologie Appliquée* **5**, 231–3.

Eysenck, H.J. (1956a). The inheritance of extraversion–introversion. *Acta Psychologica* **12**, 95–110.

Eysenck, H.J. (1956b). Diagnosis and measurement: a reply to Loevinger. *Psychological Reports* **2**, 117–18.

Eysenck, H.J. (1956c). The psychology of politics: a reply. *Psychological Bulletin* **53**, 177–82.

Eysenck, H.J. (1956d). The inheritance and nature of extraversion. *The Eugenics Review* **48**, 23–30.

Eysenck, H.J. (1956e). Reminiscence, drive and personality theory. *Journal of Abnormal and Social Psychology* **53**, 328–33.

Eysenck, H.J. (1956f). The psychology of politics and the personality similarities between fascists and communists. *Psychological Bulletin* **53**, 431–8.

Eysenck, H.J. (1957a). *Sense and nonsense in psychology*. Penguin Books, London.

Eysenck, H.J. (1957b). *The dynamics of anxiety and hysteria*. Routledge and Kegan Paul, London.

Eysenck, H.J. (1958a). The continuity of abnormal and normal behaviour. *Psychological Bulletin* **55**, 429–32.

Eysenck, H.J. (1958b). Personality tests: 1950–1955. In *Recent Progress in Psychiatry*, Vol. 3 (ed. G.W.T.H. Flemming), pp. 118–59. Churchill, London.

Eysenck, H.J. (1958c). Hysterics and dysthymics as criterion groups in the study of introversion–extraversion: a reply. *Journal of Abnormal and Social Psychology* **57**, 250–2.

Eysenck, H.J. (1959a). Scientific methodology and *The dynamics of anxiety and hysteria*. *British Journal of Medicine and Psychology* **32**, 56–63.

Eysenck, H.J. (1959b). Anxiety and hysteria—a reply to Vernon Hamilton. *British Journal of Psychology* **50**, 64–9.

Eysenck, H.J. (1959c). The Rorschach test. In *The fifth mental measurement yearbook* (ed. Oscar Buros), pp. 276–8. Gryphon Press, Highland Park, New Jersey.

Eysenck, H.J. (1959d). Learning theory and behaviour therapy. *Journal of Mental Science* **105**, 61–75.

Eysenck, H.J. (1959e). The inheritance of neuroticism: a reply. *Journal of Mental Science* **105**, 76–80.

Eysenck, H.J. (ed.) (1960a). *Behaviour therapy and the neuroses*. Pergamon Press, Oxford.

Eysenck, H.J. (1960b). What's the truth about psychoanalysis? *Reader's Digest*, January 1960, 38–43.

Eysenck, H.J. (1961). Psychoanalysis—myth or science? *Inquiry* **4**, 1–15.

Eysenck, H.J. (1962). *Know your own IQ*. Penguin Books, London.

Eysenck, H.J. (1963). Personality and cigarette smoking. *Life Science* **3**, 777–92.

Eysenck, H.J. (1964a). Review of *The act of creation*, by Arthur Koestler. *New Scientist*, 18 June 1964.

Eysenck, H.J. (1964b). *Crime and personality*. Routledge and Kegan Paul, London.

Eysenck, H.J. (ed.) (1964c). *Behaviour therapy and the neuroses*, 2nd edn. Pergamon Press, Oxford.

Eysenck, H.J. (1964d). The outcome problem in psychotherapy: a reply. *Psychotherapy: Theory Research and Practice* **1**, 97–100.

Eysenck, H.J. (1964e). Philosophers and behaviourists: a reply to Kathleen Nott. *Encounter* **23**, 53–5.

Eysenck, H.J. (1964f). The effects of psychotherapy reconsidered. *Acta Psychotherapeutica* **12**, 38–44.

Eysenck, H.J. (1965a). Letter to editor. *Times Educational Supplement*, 10 December 1965.

Eysenck, H.J. (1965b). *Smoking, health and personality*. Basic Books, New York.

Eysenck, H.J. (1965c). *Fact and fiction in psychology*. Penguin Books, London.

Eysenck, H.J. (1966a). Conditioning, introversion–extraversion and the strength of the nervous system. In *Proceedings of the 18th International Congress for Experimental Psychology*, pp. 33–44. Moscow.

Eysenck, H.J. (1966b). *Check your own IQ*. Penguin Books, London.

Eysenck, H.J. (1966c). Psychoanalysis—a necrology. *The Twentieth Century* **2**, 15–17.

Eysenck, H.J. (1967a). *The biological basis of personality*. C.C. Thomas, Springfield, Illinois.

Eysenck, H.J. (1967b). Intelligence assessment: a theoretical and experimental approach. *British Journal of Educational Psychology* **37**, 81–98.

Eysenck, H.J. (1967c). Single-trial conditioning, neurosis and the Napalkov phenomenon. *Behaviour Research and Therapy* **5**, 63–5.

Eysenck, H.J. (1968a). An experimental study of aesthetic preference for polygonal figures. *Journal of General Psychology* **79**, 3–17.

Eysenck, H.J. (1968b). A theory of the incubation of anxiety–fear responses. *Behaviour Research and Therapy* **6**, 309–21.

Eysenck, H.J. (1969a). A critique of Jensen. *New Scientist*, 1 May 1969, 228–9.

Eysenck, H.J. (1969b). Letter to the editor. *New Scientist*, 29 May 1969, 490.

Eysenck, H.J. (1969c). Environment—the new dogmatism. *The Times Educational Supplement*, 12 December 1969, 4.

Eysenck, H.J. (1969d). The rise of the mediocracy. In *The crisis in education: black paper No. 2*, pp. 34–40. Critical Quarterly Society, London.

Eysenck, H.J. (1970a). *The structure of human personality*, 3rd edn. Methuen, London.

Eysenck, H.J. (1970b). A mish-mash of theories. *Journal of Psychiatry* **9**, 140–6.

Eysenck, H.J. (1970c). The ethics of psychotherapy. *Question* **3**, 3–12.

Eysenck, H.J. (1970d). Behaviour therapy and its critics. *Journal of Behaviour Therapy and Experimental Psychiatry* **1**, 5–15.

Eysenck, H.J. (1971a). Letter to the editor. *The Sunday Times*, 27 June 1971.

Eysenck, H.J. (1971b). *Race, intelligence and education*. Maurice Temple Smith, London.

Eysenck, H.J. (1971c). *The IQ argument*. The Library Press, New York.

Eysenck, H.J. (1971d). Social attitudes and social class. *British Journal of Social and Clinical Psychology* **10**, 201–12.

Eysenck, H.J. (1971*e*). Behaviour therapy as a scientific discipline. *Journal of Consulting and Clinical Psychology* **36**, 314–19.

Eysenck, H.J. (1971*f*). The decline and fall of the Freudian empire. *Penthouse* **6**, 28–30, 70, 84, and 86.

Eysenck, H.J. (1972*a*). *Psychology is about people*. Penguin, London.

Eysenck, H.J. (1972*b*). The dangers of the new zealots. *Encounter* **39**, 79–91.

Eysenck, H.J. (1973*a*). Letter to the editor. *New Statesman* 18 May 1973, 732.

Eysenck, H.J. (1973*b*). I defied Hitler and now I defy these students. *Daily Mail*, c. early May 1973.

Eysenck, H.J. (1973*c*). Personality and the maintenance of the smoking habit. In *Smoking behaviour: motives and incentives* (ed. W.L. Dunn), pp. 113–46. Winston/Wiley, Washington.

Eysenck, H.J. (1973*d*). Letter to the editor. *The Spectator*, 5 May 1973.

Eysenck, H.J. (1973*e*). *The inequality of man*. Maurice Temple Smith, London.

Eysenck, H.J. (ed.) (1973*f*). *The measurement of intelligence*. Medical and Technical Publishers, Lancaster.

Eysenck, H.J. (1973*g*). Letter to the editor. *New Statesman*, 27 April 1973, 616.

Eysenck, H.J. (1974*a*). Races apart. Review of *Race* by John Baker. *New Society*, 14 February 1974, 298.

Eysenck, H.J. (1974*b*). Race and equality. Review of *Race* by John Baker. *Books and Bookmen*, March 1974, 17–20.

Eysenck, H.J. (1975*a*). The ethics of science and the duties of scientists. *British Association for the Advancement of Science* **1**, 23–5.

Eysenck, H.J. (1975*b*). The structure of social attitudes. *British Journal of Social and Clinical Psychology* **14**, 323–31.

Eysenck, H.J. (1976*a*). Review of *Race and IQ*, ed. Ashley Montagu. *Mankind Quarterly*, **17**, 149–50.

Eysenck, H.J. (ed.) (1976*b*). *Case studies in behaviour therapy*. Routledge and Kegan Paul, London.

Eysenck, H.J. (1976*c*). Behaviour therapy—dogma or applied science? In *Theoretical and experimental foundations of behaviour therapy* (ed. M.P. Feldman and A. Broadhurst), pp. 333–63. Wiley, London.

Eysenck, H.J. (1976*d*). Letter to the editor. *The Sunday Times*, 7 November 1976.

Eysenck, H.J. (1976*e*). *Sex and personality*. Open Books, London.

Eysenck, H.J. (1976*f*). The learning theory model of neurosis: a new approach. *Behaviour Research and Therapy* **14**, 251–61.

Eysenck, H.J. (1977*a*). Letter to the editor. *American Psychologist* **32**, 674–5.

Eysenck, H.J. (1977*b*). How the left has science by the throat. *Daily Mail*, 2 September 1977.

Eysenck, H.J. (1977*c*). Storm troopers of the left: the Australian case. *Quadrant*, November 1977.

Eysenck, H.J. (1977*d*). *Crime and personality*, 2nd edn. Routledge and Kegan Paul, London.

Eysenck, H.J. (1977*e*). *You and neurosis*. Maurice Temple Smith, London.

Eysenck, H.J. (1977*f*). Letter to the editor. *Bulletin of the British Psychological Society* **30**, 22.

Eysenck, H.J. (1977g). The case of Sir Cyril Burt. *Encounter* **48**, 19–24.

Eysenck, H.J. (1978a). Letter to the editor. *Nature* **274**, 738.

Eysenck, H.J. (1978b). Letter to the editor. *The Times*, 16 March 1978.

Eysenck, H.J. (1979a). Psychology's Jekyll and Hyde. Review of *Cyril Burt: psychologist* by Leslie Hearnshaw. *New Scientist*, 26 July 1979, 302.

Eysenck, H.J. (1979b). Letter to the editor. *New Society*, 15 February, 1979.

Eysenck, H.J. (1979c). *The structure and measurement of intelligence.* Springer Verlag, New York.

Eysenck, H.J. (1979d). Special review: behaviour therapy and the philosophers. *Behaviour Research and Therapy* **17**, 511–14.

Eysenck, H.J. (1980a). The biosocial nature of man. *Journal of Social Biology Structure* **3**, 125–34.

Eysenck, H.J. (1980b). Hans Jürgen Eysenck. In *A history of psychology in autobiography*, Vol. 7 (ed. G. Lindzey), pp. 153–87. W.H. Freeman, San Francisco.

Eysenck, H.J. (with Lindon Eaves) (1980c). *The causes and effects of smoking.* Maurice Temple Smith, London.

Eysenck, H.J. (1980d). Editorial. *Personality and Individual Differences* **1**, 1–2.

Eysenck, H.J. (1980e). Professor Sir Cyril Burt and the inheritance of intelligence: evaluation of a controversy. *Zeitschrift für Differentielle und Diagnostische Psychologie* **3**, 183–99.

Eysenck, H.J. (1980f). Psychology of the scientist: XLIV. Sir Cyril Burt: prominence versus personality. *Psychological Reports* **46**, 893–4.

Eysenck, H.J. (1981a). The nature of intelligence. In *Intelligence and learning* (ed. M.P. Friedman, J.P. Das, and N. O'Connor), pp. 67–85. Plenum Press, New York.

Eysenck, H.J. (1981b). Reply to critical notice. *Journal of Child Psychology and Psychiatry* **22**, 299–300.

Eysenck, H.J. (1982a). Neobehaviouristic (S-R) theory. In *Contemporary behaviour therapy* (ed. G.T. Wilson and C.M. Franks), pp. 205–76. Guilford Press, New York and London.

Eysenck, H.J. (1982b). *A model for intelligence.* Springer Verlag, New York.

Eysenck, H.J. (1983a). The theory of incubation: a reply to Bersh. *Behaviour Research and Therapy* **21**, 303–5.

Eysenck, H.J. (1983b). Revolution dans la theorie et la measure de l'intelligence. *La Revue Canadienne de Psycho-Education* **12**, 3–17.

Eysenck, H.J. (1983c). Revolution dans la theorie et la measure de l'intelligence: reponse a quelques critiques. *La Revue Canadienne de Psycho-Education* **12**, 144–7.

Eysenck, H.J. (1983d). *... I do. Your guide to a happy marriage.* Multimedia Publications, London.

Eysenck, H.J. (1983e). Classical conditioning and extinction. In *Perspectives on behavior therapy in the eighties* (ed. M. Rosenbaum, C. Franks, and Y. Jaffe), pp. 77-98. Springer, New York.

Eysenck, H.J. (1983f). Letter to the editor. *Encounter*, June 1983, 91–2.

Eysenck, H.J. (1983g). Forty years at the Maudsley: changes and otherwise. *Bethlem and Maudsley Gazette* **31**, 4–5.

Eysenck, H.J. (1983h). Sir Cyril Burt: polymath and psychopath. *Journal of the Association of Educational Psychology, Centenary Issue* **6**, 57–63.

Eysenck, H.J. (1984a). Scargill and the fascists of the left. *Daily Express*, 19 April 1984.

Eysenck, H.J. (1984b). The effect of race on human abilities and mental test scores. In *Perspectives on bias in mental testing* (ed. C.E. Reynolds and R.T. Brown), pp. 249–91. Plenum Press, New York.

Eysenck, H.J. (1984c). Stress, personality and smoking behaviour. In *Stress and anxiety*, Vol. 9 (ed. C.D. Spielberger, I.G. Sarason, and P.B. Defares), pp. 37–49. Hemisphere, Washington.

Eysenck, H.J. (1985a). H. Gwynne Jones, 1918–1985—an appreciation. *Behavioural Psychotherapy* **13**, 171–3.

Eysenck, H.J. (1985b). The nature of cognitive differences between Blacks and Whites. *Journal of Behaviour and Brain Science* **8**, 229.

Eysenck, H.J. (1985c). *Decline and fall of the Freudian empire*. Viking, London.

Eysenck, H.J. (1986a). Inspection time and intelligence: a historical introduction. *Personality and Individual Differences* **7**, 603–7.

Eysenck, H.J. (1986b). A critique of contemporary classification and ciagnosis. In *Contemporary directions in psychology: toward the DSM-IV* (ed. T. Millon and G.L. Klerman), pp. 73–98. Guilford Press, New York.

Eysenck, H.J. (1986c). Consensus and controversy: two types of science. In *Hans Eysenck: consensus and controversy* (ed. S. Modgil and C. Modgil), pp. 375–98. Falmer Press, Philadelphia.

Eysenck, H.J. (1986d). Nicotine's not all nasty. *The Best of Health*, February 1986.

Eysenck, H.J. (1986e). Toward a new model of intelligence. *Personality and Individual Differences* **7**, 731–6.

Eysenck, H.J. (1987a). Arousal and personality: the origins of a theory. In *Personality dimensions and arousal* (ed. J. Strelau and H.J. Eysenck), pp. 1–13. Plenum Press, New York.

Eysenck, H.J. (1987b). Personality as a predictor of cancer and cardiovascular disease, and the application of behaviour therapy in prophylaxis. *European Journal of Psychiatry* **1**, 29–41.

Eysenck, H.J. (1987c). Thomson's 'bonds' or Spearman's 'energy': sixty years on. *Mankind Quarterly* **27**, 257–74.

Eysenck, H.J. (1988a). Personality as a predictor of cancer and cardiovascular disease, and the application of behaviour therapy in prophylaxis. *British Journal of Clinical and Social Psychiatry* **6**, 4–12.

Eysenck, H.J. (1988b). Preface and behaviour therapy. In *Theoretical foundations of behaviour therapy* (ed. H.J. Eysenck and I. Martin), pp. vii–ix and 3–35. Plenum Press, New York.

Eysenck, H.J. (1988c). The role of heredity, environment, and 'preparedness' in the genesis of neurosis. In *Theoretical foundations of behaviour therapy*, (ed. H.J. Eysenck and I. Martin), pp. 379–402. Plenum Press, New York.

Eysenck, H.J. (1988d). New light on racial differences in intelligence. *MENSA Research Journal* **24**, 13–14.

Eysenck, H.J. (1989a). Ground rules for scientific debates: application to the study of racial differences [paper presented at the 10th International Conference in Human Ethology]. *Ethology and Sociobiology* **10**, 387.

Eysenck, H.J. (1989*b*). Sensitive intelligence issues. Review of *The Burt affair* by Robert Joynson. *The Spectator*, 2 September 1989.

Eysenck, H.J. (1989*c*). Refereeing in psychology journals: a reply from Hans Eysenck. *The Psychologist* **3**, 98–9.

Eysenck, H.J. (1990*a*). A sanguine veteran of psychological warfare. *Sunday Telegraph*, 11 March 1990.

Eysenck, H.J. (1990*b*). *Rebel with a cause*. W.H. Allen, London.

Eysenck, H.J. (1990*c*). Freud, science and Professor Joseph S. Jacob: a reply. *Mankind Quarterly* **31**, 183–4.

Eysenck, H.J. (1991*a*). Were we really wrong? *American Journal of Epidemiology* **133**, 416–25.

Eysenck, H.J. (1991*b*). Personality, stress and disease: an interactionist perspective. *Psychological Inquiry* **2**, 221–32.

Eysenck, H.J. (1991*c*). Reply to criticisms of the Grossarth–Maticek studies. *Psychological Inquiry* **2**, 297–323.

Eysenck, H.J. (1991*d*). Race and intelligence—an alternative hypothesis. *Mankind Quarterly* **32**, 123–5.

Eysenck, H.J. (1991*e*). Maverick psychologist. In *The history of clinical psychology in autobiography*, Vol. 1 (ed. C. Eugene Walker), pp. 39–86. Brooks/Cole, Pacific Grove, California.

Eysenck, H.J. (1991*f*). Introduction: science and racism. In *Race, intelligence and bias in academe* (ed. R. Pearson), pp. 1–55. Scott-Townsend, Washington, DC.

Eysenck, H.J. (1991*g*). *Smoking, personality and stress: psychosocial factors in the prevention of cancer and coronary heart disease*. Springer-Verlag, New York.

Eysenck, H.J. (1992*a*). Psychosocial factors, cancer and ischaemic heart disease. *British Medical Journal* **305**, 457–9.

Eysenck, H.J. (1992*b*). A hundred years of personality research, from Heymans to modern times. Undelivered lecture, University of Amsterdam, 12 February 1992.

Eysenck, H.J. (1992*c*). The definition and measurement of psychoticism. *Personality and Individual Differences* **11**, 757–85.

Eysenck, H.J. (1993*a*). Creativity and personality: suggestions for a theory. *Psychological Inquiry* **4**, 147–78.

Eysenck, H.J. (1993*b*). Reply to van der Ploeg, Vetter and Kleijn. *Psychological Inquiry* **4**, 70–3.

Eysenck, H.J. (1994*a*). The outcome problem in psychotherapy: what have we learned? *Behaviour Research and Therapy* **32**, 477–95.

Eysenck, H.J. (1994*b*). Much ado about IQ. Review of *The bell curve* by Richard Herrnstein and Charles Murray. *The Times Higher Education Supplement*, 11 November 1994.

Eysenck, H.J. (1995*a*). Burt and hero and anti-hero: a Greek tragedy. In *Burt: fraud or framed?* (ed. Nicholas J. Mackintosh), pp. 111–29. Oxford University Press, Oxford.

Eysenck, H.J. (1995*b*). *Genius: the natural history of creativity*. Cambridge University Press, Cambridge.

Eysenck, H.J. (1996). Review of *Joseph Wolpe* by Roger Poppen. *Behaviour Research and Therapy* **34**, 685–6.

Eysenck, H.J. (1997*a*). Personality and experimental psychology: the unification of psychology and the possibility of a paradigm. *Journal of Personality and Social Psychology* **73**, 1224–37.

Eysenck, H.J. (1997*b*). *Rebel with a cause* [revised and expanded]. Transaction Press, New Brunswick, New Jersey.

Eysenck, H.J. (1998). *Intelligence: a new look*. Transaction Press, New Brunswick, New Jersey.

Eysenck, H.J. (2000). Personality as a risk factor in cancer and coronary heart disease. In *Stress and health: research and clinical applications* (ed. D.T. Kenny and J.G. Carlson), pp. 291–318. Harwood, Amsterdam.

Eysenck, H.J. and Barrett, P. (1984). Psychophysiology and the measurement of intelligence. In *Methodological and statistical advances in the study of individual differences* (ed. C.R. Reynolds and V.L. Willson), pp. 1–49. Plenum Press, New York.

Eysenck, H.J. and Beech, H.R. (1971). Counter conditioning and related methods. In *Handbook of psychotherapy and behaviour change* (ed. A. Bergin and S. Garfield), pp. 543–611. Wiley, New York.

Eysenck, H.J. and Castle, M. (1970). More on *Testing negro intelligence*. The Humanist March/April 1970, 34.

Eysenck, H.J. and Coulter, T. (1972). The personality and attitudes of working-class British communists and fascists. *Journal of Social Psychology* **87**, 59–73.

Eysenck, H.J. and Evans, D. (1994). *Test your IQ*. Thorsons/Harper-Collins, London.

Eysenck, H.J. and Eysenck, M.W. (1981). *Mindwatching*. Michael Joseph, London.

Eysenck, H.J. and Eysenck, M.W. (1985). *Personality and individual differences: a natural science approach*. Plenum Publishers, New York.

Eysenck, H.J. and Eysenck, S. (1969). *Personality structure and measurement*. Routledge and Kegan Paul, London.

Eysenck, H.J. and Eysenck, S.B.G. (1971). The orthogonality of psychoticism and neuroticism: a factorial study. *Perceptual and Motor Skills* **33**, 461–2.

Eysenck, H.J. and Eysenck, S.B.G. (1976). *Psychoticism as a dimension of personality*. Hodder and Stoughton, London.

Eysenck, H.J. and Eysenck, S. (1995). Editorial: report on the present state of *Personality and individual differences*. *Personality and Individual Differences* **19**, 269–73.

Eysenck, H.J. and Frith, C.D. (1977). *Reminiscence, motivation and personality*. Plenum Publishing Corporation, New York.

Eysenck, H.J. and Gilmour, J.S.L. (1944). The psychology of philosophers: a factorial study. *Characteristics of Personality* **12**, 290–6.

Eysenck, H.J. and Kamin, L. (1981*a*). *The battle for the mind*. Macmillan/Pan, London.

Eysenck, H.J. and Kamin, L. (1981*b*). *The intelligence controversy*. John Wiley, New York.

Eysenck, H.J. and Nias, D. (1978). *Sex, violence and the media*. Maurice Temple Smith, London.

Eysenck, H.J. and Nias, D. (1982). *Astrology—science or superstition?* Maurice Temple Smith, London.

Eysenck, H.J. and O'Connor, K. (1979). Smoking, arousal and personality. In *Electrophysiological effects of nicotine* (ed. A. Redmond and C. Izard), pp. 147–57. Elsevier, Amsterdam.

Eysenck, H.J. and Prell, D.B. (1951). The inheritance of neuroticism: an experimental study. *Journal of Mental Science* **97**, 441–65.

Eysenck, H.J. and Rachman, S. (1965). *Causes and cures of neurosis*. Routledge and Kegan Paul, London.

Eysenck, H.J. and Rose, S. (1979). Race, intelligence and education. *New Scientist*, 15 March 1979, 849–52.

Eysenck, H.J. and Sargent, C. (1982). *Explaining the unexplained: mysteries of the paranormal*. Weidenfeld and Nicholson, London.

Eysenck, H.J. and Sargent, C. (1983). *Know your own PSI-Q*. Multimedia Publications, London.

Eysenck, H.J. and Sargent, C. (1986). *Are you psychic?* Prion, London.

Eysenck, H.J. and White, P.O. (1963). Personality and the measurement of intelligence. *British Journal of Educational Psychology* **33**, 197–202.

Eysenck, H.J. and Wilson, G.D. (eds.) (1973). *The experimental study of Freudian theories*. Methuen, London.

Eysenck, H.J. and Wilson, G.D. (1975). *Know your own personality*. Maurice Temple Smith, London.

Eysenck, H.J. and Wilson, G. (eds.) (1978a). *The psychological basis of ideology*. University Park Press, Baltimore.

Eysenck, H.J. and Wilson, G. (1978b). Conclusion: ideology and the study of social attitudes. In *The psychological basis of ideology* (ed. H.J. Eysenck and G. Wilson), pp. 303–12. University Park Press, Baltimore.

Eysenck, H.J. and Wilson, G.D. (1979). *The psychology of sex*. Dent, London.

Eysenck, H.J., Paul Barrett, and Lucking, S. (1986). Reaction time and intelligence: a replication study. *Intelligence* **10**, 9–40.

Eysenck, H.J., Eysenck, S.B.G., and Barrett, P. (1985). A revised version of the psychoticism scale. *Personality and Individual Differences* **1**, 21–30.

Eysenck, H.J., Gotz, K.O., Borisy, A.R., and Lynn, R. (1979). A new visual aesthetic sensitivity test: (VAST. 1. Construction and psychometric properties. *Perceptual and Motor Skills* **49**, 795–802.

Eysenck, H.J., Tarrant, M., and Woolf, M. (1960). Smoking and personality. *British Medical Journal* **11**, 1456–60.

Eysenck, H.J., Wakefield, J.A., and Friedman, A.F. (1983). Diagnosis and clinical assessment: the DSM-III. *Annual Review of Psychology* **34**, 167–93.

Eysenck, M. and Eysenck, H.J. (1980). Mischel and the concept of personality. *British Journal of Psychology* **71**, 191–204.

Fancher, R. (1983). Letter to the editor, *Encounter*, June 1983, 91.

Fancher, R. (1985). Review of *Hans Eysenck: the man and his work* by H.B. Gibson. *Journal of the History of the Behavioral Sciences* **21**, 245–7.

Fancher, R. (1990). Look at me. Review of *Rebel with a cause* by Hans Eysenck. *London Review of Books*, 28 June 1990, 22.

Fancher, R. (2003). The concept of race in the life and thought of Francis Galton. In *Defining difference: race and racism in the history of psychology* (ed. A.S. Winston), pp. 49–75. American Psychiatric Association, Washington, DC.

Farley, F. (2000). Hans J. Eysenck (1916–1997). *American Psychologist* **55**, 674–5.

Feest, U. (2005). Operationism in psychology: what the debate is about, what the debate should be about. *Journal of the History of the Behavioral Sciences* **41**, 131–49.

Feltham, C. (1996). Psychotherapy's staunchest critic: an interview with Hans Eysenck. *British Journal of Guidance and Counseling* **24**, 423–36.

Fernandez-Ballesteros, R., Zamarron, M.D., Ruiz, M.A., Sabastian, J., and Spielberger, C.D. (1997). Assessing emotional expression: Spanish adaptation of the rationality/emotional defensiveness scale. *Personality and Individual Differences* **22**, 719–29.

Fisher, R.A. (1959). *Smoking: the cancer controversy.* Oliver and Boyd, Edinburgh.

Fishman, J. (1988). The character of controversy. *Psychology Today*, December 1988.

Fletcher, C. (1965). Eysenck v. anti-smokers. *The Observer*, 27 June 1965.

Fletcher, C. (1981). Plea for the guilty. *Times Higher Education Supplement*, 16 January 1981.

Fletcher, R. (1991). *Science, ideology and the media: the Cyril Burt scandal.* Transaction, New Brunswick, New Jersey.

Flynn, J.R. (2007). *What is intelligence? Beyond the Flynn effect.* Cambridge University Press, Cambridge.

Fox, B.H. (1988).Psychogenic factors in cancer, especially its incidence. In *Topics in health psychology* (ed. S. Maes, C.D. Spielberger, R.B. Defares, and I.G. Sarason), pp. 37–55. Wiley, Chichester.

Fox, B.H. (1991). Quandaries created by the unlikely numbers in some of Grossarth-Maticek's studies. *Psychological Inquiry* **2**, 242–7.

Frank, S. (1968). To smoke or not to smoke. *True Magazine*, 15 January 1968.

Franks, C. (1954). An experimental study of conditioning as related to mental abnormality. Ph.D. thesis, University of London.

Franks, C. (1956). Conditioning and personality: a study of normal and neurotic subjects. *Journal of Abnormal and Social Psychology* **52**, 143–50.

Franks, C. (1958). Alcohol, alcoholics and conditioning: a review of the literature and theoretical considerations. *Journal of Mental Science* **104**, 14–33.

Franks, C. and Rosenbaum, M. (1983). Behavior therapy: overview and personal reflections. In *Perspectives on behavior therapy in the eighties* (ed. M. Rosenbaum, C. Franks, and Y. Jaffe), pp. 3–16. Springer, New York.

Franks, C. and Wilson, G.T. (1975). *Annual review of behaviour therapy: theory and practice*, Vol. 3. University Park Press, Baltimore.

Frearson, W. and Eysenck, H.J. (1986). Intelligence, reaction time (RT) and a new 'odd-man-out' RT paradigm. *Personality and Individual Differences* **7**, 807–17.

Freeman, H. and Kendrick, D. (1960). A case of cat phobia: treatment by a method derived from experimental psychology. *British Journal of Medicine* **11**, 497–502.

Freeman, H. and Berrios, G.E. (eds.) (1996). *150 years of British psychiatry, 1841–1991*, vol. 2. Athlone Press, London.

Freeman, J. (1997). The pugnacious psychologist [obituary for Hans Eysenck]. *The Guardian*, 8 September 1997.

Freeman, T. (1968). A psychoanalytic critique of behaviour therapy. *British Journal of Medical Psychology* **41**, 53–9.

Frenkel-Brunswik, E. (1954). Further explorations by a contributor to *Authoritarian personality*. In *Studies in the scope and method of the authoritarian personality* (ed. R. Christie and M. Jahoda), pp. 226–75. Free Press, Glencoe, Illinois.

Frentzel-Beyme, R. (1991). Levels of interest in an epidemiological approach of identifying psychomental risk factors for cancer. *Psychological Inquiry* **2**, 290–3.

Friedländer, S. (1997). *Nazi Germany and the Jews*. HarperCollins, New York.

Friedrich, O. (1972). *Before the deluge: A portrait of Berlin in the 1920's*. Harper and Row, New York.

Fulker, D. (1975). Review of *The science and politics of IQ* by Leon Kamin. *American Journal of Psychology* **88**, 505–19.

Furnham, A. (1998). Contributions to the history of psychology: CXIV. Hans Jürgen Eysenck, 1916–1997. *Perceptual and Motor Skills* **87**, 505–6.

Garssen, B. (2004). Psychological factors and cancer development: evidence after 30 years of research. *Clinical Psychology Review* **24**, 315–38.

Gelder, M., Marks, I. and Wolff, H.H. (1967). De-sensitization and psychotherapy in the treatment of phobic states: a controlled inquiry. *British Journal of Psychiatry* **113**, 53–73.

Gelwick, R.A. (1971). Personnel policies and procedures of the Waffen-SS. Ph.D. thesis, University of Nebraska at Lincoln.

Gibson, T. (1973*a*). Professor Eysenck's nose. *Anarchist Weekly* **34**, 19 May 1973.

Gibson, T. (1973*b*). Letter to the editor. *Anarchist Weekly*, 2 June 1973.

Gibson, T. (1973*c*). Letter to the editor *Anarchist Weekly*, 30 June 1973.

Gibson, T. (1975). Eysenck is not a racist—or an authoritarian. *Wildcat* **7**, 1975.

Gibson, H.B. (Tony) (1981). *Hans Eysenck: the man and his work*. Peter Owen, London.

Gilbert, M. (1987). *The Holocaust: the Jewish tragedy*. Fontana, London.

Gillham, N.W. (2001). *A life of Sir Francis Galton: from African exploration to the birth of eugenics*. Oxford University Press, Oxford.

Gillie, O. (1976). Pioneer of IQ faked his research findings. *The Sunday Times*, 24 October 1976.

Gillie, O. (1980). Burt: the scandal and the cover-up. In A balance sheet on Burt (special supplement). *Bulletin of the British Psychological Society* **33**, 9–16.

Gillie, O. (1982). The Burt scandal [letter to the editor]. *The Listener*, 6 May 1982.

Glantz, S.A., Slade, J., Bero, L.A., Hanauer, P., and Barnes, D.E. (1996). *The cigarette papers*. University of California Press, Berkeley.

Glass, G.V. (2000). Meta-analysis at 25. January 2000, p. 5, http://glass.ed.asu.edu./gene/papers/meta25.html.

Glover, E. (1955). *The technique of psychoanalysis*. International Universities Press, New York.

Glover, E. (1959). Review of *Psychotherapy by reciprocal inhibition* by Joseph Wolpe. *British Journal of Medical Psychology* **32**, 68–74.

GMC (1954). Dr. Eysenck's table talk. Review of *Uses and abuses of psychology* by Hans Eysenck. *Bethlem and Maudsley Hospital Gazette*, **1**, 121–2.

Golden, C. (1968). Cigarette cancer link is bunk. *National Enquirer*, 3 March 1968.

Gottesman, I.I., and McGuffin, P. (1996). Eliot Slater and the birth of psychiatric genetics in Great Britain. In *150 years of British psychiatry, 1841–1991*, Vol. 2 (ed. H. Freeman and G.E. Berrios), pp. 537–48. Athlone Press, London.

Gottfredson, L. (1994). Mainstream science on intelligence. *Wall Street Journal*, 13 December 1994.

Gould, S.J. (1975). Racist arguments and IQ. In *Race and IQ* (ed. A. Montagu), pp. 145–50. Oxford University Press, New York.

Gray, J. (ed. and translator) (1964). *Pavlov's typology: recent theoretical and experimental developments from the laboratory of B.M. Teplov.* Pergamon Press, London.

Gray, J. (1997a). Foreword to *The scientific study of human nature: tribute to Hans J. Eysenck at eighty* (ed. H. Nyborg), xi–xiii. Elsevier, Oxford.

Gray, J. (1997b). Obituary: Hans Jürgen Eysenck (1916–97). *Nature* **389**, 794.

Green, R.T. and Stacey, B.G. (1962). The T concept. *Nature* **196**, 94.

Green, R.T. and Stacey, B.G. (1964). Was Torquemada tenderminded? *Acta Psychologica* **22**, 250–71.

Gross, A.G. (1990). *The rhetoric of science.* Harvard University Press, Cambridge, Massachusetts.

Grossarth-Maticek, R. (1977). Social scientific aspects in the aetiology of organic diseases: perspective, method, and results of a prospective study. Habilitation thesis, University of Heidelberg. [Translated by Manfred Amelang, University of Heidelberg, 1977. Original German title: Veröffentlichung im Rahmen des interdisziplinären Forschungsprojektes sozialwissenschaftlicher Onkologie.]

Grossarth-Maticek, R. (1980a). Psychosocial predictors of cancer and internal diseases: an overview. *Psychotherapy and Psychosomatics* **33**, 122–8.

Grossarth-Maticek, R. (1980b). Social psychotherapy and course of the disease: first experiences with cancer patients. *Psychotherapy and Psychosomatics* **33**, 129–38.

Grossarth-Maticek, R. and Eysenck, H.J. (1989). Is media information that smoking causes illness a self-fulfilling prophecy. *Psychological Reports* **65**, 177–8.

Grossarth-Maticek, R. and Eysenck, H.J. (1990). Personality, stress and disease: description and validation of a new inventory. *Psychological Reports* **66**, 355–73.

Grossarth-Maticek, R. and Eysenck, H.J. (1991). Creative novation behaviour therapy as a prophylactic treatment for cancer and coronary heart disease. I. Description of treatment. *Behaviour Research and Therapy* **29**, 1–16.

Grossarth-Maticek, R., Eysenck, H.J., and Barrett, P. (1993). Prediction of cancer and coronary heart disease as a function of questionnaire administration. *Psychological Reports* **73**, 943–59.

Grossarth-Maticek, R., Eysenck, H.J., and Boyle, G.J. (1995). Method of test administration as a factor in test validity: the use of a personality questionnaire in the prediction of cancer and coronary heart disease. *Behaviour Research and Therapy* **33**, 705–10.

Grossarth-Maticek, R., Eysenck, H.J., and Vetter, H. (1988). Personality type, smoking habit and their interactions as predictors of cancer and coronary heart disease. *Personality and Individual Differences* **9**, 479–95.

Grossarth-Maticek, R., Eysenck, H.J., and Vetter, H. (1989). The causes and cures of prejudice: an empirical study of the frustration–aggression hypothesis. *Personality and Individual Differences* **10**, 547–58.

Guardian (1973). People who don't want to know [editorial]. *The Guardian*, 10 May 1973.

Gudjonsson, G. (1997). Crime and personality. In *The scientific study of human nature: tribute to Hans J. Eysenck at eighty* (ed. H. Nyborg), pp. 142–64. Elsevier, Oxford.

Gurjeva, L.G. (2001). James Sully and scientific psychology. In *Psychology in Britain: historical essays and personal reflections* (ed. G.C. Bunn, A.D. Lovie, and G.D. Richards), pp. 72–94. BPS Books, Leicester.

Guthrie, R.V. (1976). *Even the rat was white: a historical view of psychology*. Allyn and Bacon, Needham Heights, Massachusetts.

Hacking, I. (1995). *Rewriting the soul: multiple personality and the sciences of memory*. Princeton University Press, Princeton.

Hagan, P. (2003). Review queries usefulness of peer review. *The Scientist*, 28 January 2003.

Haggbloom, S.J., Warnick, R., Warnick, J.E., Jones, V.K., Yarbrough, G.L., Russell, T.M., Borecky, C.M., McGahhey, R., Powell III, J.L., Beavers, J., and Monte, E. (2002). The 100 most eminent psychologists of the 20th century. *Review of General Psychology* **6**, 139–52.

Hall, J. (2007a). The emergence of clinical psychology in Britain from 1943 to 1958. Part I: Core tasks and the professionalisation process. *History and Philosophy of Psychology* **9**(1), 29–55.

Hall, J. (2007b). The emergence of clinical psychology in Britain from 1943 to 1958. Part II: Practice and research traditions. *History and Philosophy of Psychology* **9**(2), 1–33.

Hall, S.B. (1955). Psychotherapy: misapprehensions and realities. *British Journal of Medical Psychology* **26**, 295–9.

Halsey, A.H. (1971). Discriminations. Review of *Race, intelligence and education* by Hans Eysenck. *The Guardian*, 17 June 1971.

Hamilton, V. (1959a). Eysenck's theories of anxiety and hysteria—a methodological critique. *British Journal of Psychology* **50**, 48–63.

Hamilton, V. (1959b). Theories of anxiety and hysteria—a rejoinder to Hans Eysenck. *British Journal of Psychology* **50**, 276–9.

Hamilton, V. (1964). Techniques and methods in psychological assessment: a critical appraisal. *Bulletin of the British Psychological Society* **17**, 27–36.

Hanley, C. and Rokeach, M. (1956). Care and carelessness in psychology. *Psychological Bulletin* **53**, 183–6.

Harkness, J.M. (2006). The US Public Health Service and smoking in the 1950s: the tale of two more statements. *Journal of the History of Medicine and Allied Sciences* **62**, 171–212.

Hartshorne, E. (1937). *The German universities and National Socialism*. George Allen and Unwin, London.

Hastings, M. (1981). Putting their eggheads together. *New Standard*, 28 April 1981.

Hayward, R. (2001). 'Our friends electric': mechanical models of mind in postwar Britain. In *Psychology in Britain: historical essays and personal reflections* (ed. G.C. Bunn, A.D. Lovie, and G.D. Richards), pp. 290–308. BPS Books, Leicester.

Hearnshaw, L. (1964). *Short history of British psychology, 1840–1940*. Methuen, London.

Hearnshaw, L. (1979). *Cyril Burt: psychologist*. Hodder and Stoughton, London.

Henderson, D. and Gillespie, R.D. (1943). *Textbook of psychiatry*. Oxford University Press, Oxford.

Hendrickson, D.E., and Hendrickson, A.E. (1980). The biological basis of individual differences. *Personality and Individual Differences* **1**, 3–33.

Herman, E. (1995). *The romance of American psychology: political culture in the age of experts*. University of California Press, San Francisco.

Hewitt, J.K., Eysenck, H.J., and Eaves, L. (1977). The structure of social attitudes after 26 years: a replication. *Psychological Reports* **40**, 138–88.

Hilberg, R. (1985). *The destruction of the European Jews*. Holmes and Meier, New York.

Hildebrand, H.P. (1958). A factorial study of introversion–extraversion. *British Journal of Psychology* **49**, 1–11.

Hilgartner, S. (1990). The dominant view of popularization: conceptual problems, political uses. *Social Studies of Science* **20**, 519–39.

Hill, G. (1980). Review of *The causes and effects of smoking* by Hans Eysenck (with Lindon Eaves). *The Times*, 12 December 1980.

Hilts, P.J. (1997). *Smokescreen: the truth behind the tobacco industry cover-up*. Addison-Wesley, Reading, Massachusetts.

Hinshelwood, R.D. (1995). Psychoanalysis in Britain: points of cultural access, 1893–1918. *International Journal of Psychoanalysis* **76**, 135–51.

Hirsh, J. (1981). To 'unfrock the charlatans'. *Sage Race Relations Abstracts* **6**, 20.

Hobsbawm, E. (2002). *Interesting times: a twentieth century life*. Allen Lane, London.

Honan, W.H. (1997). Hans J. Eysenck, 81, a heretic in the field of psychotherapy. *New York Times*, 10 September 1997.

Howarth, E. (1973). A hierarchical oblique factor analysis of Eysenck's rating study of 700 neurotics. *Social Behaviour and Personality* **1**, 81–7.

Howarth, E. (1976). A psychometric investigation of Eysenck's personality inventory. *Journal of Personality Assessment* **40**, 173–85.

Hudson, L. (1971). Science and popularisation. Review of *Race, intelligence and education* by Hans Eysenck. *New Society*, 1 July 1971, 29–30.

Hull, C.L. (1952). *A behavior system*. Yale University Press, New Haven, Connecticut.

Hunt, J.McV. (ed.) (1944). *Personality and the behavior disorders*, Vols. 1 and 2. Ronald Press, New York.

Hyman, H. and Sheatsley, P. (1954). *The authoritarian personality*: a methodological critique. In *Studies in the scope and method of the authoritarian personality* (ed. R. Christie and M. Jahoda), pp. 50–122. Free Press, Glencoe, Illinois.

Institute of Psychiatry (1952–53). Annual Report, 1952–1953, Maudsley Hospital.

Institute of Psychiatry (1956–57). Annual Report, 1956–1957, Maudsley Hospital.

Izbicki, J. (1973). Charges urged over Eysenck attack at LSE. *Daily Telegraph*, 10 May 1973.

Jenner, F.A. (1991). Erwin Stengel: a personal memoir. In *150 years of British psychiatry, 1841–1991*, Vol. 1 (ed. G.E. Berrios and H. Freeman), pp. 436–44. Royal College of Psychiatrists, London.

Jensen, A.R. (1957). Authoritarian attitudes and personality maladjustment. *Journal of Abnormal and Social Psychology* **54**, 303–11.

Jensen, A.R. (1969). How much can we boost IQ and scholastic achievement? *Harvard Educational Review* **39**, 1–123.

Jensen, A.R. (1972). *Genetics and education*. Methuen, London.

Jensen, A.R. (1974). Kinship correlations reported by Sir Cyril Burt. *Behavior Genetics* **4**, 24–5.

Jensen, A.R. (1976). Letter to the editor. *The Times*, 9 December 1976.

Jensen, A.R. (1980). *Bias in mental testing*. Free Press, New York.

Jensen, A.R. (1986). The theory of intelligence. In *Hans Eysenck: consensus and controversy* (ed. S. Modgil and C. Modgil), pp. 89–102. Falmer Press, Philadelphia.

Jensen, A.R. (1987). Individual differences in the Hick paradigm. In *Speed of information processing and intelligence* (ed. P.A. Vernon), pp. 101–75. Ablex, Norwood, New Jersey.

Jensen, A.R. (1997a). Eysenck as teacher and mentor. In *The scientific study of human nature: tribute to Hans J. Eysenck at eighty* (ed. H. Nyborg), pp. 543–59. Elsevier, Oxford.

Jensen, A.R. (1997b). The psychometrics of intelligence. In *The scientific study of human nature: tribute to Hans J. Eysenck at eighty* (ed. H. Nyborg), pp. 221–39. Elsevier, Oxford.

Jensen, A.R. (2000). Hans Eysenck's final thoughts on intelligence. Review of *Intelligence: a new look* by Hans Eysenck. *Personality and Individual Differences* **28**, 191–4.

Jones, G. (1956). The application of conditioning and learning techniques to the treatment of a psychiatric patient. *Journal of Abnormal and Social Psychology* **52**, 414–19.

Jones, G. (1960). Individual differences in inhibitory potential. *British Journal of Psychology* **51**, 220–5.

Jones, G. (1984). Behaviour therapy—an autobiographic view. *Behavioural Psychotherapy* **12**, 7–16.

Jones, K. (1991). The culture of the mental hospital. In *150 years of British psychiatry, 1841–1991*, Vol. 1 (ed. G.E. Berrios and H. Freeman), pp. 17–28. Royal College of Psychiatrists, London.

Kamin, L.J. (1974). *The science and politics of IQ*. Erlbaum, Potomac, Maryland.

Kaplan, J., and Bennett, T. (2003). Use of race and ethnicity in biomedical publications. *Journal of the American Medical Association* **289**, 2709–12.

Kaplan, M.A. (1998). *Between dignity and despair: Jewish life in Nazi Germany*. Oxford University Press, New York.

Karon, B.P., and Saunders, D.R. (1958). Some implication of the Eysenck–Prell study of 'The inheritance of neuroticism': a critique. *Journal of Mental Science* **104**, 350–8.

Kazdin, AE. (1978). *History of behaviour modification: experimental foundations of contemporary research*. University Park Press, Baltimore.

Kendrick, D.C. (1960). Effects of drive and effort on inhibition with reinforcement. *British Journal of Psychology* **51**, 211–19.

Kendrick, D.C. (1981). Neuroticism and extraversion as explanatory concepts in clinical psychology. In *Dimensions of personality: papers in honour of H.J. Eysenck* (ed. R. Lynn), pp. 253–62. Pergamon Press, Oxford.

Kenna, J.C. (1966). Some aspects of the development of psychology departments in British universities. Unpublished manuscript, December 1966.

Kenny, M.G. (2002). Toward a racial abyss: eugenics, Wickliffe Draper, and the origins of the Pioneer Fund. *Journal of the History of the Behavioural Sciences* **38**, 259–83.

Kenny, M.G. (2006). A question of blood, race, and politics. *Journal of the History of Medicine and Allied Sciences* **61**, 456–91.

Kimble, G.A. (1949). An experimental test of a two factor theory of inhibition. *Journal of Experimental Psychology* **39**, 15–23.

Kissen, D.M. and Eysenck, H.J. (1962). Personality in male lung cancer patients. *Journal of Psychosomatic Research* **6**, 123–7.

Kitchin, C.H. (1966). *You may smoke*. Award Books, New York.

Kline, P. (1981). No smoking. Review of *The causes and effects of smoking* by Hans Eysenck (with Lindon Eaves). *London Review of Books*, 19 February–4 March 1981.

Knopfelmacher, F. (1977). The Eysenck–Jensen scandal. *Nation Review*, 22–28 September 1977.

Knorr-Cetina, K. (1981). *The manufacture of knowledge: an essay on the constructivist and contextual nature of science.* Pergamon Press, Oxford.

Koestler, A. (1976). *The thirteenth tribe: the Khazar empire and its heritage.* Hutchinson, London.

Kurz, F. (2002). Akademisches Schattenreich. *Der Spiegel*, **37**, 9 October 2002.

Ladd, B. (1997). *The ghosts of Berlin: confronting German history in the urban landscape.* University of Chicago Press, Chicago.

Lancet (1933). Eugenics in Germany [unsigned editorial; later found to be by Aubrey Lewis]. *The Lancet* **ii** (August 1933), 297–8.

Lancet (1947). Review of *Dimensions of personality* by Hans Eysenck. *The Lancet* **249**, 713.

Lancet (1954). Review of *Uses and abuses of psychology* by Hans Eysenck. *The Lancet* **263**, 348.

Lane, C. (1944). The tainted sources of *The bell curve*. *The New York Review of Books*, 1 December 1994, 14–19.

Lazarus, A. (1958). New methods in psychotherapy: a case study. *South African Medical Journal* **33**, 660–3.

Lazarus, A. (1986). On sterile paradigms and the realities of clinical practice. In *Hans Eysenck: consensus and controversy* (ed. S. Modgil and C. Modgil), pp. 247–57 and 260–1. Falmer Press, Philadelphia.

Lee, P.N. (1991). Personality and disease: a call for replication. *Psychological Inquiry* **2**, 251–3.

Levine, G. (2002). *Dying to know: epistemology and narrative in Victorian England.* University of Chicago Press, Chicago.

Levis, D.J. and Malloy, P.F. (1982). Research in infrahuman and human conditioning. In *Contemporary behaviour therapy* (ed. G.T. Wilson and C.M. Franks), pp. 65–118. Guilford Press, New York and London.

Lewis, A. (1947). Foreword to *Dimensions of personality*, by Hans Eysenck, p. vii. Routledge and Kegan Paul, London.

Lewontin, R.C. (1970). Race and intelligence. *Science and Public Affairs, Bulletin of Atomic Scientists*, March 1970, 2–8.

Littlewood, R. (1995). *Mankind Quarterly* again. *Anthropology Today* **11**, 17–18.

Loehlin, J. (1986). H.J. Eysenck and behaviour genetics: a critical view. In *Hans Eysenck: consensus and controversy* (ed. S. Modgil and C. Modgil), pp. 49–57. Falmer Press, Philadelphia.

Loevinger, J. (1955). Diagnosis and measurement: a reply to Eysenck. *Psychological Reports* **1**, 277–8.

Lombardo, G. and Foschi, R. (2003). The concept of personality in 19th century French and 20th century American psychology. *History of Psychology* **6**, 123–42.

Lombardo, P.A.(2002). 'The American breed': Nazi eugenics and the origins of the Pioneer Fund. *Albany Law Review* **65**, 743–830.

London, P. (1972). The end of ideology in behaviour modification. *American Psychologist* **27**, 913–20.

London, P. (1983). Science, culture, and psychotherapy: the state of the art. In *Perspectives on behavior therapy in the eighties* (ed. M. Rosenbaum, C. Franks, and Y. Jaffe), pp. 17–32. Springer, New York.

London School of Economics Social Science Society (1973). Letter to the editor. *New Statesman*, 11 May 1973, 693.

Lord, F.M., and Novick, M.R. (1968). *Statistical theories of mental test scores*. Addison-Wesley, London.

Lovie, A.D. (1983). Images of man in early factor analysis—psychological and philosophical aspects. In *Studies in the history of psychology and the social sciences* (ed. S. Bem, H. van Rappard, and W. van Hoorn), pp. 235–47. Leiden University, Leiden.

Lovie, A.D. and Lovie, P. (1993). Charles Spearman, Cyril Burt, and the origins of factor analysis. *Journal of the History of the Behavioral Sciences* **29**, 308–21.

Lovie, P. and Lovie, A.D. (1995). The cold equations: Spearman and Wilson on factor indeterminacy. *British Journal of Mathematical and Statistical Psychology* **48**, 237–53.

Low, I. (1978). Pursuing truth. *New Scientist*, 6 April 1978.

Lubin, A. (1950). A note on 'criterion analysis'. *Psychological Review* **57**, 54–7.

Luborsky, L. (1954). A note on Eysenck's article: 'The effects of psychotherapy: an evaluation'. *British Journal of Psychology* **45**, 129–31.

Lykken, D.T. (1959). Turbulent complication. Review of *The dynamics of anxiety and hysteria* by Hans Eysenck. *Contemporary Psychology* **4**, 377–9.

Lykken, D.T. (1991). What's wrong with psychology anyway? In *Thinking clearly about psychology: matters of public interest*, Vol. 1 (ed. D. Cicchetti and W.M. Grove), pp. 3–39. University of Minnesota Press, Minneapolis.

Lynn, R. (ed.) (1981). *Dimensions of personality: papers in honour of H.J. Eysenck*. Pergamon Press, Oxford.

MacKinnon, D.W. (1944). The structure of personality. In *Personality and the behavior disorders*, Vol. 1 (ed. J.McV. Hunt), pp. 3–48. Ronald Press, New York.

MacKinnon, D. (1953). Fact and fancy in personality research. *American Psychologist* **8**, 138–46.

Mackintosh, N.J. (1981). Psychological sound and fury. Review of *The battle for the mind* by Hans Eysenck and Leon Kamin. *The Times Educational Supplement*, 13 March 1981.

Mackintosh, N.J. (ed.) (1995). *Burt: fraud or framed?* Oxford University Press, Oxford.

Mackintosh, N.J. (1998). *IQ and human intelligence*. Oxford University Press, Oxford.

Maddox, J. (1961). The influence of Eysenck. *The Guardian*, 18 April 1961.

Maller, J.B. (1933). Studies in character and personality in German psychological literature. *Psychological Bulletin* **30**, 209–32.

Marsh, P. (1979). Fascist mythology. Review of *Fascists: a social psychological view of the National Front* by Michael Billig. *New Society*, 18 January 1979.

Martin, I. (2001). Hans Eysenck at the Maudsley—the early years. *Personality and Individual Differences* **31**, 7–9.

Marusic, A., Gudjonsson, G., Eysenck, H.J., and Starc, R. (1999). Biological and psychosocial risk factors in ischaemic heart disease: empirical findings and a biopsychosocial model. *Personality and Individual Differences* **26**, 285–304.

Mather, I. (1973). They call ME a racialist ... But these 'students' are the new Hitlers. Interview with Hans J. Eysenck, *Daily Mail*, 10 May 1973.

Matthews, G. and Deary, I. (1998). *Personality traits*. Cambridge University Press, Cambridge.

Matthews, G. and Gilliland, K. (1999). The personality theories of H.J. Eysenck and J.A. Gray: a comparative review. *Personality and Individual Differences* **26**, 583–626.

Mayes, R. and Horwitz, A.V. (2005). DSM-III and the revolution in the classification of mental illness. *Journal of the History of the Behavioral Sciences* **41**, 249–67.

Maynard, D.W. and Schaeffer, N.C. (2000). Toward a sociology of social scientific knowledge: survey research and ethnomethodology's asymmetric alternates. *Social Studies of Science* **30**, 323–70.

McConaghy, N. (1970). Review of *The biological basis of personality* by Hans Eysenck. *Australian and New Zealand Journal of Psychiatry* **4**, 113.

McCourt, K., Bouchard, T.J., Lykken, D.T., Tellegen, A., and Keynes, M. (1999). Authoritarianism revisited: genetic and environmental influences examined in twins reared apart and together. *Personality and Individual Differences* **27**, 985–1014.

McGuire, R.J., Mowbray, R.M., and Vallance, R.C. (1963). The Maudsley personality inventory used with psychiatry patients. *British Journal of Psychology* **54**, 157–66.

McIlraith, S. (1965). Revolutionary new treatment for sexual aberration. *People*, 5 May 1965.

Meakin, C. (1969). Letter to the editor. *New Scientist*, 22 May, 1969, 429.

Medawar, P. (1991). Is the scientific paper a fraud? In Peter Medawar, *The threat and the glory: reflections on science and scientists*, pp. 228–33. Oxford University Press, Oxford.

Meehl, P.E. (1954). *Clinical versus statistical prediction: a theoretical analysis and a review of the evidence*. University of Minnesota Press, Minneapolis.

Melvin, D. (1955). An experimental and statistical study of two primary social attitudes. Ph.D. thesis, University of London.

Meredith, P. (1973). Eye synch. *The Guardian*, 26 May 1973.

Meyer, V. (1957). The treatment of two phobic patients on the basis of learning principles. *Journal of Abnormal and Social Psychology* **55**, 261–6.

Micale, M.S. (1995). *Approaching hysteria: disease and its interpretations*. Princeton University Press, Princeton.

Mildenberger, F. (2007). Kraepelin and the 'urnings': male homosexuality in psychiatric discourse. *History of Psychiatry* **18**, 321–35.

Miller, D.J. and Hersen, M. (eds.) (1992). *Research fraud in the behavioural and biomedical sciences*. John Wiley, New York.

Modgil, S. and C. Modgil (eds.) (1986). *Hans Eysenck: consensus and controversy*. Falmer Press, Philadelphia.

Mollon, J.D. History of the EPS: Meetings. http://www.eps.ac.uk/society/meetings.html.

Monks, J. (1977). Campus drowns out free speech. *The Australian*, 16 September 1977.

Montagu, A. (ed.) (1975). *Race and IQ*. Oxford University Press, New York.

Montagu, A. (1978). The ethics of book reviewing. *Current Anthropology* **19**, 385.

Monte, C.F. (1991). *Beneath the mask: an introduction to theories of personality*. Holt, Rinehart, and Winston, New York.

Moore, J. (1996). Metabiographical reflections on Charles Darwin. In *Telling lives in science: essays on scientific biography* (ed. M. Shortland and R. Yeo), pp. 267–81. Cambridge University Press, Cambridge.

Morris, E. (1999). *Dutch: a memoir of Ronald Reagan*. Random House, New York.

Mowrer, O.H. and Mowrer, W.M. (1938). Enuresis: a method for its study and treatment. *American Journal of Orthopsychiatry* **8**, 436–59.

Mungo, C.J. (1971). Letter to the editor. *The Times*, 28 June 1971.

Murphy, G. and Jensen, F. (1932). *Approaches to personality*. Coward-McCamus, New York.

Napalkov, A.V. (1963). Information process of the brain. In *Progress in brain research*, Vol. 2 (ed. N. Weiner and J.P. Schade), pp. 182–6. Elsevier, Amsterdam.

Neisser, U. (ed.) (1998). *The rising curve: long term gains in IQ and related measures*. American Psychological Association, Washington, DC.

Neisser, U. (2004). Serious scientists or disgusting racists? *Contemporary Psychology* **49**, 5–7.

Neisser, U., Boodoo, G., Bouchard, J.R. Jr, Boykin, A.W., Brody, N, Ceci, S.J., Halpern, D.F., Loehlin, J.C., Perloff, R., Sternberg, R.J. and Urbina, S. (1996). Intelligence: knowns and unknowns. *American Psychologist* **51**, 77–101.

Neue Anthropologie (1976). Interview mit Hans-Jürgen Eysenck. *Neue Anthropologie*, January/March, 1976, 16–17.

New Scientist (1969). Intelligence and 'race' [editorial]. *New Scientist* 1 May 1969, 219.

New Scientist (1980). Smoking out censorship [editorial]. *New Scientist* **88** (18/25 December 1980), 756.

Nias, D. (1981). Humour and personality. In *Dimensions of personality: papers in honour of H.J. Eysenck* (ed. R. Lynn), pp. 287–313. Pergamon, Oxford.

Nias, D. (1997). Psychology and medicine. In *The scientific study of human nature: tribute to Hans J. Eysenck at eighty* (ed. H. Nyborg), pp. 92–108. Elsevier, Oxford.

Nicholson, I. (2003). *Inventing personality: Gordon Allport and the science of selfhood*. APA Press, Washington, DC.

Nicholson, J. (1981). A score draw. *New Society*, 12 March 1981.

Nigniewitzky, R.D. (1955). A statistical study of rigidity as a personality variable. MA thesis, University of London.

Nigniewitzky, R.D. (1956). A statistical and experimental study of rigidity in relation to personality and social attitudes. Ph.D. thesis, University of London.

Nisbett, J. (1971). Review of *Race, intelligence and education* by Hans Eysenck. *The Times Educational Supplement*, 18 June 1971.

Nyborg, H. (ed.) (1997). *The scientific study of human nature: tribute to Hans J. Eysenck at eighty*. Pergamon, Oxford.

Nyborg, H. (2003). The sociology of psychometric and bio-behavioral sciences: a case study of destructive social reductionism and collective fraud in 20th century academia. In *The scientific study of general intelligence: tribute to Arthur R. Jensen* (ed. H. Nyborg), pp. 441–502. Pergamon, Amsterdam.

O'Connor, N. (1952). Review of *The scientific study of personality* by Hans Eysenck. *Bulletin of the British Psychological Society* **3**, 115.

Oxford Mail (1963). Freud ousted by Pavlov: professor. *Oxford Mail*, 29 August 1963.

Page, B., Pringle, P., and Fay, S. (1971). The fallibility of H.J. Eysenck. *The Sunday Times*, 20 June 1971.

Page, E.B. (1972). Behavior and heredity. *American Psychologist* **27**, 660–1.

Parry, M. (1973). Bully boys on the campus. *Yorkshire Post*, 11 July 1973.

Pavlov, I.P. (1927). *Conditioned reflexes: an investigation of the physiological activity of the cerebral cortex* [translator G.V. Anrep]. Oxford University Press, Oxford.

Pavlov, I.P. (1955). *Selected works* [translator S. Belsky]. Foreign Languages Publishing House, Moscow.

Payne, R.W (1955). L'utilité du test de Rorschach en psychologie clinique. *Revue de Psychologie Appliquée* **3**, 255–64.

Payne, R.W. (1957). Experimental method in clinical psychological practice. *Journal of Mental Science* **103**, 189–96.

Payne, R.W. (2000). The beginnings of the clinical psychology programme at the Maudsley Hospital, 1947–1959. *Clinical Psychology Forum* **145**, 17–21.

Payne, R.W. and Jones, H.G. (1957). Statistics for the investigation of individual cases. *Journal of Clinical Psychology* **13**, 115–21.

Pearce, M. (1977). Professors, protestors and police. *Farrago*, 16 September 1977.

Pearson, J.S. and Kley, I.B. (1957). On the application of genetic expectancies as age-specific base rates in the study of human behaviour disorders. *Psychological Bulletin* **54**, 406–20.

Pearson, J.S. and Kley, I.B. (1958). Discontinuity and correlation: a reply to Eysenck. *Psychological Bulletin* **55**, 432–5.

Pelosi, A.J. and Appleby, L. (1992). Psychological influences on cancer and ischaemic heart disease. *British Medical Journal* **304**, 1295–8.

Pelosi, A.J. and Appleby, L. (1993). Personality and fatal diseases. *British Medical Journal* **306**, 1666–7.

Penrose, L. (1971). Negro intelligence. *The Friend*, 10 September 1971.

Personality and Individual Differences (2001). Bibliography: Hans Eysenck, Ph.D., D.Sc., 1939–2000 [Sybil Eysenck et al.]. *Personality and Individual Differences* **31**, 45–99.

Pickering, A. (1995). *The mangle of practice: time, agency and science.* University of Chicago Press, Chicago.

Pigliucci, M. (2001). *Phenotypic plasticity: beyond nature and nurture.* Johns Hopkins University Press, Baltimore.

Pigliucci, M. and Kaplan, J. (2003). On the concept of biological race and its applicability to humans. *Philosophy of Science* **70**, 1161–72.

Pilgrim, D. and Treacher, A. (1992). *Clinical psychology observed.* Tavistock, London.

Pilkington, G.W. and McKellar, P. (1960). Inhibition as a concept in psychology. *British Journal of Psychology* **51**, 194–201.

Pinard, J.W. (1932). Tests of perseveration. 1. Their relation to character. *British Journal of Psychology* **23**, 5–19.

Plomin, R. (1994). *Genetics and experience: the interplay between nature and nurture.* Sage, Thousand Oaks, California.

Plomin, R. and Loehlin, J. (1989). Direct and indirect IQ heritability estimates. *Behavior Genetics* **19**, 331–42.

Plomin, R., DeFries, J.C., McClearn, G.E., and McGuffin, P. (2001). *Behavioral genetics*. Freeman, New York.

Porter, T.M. (1995). *Trust in numbers: the pursuit of objectivity in science and public life*. Princeton University Press, Princeton.

Porter, T.M. (2004). *Karl Pearson: the scientific life in a statistical age*. Princeton University Press, Princeton.

Portes, A. (1971a). On the emergence of behaviour therapy in modern society. *Journal of Consulting and Clinical Psychology* **36**, 303–13.

Portes, A. (1971b). Behaviour therapy and critical speculation. *Journal of Consulting and Clinical Psychology* **36**, 320–4.

Pressman, J.D. (1998). *Last resort: psychosurgery and the limits of medicine*. Cambridge University Press, Cambridge.

Prins, A. (1998). Ageing and expertise: Alzheimer's disease and the medical professions, 1930–1990. Ph.D. thesis, University of Amsterdam.

Proshansky, H.M. (1973). Behavior and heredity: statement by the Society for the Psychological Study of Social Issues. *American Psychologist* **28**, 620–1.

Rachman, J. (S.) (2000). Joseph Wolpe (1915–1997). *American Psychologist* **55**, 431–2.

Rachman, S. (1981). H.J. Eysenck's contribution to behaviour therapy. In *Dimensions of personality: papers in honour of H.J. Eysenck* (ed. R. Lynn), pp. 315–30. Pergamon Press, Oxford.

Rachman, S. and Eysenck, H.J. (1966). Reply to a 'critique and reformation' of behaviour therapy. *Psychological Bulletin* **65**, 165–9.

Rafter, N.H. (2006). H.J. Eysenck in Fagin's kitchen: the return to biological theory in 20th century criminology. *History of the Human Sciences* **19**, 37–56.

Raine, A. (1997). Classical conditioning, arousal and crime: a biosocial perspective. In *The scientific study of human nature: tribute to Hans J. Eysenck at eighty* (ed. H. Nyborg), pp. 122–41. Elsevier, Oxford.

Ray, J. (1986). Eysenck on social attitudes: a historical critique. In *Hans Eysenck: consensus and controversy* (ed. S. Modgil and C. Modgil), pp. 155–73. Falmer Press, Philadelphia.

Rechtschaffen, A. (1958). Neural satiation, reactive inhibition, and introversion–extraversion. *Journal of Abnormal and Social Psychology* **57**, 283–91.

Rego, B. (2009). The polonium brief: a hidden history of cancer, radiation, and the tobacco industry. *Isis* **100**, 453–84.

Reid, R.L. (1960). Inhibition—Pavlov, Hull, Eysenck. *British Journal of Psychology* **51**, 226–32.

Revelle, W. (1995). Personality processes. *Annual Review of Psychology* **46**, 295–328.

Richards, B. (1983). Clinical psychology, the individual and the welfare state. Ph.D. thesis, North East London Polytechnic.

Richards, G. (1997). *'Race', racism and psychology: towards a reflexive history*. Routledge, London.

Richards, G. (2000). Britain on the couch: the popularisation of psychoanalysis in Britain, 1918–1940. *Science in Context* **13**, 183–230.

Richards, G. (2001). Edward Cox, the Psychological Society of Great Britain (1875–1879) and the meaning of an institutional failure. In *Psychology in Britain: historical essays and personal reflections* (ed. G.C. Bunn, A.D. Lovie, and G.D. Richards), pp. 33–53. BPS Books, Leicester.

Richards, G. (2003). 'It's an American thing': the 'race' and intelligence controversy from a British perspective. In *Defining difference: race and racism in the history of psychology* (ed. A.S. Winston), pp. 137–70. American Psychological Association, Washington, DC.

Richards, G. (2004). Eysenck, Hans Jürgen (1916–1997), psychologist. In *Oxford dictionary of national biography* (ed. C. Matthew and B. Harrison). Oxford University Press, Oxford.

Richardson, M. (1954). Review of *Uses and abuses of psychology* by Hans Eysenck. *The New Statesman and Nation*, 23 January 1954.

Roback, A.A. (1927). *A bibliography of character and personality*. Sci-art, Cambridge.

Robinson, D.L. (2001). How brain arousal systems determine different temperament types and the major dimensions of personality. *Personality and Individual Differences* **31**, 1233–59.

Rogers, K.H. (1935). The study of personality. *Journal of Abnormal and Social Psychology* **29**, 357–66.

Rogers, T.B. (1995). *The psychological testing enterprise: an introduction*. Brooks/Cole, Pacific Grove, California.

Roiser, M. and Willig, C. (2002). The strange death of the authoritarian personality: 50 years of psychological and political debate. *History of the Human Sciences* **15**, 71–96.

Rokeach, M. (1948). Generalised mental rigidity as a factor in ethnocentrism. *Journal of Abnormal and Social Psychology* **43**, 259–78.

Rokeach, M. (1960). *The open and closed mind*. Basic Books, New York.

Rokeach, M. and Hanley, C. (1956). Eysenck's tender-mindedness dimension: a critique. *Psychological Bulletin* **53**, 169–76.

Rose, N. (2007). *The politics of life itself: biomedicine, power, and subjectivity in the twenty-first century*. Princeton University Press, Princeton, NJ.

Rose, S. (1978). Letter to the editor. *Nature* **274**, 738.

Rose, S. (1979). Letter to the editor. *New Society*, 15 March 1979.

Rosenberg, N., Pritchard, J.K., Weber, J.L., Cann, H.M., Kidd, K.K., Zhivokovsky, L.A., and Feldman, M.W. (2002). Genetic structure of human populations. *Science* **298**, 2381–5.

Rosenzweig, S. (1954). A transvaluation of psychotherapy—a reply to Hans Eysenck. *Journal of Abnormal and Social Psychology* **127**, 330–43.

Rosner, R. (2005). Psychotherapy research and the National Institute of Mental Health, 1948–1980. In *Psychology and the National Institute of Mental Health: a historical analysis of science, practice and policy* (ed. W.E. Pickren and S.F. Schneider), pp. 113–50. American Psychological Association, Washington, DC.

Rushton, J.P. (2001). A scientometric appreciation of H.J. Eysenck's contribution to psychology. *Personality and Individual Differences* **3**, 17–39.

Rushton, J.P. (2002). The Pioneer Fund and the scientific study of human differences. *Albany Law Review* **66**, 207–62.

Rushton, J.P. and Jensen, A.R. (2005). Thirty years of research on race differences in cognitive ability. *Psychology, Public Policy and Law* **11**, 235–94.

Rust, J. (1975). Cortical evoked potential, personality and intelligence. *Journal of Comparative and Physiological Psychology* **89**, 1220–6.

Rutherford, A. (2003). Skinner boxes for psychotics: operant conditioning at Metropolitan State Hospital. *The Behavior Analyst* **26**, 267–79.

Rutter, M. (2001). The emergence of developmental psychopathology. In *Psychology in Britain: historical essays and personal reflections* (ed. G.C. Bunn, A.D. Lovie, and G.D. Richards), pp. 422–32. BPS Books, Leicester.

Rutter, M. (2006). *Genes and behavior: nature–nurture interplay explained.* Blackwell, Malden, Massachusetts.

Salter, A. (1952). *The case against psychoanalysis.* Holt, New York.

Sanavio, E. (ed.) (1999). *Behaviour therapy and cognitive behaviour therapy today: essays in honour of Hans J. Eysenck.* Elsevier, Oxford.

Sanford, N. (1953). Psychotherapy. *Annual Review of Psychology* **4**, 317–42.

Sang, J.H. (1971). Letter to the editor. *New Society*, 1 July 1971, 30.

Scarr-Salapatek, S. (1971). Unknowns in the IQ equation. *Science* **174**, 1223–8.

Schoenthaler, S.J. and Eysenck, H.J. (1997). Raising IQ level by vitamins and mineral supplementation. In *Intelligence, heredity, and environment* (ed. R.J. Sternberg and E. Grigorenko), pp. 363–92. Cambridge University Press, Cambridge.

Schoenthaler, S.J., Amos, P., Eysenck, H.J., Peritz, E., and Yudkin, J. (1991). Controlled trial of vitamin–mineral supplementation: effects on intelligence and performance. *Personality and Individual Differences* **12**, 351–62.

Scull, A. (2004). The insanity of place. *History of Psychiatry* **15**, 417–36.

Scull, A. (2005). *Madhouse: a tragic tale of megalomania and modern medicine.* Yale University Press, New Haven, Connecticut.

Segerstråle, U. (2001). *Defenders of the truth: the sociobiology debate.* Oxford University Press, Oxford.

Seltzer, C.C. (1967). Constitution and heredity in relation to tobacco smoking. *Annals of the New York Academy of Sciences* **142**, 322–30.

Seltzer, C.C. (1972). Critical appraisal of the Royal College of Physician's report on smoking and health. *The Lancet* **300**, 243–8.

Shapiro, D. and Shapiro, D. (1970). The 'double standard' in evaluations of psychotherapies. *Bulletin of the British Psychological Society* **30**, 209–10.

Shapiro, M. (1951). An experimental approach to diagnostic testing. *Journal of Mental Science* **97**, 748–64.

Shapiro, M. (1955). Training of clinical psychologists at the Institute of Psychiatry. *Bulletin of the British Psychological Society* **26**, 15–20.

Shapiro, M. (1957). Experimental method in the psychological description of the individual psychiatric patient. *International Journal of Social Psychiatry* **3**, 89–102.

Shapiro, M. and Nelson, E.H. (1955). An investigation of an abnormality of cognitive function in a co-operative young psychotic: an example of the application of the experimental method to the single case. *Journal of Clinical Psychology* **11**, 344–51.

Shepherd, M. (1977). A representative psychiatrist: the career and contributions of Sir Aubrey Lewis. *American Journal of Psychiatry* **134**, 7–13.

Shields, J. (1962). *Monozygotic twins: brought up apart and brought up together.* Oxford University Press, Oxford.

Shils, E.A. (1954). Authoritarianism, right and left. In *Studies in the scope and method of the authoritarian personality* (ed. R. Christie and M. Jahoda), pp. 24–49. Free Press, Glencoe, Illinois.

Showalter, E. (1987). *The female malady: women, madness and female culture*. Virago Press, London.

Shuey, A.M. (1966). *The testing of Negro intelligence*, 2nd edn. Social Science Press, New York.

Sigal, J.J., Star, K.H., and Franks, C.M. (1958*a*). Hysterics and dysthymics as criterion groups in the study of introversion–extraversion. *Journal of Abnormal and Social Psychology* **57**, 143–8.

Sigal, J.J., Star, K.H., and Franks, C.M. (1958*b*). Hysterics and dysthymics as criterion groups in the study of introversion–extraversion: a rejoinder to Eysenck's reply. *Journal of Abnormal and Social Psychology* **57** 381–2.

Simon, B. (1969). Intelligence does not depend on race, class or colour. *Morning Star*, 25 August 1969.

Skinner, B.F., Solomon, H., and Lindsley, O. (1953). *Studies in behavior therapy*, status report no .1, Naval Research Contract N5 ori-7662.

Slater, E. (1971). Autobiographical sketch. In *Man, mind and heredity* (ed. J. Shields and I. Gottesman), pp. 1–23. Johns Hopkins Press, Baltimore.

Smith, M., Glass, G., and Miller, T. (1980). *The benefits of psychotherapy*. Johns Hopkins Press, Baltimore.

Smith, R. (1992). *Inhibition: history and meaning in the sciences of mind and brain*. Free Association, London.

Smith, R. (1997). *The Fontana history of the human sciences*. Fontana, London.

Smith, R. (2001). Physiology and psychology, or brain and mind, in the age of C.S. Sherrington. In *Psychology in Britain: historical essays and personal reflections* (ed. G.C. Bunn, A.D. Lovie, and G.D. Richards), pp. 223–42. BPS Books, Leicester.

Smith, R.E. (1969). The other side of the coin. Review of *The biological basis of personality* by Hans Eysenck. *Contemporary Psychology* **14**, 628–30.

Snyderman, M. and Rothman, S. (1988). *The IQ controversy: the media and public policy*. Transaction Books, New Brunswick, New Jersey.

Söderqvist, T. (1996). Existential projects and existential choice in science: science biography as an edifying genre. In *Telling lives in science: essays on scientific biography* (ed. M. Shortland and R. Yeo), pp. 45–84. Cambridge University Press, Cambridge.

Solomon, J. (1998). *Objectivity in the making: Francis Bacon and the politics of inquiry*. Johns Hopkins University Press, Baltimore.

Soyland, A.J. (1994). *Psychology as metaphor*. Sage, London.

Spearman, C.E. (1904). 'General intelligence' objectively determined and measured. *American Journal of Psychology* **15**, 201–93.

Spiegel, D. (1991). Second thoughts on personality, stress, and disease. *Psychological Inquiry* **2**, 266–8.

Spiegel, D., Bloom, J.R., Kraemer, H.C., and Gottheil, E. (1989). Effect of psychosocial treatment on survival of patients with metastatic breast cancer. *The Lancet* **334**, 888–901.

Spielberger, C. (1986). Smoking, personality and health. In *Hans Eysenck: consensus and controversy* (ed. S. Modgil and C. Modgil), pp. 305–15. Falmer Press, Philadelphia.

Stagner, R. (1937). *Psychology of personality*. McGraw-Hill, New York.

Stallman, H. (ed.) (1966). *Das Prinz-Heinrichs-Gymnasium zu Schöneberg, 1890–1945*. Privately published, Berlin.

Stankov, L. (1998). H. Eysenck on intelligence: biological correlates and polemics. *Psihologija* **31**, 257–70.

Stebbing, S. (1930). *A modern introduction to logic*. Methuen, London.

Stebbing, S. (1937). *Philosophy and the physicists*. Dover Publications, New York.

Stephen, J., Rahn, M., Verhoef, M., and Leis, A. (2007). What is the state of the evidence on the mind–cancer survival question, and where do we go from here? *Supportive Care in Cancer* **15**, 923–30.

Stephenson, W. (1939). Methodological considerations of Jung's typology. *Journal of Mental Science* **85**, 185–205.

Stephenson, W., Mackenzie, M., Simmins, C.A., Kapp, D.M., Studman, G.L., and de B. Hubert, W.H. (1934). Spearman factors and psychiatry. *British Journal of Medical Psychology* **14**, 101–35.

Stolley, P.D. (1991*a*). When genius errs: R.A. Fisher and the lung cancer controversy. *American Journal of Epidemiology* **133**, 416–25.

Stolley, P.D. (1991*b*). Author's response to 'How much retropsychology?'. *American Journal of Epidemiology* **133**, 428.

Stone, W.F. (1974). *The psychology of politics*. Free House, London.

Stone, W.F., Lederer, G., and Christie, R. (eds.) (1993). *Strength and weakness: the authoritarian personality today*. Springer Verlag, New York.

Storms, L.H. and Sigal, J.J. (1958). Eysenck's personality theory with special reference to *The dynamics of anxiety and hysteria*. *British Journal of Medical Psychology* **31**, 228–46.

Storms, L.H. and Sigal, J.J. (1959). Misconceptions in 'Scientific methodology and *The Dynamics of anxiety and hysteria*'. *British Journal of Medical Psychology* **32**, 64–7.

Stott, D.H. (1983). *Issues in the intelligence debate*. NFER-Nelson, Windsor, Berkshire.

Strelau, J. and Zawadzki, B. (1997). Temperament and personality: Eysenck's three superfactors. In *The scientific study of human nature: tribute to Hans J. Eysenck at eighty* (ed. H. Nyborg), pp. 68–91. Elsevier, Oxford.

Strupp, H. (1963). The outcome problem in psychotherapy revisited. *Psychotherapy: Theory Research and Practice* **1**, 1–13.

Strupp, H. and Howard, I.I. (1992). A brief history of psychotherapy research. In *History of psychotherapy: a century of change* (ed. D.K. Freedheim), pp. 309–34. American Psychological Association, Washington, DC.

Studman, G.L. (1935). The factor theory in the field of personality. *Character and Personality* **4**, 34–43.

Stürmer, T., Hasselback, P., and Amelang, M. (2006). Personality, lifestyle, and risk of cardiovascular disease and cancer: follow-up of a population based cohort. *British Medical Journal* **332**, 1359.

Sunday Express (1973). Bring them to justice [editorial]. *Sunday Express*, 13 May 1973.

Sutherland, J.D. (1951). The Tavistock Clinic and the Tavistock Institute of Human Relations. *Quarterly Bulletin of the British Psychological Society* **2**, 105–11.

Sweet, C. (1955). Mr. Eysenck and the state of social science. *Clare Market Review* **50**, 34–9.

Sydney Morning Herald (1977). Shameful, dangerous [editorial]. *The Sydney Morning Herald*, 17 September 1977.

Talley, C., Kushner, H.I., and Sterk, C.E. (2004). Lung cancer, chronic disease epidemiology, and medicine, 1948–1964. *Journal of the History of Medicine and Allied Sciences* **59**, 329–74.

Tang, H., Quertermous, T., Rodriguez, B., Kardia, S.L.R., Zhu, X., Brown, A., Pankow, J.S., Province, M.A., Hunt, S.C., Boerwinkle, E., Schork, N.J., and Risch, N.J. (2005). Genetic structure, self-identified race/ethnicity, and confounding in case-control association studies. *American Journal of Human Genetics* **76**, 268–75.

Tate, C. and Audette, D. (2001). Theory and research on 'race' as a natural kind variable in psychology. *Theory and Psychology* **11**, 495–520.

Temoshok, L. (1991). Assessing the assessment of psychosocial factors. *Psychological Inquiry* **2**, 276–80.

Temple Smith, M. (1971). Letter to the editor. *The Times*, 30 June 1971.

Thurstone, L.L. (1938). *Primary mental abilities*, Psychometric Monographs, no .1. University of Chicago Press, Chicago.

Times Educational Supplement (1969). Black paper two savages the progressives. *The Times Educational Supplement*, 16 October 1969.

Times Literary Supplement (1954). Review of *Uses and abuses of psychology* by Hans Eysenck. *The Times Literary Supplement* 19 February 1954.

Timmermann, C. (2007). As depressing as it was predictable? Lung cancer, clinical trials, and the Medical Research Council in postwar Britain. *Bulletin of the History of Medicine* **81**, 312–34.

Tizard, J. (1977). The Burt affair. *University of London Bulletin*, May 1977, 4–7.

Todes, D.P. (2002). *Pavlov's physiology factory: experiment, interpretation, laboratory enterprise*. Johns Hopkins, Baltimore.

Tucker, W.H. (1994). *The science and politics of racial research*. University of Illinois Press, Urbana.

Tucker, W.H. (2002). *The funding of scientific racism*. University of Illinois Press, Urbana.

Tucker, W.H. (2007). Burt's separated twins: the larger picture. *Journal of the History of the Behavioral Sciences* **43**, 81–6.

Turner, J. (1981). Mind if I smoke? *Books*, 1 January 1981.

Turner, T. (1991). 'Not worth powder and shot:' the public profile of the Medico-Psychological Association, c .1851–1914. In *150 years of British Psychiatry, 1841–1991*, Vol. 1 (ed. G.E. Berrios and H. Freeman), pp. 3–16. Royal College of Psychiatrists, London.

Turner, T. (1996). James Crichton-Browne and the anti-psychoanalysts. In *150 years of British psychiatry, 1841–1991*, Vol. 2 (ed. H. Freeman and G.E. Berrios), pp. 144–55. Athlone Press, London.

University of Sydney (1978). Statement by committee of enquiry. The disruption of meetings with Professor Eysenck. *The University of Sydney News*, 6 March 1978, 224–77.

van der Ploeg, H. (1991). What a wonderful world it would be: a reanalysis of some of the work of Grossarth-Maticek. *Psychological Inquiry* **2**, 280–5.

van der Ploeg, H. and Kleijn, W.C. (1993). Some further doubts about Grossarth-Maticek's data base. *Psychological Inquiry* **4**, 68–9.

van der Ploeg, H. and Vetter, H. (1993). Two for the price of one: the empirical basis of the Grossarth-Maticek interviews. *Psychological Inquiry* **4**, 65–6.

Vandenbroucke, J.P. (1991). How much retropsychology? and reply to Eysenck. *American Journal of Epidemiology* **133**, 426–7.

Velden, M. (1999). Vexed variations. Review of *Intelligence: a new look* by Hans Eysenck. *The Times Literary Supplement*, 16 April 1999, 34.

Venables, P. (1970). Review of *The biological basis of personality* by Hans Eysenck. *Quarterly Journal of Experimental Psychology* **22**, 70–1.

Vernon, P.A. (1997). Behavioural genetic and biological approaches to intelligence. In *The scientific study of human nature: tribute to Hans J. Eysenck at eighty* (ed. H. Nyborg), 240–58. Pergamon, Oxford.

Vernon, P.E. (1933). The American v. the German methods of approach to the study of temperament and personality. *British Journal of Psychology* **24**, 156–75.

Vernon, P.E. (1962). *The structure of human abilities*. Methuen, London.

Verrall, R. and Reed-Herbert, A. (1978). Letter to the editor. *The Times*, 20 March 1978.

Vetter, H. (1991). Some observations on Grossarth-Maticek's data base. *Psychological Inquiry* **2**, 286–7.

Vetter, H. (1993). Further dubious configurations in Grossarth-Maticek's psychosomatic data. *Psychological Inquiry* **4**, 66–7.

Vital, D. (1999). *A people apart: the Jews in Europe, 1789–1939*. Oxford University Press, Oxford.

Vogel, F., Kruger, J., Schalt, E., Schnobel, R., and Hassling, L. (1987). No consistent relationships between oscillations and latencies of visual evoked EEG potentials and measures of mental performance. *Human Neurobiology* **6**, 173–82.

Waddington, K. (1998). Enemies within: post-war Bethlem and the Maudsley Hospital. In *Cultures of psychiatry and mental health care in postwar Britain and the Netherlands* (ed. M. Gijswijt-Hofstra and R. Porter), pp. 185–202. Rodopi, Amsterdam.

Wakefield, J.C. (1997). Diagnosing DSM-IV—part II: Eysenck (1986) and the essentialist fallacy. *Behaviour Research and Therapy* **35**, 651–65.

Watt, D. (1973). Bad year for the universities. *The Spectator*, 18 August 1973.

Webb, E. (1915). Character and intelligence. *British Journal of Psychology, Monograph Supplement* **1**, no. 3.

Wegner, B. (1982). *Hitler's politische Soldaten: die Waffen SS, 1933–1945*. Schöningh, Paderborn.

Weidman, N. (1997). Heredity, intelligence and neuropsychology; or, why the bell curve is good science. *Journal of the History of Behavioral Sciences* **33**, 141–4.

Weitzman, B. (1967). Behaviour therapy and psychotherapy. *Psychological Review* **74**, 300–17.

Wessely, S. (1993). Is cancer all in the mind? *The Times*, 22 June 1993.

Westfall, J. (1973). Letter to the editor. *Anarchist Weekly*, 26 May 1973.

Weyher, H.F. (1999). The Pioneer Fund, the behavioural sciences, and the media's false stories. *Intelligence* **26**, 319–36.

Williams, A. (1973). Letter to the editor. *Anarchist Weekly*, 16 June 1973.

Wilson, D. (2001). A 'precipitous *dégringolade*'? The uncertain progress of British comparative psychology in the twentieth century. In *Psychology in Britain: historical essays and personal reflections* (ed. G.C. Bunn, A.D. Lovie, and G.D. Richards), pp. 243–66. BPS Books, Leicester.

Winston, A. (1998). Science in the service of the far right: Henry E. Garrett, the IAAEE, and the Liberty Lobby. *Journal of Social Issues* **54**, 179–210.

Winston, A.S., Butzer, B., and Ferris, M.D. (2003). Constructing difference: heredity, intelligence, and race in textbooks, 1930–1970. In *Defining difference: race and racism in the history of psychology* (ed. A.S. Winston), pp. 199–229. American Psychological Association, Washington, DC.

Winter, D.G. (2001). A tough look at tough-mindedness. *Contemporary Psychology* **46**, 486–9.

WLG (1953). Review of *The scientific study of personality* by Hans Eysenck. *British Journal of Psychology (Statistical Section)* **5**, 208–12.

WLG (1954). A reply. *British Journal of Psychology (Statistical Section)* **6**, 46–52.

Wober, M. (1971). Race and intelligence. *Transition* **40**, December 1971, 17–26.

Wolfe, A. (2005). 'The authoritarian personality' revisited. *The Chronicle Review* **52**, B12.

Wolfle, D. (1942). Factor analysis in the study of personality. *Journal of Abnormal and Social Psychology* **37**, 393–7.

Wolpe, J. (1952). Objective psychotherapy of the neuroses. *South African Medical Journal* **26**, 825–9.

Wolpe, J. (1954). Reciprocal inhibition as the main basis of psychotherapeutic effects. *Archives of Neurology and Psychiatry* **72**, 205–26.

Wolpe, J. (1965). Letter to the editor. *British Medical Journal* **250** (19 June 1965), 1609.

Wynne, B. (1991). Knowledge in context. *Science Technology and Human Values* **16**, 11–121.

Wynne, B., Wilsdon, J., and Stilgoe, J. (2005). *The public value of science*. Demos, London.

Yates, A.J. (1958). The application of learning theory to the treatment of tics. *Journal of Abnormal and Social Psychology* **56**, 175–82.

Zangwill, O. (1951). *Introduction to psychology*. Methuen, London.

Zentralinstitut für Wissenschaftliche Forschung, Freien Universität Berlin (1995). *Gedenkbuch Berlins der Jüdischen Opfer des Nationalsozialismus*, Berlin edition. Hentrich, Berlin.

Index

Note: *Italic* page numbers denote illustrations

abnormal-normal continuum 95–8
Adorno, Theodor 244–5, 247, 248
aesthetics research 55–8, 65–7
Allport, Gordon 81, 83
Amelang, Manfred 379, 390, 392–3, 395, 398, 400
American tobacco industry 368–74, 401–8
animal experimentation 121, 131–2, 233–4, 277–8, 363
anti-Semitism 25–36, 244–5, 308–9, 357–8
anxiety
  due to conditioned learning 225–6
  neurotic anxiety 97, 205, 207
  see also *Dynamics of anxiety and hysteria, The* (Eysenck)
Appleby, Louis 391–2, 393
arousal model/theory, Gray's and Eysenck's 152–7
attitude research 242–7, 311–13
Australia tour (1977) 1, *316*, 318–19
Authoritarian personality, The (Adorno et al) 244–6
authoritarianism 241–2, 243, 248, 312
  left-wing 252, 255, 270, 309, 312–13
aversion therapy, criticisms of 227–9

Babington Smith, Bernard 59–60
Baker, John 304, 305
Barrett, Paul 165, 339–42, 411
Bartlett, Frederic 47, 48, 50
Beacon interview 304, 320–1
Behaviour Research and Therapy (journal) 152, 215, 426
behaviour therapy 181–2
  cognitive factors 222–5, 230, 234–5
  controversy over 208–12
  early techniques 202–8
  Eysenck's last theories 233–5
  and Eysenck's out-dated ideas 235–7
  Eysenck's role in development of 238–40
  importance of theory 225–7
  methods of working 237–8
  political criticisms 227–30
  science and techniques of 230–3
  successes of 212–17
Behaviour therapy and the neuroses (Eysenck) 212
bell-and-pad method for bed-wetting 204

bell curve debate 345, 347, 356
Berlin, Eysenck's early life and education in 14–24
Bethlem Royal Hospital 107, 119, 120, 121
'big five' model, Costa and McCrae 115, 177
Billig, Michael 320–5
biographer's standpoint 6–7
Biological basis of personality, The (Eysenck) 117, 154–7, 176
biology of personality 117–18
  first model 129–32
    critiques of 140–8
    Eysenck's publication rate 135–40
    student researchers 132–5
  IoP research department 118–21
  Pavlov's work, Eysenck's variation on 124–9
  re-emergence of 122–4
  second model 148–58
    decline in research 161–3
    experimental and correlational psychology, reconciliation of 173–9
    personality dimensions, uncertainty of 158–60
    scientific consensus building 169–72
    trust issues 164–9
Blewett, D. B. 275
body sway test and suggestibility 84, 130, *150*
Breger, Louis 222–3
British Psychological Society (BPS) 168–9
  belated recognition of Eysenck 425n
  Eysenck's attempt to reform medical section 197–9
  intelligence assessment symposium 279
  professionalization 220–1
  symposium on behaviour therapy 218
  symposium on inhibition 144–5
Broadbent, Donald 148, 168, 174–5
Broadhurst, Peter 131, 277–8
Burt, Cyril 37, 42–3, 45–6
  attacks on reputation of 313–19, 328–9
  end of Eysenck's apprenticeship with 67–70
  Eysenck's disagreements with 63–4
  Eysenck's wartime letters to 61–3
  factor analysis work 52–3, 56–7, 66–7, 102–3
  general emotionality factor 82, 83
  intelligence testing work 51
  mentoring Eysenck 53–5, 59–61, 65

Burt, Cyril (*continued*)
  and review of Eysenck's work by WLG 68, 111–12
  suporting Jensen 283–4

Cambridge University, psychology at 46–50, 148, 174–5
cancer and personality *see* smoking, cancer and personality
Causes and effects of smoking, The (Eysenck) 374–8
Chomsky, Noam 222
Christie, Richard 248–51, 254–7, 262–5
Claridge, Gordon 151, 155, 157, 177, 409
clinical psychology 181–2
  behaviour therapy
    cognitive challenge to 222–5, 230, 234–5
    controversy over 208–12
    criticisms of 227–30
    early techniques 202–8
    Eysenck's last theories 233–5
    and Eysenck's out-dated ideas 235–7
    Eysenck's role in development of 238–40
    importance of theory 225–7
    methods of working 237–8
    science and techniques of 230–3
    successes of 212–17
  clinical course at the Maudsley 182–7
  Eysenck-Shapiro split 200–2
  Eysenck's vision for 187–9
  psychoanalysis/psychotherapy
    anti-Freudianism 217–22
    Eysenck's attack on 189–96
  reform of BPS medical section 197–9
Clockwork orange, A (Burgess) 227, 228–9
cognitive approaches 230, 234–5
cognitive neuroscience 178–9
cognitive performance research 174–5
colour preferences, research on 64–5
Communist Party of England (Marxist-Leninist) (CPE-ML) protest 300–1
communists and tough-mindedness 246–8, 268–9
  Christie's analysis of Coulter's data 254–5
  Coulter's work, Eysenck's write-up of 308–10
  critique of T scale 270
  Rokeach and Hanley critique 251–4
conditioned learning
  and bed-wetting 203–4
  introvert-extravert differences 129–31
  and reciprocal inhibition 205
  theory and practice 225–7
  see also Pavlov, Ivan, work of
Conditioned reflexes (Pavlov) 123, 126, 185
conservatism-radicalism (R) factor 243, 249, 308
continuity issue 111–13

Controversy (BBC programme) 296, 305
correlational psychology 43–6
cortical arousal 152–7, 377
cortical excitation 125–6, 129–30
cortical inhibition 124–30
Coulter, Thelma 247–8, 254–5, 308–10
Council for Tobacco Research (CTR) 369–74
creative novation therapy 234, 382–3, 388, 391
Crime and personality (Eysenck) 139, 161
criterion analysis 105–6, 108–9, 113
Crown, Sidney 220

Daily Mail 302, 304, 307, 319
Darlington, C. D. 277, 278, 294
Darwin, Charles 42
Davies, Margaret (first wife) 30, 36, 52–3, 55, 58, 62, 200
Deary, Ian 113–14
de-sensitization treatment 205, 212, 216, 220, 224–5, 226
dichotomous typologies 92
Dimensions of personality (Eysenck)
  American researchers 104–5
  and factor analysis 101–4
  foundations of 113–15
  and MacKinnon's typologies 89–95
  normal-abnormal continuum 95–8
  threat to psychiatric diagnostic categories 100–1
Dynamics of anxiety and hysteria, The (Eysenck) 129–32, 140
  critiques of 140–8, 166–7

Eaves, Lindon 234, 311, 343–4, 372, 374–5
educational practices, Eysenck's views on 284–5
EEG measures of intelligence 165, 333, 338–41
'Effects of psychotherapy, The' (Eysenck), article 191–6
emotionality, Burt's factorial work 82–3
environmental factors and rise in IQ 347–8
eugenics 48–9, 76, 272–3
evoked potentials (EPs) and IQ scores 333, 335, 336, 338–41
experimental and correlational psychology, reconciliation of 173–9
Experimental Psychology Group/Society 110, 169, 175
extraversion-introversion factor *see* introversion-extraversion factor
extraverts
  excitation and inhibition 128–32
  and hysteria 144–5
  individual differences in performance 154
  in the prison population 161
  smoking habits 366, 377

eye-blink conditioning experiments,
    Franks 128, 132, 141, 142, 143, 167n
Eysenck, Eduard (father) 14, 15–16, 17–18,
    23, *24*, 29–30, 34
Eysenck, Hans 3–4, *98–9*, *213*, *415*
    adversarial style of 263–4, 415–19
    Berlin University, attempt to enter 22–5
    biographies of Eysenck 4–6
    birth and family background 14–18
    childhood and secondary education 19–21
    competitiveness 10–11, 169–70, 258,
        415–20
    death 409–10
    detachment of 111, 118, 166, 167, 249,
        410–15
    divorce from Margaret 200
    doctoral thesis 57–8, 62
    employment difficulties, post-war 62–3
    employment, first job 70, 76
    Jewishness, denial of 31–2, 304
    marriage to Margaret 52
    modus operandi of 138–40, 164–9, 410–11
    opinions about 1–2, 420–2
    passive-aggressive quality 257–61
    political views 264–5, 357–8
    recognition of 4, 425–6
    retirement, official 335–6
    self-belief of 422–3
    undergraduate years 51–3
    war years 58–63, 70–1
    see also publications
Eysenck, Margaret (first wife) 30, 36,
    52–3, 55, 58, 62, 200
Eysenck, Michael (son) 2, 6, 30, 36, 52,
    157, 291, 359
Eysenck Personality Inventory (EPI) 159–60
Eysenck Personality Questionnaire
    (EPQ) 159–60
Eysenck, Ruth (mother) 14–16, 17, 25
Eysenck, Sybil (second wife) 5, 35–6,
    149, 160, 200, 306

F scale, 'fascist potential' 244, 247–8,
    254–5, 308–9
factor analysis 44–6
    1944 personality factors 87–9
    and personality dimensions 100–5
Factors of the mind, The (Burt) 66
Facts of life, The (Darlington) 277
fascism
    California F scale 244–5
    ideology of National Front 320
fascists and tough-mindedness 246–8, 268–9
    Christie's analysis of Coulter's data 254–5
    Coulter's work, Eysenck's write-up
        of 308–10
    critique of T scale 270
fears and phobias 205, 220, 225–6, 233

'Flynn effect' 347
Franks, Cyril 130–1, 141–4, 147, 148, 416
Frentzel-Beyme, Rainer 390–1
Freudian theories, attacks on 39, 183, 191–6,
    197, 206, 217–23
Front *see* National Front
Fulker, David 327–8, 346–7
funding for research 119, 121
    from Big Tobacco 338, 369–70, 401–8
    Pioneer Fund 349–50
Furneaux, Desmond *150*, 279–80, 411
Furnham, Adrian 418, 419

Galton, Francis 42, 48–9, 272–3
Gelder, Michael 211, 223–5
general ability factor, Spearman's g 45–6, 57,
    279, 327
general aesthetic factor 56, 64
Genes, culture and personality (Eaves, Eysenck
    and Martin) 343–4
genetics of personality 272–8, 343–8
    research on smokers and non-smokers 364
    and variation in political
        convictions 311–12
Germany
    childhood in 13–21
    departure from 21–5
    rejection of 33–5
Gibson, Tony (first biographer) 38, 62,
    250–1, 304, 310–11, 414, 420–1
Gillie, Oliver 313, 317–18
Glass, Max (step-father) 16, 17
Glover, Edward 219
Gray, Jeffrey 126–7, 167, 175, 279
    biological model of personality 156
    comments about Eysenck 359, 406, 409,
        412, 416, 417, 422
    translation of Pavlov's work 152–3
Green, Robert 270
Grossarth-Maticek, Ronald 234, 361–2,
    377–400, 404–5, 408
Guardian, The 260, 293, 301, 410, 421

Hamilton, Vernon 140–2, 147, 167
Hanley, Charles 251–4, 256–7
Hans Eysenck: the man and his work
    (Gibson) 5
Hearnshaw, Leslie 51, 68, 319, 328
Heidelberg study 378–96
Hendrickson paradigm, EP and IQ 333–5,
    338–43
heritability of intelligence 289–91, 329, 345–6
heritability of personality 274–8
Hildebrand, Peter 131, 141–2, 147, 158,
    173–4
Himmelweit, Hilde *150*, 184–5
Hobsbawm, Eric 19–20, 30, 33
Howarth, Edgar 158–9

Hull, C. L. 127–8
humour research 62, 64
hysteria *see Dynamics of anxiety and hysteria, The* (Eysenck)

incubation model/theory 225–6, 233–4
individual differences, UCL 42–6
Inequality of man, The (Eysenck) 298–9
inheritance of personality 274–8
inhibition 125–30, 144–5, 151–4
  reciprocal inhibition 205, 207, 209, 219, 239
inspection time research 336–8
Institute of Psychiatry (IoP) 120–1
intelligence
  heritability of 272–8
  and nutrition 347–8
  and processing speed 175, 279, 337
  and race *see* race and IQ controversy
  Spearman's general ability factor 43–6
  testing 43–4, 45, 51, 283–5, 331–4
Intelligence: the battle for the mind (Eysenck and Kamin) 329
Intelligence: a new look (Eysenck) 342–3, 353
introversion-extraversion factor 66–7, 83, 96–7
  and conditionability 130–1
  excitation and inhibition 129–30
  Gray's contribution 152–4, 156
  inhibition differences 128–9
  and neurotic disorders 131–2, 141–2, 143–4
  scales measuring 159–60
  see also extraverts
IQ and race controversy *see* race and IQ controversy
IQ argument, The (Eysenck) *see* Race, intelligence and education (Eysenck)

Jacob, Ed (legal advisor to R. J. Reynolds) 373–4, 402
Jensen, Arthur 139, 271–2, 297, *316*, 352
  article on race and IQ 280, 281–2
  Australia tour 318–19
  and the Burt affair 313–18
  and Eysenck's race and IQ book 286, 288
  personality research 280
  petition supporting 298
  support from Burt and Eysenck 283–4
Jews
  Eysenck's denial of Jewish heritage 26, 30, 31–2, 304
  fate of Eysenck's grandmother 35–6
  persecution by Nazis 16–17, 26–9
  and political extremes research 308, 309
  prejudice studies 244–5
Jones, Gwynne 145, 198–9, 203, 207, 209, 229
Jones, Wynn 61, 64, 78

Kamin, Leon 313, 325, 329–30, 334, 359
Kendrick, Donald 145, 214, 233
kinaesthetic after-effects 128
Kissen, David 365
Koestler, Arthur 31–2, 265

Laing, R. D. 227–8
Lazarus, Arnold 226, 239
Leach, Penelope 261
learning theories
  application to therapy 204–5
  and behaviour theory 222–3
  and clinical psychology 212–17
  and de-sensitization 226
  and Wolpe's reciprocal inhibition 205–6, 207
  see also conditioned learning
Lee, P. N. 387, 389, 390, 404
Lewis, Aubrey 70, 75–7, 84–6, 108
limbic system 153–4
litigation-oriented research 369–74, 403
London-Oxbridge differences 46–51
London School of Economics (LSE) incident 299–301, 304, *315*
Lubin, Ardie 109
Lykken, David 145
Lynn, Richard 284–5, 349, 351

MacKinnon, Donald. 80, 84, 89–93
Mackintosh, Nicholas 329, 342, 345
Mankind Quarterly (journal) 322–3, 349
Mapother, Edward 75–6
Marks, Isaac 211, 223–5, 228
Martin, Irene 133, 136, 137, 162, 186, 215
Maudsley Hospital *149*
  behaviour therapy at 202–4, 210–11
  BPS membership reforms 197–9
  clinical course at 181–7, 215–16
  clinical psychology at 202–4
  establishment of 74–7
  Eysenck's positions at 77–80, 118–21
  failure to bring Wolpe to 211–12
  provocation of psychiatrists at 208–10
  see also Mill Hill Emergency Hospital
Maudsley personality inventory (MPI) 136, 142–4, 159
Maxwell, Robert 152
McConaghy, Neil 156–7
McDougall, William 43, 46, 88–9, *92*, 122
McGaugh, James 222–3
McGuire, Ralph 159
Medawar, Peter 165–6
media publicity 11, 214, *314*, *316*
  and aversion therapy 227–8
  popular Pelican series 265–9
  race and IQ controversy 283, 296–7, 298
  dealing with 301–8

smoking and health 376–7
  Grossarth-Maticek claims 396–8
Melvin, D. 250, 252–3, 254, 255, 256
Meyer, Adolf, psychiatrist 76–7
Mill Hill Emergency Hospital 3, 70, 75, 77–80
  Howarth's re-analysis of Eysenck's data 158–9
  psychiatric casualties of war 95–8
  research work at 84–7
  see also Maudsley Hospital
Model for intelligence, A (Eysenck) 334–5
Mollon, John 146–7, 175
Montagu, Ashley, Eysenck's critique of *Race and IQ* 322–3

National Front 283, 319–23
natural sciences, Eysenck's idealization of 8, 38–9
nature-nurture dichotomy 273–4, 311–12, 319, 326, 343–8, 355
Nazi party/Nazism
  and Eysenck's father 17–18
  Eysenck's rejection of 21–5, 33–4, 303–4
  impact on media and film industry 16–17
neurosis and suggestibility 84
neuroticism 88, 93, 97–8, 100–1
  and cancer 364–5, 375, 377, 382
  heritability of 274–5
  and limbic system activation 153
  measures of 159–60
  neurotic types 92
  normal distribution of 96–7
New Scientist (journal) 283, 329
nicotine, research on effects of 338, 368, 371–2, 405–7
normal-abnormal continuum 95–8
normal personality types, MacKinnon's twofold typology of 91
nutrition and rise in IQ 347–8

objectivity, scientific 7–10, 420
Oxbridge-London differences 46–51

Parry, John 55, 61, 63
Pavlov, Ivan, work of 123, 124–6
  ambiguity of Russian terminology 126–7
  Eysenck's interpretation 127–8
  Gray's translation 152
  Teplov's extensions 152–3
Pavlov's typology (Gray) 152
Payne, Bob 26, 29, 30, 32, 186, 201–3, 204
Pearson, Karl 41–2, 44–5
Pearson, Roger 323
Pelosi, Tony 391–2, 393, 397, 399
Penguin paperbacks, popularity of 265–9
Penrose, Lionel 294–5

Personality and Individual Differences (PAID), journal 331, 426
personality dimensions 73
  1944 personality factors 87–9
  continuity between normal and abnormal 95
  Eysenck's work in clinical psychiatry 77–80
  factor analysis 100–5
  Maudsley Hospital 74–7
  Mill Hill research programme 84–7
  origins of *Dimensions of Personality* 113–15
  Scientific study of personality (1952) 105–13
  sourced from MacKinnon's two-fold typologies 89–93
  temperament studies 80–4
  uncertainty of 158–60
personality traits and types, twofold typologies 91, 92
phobias, treatment of 225–6
physics, Eysenck's idealization of 38, 39–40
Pioneer Fund 338, 349–50, 406
political attitudes research
  in America 243–4
  Authoritarian personality, The (Adorno et al) 244–6
  Jews excluded from analyses 308, 309
  political extremes research 246–8
  Rokeach and Christie attacks 248–57, 262–5
  post-war research 242–3
  prejudice studies 244–5, 264, 319
popularization of psychology 265–9
prejudice, studies in 244–5, 264, 319
Prell, Donald 274–7
preparedness notion, Seligman 233–4
processing speed 175, 279, 337
professional associations, Eysenck's involvement with 168–9
projective tests, Eysenck's attacks on 147, 201
psychiatry
  British psychoanalysts shielded by 220
  diagnostic issues 100–1, 201
  at the Maudsley hospital 74–80, 119–21
  psychiatric patients, signs and symptoms 85, *86*
psychoanalysis 183
  Eysenck's attack on 189–91, 197–9, 218–19
Psychological Inquiry, Grossarth-Maticek's studies 389–91, 393
psychological testing 49–51, 159–60
psychological types 385–7
Psychology of politics, The (Eysenck) 170, 247, 266, 269, 307
  Rokeach and Christie criticisms of 248–57, 262–5, 269
Psychology is about people (Eysenck) 298–9
psychosomatic research 365, 378–81, 386

psychotherapy
  APA training recommendations 188, 189
  effectiveness of 191–6
  and health outcomes 388–9
  in the US 189–90
  see also behaviour therapy; psychoanalysis
psychoticism 107–8, 160
publications 9
  during the 1950s and 1960s 137–8
  during retirement 336–7
  first article 54
  high output rate 138–9
  lack of detail in 166, 257
  Pelican series 265–9
  post-doctoral (1940s) 63–7

R scale, radicalism-conservatism 249, 308
Rabbitt, Patrick 174–6
race and IQ controversy 1, 162, 271–2
  British perspective 282–5
  the Burt affair 313–19
  Coulter's thesis 308–10
  eugenic movement 272–3
  genetics of personality and IQ 274–8
  intelligence research 331–4
  intelligence testing and education 279–82
  Jensen, support for 348–9
  Kamin-Eysenck debate 329–30
  LSE incident 299–301
  media manipulation 301–4
  Model for intelligence, A (1983) 334–6
  National Front issues 319–23
  nature versus nurture 273–4, 343–8, 358
  Pioneer funding 349–50
  political implications 324–6
  post-retirement work 336–43
  public hounding, effects of 306–7
  public image politics 356–60
  publicity, avoidance of 305–6
  Race, intelligence and education (1971) 286–91
    rationale and social consequences 291–3
    reaction to 293–9
  racial science 353–6
  Structure and measurement of intelligence, The (1979) 326–9
Race and IQ (Montagu) 322
Race (Baker) 304, 305
Race, intelligence and education (Eysenck) 271, 286–7, *315*
  rationale and social consequences 291–3
  reaction to 293–9
  US edition, IQ argument 287–90
Rachman, Stanley J. ('Jack') 205–6, 211, 216–17, 223–4
radicalism-conservatism (R) factor 243, 249, 308
reaction-time research 332–3, 336–8

Rebel with a cause (Eysenck's autobiography) 4–5
reciprocal inhibition 205–6, 207, 209, 219, 239
Reid, R. L. 145
Reynolds, R.J. (tobacco company) 338, 373–4, 383–5, 402
Rokeach, Milton 248–57, 262–3, 264–5, 312
Ronald Grossarth-Maticek (RGM) questionnaire 380–1
Rorschach test 147, 181, 187, 201
Rose, Steven 321, 329
Rostal, Sybil (Eysenck's second wife) 5, 35–6, 160, 200, 306
Royal Medico-Psychological Association (RMPA), Eysenck's presentation at 208–10
Rushton, J. Philippe 299, 305, *315*, 349, 351
Russell, Roger 121, 133
Rutter, Michael 146, 418, 426

Sanford, Nevitt 194, 258
Scarr-Salapatek, Sandra 288, 289, 293, 295
Schutzstaffel (SS), refusal to join 22–3, 25
science
  as a competitive game 10–12, 257–61
  and intellectual exchange 136–7, 176
  practice of 166–8
  trust issues 164–6
Scientific study of personality, The (Eysenck) 106–7, 122
  criticisms of 108–13
  and psychoticism dimension 73, 107–8
  WLG review of 68, 111–12
Seligman, Martin 233–4
Sense and nonsense in psychology (Eysenck) 266, 268–9
Shapiro, Monte 203, 204, 206–7, *213*, 231
  contribution to clinical psychology 238–9
  disagreements with Eysenck 199, 200, 202
  heading clinical section at the Maudsley 135, 184–7
  methods of practice 200–1
  retirement 216
Shils, Edward 246
Shuey, Audrey 281, 283
Sigal, J. J. 140–4, 147, 167
Skinner, B. F. 239
smoking, cancer and personality 361–2
  absence of news headlines 396–8
  Causes and effects of smoking, The (Eysenck) 374–8
  Eysenck's reaction to criticism 393
  Grossarth-Maticek's work
    creative novation behaviour therapy 388
    Eysenck's influence 385–7
    scepticism of 387–92
    tobacco company involvement 383–5

uncertain status of 398–400
Heidelberg connection 378–81
implications of Eysenck's
    involvement 393–6
reconciling differences and doubts 381–3
replication attempts 392
smoking-cancer research 363–8
tobacco industry sponsorship 368–74,
    401–8
Smoking, health and personality
    (Eysenck) 365–8, 370, 374
Smoking, personality and stress
    (Eysenck) 405
social attitude research 242–6
Spearman, Charles 43–6, 50, 78
Spielberger, Charles 372–3, 385, 400
Stacey, Barrie 270
Stebbing, Susan 39, 55, 106
Stephenson, William 45–6, 53, 67, 78
Storms, L. H. 140–3, 147, 167
stress-cancer link 377, 387–8, 400–1, 405, 408
string length and IQ 334–5, 339–40
Structure of human personality, The
    (Eysenck) 266
Structure and measurement of intelligence, The
    (Eysenck) 327–8
student researchers
    Eysenck's control of 139–40
    Eysenck's influence on 419–20
    IoP psychology department 132–5
    mid-1970s decline in 161–2
    unpublished work, access to 256
Studman, Grace 78, 82
suggestibility and neurosis 84–5, 87
Sunday Times, The 287, 291, 293, 294,
    313, 317
systematic de-sensitization 205, 212, 216, 220,
    224–5, 226

T scale, tough-mindedness 246–8, 249,
    251–5, 270, 309
Tatchell, Peter 228
Temoshok, Lydia 386, 389
temperament and personality 66–7, 80–3,
    101, 178
Temple Smith, Maurice 287

testing, psychological 49–51, 159–60
Thurstone, Louis Leon 45–6, 54
Times, The 294, 317, 322
tobacco industry
    litigation-oriented research 369–74, 403
    sponsoring research 338, 369–70, 401–8
Tobacco Research Council (TRC) 364–5
tough- vs. tender-minded dimension 242–3,
    246–9, 251–5, 308–9, 312–13
Trust in Numbers (Porter) 8, 39n
trust issues 164–9, 420–2
    Eysenck's 'three steps to trust' 381–2, 395
twin studies 274–5, 276–7, 329–30, 343–5

University College, London (UCL) 37, 41–3
    Cambridge University as rival to 46–51
    doctoral work at 53–8
    enrolling as a psychologist 37–41
    evacuation of during the war 58
    part-time teaching post at 62
    taking over chair from Burt 69
    undergraduate years at 51–3
US tobacco industry 368–74, 401–8
Uses and abuses of psychology (Eysenck) 204,
    266, 279, 280

van der Ploeg, H. 384, 385, 390
Vernon, Philip 70, 82, 83–4, 288
Vetter, Hermann 390, 394, 399–400

war years 70–1
    air raids 63, 67
    internment threat 61–2
    trials and uncertainties 58–63
Webb, E., 'will' factor 82, 83, 88
Werner, Antonia (maternal grandmother) 15,
    35–6
Werner, Ruth (Eysenck's mother) 14–16,
    17, 25
WLG, review of Eysenck's work by 68,
    111–12
Wolpe, Joseph 205–12, 239–40

Yates, Aubrey 212, 238

Zangwill, Oliver 50, 110, 174